AF615074

Aquatic Microbiology

Aquatic

Microbiology

An ecological approach

Edited by
Timothy Edgcumbe Ford
Department of Environmental Health
Harvard School of Public Health
Boston, Massachusetts

Boston
BLACKWELL SCIENTIFIC PUBLICATIONS
Oxford London Edinburgh
Melbourne Paris Berlin Vienna

To my parents, Sheila and Stanley Ford

Editorial offices:
238 Main Street, Cambridge
Massachusetts 02142, USA
Osney Mead, Oxford OX2 0EL, England
25 John Street, London WC1N 2BL
England
23 Ainslie Place, Edinburgh EH3 6AJ
Scotland
54 University Street, Carlton
Victoria 3053, Australia

Other editorial offices:
Librairie Arnette SA
2 rue Casimir-Delavigne
75006 Paris
France

Blackwell Wissenschafts-Verlag
Meinekestrasse 4
D-1000 Berlin 15
Germany

Blackwell MZV
Feldgasse 13
A-1238 Wien
Austria

First published 1993

Set by Excel Typesetters Co., Hong Kong
Printed and bound in the United States of America
by Braun-Brumfield, Ann Arbor, Michigan

93 94 95 96 5 4 3 2 1

DISTRIBUTORS

USA
Blackwell Scientific Publications, Inc.
238 Main Street, Cambridge
Massachusetts 02142
(*Orders*: Tel: 800 759-6102
617 876-7000)

Canada
Oxford University Press
70 Wynford Drive
Don Mills
Ontario M3C 1J9
(*Orders*: Tel: 416 441-2941)

Australia
Blackwell Scientific Publications Pty Ltd
54 University Street
Carlton, Victoria 3053
(*Orders*: Tel: 03 347-5552)

Outside North America and Australia
Marston Book Services Ltd
PO Box 87
Oxford OX2 0DT
(*Orders*: Tel: 0865 791155
Fax: 0865 791927
Telex: 837515)

Library of Congress
Cataloging-in-Publication Data

Aquatic microbiology: an ecological approach/
edited by Timothy Edgcumbe Ford.
p. cm.
Includes bibliographical references and index,
ISBN 0–86542–225–7
1. Aquatic microbiology,
2. Aquatic ecology.
I. Ford, Timothy, Edgcumbe.
QR105.A73 1993
576′.192 – dc20

Contents

Contributors

Michael Boyle
Microbial Ecology Laboratory, Harvard University, Cambridge, Massachusetts

Thomas E. Cappenberg
Netherlands Institute of Ecology, Centre for Limnology, Nieuwersluis, The Netherlands

Nina F. Caraco
Institute of Ecosystem Studies, Mary Flagler Cary Arboretum, Millbrook, New York

Jonathan J. Cole
Institute of Ecosystem Studies, Mary Flagler Cary Arboretum, Millbrook, New York

Hugh W. Ducklow
Horn Point Environmental Laboratory, Cambridge, Maryland

Klas B.G. Flärdh
Department of General and Marine Microbiology, University of Göteborg, Göteborg, Sweden

Timothy E. Ford
Department of Environmental Health, Harvard School of Public Health, Boston, Massachusetts

William C. Ghiorse
Section of Microbiology, Cornell University, Ithaca, New York

John E. Hobbie
The Ecosystems Center, Marine Biological Laboratory, Woods Hole, Massachusetts

Robert W. Howarth
Cornell Center for the Environment, Cornell University, Ithaca, New York

Louis A. Kaplan
Stroud Water Research Center of the Academy of Natural Sciences, Avondale, Pennsylvania

David L. Kirchman
Marine Biology/Biochemistry Program, College of Marine Studies, University of Delaware, Lewes, Delaware

Staffan Kjelleberg
Department of General and Marine Microbiology, University of Göteborg, Göteborg, Sweden

Allan E. Konopka
Department of Biological Sciences, Purdue University West Lafayette, Indiana

Donn J. Kushner
Department of Microbiology, University of Toronto, Toronto, Canada

Maurice A. Lock
School of Biological Sciences, University of Wales, Bangor, United Kingdom

Eugene L. Madsen
Section of Microbiology, Cornell University, Ithaca, New York

James S. Maki
Department of Biology, Marquette University, Milwaukee, Wisconsin

Robert V. Miller
Department of Microbiology and Molecular Genetics, Oklahoma State University, Stillwater, Oklahoma

David J.W. Moriarty
CSIRO Marine Laboratories, Cleveland, Queensland, Australia

J. Denis Newbold
Stroud Water Research Center of the Academy of Natural Sciences, Avondale, Pennsylvania

Thomas Nyström
Department of General and Marine Microbiology, University of Göteborg, Göteborg, Sweden

Hans W. Paerl
Insititute of Marine Sciences, University of North Carolina, Morehead City, North Carolina

John H. Paul
Department of Marine Science, University of South Florida, St Petersburg, Florida

Fuh-Kwo Shiah
Horn Point Environmental Laboratory, Cambridge, Maryland

Cornelis L.M. Steenbergen
Netherlands Institute of Ecology, Centre for Limnology, Nieuwersluis, The Netherlands

Jean-Pierre R.A. Sweerts
Manager, Water Quality Studies, Delft Hydraulics, Delft, The Netherlands

Peter Westermann
Department of General Microbiology, University of Copenhagen, Copenhagen, Denmark

Preface and acknowledgments

Research at the microbiological level often seems to limit our understanding of the ecosystem as a whole. We delineate specialized areas for study. This has inevitably resulted in excellent research on specific processes in freshwater, marine, and anthropogenic systems that fails to recognize the many important interconnections between them.

In an ecological approach to aquatic microbiology, we have tried to recognize the interconnections between environments and to highlight dominant processes and their rates. The reader will note that many chapters address similar processes within their respective environments. The characteristics of these ecosystems are, however, dominated by the rates of the processes. Changes along the freshwater–estuarine–marine continuum are the result of interaction between physicochemical and biological processes. The resultant decrease in the rate of a specific process does not mean it is no longer an important component, rather that it no longer dominates. For example, it may still provide a key electron source or sink for other processes. Conversely, dominance of a process within a particular ecosystem may have little effect on global fluxes. This is a particularly important concept in terms of localized anthropogenic effects (e.g., sewage outfall). In Chapter 1, we have tried to explain further our perspective on the relationships between microbial ecosystems.

This book brings together authors from many different fields in aquatic microbiology. The authors have been asked to discuss their research from an ecosystem perspective. This book is designed to present the most recent advances in aquatic microbial research in a manner that will appeal to both researchers and advanced students. The ecosystem perspective will also enable the reader less familiar with aquatic microbiology to understand the relevance of research in this complex field to the environmental and indeed, global perturbations of current concern.

Chapters 1 and 2 are essentially introductory. Chapter 1 presents our perspectives on the ecology of aquatic microbes and Chapter 2 addresses the advances and limitations of the methodology. The central part of the book roughly follows the continuum thought. Chapters 3–5 are concerned with lake ecosystems from general to more specific theories. Chapters 6 and 7 address rivers and streams; Chapter 8 addresses ground water and Chapter 9, freshwater wetlands. Chapters 10–14 follow a logical progression from salt marsh/estuary to open ocean nutrient cycles. The final part of the book, Chapters 15–19, covers extreme and anthropogenic processes, with Chapter 19 addressing genetic stability of engineered microorganisms.

I am particularly grateful to John Hobbie for his support and contributions to this book. In addition, each chapter has been externally reviewed by at least two authorities in their respective fields.

I am extremely grateful to the following reviewers who have contributed significantly to the quality of each chapter: Felix Baerlocher, David Balkwill, Russell Bell, Angelo Carlucci, Ronald Crawford, Hugh Ducklow, Jed Fuhrman, Malte Hermansson, Mark Hines, John Hobbie, Barbara Javor, Louis Kaplan, Gary King, David Kirchman, Staffan Kjelleberg, Bruce Logan, Jim Maki, Kevin Marshall, Lutz Meyer-Reil, Ralph Mitchell, Richard Morita, Pat Mulholland, David Nedwell, Betty Olson, Ron Oremland, Mike Pace, John Paul, Carlos Pedros-Alio, Don Reasoner, Joan Rose, Gary Sayler, Ray Seidler, Steve Smith, Jim Staley, David Strayer, Bo Thamdrup, E. Michael Thurman, Bess Ward, Brian Whitton, and John Wilson.

1

John E. Hobbie & Timothy E. Ford

A perspective on the ecology of aquatic microbes

1.1 Introduction

This multiauthored book was written for advanced students in ecology or microbiology as an overview of current understanding and research in aquatic microbial ecology. The growth of knowledge regarding aquatic microbes has been so enormous, encompassing habitats from marshes to the deep sea, that a multiauthored approach is perhaps the only reasonable option. However, authors bring their own viewpoints and ideas about the important concepts and organizing themes to each chapter. The result is a good summary of the field that may lack a sense of unity and completeness. Even advanced students may have difficulty in organizing the information into a coherent whole.

Given the breadth of the field, it is likely that students and scientists from outside aquatic microbiology lack the necessary perspective for critical evaluation of the field. In this context, perspective is the "capacity to view things in their true relations or relative importance" (*Webster's Seventh New Collegiate Dictionary 1970*). Perspective takes specialized knowledge and experience. Understanding and interpretation of data requires continual questioning of methods, asking about artifacts of measurement, and trying to determine the importance of each finding. Because these questions about the true state of microbes in nature and about the importance of various data are difficult for students and scientists outside this field to formulate, we consider it worthwhile to present a perspective in this chapter.

1.2 Perspectives

1.2.1 What is meant by the ecology of aquatic microbes?

The subject of this book is aquatic microbes in their natural settings, i.e., their ecology. This includes:

1 understanding what types are where;
2 how many there are;
3 what they are doing;
4 how they survive;
5 what controls their activity and numbers;

6 what role they play in the large cycles of carbon, trace gases, and nutrients.

This is certainly a broad set of questions, so broad that it is often difficult to find the dividing line between ecology and physiology. One clear-cut example of the difference occurs in studies of *Sulfolobus* which grows in sulfurous hot springs at a pH of 2–3 and a temperature of 80°C (discussed in Vestal & Hobbie 1988). Studying the internal mechanisms of survival, such as the role of the diphytanyl diether-linked lipids in membrane stability at these temperatures (see Chapter 15), is physiology. Studying *Sulfolobus* in a hot spring, e.g., the rate and control of sulfur oxidation and acid production, is ecology. From these examples it should be clear that the subject matter of ecology is defined by the questions asked.

In studies of aquatic microbes, ecology differs from physiology not so much in the type of measurements made as in the reason or justification for making those measurements. In ecological studies, the ultimate explanation must be tied back to the microbe in nature. Even if careful measurements of the response of a microbe to temperature are made in the laboratory, we still need to know if the microbes respond in the same way in the field, a complex habitat where they often must cope with too little substrate for optimal growth. Any study that just reports laboratory-derived data is not truly an ecological study.

1.2.2 A methods perspective: microbiology under very difficult conditions

Much of our understanding of microbial ecology is quite recent. This has been due in part to the great difficulty in making accurate measurements of microbes and their activities in nature. In fact, the field of ecological microbiology has lagged far behind other areas of microbiology for just this reason. How could this lag develop when microbiology as a discipline has been so successful? One reason may be that the techniques that worked so well for the laboratory study of microbial physiology or for the study of microbially caused diseases are unsuitable for studies of microbial ecology. For example, the standard plate technique that is successfully used for isolating microbes present in nutrient-rich environments, such as diseased tissues or sewage, simply does not work for the extremely dilute waters of lakes or oceans. Often less than 1% of the bacteria from these environments will grow on nutrient-rich media. Buck (1979) points out that as late as 1966 these methods were still being seriously discussed at meetings. No other techniques were available to address the question of "what organisms are out there" until the recent advent of molecular probes (see Chapter 2).

Many other techniques that work well in the laboratory or in very nutrient-rich environments do not work as well under natural conditions. For example, microbial activity or respiration in cultures may be measured as a change in oxygen concentration over time. However, this technique is not sensitive enough to measure changes under natural conditions except by long incubations, e.g., lake water samples must be held in closed bottles for several days. When multiday measurements of this type are carried out, some of the microbes change their rates of activity, some grow, and some die; the measured respiration rate does not correspond to the actual rate in nature.

The response of lake bacteria in a water bottle points out two principles.

1 Bacteria are ubiquitous and there are billions in every liter of natural water or cubic centimeter of soil or sediment; hundreds or thousands of species are likely to be growing under any given condition. These forms that grow may or may not have been active before the sample was collected.

2 Microbes respond rapidly to changed conditions in their environment; some become active and start to grow within minutes or hours.

Both of these principles work together so that there is always microbial growth and microbial activity when measurements are made, and microbiologists always obtain good graphs of seasonal cycles from their field measurements. However, both the activity and biomass measurements can be artifacts of the experimental methods employed.

To some degree, the whole field is still methods limited. There are many successful methods but we still lack some crucial techniques that would allow a complete picture. Much of the effort in the field has been spent on investigations of the physiology and potentialities of many types of bacteria in the laboratory, yet in the field we do not even know the identity of the species carrying out many of the important processes, or the correct rate of growth of bacteria in water and sediments.

1.2.3 Immense progress in the past two decades

To counteract the gloomy picture of difficult conditions, techniques that do not work in the field, and methods limitation, we would also emphasize the tremendous strides made in the field during the past two decades. We now know how many microbes exist in habitats ranging from lakes to the depths of the ocean. We have a good idea about the relative growth rate of microbes in many environments, and we understand most of the steps in the microbial processing of carbon compounds, of nutrients such as nitrogen, and of elements such as sulfur. These giant steps are recent and they form the material reported in this book.

New methods have resulted in exciting new insights. Descriptions of these are scattered throughout the chapters; Chapter 2 specifically addresses five recent technologies that are particularly important in measuring microbial numbers, species, and biomass (see Section 2.1).

1.2.4 The perspective of processes

The different habitats of aquatic microbes – from the surfaces of particles in the ocean to the water column of oligotrophic lakes is the organizational thread running through this book. Because this book is made up of contributions from many authors, there are many different special interests and different formats for dealing with the various habitats.

The differences in presentation may obscure the point that the same processes and the same microbes (or types of microbes) occur in every habitat; it is the relative importance of any particular process or the controls on a process that may differ in different habitats. These differences often arise because of physical factors, such as temperature and light, and chemical factors, such as oxygen and the presence of hydrogen sulfide or different substrate qualities. For example, anaerobic processes are discussed in terms of redox gradients in eutrophic lakes (see Chapter 4), freshwater swamps (see Chapter 9), and salt-marsh sediments (see Chapter 11). Despite differences in the way processes are controlled from habitat to habitat, they are basically similar and the ecology of the microbes carrying them out is often similar no matter where they are found. Even a major environmental factor, such as the salt in sea water versus the absence of salt in fresh water, appears to have little or no effect on the ecology of aquatic bacteria. Wood (1965) stated that "Salinity itself does not affect bacterial processes, but merely regulates the strains of bacteria responsible for them."

For a list of the various microbial processes, the relevant biochemical pathways, and microbial energetics, students should consult a basic microbiology textbook. Fenchel & Blackburn (1979) have an excellent review of many processes from the ecological viewpoint in a chapter titled "The requirements of the bacterial cell." They point out that the energy-yielding processes for bacteria are either chemotrophic oxidation–reduction reactions or phototrophic reactions. For phototrophic organisms (plants and photosynthetic prokaryotes), there are oxygenic and anoxygenic reactions. For chemotrophic reactions, there are only oxidations of H_2, CHO, CH_4, HS^-, inorganic nitrogen compounds, and ferrous iron. Stanier *et al.*'s (1986) chapter on the "Effect of the Environment on Microbial Growth" provides a good summary of membrane transport of nutrients in relation to biochemical and physicochemical processes.

1.2.5 The perspective of scale

Initially, it might appear that there are few links between the current interest in ecological scaling and microbes, as microbes exist and operate on a single spatial scale of a few microns. However, it is not the microbes but the ecological questions that exist at a number of scales. At the global end of the scale fall questions about the total quantity of microbial production of the greenhouse gas methane from marshes and bogs. At the microscopic end of the scale fall questions of enzyme induction on the surface of bacteria. The perspective of scale is one way the entire field can be organized. We will separate scale into the following areas:

1 global;
2 regional;
3 habitat;
4 community;
5 whole organism;
6 physiological and molecular considerations.

It should be stressed that the questions asked often determine the type of measurements needed for the answers. For example, questions about the ecology of denitrifying bacteria in a biofilm clearly require identification of species. The techniques used might be to culture microbes from various layers or to identify them in a slice of the film with a fluorescent antibody method. In contrast, questions about the controls of trace gas release from an estuary will not require identification of the species. Instead, the question might be answered by field measurements of gas production in different parts of the estuary at different times of the year.

Global scale

At the level of the whole globe, which for aquatic ecologists means the entire world ocean, all lakes, rivers, freshwater and salt marshes, and even the earth ground water, the questions are concerned with the large-scale cycles of carbon, nitrogen, sulfur, and other nutrients. Typical questions include: Will microbial degradation slow if increased carbon dioxide in the atmosphere changes the chemistry of plants? and, What is the contribution of oceanic nitrogen fixation to the nitrogen requirement of marine photosynthesis?

Regional scale

The regional scale of questions includes those about processes or controls occurring in broad categories of habitats, such as tropical rivers, arctic lakes, and temperate tidal marshes. We could define this scale further to include, for example, oligotrophic northern lakes and eutrophic temperate coastal oceans. Many of the questions concern processes of nutrient cycling, e.g., carbon, nitrogen, and phosphorus, their operation and control in these different types of habitat. Typical questions address such diverse topics as:

1 What controls bacterial numbers in northern oligotrophic lakes: substrate or predators? (See Chapter 5.)

2 What is the rate of methane production in temperate freshwater marshes? (See Chapter 9.)

3 Which hypothesis (stripping, dilution, or another) explains measured fluxes at the sediment–water interface in temperate streams? (See Chapter 7.)

4 What is the role of hydrodynamics in regulating microbial populations in temperate estuarine waters? (See Chapter 11.)

Habitat scale

Questions at the habitat scale of individual lakes, estuaries, sections of oceans, such as the New York Bight, and individual marshes, often concern microbes as parts of whole ecosystems. For example, how much of the dissolved organic carbon (DOC) that enters Lake Michigan is used by microbes? Or, how important to total carbon flux is the chemosynthetic production at depth in the Black Sea? In many cases, enough is known about the whole system that microbial rates can be checked against other measurements. Thus, we can compare the annual bacterial production rate of a small lake to the measured inputs of organic matter through inflow and algal photosynthesis. If our techniques are accurate then these measurements should balance.

Many of the societal environmental problems occur at this scale. Chapter 18 discusses localized ecosystem effects of waste and cooling water discharge and Chapter 4 discusses the response of microbes to eutrophication or overfertilization.

Community scale

Interacting microbial species form communities. These community interactions take place at scales of tens of microns. Appropriate questions may concern predation of one microbe upon another or perhaps the transfer of nutrients between microbes. For example, Chapter 4 describes the chain of microbes and processes in sediments that begins with anaerobic hydrolysis, depolymerization, and fermentation. Products from fermentation are consumed by various bacteria carrying out denitrification, sulfate reduction, and methanogenesis. The location of the various organisms within this decomposer community can depend upon factors such as the presence of plant roots, sulfate, or the thickness of the oxygenated layer of the sediments. The authors identify the crucial questions as the controls of the initial exoenzymatic hydrolysis. Chapter 6 addresses questions on structure and function of river biofilms. The biofilm provides a relatively nutrient-rich and protective environment where immobilization of extracellular enzymes could provide degradation pathways for successive generations of microbes. In the water column of the oceans, the organic aggregate material may fulfill the same function in providing protection and sites for nutrient concentration (discussed in Chapter 13).

Whole organism

Most microbiologists study individual microbes or cultures of microbes grown from an individual. The information derived from these studies is really about the organism and its immediate surroundings. That is, on a scale of one to several microns for most aquatic microbes. With this approach, it is the species (or strain) that is being studied: its presence, capabilities, and physiological controls. From the ecological perspective, distribution studies of individual species are limited in that the final result is the knowledge that a particular type of organism with a particular physiological capability was present. While this is important information, we know that presence does not necessarily equal activity. Unfortunately there are no current methods that will allow measurements of rates at this level. This is particularly relevant to groundwater microbiology and is discussed in detail in Chapter 8.

The appropriate questions at the whole-organism level center around the determination of what species are present in various habitats. Once the species name or type is determined, then a great deal of ecological information may be transferable from prior laboratory studies. For example, the layering of various species of bacteria in an algal mat will inform us about the energy and substrate transfers within this community. Another type of question deals with the transfer of maximum activity rates from laboratory experimentation to the field. Watson (1965) determined the maximum rate of nitrification per cell under optimum laboratory conditions, then multiplied this rate by the number of nitrifying bacteria found in the oceanic water column to determine the maximum possible natural rate. In this particular situation, the laboratory-derived rate turned out to be far less than the rate determined in the field. From this we can conclude that other organisms must be carrying out nitrification, or that environmental factors absent in laboratory studies accelerate this process.

Physiological and molecular level

The appropriate questions at this micron or submicron scale concern the actual physiological, biochemical, and functional activities at work in nature. For example, it is well known from laboratory studies that bacteria are able to use organic matter escaping from photosynthesizing algae. However, it is difficult to measure the rates of this use in nature and to answer the questions listed below.

1 Is the organic matter produced by algae the dominant substrate for heterotrophic planktonic bacteria?

2 How close is the linkage in time between algal photosynthesis and bacterial growth, and are heterotrophic bacteria active only at the same time of day that the algae photosynthesize?

To answer these questions requires activity measurements on individual cells or examination of specific properties peculiar to bacteria. One approach is through extremely laborious techniques involving autoradiographic measurements of the amount of ^{14}C transferred from algae to bacteria. Alternatively, as specific substrates are produced by the algae, they are likely to act as promoters or repressors for specific enzyme function. Measurement of changes in regulatory molecules within bacteria, such as messenger RNA, can therefore be used to describe nutritional status.

1.3 Research needs

In this book, the ecological viewpoint of aquatic microbes will be stressed continually. Physiology and molecular biology have been brought in where appropriate (particularly in discussions of survival mechanisms in Chapters 12, 15, and 16). Although a considerable amount of material has been presented and it is obvious that we know a great deal, it appears that we still lack much of the basic information that will enable us to have a unified view of aquatic microbial ecology. In the following section, we attempt to present the missing ecological information. This falls into the following categories in natural systems:

1 species and biomass (abundance);
2 microbial processes and their rates;
3 microbial growth rates and efficiencies.

1.3.1 Species and biomass

It is amazing that microbial ecology has progressed as far as it has with so little information on which major species are active in aquatic ecosystems. We do have long lists of species isolated from various waters and sediments, but because microbes can survive in an inactive state for years, their mere presence does not equal activity (see Section 8.3.4). It is also true that in a water sample the numbers of individual bacterial cells per milliliter as determined by culture techniques are usually hundreds of times less than the numbers determined by electron microscopy or by epifluorescence. Furthermore, the bacteria that grow in culture are not necessarily the ones that are active in the environment. Here the molecular probe techniques that are just beginning to come into use will be important. At a minimum, probes developed from cells already in culture will tell us how many of these culturable species are present in natural samples. As techniques improve, they may eventually lead to the identification of all of the forms present. It should also be possible to use the polymerase chain reaction to identify genes present in the uncultured natural populations and, from this, gain clues as to the possible culture requirements of the cells. For example, Giovannoni *et al.* (1990) found DNA sequences similar to those of members of the α-purple eubacterial phylum in unculturable microbes from the Sargasso Sea. They proposed that culture media suitable for sulfide oxidizers might be used for isolating these bacteria.

What kind of ecological insights will be produced when we are finally able to identify all of the microbial species in a water sample from nature? For some species, such as sulfate reducers, it is likely that identification is sufficient to tell us a great deal about their ecology. For others, identification will tell us little about their interactions in nature. For example, in the case of the heterotrophic oxidizers of organic matter in the plankton, a number of species appear to carry out the same general transformations in culture studies. The analogy may be with the algae of the oceanic and limnic plankton. Many have been identified for decades and a large number have been studied in culture. They all compete for the low concentrations of nitrogen and phosphorus in the upper waters of oceans and appear to carry out many of the same processes in culture. Identification has not helped to complete our understanding of their ecology. The abundance of algal species in the ocean with apparently similar niches has been called the "paradox of the plankton" (Hutchinson 1961). While we do not yet know if the same paradox will appear for the planktonic bacteria, it is useful to bear in mind that identification is only one of the steps towards the ecological understanding we seek.

The exact biomass of microbes present in natural waters and sediments is another difficult problem. This quantity becomes important when we need to calculate the growth rate of microbes or quantify any measurement of the microbial cycling of carbon, nitrogen, phosphorus, or sulfur. A number of different methods have been proposed for biomass estimation but the assumptions and errors remain large. If a cell component is used to estimate biomass, such as the muramic acid of bacterial cell walls or the ergosterol of fungi, then a value for the conversion of muramic acid or ergosterol to carbon or cell volume must be used; this conversion value is difficult to apply as it may vary with the morphology and physiology of the cell. If the numbers of cells are counted by direct microscopy, then the volume must be calculated and the carbon conversion factor applied. In this case the small size of the microbes creates the problem. A slight error in the estimation of the radius of a bacterial cell will cause a large error in the volume estimate because the radius is cubed to calculate the volume of a sphere. Although we have an approximate idea of biomass, these and other problems still create too much uncertainty for ecological calculations, such as growth efficiency.

Two promising techniques are now under development. We have long known that the flow-through cell sorter can be used for determining various properties of cells; these cells were always an order of magnitude larger than natural bacteria. Recently, Robertson & Button (1989) have described a modified cell sorter for the simultaneous measurement of DNA content (fluorescent staining of nucleic acids) and cell volume (by light scattering). (Cell volume and DNA content for bacteria in a freshwater lake and a marine bay are compared in Chapter 2.) Another method makes use of the constant relationship between the leucine content of bacteria and the total protein and total carbon of cells (Simon & Azam 1989). In this method, the particles in the plankton are filtered through a series of filters and the amino acids measured for the smallest size fractions. The result is total bacterial carbon.

For most ecological purposes, the biomass measurements are not enough; we also need to measure the rate of turnover of this mass. Several mechanisms for turnover are known, such as removal by grazing or viral attack, but in most systems we do not yet understand which dominates. To measure grazing by ciliates or microflagellates, the ideal method would use the disappearance of the natural bacteria. These bacteria are growing while the measurements are being made, so we have turned to the loss of nongrowing particles for the measurement. Some of these, such as the inert fluorescent particles used to measure uptake into flagellates, may be so unnatural that there is discrimination against them by these protozoans. Real bacteria would be better, but nonreproducing buds of coliform bacteria or stained but heat-treated natural bacteria (Sherr *et al.* 1987) may also have surface properties quite different from natural bacteria. Another problem is how best to count the number of flagellates and ciliates in natural waters. Existing techniques are laborious and need improvement.

Viruses have also been proposed as controlling agents of bacterial biomass (see Chapter 19), but we need to learn more about the number of viruses present and the rate of infection before their true effect is known.

1.3.2 *Microbial processes*

There has been a great deal of work on the pathways of the various microbial metabolic processes. While there are always interesting details to be learned about any process, it is fair to say that the broad details of the possible pathways are known but that we know much less about the rates of these processes in nature. For example, the process of denitrification occurs in all sediments and some water columns. The reactions are well known, as are their requirements for nitrate and anaerobic conditions. Denitrification is given the major credit for removing the nitrogen that enters lakes and coastal waters, yet this credit has been based on little data. Not only do we need many more measurements of rates, but also improved methods of measurement. One promising method is the use of microelectrodes to measure oxygen and other gases and ions at intervals of tens of microns within biofilms or sediments (their potential use in freshwater swamps is discussed in Chapter 9). Once the gradients are known, then more accurate fluxes may be calculated. In one application, Nielsen *et al.* (1990) were able to measure both denitrification and photosynthesis in stream sediments.

In order to completely understand the ecology of aquatic microbes, we must go beyond the measurement of rates and determine their controls. Each process is likely to have different controlling factors in different systems, but one control that cuts across many systems is the quality of the organic substrates. The word "quality" is taken here to mean the nutritive value of the material for microbes; it may be difficult to define biochemically but it is obvious that fresh algal detritus is a high-quality substrate and decomposes rapidly. After some weeks, the best substrates have been removed from the detritus and decomposition is slowed; the detritus is now a low-quality nutrient source. In work on a coastal bay in Massachusetts, Banta (1991) found that denitrification in the bay occurred almost exclusively during the few weeks after the detritus from the spring algal bloom reached the sediments. This type of effect is well known but we currently have no standard measure of substrate quality.

In Chapter 7, measurement of the quality of DOC is considered the primary need for future research. The authors conclude that microbial ecologists must devise ways to measure the biological lability of this DOC and suggest that perfusion experiments with stream-bed substrates are one approach.

Finally, there is the problem of the regulation by microbes of their own activity and life histories. These are difficult questions to study in the laboratory and have yet to be addressed in natural populations. Some of the questions of regulation concern microbial growth. For example, do bacteria of the plankton regulate their own numbers?

There is evidence that bacteria will grow to a certain population density when introduced into filtered natural water. When the same water is refiltered, bacteria regrow to about the same abundance. Why? Another set of questions concerns activity. For example, does the heterotrophic activity of microbes change over the day as changes in the light within a water column change the amount of photosynthetically produced carbon? We have demonstrated a close dependence of heterotrophic activity on algal productivity in an artificially enriched arctic river (Peterson *et al.* 1985), but the evidence for a similar dependence in most unperturbed aquatic systems is weak.

1.3.3 Growth rates and efficiencies

The crucial information still missing from almost all ecological studies is the growth rate of the microbes. This information, plus a knowledge about the efficiency of carbon conversion, will allow us to bring together a quantitative map of all the transformation and growth rates in a microbial system. When this is done, our understanding may then be checked against whole-system measurements, such as the input and output of organic carbon or the total respiration of the system. These whole-system measurements (assuming they are likely to be more accurate than our small-scale measurements) may be looked at as setting the constraints or limits for the entire system. The situation now is very different, and there are many reported estimates of bacterial productivity that exceed the input rates for the entire system. While incomplete information is often to blame, why is it that the bacterial production estimates are almost always greater than the estimates of algal production? Could it be due to faulty techniques and/or interpretation? Certainly we need a few studies in contained ecosystems to measure all the inputs, outputs, and microbial processes. These studies should last for reasonably long periods of time (weeks or months) and take place in relatively large systems (e.g., the $12\,m^3$ permanent tanks at the University of Rhode Island). Not surprisingly, discussion of growth rates and efficiencies occurs in a number of chapters (particularly Chapters 10, 12, and 13).

To reach the point where we can carry out a complete system analysis, more work is needed on methods for accurately measuring productivity of all microbes in nature and especially the heterotrophs. While we may generally know the broad limits of the rates of growth, the resolution is not good enough to use the numbers in calculations of budgets, although this view is not shared by all aquatic microbiologists. No matter which view is correct, all would agree that a demonstration of the accuracy of the various methods against some absolute standard is needed.

One example of the difficulty of the growth measurement is the case of the thymidine incorporation technique. It is reasonable to assume that the incorporation of a precursor to DNA, thymidine, may be proportional to the total growth of a bacterial cell. A method has been developed that uses very low levels of added ^{3}H-thymidine, in the nanomolar range, and measures the rate of incorporation during an incubation period that does not last more than a few minutes. The rates of incorporation into DNA in the cells have been shown to change over orders of magnitude in parallel with the amount of added substrate, so we do know that, in general, the growth rate is being measured. However, we need greater accuracy. Various workers have calibrated the thymidine incorporation rate in a bottle experiment against the changes in numbers of bacteria over many hours or a few days. The result is a conversion factor; the major problem lies in the possibility that the physiological processes of the bacteria may be different in the calibration bottle compared to the natural system. After all, the argument goes, the bacteria produce a certain amount of thymidine themselves and if the endogenous production increases during this incubation then a large amount of bacterial thymidine would be produced with a relatively small amount of ^{3}H-thymidine incorporation.

In order to put microbial production into an energy or carbon budget calculation, we also need to know the efficiency of conversion of organic matter to microbial mass. This growth efficiency, on individual compounds, such as sugars, or on complex artificial media, can be measured in the laboratory, where values of 40–60% efficiency are reported. Growth efficiencies can also be measured in long-term growth experiments using natural waters, where values as low as 10% are reported. Neither of these measurements may accurately reflect the natural situation but they may give the extremes. Once again, these values are not helpful for any budgets or calculations. One problem with the long-term growth measurements, such as those of Tranvik (1988), is the possibility that the microbes of the plankton live mainly on the high-quality DOC leaking from photosynthesizing algae. When the water from lakes or oceans is filtered and the algae removed, then the bacteria switch to using the DOC of lower quality that remains after the initial bacterial degradation. The growth efficiency of bacteria on the high-quality DOC may be 50% while the efficiency on the low-quality DOC is only 10%. In nature some mixture of these two processes is undoubtedly occurring but how can this mixed reaction be measured?

1.4 Conclusion

Aquatic microbiology from an ecological perspective appears to be relatively immature. While there is a great deal of information about microbes in oceans, lakes, marshes, and streams, we lack key integrating information. We need information on the species carrying out the microbial processes, the controls of these processes, the importance of grazing by protozoans or infection by viruses as controls of the microbial biomass, and microbial growth rates in nature. Although there has been tremendous progress, key techniques for measurement are still needed before we can have faith in the results and begin to apply the techniques over a range of habitats. These techniques will allow a quantitative view of the ecology of aquatic bacteria rather than the largely qualitative and descriptive view we have today.

References

Banta G. (1991) *Decomposition and Nitrogen Cycling in Coastal Marine Sediments.* PhD thesis, Boston University Marine Program.

Buck J.D. (1979) The plate count in aquatic microbiology. In Costerton J.W. & Colwell R.R. (eds) *Native Aquatic Bacteria: Enumeration, Activity, and Ecology*, pp. 19–28. American Society for Testing and Materials, Philadelphia.

Fenchel T. & Blackburn T.H. (1979) *Bacteria and Mineral Cycling.* Academic Press, New York.

Giovannoni J., Britschgi T.B., Moyer C.L. & Field K.G. (1990) Genetic diversity in Sargasso Sea bacterioplankton. *Nature* **345**, 60–63.

Hutchinson G.E. (1961) The paradox of the plankton. *American Naturalist* **95**, 137–146.

Nielsen L.P., Christensen P.B., Revsbech N.P. & Sorensen J. (1990) Denitrification and photosynthesis in stream sediment studied with microsensor and wholecore techniques. *Limnol. Oceanogr.* **35**, 1135–1144.

Peterson B.J., Hobbie J.E., Hershey A.E., Lock M.A., Ford T.E., Vestal J.R., McKinley V.L., Hullar M.A.J., Miller M.C., Ventullo R.M. & Volk G.S. (1985) Transformation of a tundra river from heterotrophy to autotrophy by addition of phosphorus. *Science* **229**, 1383–1386.

Robertson B.R. & Button D.K. (1989) Characterizing aquatic bacteria according to population, cell size, and apparent DNA content by flow cytometry. *Cytometry* **10**, 70–76.

Sherr B.F., Sherr E.B. & Fallon R.D. (1987) Use of monodispersed, fluorescently labeled bacteria to estimate *in situ* protozoan bactivory. *Appl. Environ. Microbiol.* **53**, 958–965.

Simon M. & Azam F. (1989) Protein content and protein synthesis rates of planktonic marine bacteria. *Mar. Ecol. Prog. Ser.* **51**, 201–213.

Stanier R.Y., Ingraham J.L., Wheelis M.L. & Painter P.R. (1986) *The Microbial World*, 5th edn. Prentice-Hall, New Jersey.

Tranvik L.J. (1988) Availability of dissolved organic carbon for planktonic bacteria in oligotrophic lakes of differing humic content. *Microb. Ecol.* **16**, 311–322.

Vestal J.R. & Hobbie J.E. (1988) Microbial adaptations to extreme environments. In Lynch J.M. & Hobbie J.E. (eds) *Micro-organisms in Action: Concepts and Applications in Microbial Ecology*, 2nd edn, pp. 193–206. Blackwell Scientific Publications, Oxford, UK.

Watson S.W. (1965) Characteristics of a marine nitrifying bacterium, *Nitrosocystis oceanis* sp. n. *Limnol. Oceanogr.* **10** (Suppl.), R274–R289.

Wood E.J.F. (1965) *Marine Microbial Ecology.* Chapman and Hall, London.

2

John H. Paul

The advances and limitations of methodology

2.1 Introduction

Our understanding of microbial processes in the environment is only as good as our methods. Perhaps in no other scientific field have dogmas been so completely reversed by new information resulting from improvements in methodology. One such classic example is our understanding of the abundance of bacteria in marine and freshwater environments. Enumeration of bacteria by plate counting had resulted in aquatic bacterial abundances of 10^1–$10^4\,ml^{-1}$ (Sieburth 1971; Gunderson *et al.* 1972), implying that bacterial biomass was insignificant in most aquatic environments. The advent of epifluorescence microscopy and the direct counting of bacteria (Francisco *et al.* 1973; Hobbie *et al.* 1977) revealed that bacterial abundance exceeded plate-count estimates by several orders of magnitude. These observations, coupled with the advent of methods to measure bacterial production (Karl 1981; Fuhrman & Azam 1982) resulted in the Microbial Loop hypothesis (Azam *et al.* 1983), which has elevated the significance of heterotrophic bacteria to the same order of magnitude as primary producers in terms of amount of carbon cycled.

In this chapter I will look at the five technologies which are currently revolutionizing our understanding of the role of microorganisms in the environment:

1 immunology and immunocytochemistry;
2 flow cytometry;
3 image analysis technology;
4 nucleic acid technology (NAT);
5 transmission electron microscopy (TEM).

In some cases, the technology is not new, e.g., TEM, but its application to the study of microorganisms in the environment is novel. The list is not intended to be a list of all new methods in microbial ecology; these technologies were chosen because they have proved themselves by providing new insights in aquatic microbiology.

2.2 Immunology and immunocytochemistry

2.2.1 Detection of unspecified cell surface antigens

The application of immunological methods to microbial ecology involves the production of antibodies in mammals after injection with antigens, i.e., microbial cells or parts thereof. Thus, cultivation of the target organism (or a close seriological relative) is required (Bohlool & Schmidt 1980). The production of antibodies is most often accomplished by polyclonal technology, which results in a heterogeneous collection of antibody molecules that react to various antigenic sites on the target cell surface. If an antibody-producing lymphocyte is fused with a melanoma or other cancer cell, a monoclonal antibody is produced. The monoclonal antibody is thus specific for one antigenic site. In application to most problems in microbial ecology, the desired antibodies (monoclonal or polyclonal) are conjugated with a fluorescent dye, such as rhodamine or fluorescein isothiocyanate, yielding a fluorescent antibody which can be used to optically detect the target organism against a background of nontarget organisms. The technology is most often used in the "sandwich" or indirect immunofluorescence assay, where the primary antibody (nonfluorescent) reacts to the target cells, and a second antibody, directed against the first, contains the fluorescent conjugate.

The utility of immunological techniques in aquatic microbiology was demonstrated elegantly by the early work of Ward and colleagues on marine nitrifying bacteria (Ward 1982; 1984; Ward *et al.* 1982; 1984). This group of bacteria was perhaps best suited for immunological studies because of relatively few taxa (and thus serogroups) of isolated strains involved in nitrification (Ward & Carlucci 1985).

More recently, immunofluorescence technology has been successfully used to identify types of *Synechococcus* occurring in the marine environment (Campbell & Carpenter 1987; Campbell 1988; Campbell & Iturriaga 1988). *Synechococcus* cells are minute prokaryotic picoplankton that possess little morphological variation that might aid in identification of strains. Unlike many prototrophic bacteria, a number of *Synechococcus* strains have been isolated, and immunological studies of natural populations suggest that those isolated reflect the dominant serotypes found in at least some oceanic environments. For example, antisera raised to *Synechococcus* clone WH7803 cross-reacted with other tropical isolates (Campbell & Iturriaga 1988) and could be used to enumerate *Synechococcus* in tropical and subtropical oceanic environments by indirect immunofluorescence microscopy (Campbell & Carpenter 1987). In fact, 85% of the *Synechococcus*-type cells seen in one tropical environment crossreacted with anti-WH7803 antibodies. Combining immunofluorescence and single-cell microfluorometry has enabled determination of the phycourobilin/phycoerythrobilin chromophore content of phycoerythrin in natural *Synechococcus* populations (Campbell & Iturriaga 1988).

Eukaryotic ultraplankton, minute cells of diverse taxa with little morphological variation, have also been successfully characterized and identified by immunofluorescence (Shapiro 1988). Studies using antisera to seven clonal ultraplankton from six major algal groups indicated that the antibodies were species-specific, enabling detection of the same species isolated from different parts of the oceans (Shapiro *et al.* 1989a,b).

In the examination of unidentified clones of open oceanic eukaryotic ultraplankton, classification of clones by pigment analysis was verified by immunological methods (Shapiro *et al.* 1989b). This approach, when used in conjunction with epifluorescence microscopy and/or flow cytometry, should enable classification to major taxonomic division and perhaps species, of dominant eukaryotic ultraplankton in the oceans.

2.2.2 Detection of specified antigen targets

The immunological approach has been successfully used to detect marine toxins, particularly those of dinoflagellate origin including brevitoxin (Baden *et al.* 1988), ciguatoxin (Hokama *et al.* 1988), saxitoxin, and gonyatoxin (Guire *et al.* 1988).

Antisera to enzymes catalyzing important ecological functions have been used to detect these activities and enumerate microorganisms containing these target enzymes. Orellana *et al.* (1988) made antibodies to ribulose-1,5-bisphosphate carboxylase oxygenase (Rubisco) and the major light harvesting pigments of dinoflagellates, the peridinin/chlorophyll *a* binding protein. Rubisco is the enzyme involved in the fixation of carbon dioxide in nearly all autotrophic organisms on earth, and has a relatively conserved primary structure. The antibodies were used in indirect immunofluorescence analysis of phytoplankton cultures by epifluorescence microscopy and flow cytometry to detect regulation of Rubisco in response to medium nitrogen concentration.

Anti-Rubisco antibodies have also been used to locate carbon dioxide-fixing symbionts in bivalve tissues (Cavanaugh *et al.* 1988).

Paerl *et al.* (1989) used colloidal gold-conjugated antibodies to detect nitrogenase in the pelagic nitrogen-fixing cyanobacterium *Trichodesmium thiebautii*, an organism that has only recently been cultivated. *Trichodesmium* possesses no cellular differentiation or obvious mechanism to protect its nitrogenase from oxygen inactivation. TEM of sections of the immunogold-antibody treated *Trichodesmium* filaments indicated that all cells within a filament contained approximately equal quantities of nitrogenase with no indication of intracellular localization. This work implies that the degree of colony formation by individual *Trichodesmium* filaments regulates the production of nitrogenase, and that nitrogen fixation must occur during quiescent sea states (Paerl *et al.* 1989).

Currin *et al.* (1990) raised antibodies to the FeS protein of nitrogenase of *Rhodospirillum rubrum*. Immunoblots and epifluorescence microscopy indicated that these antibodies crossreacted with marine and freshwater diazotrophic bacteria and cyanobacteria. This technology should be extremely useful in determination of the nitrogen fixers in mixed microbial assemblages, such as microbial mats, detrital particles, and other complex microbial communities.

Immunological techniques have also been used to detect gene expression in individual cells. The regulation of denitrification enzymes in response to oxygen tension and electron acceptor concentrations has been studied by rocket immunoelectrophoresis (Korner & Zumft 1989). Antisera to nitrate reductase, nitrite reductase, and N_2O reductase were used to determine the precise oxygen tensions that enabled synthesis of these enzymes.

The combination of nucleic acid technology (NAT) and immunology was elegantly employed by Zehr *et al.* (1990) for the study of nitrogenase regulation in *Trichodesmium* in the tropical oceans. After successfully amplifying and cloning a portion of the *nifH* gene (see Section 2.5.4), the gene product was synthesized using an *in vivo* translation system. The protein generated was used to raise antisera to the *Trichodesmium* nitrogenase. Using these antibodies, diel regulation of nitrogenase in natural communities of *Trichodesmium* was observed (Capone *et al.* 1990).

2.3 Flow cytometry

Flow cytometers or fluorescence activated cell sorters are instruments that were originally developed for use in the study of cell biology by analyzing optical parameters of many individual cells in a short space of time. Most flow cytometers operate by aspirating a sample by use of a sheath fluid of similar index of refraction as the sample. The sample flows through an orifice and across the path of a light source (usually a laser), and measurements of optical parameters (typically light scatter and fluorescence) are made on each cell by photodetectors. The signals received are processed by a computer and, if selected, can be the basis for automatically sorting and collecting cells of the predefined optical properties.

The use of flow cytometry in microbial ecology has tremendous potential which is only starting to be realized. The attraction of the use of flow cytometry is the rapid analysis of relatively large sample sizes, i.e., 10^4–10^6 cells per minute, by fluorescence and light scatter in a nondestructive manner with the potential to sort and separate based upon these optical parameters. Commercially available flow cytometers work best with cells $>0.5\ \mu m$ and at concentrations of $\sim 10^6\ ml^{-1}$, but modifications have been made to detect smaller and less abundant cells in natural water samples.

Perhaps the most obvious utility of flow cytometers is the analysis

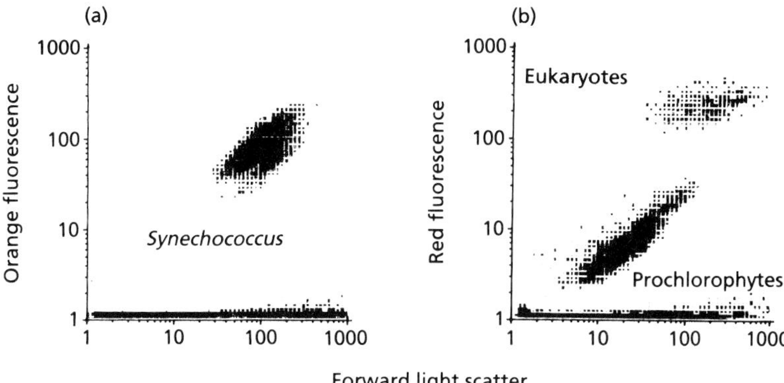

Figure 2.1 Flow cytometric signatures of (a) *Synechococcus*, and (b) prochlorophytes and eukaryotic phytoplankton from a sample taken near Bermuda in 1989. Orange fluorescence, an indication of phycoerythrin content, or red fluorescence, an indication of chlorophyll content, are plotted against forward angle light scatter. In (b) electronic gating has prevented cells that contain orange fluorescence from being plotted, thus segregating against *Synechococcus*-type cells. From Olson *et al.* (1990).

of natural phytoplankton populations by autofluorescence. This technology has been successfully used to enumerate and separate cyanobacterial phototrophs from other phytoplankton (Yentsch *et al.* 1983; Olson *et al.* 1985; Li & Wood 1988) based on phycoerythrin (orange) and chlorophyll (red) fluorescence. Fine differences in types of phycoerythrin, i.e., phycourobilin or phycoerythrobilin chromophores, in laboratory clones (Wood *et al.* 1985) and natural marine populations of *Synechococcus* (Olson *et al.* 1988) could be detected by flow cytometry.

Perhaps the most noteworthy finding resulting from the application of flow cytometry to environmental microbiology was the discovery of *Prochloron*-type phototrophs in the oceans (Chisholm *et al.* 1988). Prochlorophytes are prokaryotic organisms that contain chlorophyll *b* and are thought to be the progenitors of green algae and higher plants. The unique fluorescence signature and small size ($<0.8\ \mu m$) differentiated these organisms from *Synechococcus*-type cells (Figure 2.1). The abundance of these organisms ($\sim 10^5\ ml^{-1}$) and their location in the water column, typically below the *Synechococcus*-type picoplankton, indicated that these organisms were an extremely important component of the open oceanic photoautotrophic community. Improvements in sensitivity in flow cytometry enabled detection of prochlorophytes over the entire euphotic zone, even though the fluorescence of near-surface prochlorophyte populations was extremely faint (Olson *et al.* 1990). The median depth of the prochlorophyte population was correlated to the median depth of the nitracline (Figure 2.2).

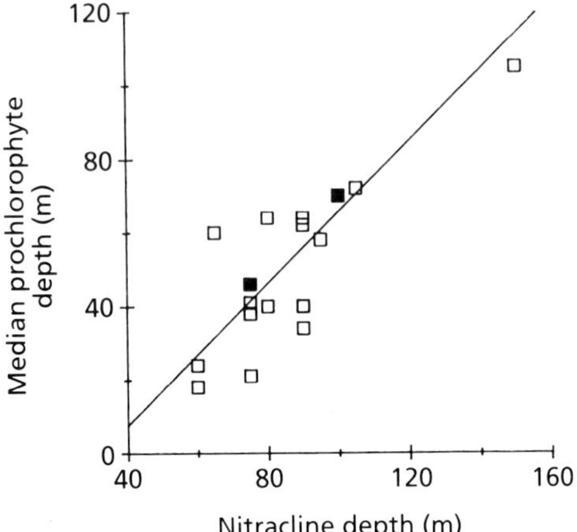

Figure 2.2 Median depth of the prochlorophyte populations as a function of nitracline depth in the North Atlantic for samples taken in May 1988 (solid squares) and in May 1989 (clear squares). From Olson *et al.* (1990).

The capability to stain DNA with specific fluorochromes (Olson *et al.* 1983) has enabled a variety of questions concerning the control of growth in phytoplankton to be investigated (Vaulot & Chisholm 1987; Armbrust *et al.* 1990). For example, control of the cell cycle and periods of DNA synthesis have been investigated in marine *Synechococcus* strains with this technology (Armbrust *et al.* 1989). Binder & Chisholm (1990) demonstrated that there were multiple genome copies in *Synechococcus* PCC 6301, even in slowly growing cells. They concluded that DNA synthesis was initiated asynchronously in this organism.

Because flow cytometry was developed in the biomedical fields for the study of mammalian cells, the technology was not designed to detect the much fainter fluorescence associated with prokaryotic cells. Advances in flow cytometry technology have now enabled detection of cultured bacteria (Van Dilla *et al.* 1983; Amann *et al.* 1990a; Sanders *et al.* 1990), and the much smaller cells in environmental samples (Robertson & Button 1989). By using high-intensity ultraviolet (UV) excitation, and optimizing optics and electronics, Robertson and Button found they could detect and enumerate natural populations of bacteria from marine and freshwater environments in Alaska. Forward angle light scatter (FALS) was found to be linearly proportional to cell volume, enabling determination of mean cell volume for any given population of cells. A limitation of the method was the requirement for use of high-intensity UV wavelengths, which greatly shortened the life of the laser.

Sanders *et al.* (1990) improved upon the method of Van Dilla *et al.* (1983) for guanine+cytosine (G+C) content determination in bacteria by flow cytometry. By staining cells with both chromomycin A3, which stains only GC-regions of DNA, and Hoechst 33258, which stains adenine+thymine (AT) regions of DNA, it was found that G+C content could be determined by the ratio of the fluorescence of these two stains. The major advantage of flow cytometry over standard methods for determining G+C content was obviating the need for highly purified DNA from the strains to be analyzed and the capability to analyze mixtures of bacteria.

Flow cytometry may be used in the analysis of phylogenetic diversity of natural microbial populations. Amann *et al.* (1990b) investigated discrimination of phylogenetically diverse bacteria by flow cytometry of cells labeled with fluorescent oligonucleotides specific for recombinant RNA (rRNA) targets (see Section 2.5.3). The methyl-rhodamine conjugated oligonucleotides were designed to hybridize to eubacteria, eukaryotes, sulfate-reducing bacteria, and the genus *Desulfobacter*. Fluorescence was used in conjunction with FALS to differentiate *Desulfovibrio gigas* from *Escherichia coli* by staining with the sulfate-reducing bacteria probe, even in the presence of excess *E. coli*. Use of multiple-fluorescent probes to hybridize to different sites of the same rRNA molecule resulted in enhanced fluorescence and detection. The future application of this technology may be to follow population dynamics by changes in staining with specific oligonucleotides. A limitation of the method as performed was the use of methyl-rhodamine as fluorochrome, the fluorescence of which overlaps with chlorophyll autofluorescence. These preliminary studies were performed with organisms in culture, and, with refinement, this technology should be applicable to complex assemblages in aquatic environments.

2.4 Image analysis technology

The evolution of the bacterial direct count technology (Francisco *et al.* 1973; Hobbie *et al.* 1977) has revolutionized our understanding of the abundance and size of microorganisms in aquatic environments. Direct visualization of microbial cells enables sizing which is required for biomass estimation, which can be used for determination of carbon and energy flows through microbial populations. To reduce the tedium and subjectivity of these techniques, image analysis systems have been employed in conjunction with epifluorescence microscopy, and scanning electron microscopy (Krambeck *et al.* 1981), for bacterial enumeration and sizing (Sieracki *et al.* 1985; Bjornsen 1986; David & Paul 1989).

Salient components of all image analysis systems used in microbial ecology are:

1 an epifluorescence microscope;
2 a camera detection system;

3 an image processor to digitize the signal;
4 software and/or hardware to analyze the digitized image.

The enumeration of bacteria with such systems (even on shipboard) can be rapidly accomplished (Sieracki *et al.* 1985; Bjornsen 1986; David & Paul 1989). Problems arise in counting when detrital particles are abundant in samples or when bacteria clump together.

The measurement of microorganisms for biovolume and biomass determination present problems not encountered when simply counting digitized images. These include converting a two-dimensional image into a three-dimensional image and errors in size caused by "haloing" associated with fluorescent images. Recently, algorithms have been developed to mathematically and electronically determine the edge of an image, termed segmentation (Sieracki *et al.* 1989a). Conversion of two-dimensional images into volumes was first accomplished by use of simple formulae, i.e., assuming particles are spheres, prolate spheroids, or cylinders with spheres on the ends (Sieracki *et al.* 1985; David & Paul 1989). Because volumes are based upon the cube of the linear dimension, small errors in measurement or assumption in shape can result in large errors in biovolume. Recently, Sieracki *et al.* (1989b) devised a method to integrate two-dimensional images into three-dimensional shapes by assuming symmetry in the third dimension, which resulted in greater fidelity to actual volumes.

Image analysis technology has also been employed to estimate phytoplankton growth rates based on cellular DNA content determined by microfluorometry (Carpenter & Chang 1988; Chang & Carpenter 1988). The fluorescence intensity of 4,6-diamidino-2-phenylindole (DAPI)-stained nuclei of the dinoflagellate *Heterocapsa triquetra* was shown to be proportional to nuclear DNA content. Growth rates via *in situ* nuclear DNA content agreed with those determined by cell counts (Chang & Carpenter 1988).

Recently, real color image analysis has been used in microbial ecology (Sieracki & Viles 1990). In real color image analysis, the image is divided into three color planes (blue, red, and green). Bacterial cells stained with DAPI fluoresce blue, and can be differentiated from detritus, which stains yellowish-green. Heterotrophs can be differentiated from autotrophs because autotrophs fluoresce when illuminated with blue light (450–490 nm). Chlorophyll will fluoresce red, and phycoerythrin (the dominant pigment in cyanobacterial picoplankton) will fluoresce orange. Differentiation is accomplished by looking at changes in emission brightness in red, green, and blue using blue and UV excitation (Figure 2.3) (Sieracki & Viles 1990).

Many of the applications of flow cytometry that require differentiation of cell types based upon autofluorescence or fluorochrome

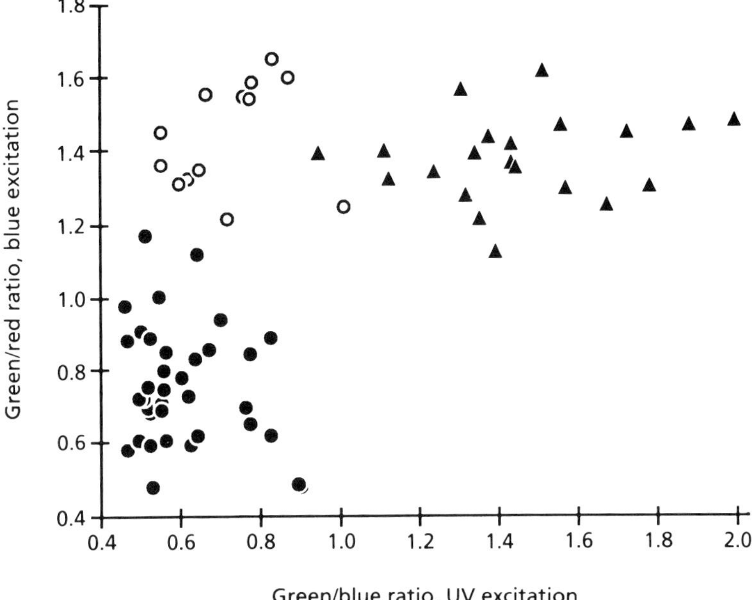

Figure 2.3 Discrimination of detrital particles (triangles), heterotrophic (clear circles), and phototrophic (solid circles) nanoplankton cells from Chesapeake Bay on the basis of different color fluorescence emissions under blue and violet excitation in samples stained with both DAPI and proflavine, using color image analysis. The ratios of the average brightness of each cell in the red, green, and blue color channels were used. DAPI-stained DNA in living cells appears blue under UV excitation resulting in a lower green to blue ratio than detritus. Both nucleus and cytoplasm of heterotrophs and phototrophs appear green under blue excitation. The presence of chlorophylls or phycoerythrin gives these cells a red fluorescence. From Sieracki & Viles (1990).

staining could be accomplished by color image analysis. Phylogenetic probing using 16S rRNA probes (see Section 2.5.3) could also be accomplished with this technology. An advantage of color image analysis systems is the real size and shape information that would be obtained for each cell detected. The ability to discriminate a range of cell size "bins" and fluorescence characteristics could enable differentiation of many subpopulations of microorganisms compared to that separable by flow cytometry. Limitations of color image analysis are its expense, lower resolution, and sensitivity compared to black and white image analysis. A general limitation to image analysis is the tendency for cells to fade (termed photooxidation) under illumination.

The future of image analysis may be tied to the introduction of cooled, charge-coupled device (CCD) cameras (Hiraoka *et al.* 1987). There is a linear response with the CCD camera over a much greater range of light intensities (termed dynamic range) than a video camera. CCD cameras can discriminate 4096 grey levels, compared to 256 for a video camera, resulting in greater resolution and ease in image segmentation. High-resolution CCD cameras (1000 × 1000 pixels and some even 2000 × 2000) will be vast improvements over the 256 × 256 video cameras (Sieracki & Webb 1991). Lastly, high-performance three-chip color CCD cameras are now being made available. As this technology develops, greater developments in the application to aquatic microbiology should be expected.

2.5 Nucleic acid technology

2.5.1 Nucleic acid extraction and purification

As with immunological methods and flow cytometry/image analysis technology, the goal of NAT is the characterization of natural populations without the need to isolate or cultivate individual community members. In most characterizations based on NAT there is a requirement for purification of nucleic acids from a dilute or sparse microbial population. When the analysis requires simply probing – nucleic acid hybridization (NAH) – of the community DNA, purification is usually not essential. However, when DNA or RNA is to be used for further molecular studies, purity is essential.

Extraction of nucleic acids from soil or sediment has the advantage over water column environments in that bacterial populations are more numerous (per unit volume) by several orders of magnitude. The disadvantage of extraction from soil or sediment is the contaminants, humic materials, and plant pigment degradation products from which the DNA must be purified. Two approaches to this problem have been successfully employed, including separation of the bacteria from soil particles prior to lysis (Holben *et al.* 1988) or direct extraction of total DNA from soil or sediment (Ogram *et al.* 1987; Steffan *et al.* 1988). An integral component of sediment, and occasionally water column (Paul *et al.* 1990), nucleic acid purification is the use of polyvinyl polypyrrolidone (PVPP) (Holben *et al.* 1988; Steffan *et al.* 1988; Weller & Ward 1989; Paul *et al.* 1990). This insoluble polymer binds humic and polyphenolic materials, preventing their copurification with DNA which occurs when using standard nucleic acid purification protocols, i.e., phenyl chloroform extraction and alcohol precipitation.

All water column DNA concentration techniques use some form of filtration. Collection of cells on 0.22 μm membrane filters followed by standard lysis procedures (Fuhrman *et al.* 1988; Paul *et al.* 1990) was reportedly more efficient than hollow fiber filtration, resulting in recoveries estimated at 25–50% for 8 to 40 liters of seawater. Sterivex® filter cartridges have been used to concentrate bacteria followed by extraction of nucleic acids in the same cartridges (Sommerville *et al.* 1989). This technology was estimated to be between 17 and 40% efficient but only enabled sampling of between 0.3 and 1 liter of water. Tangential flow filtration has been used to concentrate picoplankton from 1700 to 5000 liters of oligotrophic ocean water with a 37% efficiency (Giovannoni *et al.* 1990b). This procedure resulted in milligram quantities of DNA which were used for 16S rRNA analysis (see below).

Filtration has been used to collect viruses and viral nucleic acids for probing (Jiang *et al.* 1986; 1990) from large volumes (189 liters) of estuarine water.

2.5.2 Use of nucleic acid hybridization in environmental microbiology – non-rRNA targets

NAH, i.e., molecular probing, has been widely used to detect genes of specific bacterial metabolic potential, including heavy metal detoxification, aromatic/xenobiotic degradation, and pathogenic bacteria in a variety of samples (for earlier work, see Hazen & Jimenez 1988). In most cases this technology involves a probe to a target gene sequence, varying from oligonucleotides to several kilobases to entire organism genomes, that is labeled chemically or radiochemically for use in hybridization to denatured, immobilized DNA or RNA. The procedure is most often used with isolates, either by probing nucleic acids partially purified from cultures (Morris *et al.* 1987; Bermudez & Hazen 1988; Barkay *et al.* 1989a; Barkay & Liebert 1990) or for "first" isolates by colony hybridization (Barkay *et al.* 1985; Sayler *et al.* 1985; Walia *et al.* 1990).

Fredrickson *et al.* (1988) combined the most probable number format with NAH in a microtiter plate assay for detection of Tn5 in rhizobia or pseudomonads added to soil. Detection limits were 10–100 cells g^{-1} soil.

Perhaps the greatest attraction of NAH is to detect organisms present in the environment by DNA community extraction and probing without the need for cultivation. Extraction of community DNA for NAH has been employed to detect pathogenic viruses (Jiang *et al.* 1986; 1990), *nifH* genes in sediments (Ogram *et al.* 1987), *Bradyrhizobium* in soil (Holben *et al.* 1988), polychlorinated biphenyl-degrading bacteria in contaminated soils (Walia *et al.* 1990), *mer* genes in aquatic bacteria (Barkay *et al.* 1989b), and plasmid-bearing *Frankia* in crushed root nodules (Simonet *et al.* 1988).

The utility of probing nucleic acid extracts of environmental samples may be greatest for detection of known strains, including genetically engineered microorganisms, added to the environment for agricultural, insecticidal, and bioremediation purposes (Holben *et al.* 1988). In this application, stringency of hybridization can be kept high and background hybridization detected and corrected by using uninoculated controls. The limitations of NAH can be best seen by detection of target genes in ambient communities (Barkay *et al.* 1989a,b; Walia *et al.* 1990). Here we see the conflict between conservation of gene sequence, stringency of hybridization, and nonspecific (background) hybridization. A common observation with this approach is that there are many more bacteria possessing the target phenotype than the target genotype (Barkay *et al.* 1989a,b). Where several nonhybridizing mercury-reducing isolates were probed under lower stringency, all hybridized to the *merA* (TN501) probe (Barkay & Liebert 1990), implying a common molecular mechanism for Hg^{2+} resistance and volatilization. Lowering stringency in probing environmental nucleic acid extracts is risky because not only will the less homologous target operons be detected, but undoubtedly a host of other nonhomologous genes.

The second problem can be the detection of the target gene in the absence of phenotypic activity (Walia *et al.* 1990). Therefore, the presence or absence of hybridization to DNA from a natural population should be interpreted with caution unless a diversity of controls are available. The direct hybridization approach in some applications is being supplanted by the more specific polymerase chain reaction (PCR) approach.

A novel approach to community genome analysis has been taken by Lee & Fuhrman (1990; 1991a). In place of using defined probes from plasmids, chromosomes, transposons, or specific oligonucleotides, they used purified community DNA as a probe for similar sequences from the same or different environments. Because no defined gene sequence is used as a probe, only information on the relative similarities of the populations can be obtained, in pairwise comparisons. Using this approach, two summer samples of bacterial populations in Long Island Sound were found to be over 95% similar, but only 40% similar to fall populations (Lee & Fuhrman 1990). Coral reef bacteria were <10% similar to those of all other environments. In open ocean water, deep populations (500–1000 m) were 30–90% similar, but totally different (<5%) from surface bacteria (Lee & Fuhrman 1991a). Bacteria from the different ocean basins were <16% similar (Lee & Fuhrman 1991b). These results indicate that spatial and temporal variations strongly affect population composition. Such technology possesses the capability to determine the time and space scales for changes in microbial community composition. However, this technology will not enable determination of how these populations are changing, what components are different, and why the population is changing.

2.5.3 16S rRNA technology

The advent of analysis of rRNA sequences has revolutionized bacterial phylogeny and may well revolutionize microbial ecology. The 16S-like rRNAs have been particularly useful, owing to their large size (1500–2000 nucleotides), in providing quantitative evolutionary relationships between microorganisms (Olsen *et al.* 1986; Woese 1987). Additionally, when rRNA molecules are the target in hybridization studies the potential sensitivity is much greater than detection of DNA sequences, since actively growing cells can contain 10^4 ribosomes per cell (Ingraham *et al.* 1983).

Two approaches have been taken in application of 16S rRNA technology to microbial ecology:

1 detection of bacterial types by NAH;

2 sequence analysis of "clones" taken from the natural population. (Here clone refers to the isolation of a target rRNA gene sequence into one of the vectors used in molecular biology.)

The utility of the probing approach was shown by Giovannoni *et al.* (1988) using kingdom-specific probes targeted for conserved 16S rRNA sequences found in all living cells (universal probes) and sequences found in eubacterial, eukaryotic, and archaebacterial kingdoms. Unlike the NAH technology described above, the stringencies for hybridization were well defined for the oligonucleotide probes described. A control for nonspecific hybridization was the complementary strand oligonucleotides for the universal probe. The kingdom-probing approach was also used to demonstrate that the majority of the cellular DNA in the oligotrophic surface waters was eubacterial (Figure 2.4) (Giovannoni *et al.* 1990b). The specificity of 16S rRNA probes could be designed to differentiate closely related species and strains (Stahl *et al.* 1988).

By linking these probes to fluorescent dyes, questions concerning phylogenetic diversity could be answered by epifluorescence microscopy (DeLong *et al.* 1989b; Amann *et al.* 1990a) and flow cytometry (see above) (Amann *et al.* 1990b). The specificity of the fluorescent probes, hybridization, and washing conditions could be adjusted to enable single mismatch discrimination between probe and nontarget sequences (Amann *et al.* 1990a). This approach was also used to detect uncultured bacterial symbionts (*Holospora* spp.) in the nuclei of protozoans (Amann *et al.* 1991) and symbiotic cellulolytic, nitrogen-fixing bacteria in shipworms (Distel *et al.* 1991).

The sequence analysis approach has provided conclusive proof that isolation and cultivation has not adequately described the microbial diversity found in the environment. The sequence approach calls for cloning of 16S rRNA genes from community DNA. A limitation of the approach is that the investigator usually does not see from which organism the genes arise, thereby losing all other phenotypic information, i.e., morphology, pigment composition, physiology, etc. Additionally, it is not known if the sequence was derived from living cells, dead cells, or extracellular DNA.

The classic approach to cloning 16S genes is to make a shotgun

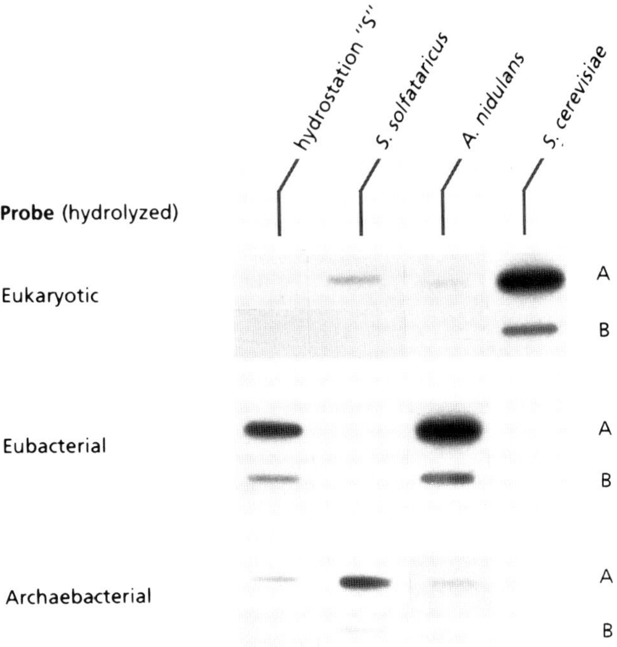

Figure 2.4 Probing of purified DNA from hydrostation sulfur from the Sargasso Sea, *Sulfolobus solfataricus, Anacystis nidulans* (*Synechococcus* sp.), and *Saccharomyces cerevisiae* with hydrolyzed 16S or 18S RNA from *S. cerevisiae* (eukaryotic probe), *Oceanospirilum linum* (eubacterial probe) or *S. solfataricus* (archaebacterial probe). Row A is 1 μg DNA, Row B is 0.1 μg DNA. The strong hybridization occurring with the eubacterial probe suggests that most of the DNA in the oligotrophic oceanic sample is of eubacterial origin. From Giovannoni *et al.* (1990b).

library from DNA isolated from the environment. Although this method has been successfully used by several investigators (DeLong *et al.* 1989b), a drawback is the relatively large amounts of DNA required, since only a small percentage of the clones will contain 16S sequences of interest. The second approach is to isolate clones from a specific part of the 16S molecule. In the method of Weller and Ward (Weller & Ward 1989; Ward *et al.* 1990a,b), the rRNA is isolated from the population and converted to complementary DNA (cDNA) by use of a primer specific for a 16S sequence. The cDNA – termed ribosomal complementary DNA (rcDNA) – is cloned into a vector and is then available for sequencing. Because the sequence is derived initially from RNA and not DNA, the chances are greater that these sequences reflect living, actively growing organisms, than those obtained from cloning of DNA.

The third method involves amplification of sequences by the PCR (Giovanonni *et al.* 1990a,b). The amplified 16S sequences are then cloned into vectors. As with the rcDNA methods, only rRNA gene clones are obtained.

That this methodology detects organisms that are truly not cultivatable is perhaps best illustrated by the work of Ward *et al.* (1990a,b) on the well-characterized Octopus Springs microbial mat. This cyanobacterial mat community was thought to be composed of primarily two organisms, *Synechococcus lividens* and the photosynthetic bacterium *Chloroflexus aurantiacus*. Sequence analyses of eight unique rRNA clones from this mat were found to be different from those of any organism previously isolated from the mat or from any other geothermal environment (Ward *et al.* 1990a,b). This work illustrates that even simple microbial communities intensively investigated by classical techniques may have important microbial components never before isolated.

Giovannoni *et al.* (1990a,b) sequenced 12 clones from amplified 16S rRNA genes from picoplankton of the Sargasso Sea. These clones were divided into two clusters, one termed Sar 7, which was phylogenetically related to cyanobacteria, prochlorophytes, and chloroplasts, and a second cluster in the α-purple division, termed the Sar 11 cluster. These organisms were not closely related to any organisms currently in culture, being derived from somewhat distant ancestors to known forms (the closest being *Agrobacterium tumefaciens*). These organisms were thought to be representative of the oligotrophic bacterioplankton of the Sargasso Sea, and were shown to comprise ~12.5% of the ribosome DNA gene abundance of this environment. Further work by DeLong and Schmidt (DeLong *et al.* 1989a; DeLong 1991; Schmidt *et al.* 1991) in oligotrophic Pacific surface waters near Hawaii yielded organisms closely related to those in the Sar 11 cluster (ALO 21, ALO 38), some organisms related to the Sar 7 cluster, and some isolates related to heterotrophic marine bacteria in culture (*Alteromonas haloplanktis* and *Vibrio harveyi*) (Figure 2.5). Additional work by the Giovanonni lab on sequence analysis of clones obtained from the Sargasso Sea indicated other sequences similar to the Sar 11 cluster (Sar 95), a strain somewhat related to *Erythrobacter* (Sar 83), a strain distantly related to *Oceanospirilum linum* (Sar 92), and some oxygenic photoautotrophs (Sar 100 and Sar 139), in the Sar 7 cluster (Britschgi & Giovanonni 1991). This work demonstrates the circumglobal phylogenetic similarity in noncultivatable oceanic oligotrophic picoplankton.

A limitation of the technique of rRNA sequence analysis is the tremendous work required to process one sample. The technology as it exists today does not lend itself to intensive time series or profile sampling. Improvements in sequencing technology and cloning may

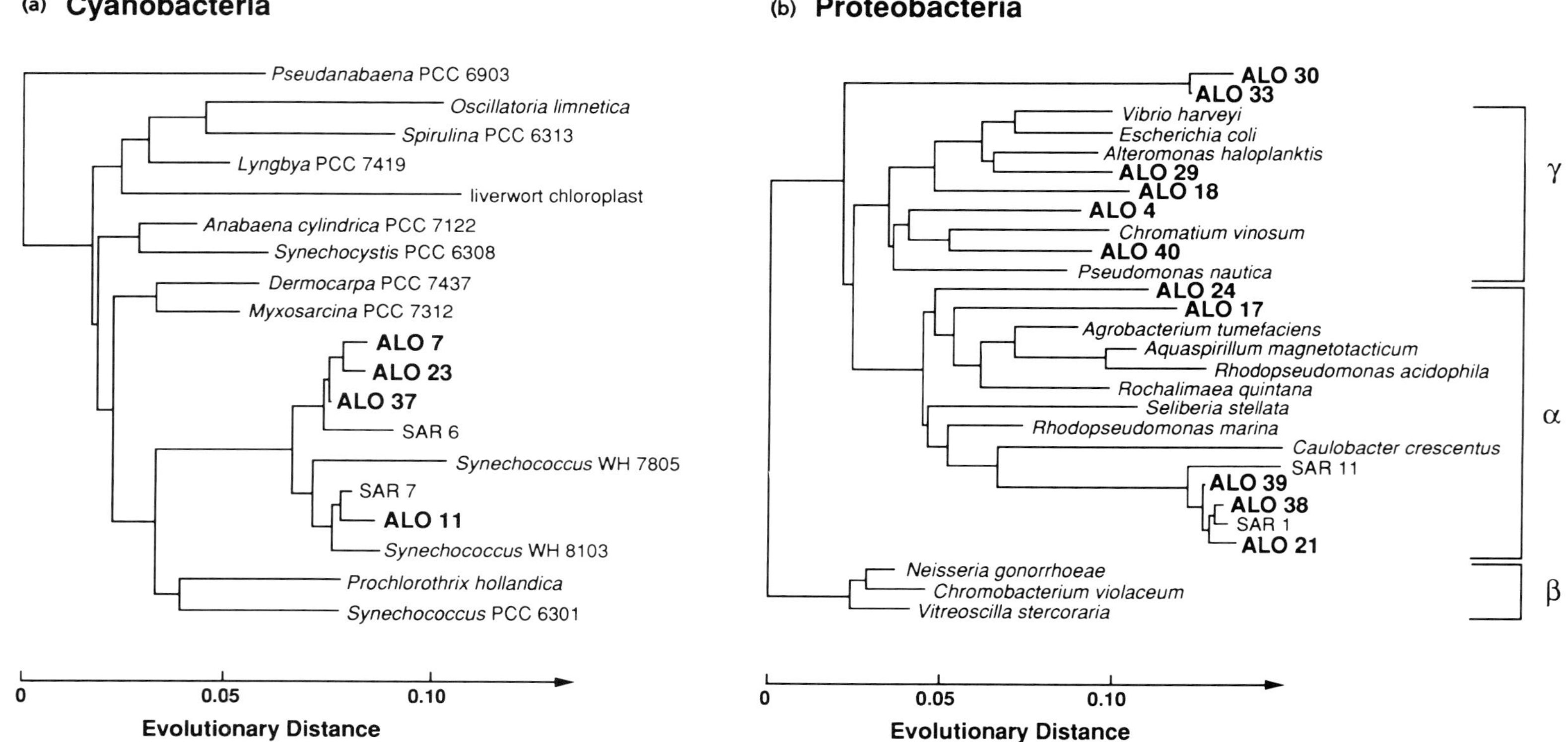

Figure 2.5 Evolutionary relationship of bacterioplankton population constituents of the Sargasso Sea (numbered clones with SAR prefix) and the central North Pacific (numbered clones with ALO prefix). Pairwise evolutionary distances were estimated by comparing partial small subunit rRNA sequences. From Schmidt *et al.* (1991).

lead to the automation needed to provide analysis of the intensive spatial and temporal sampling needed to truly investigate the dynamics of oceanic microbial communities.

2.5.4 Polymerase chain reaction

The PCR is a methodology employed to enzymatically multiply (amplify) specific genes *in vitro* (Saiki *et al.* 1985; 1988). The procedure is based upon thermal denaturation (at 94–100°C) of DNA in a sample, followed by hybridization of oligonucleotide primers (annealing) to opposite strands of the DNA. A thermostable DNA polymerase is then used to synthesize a new strand of the DNA template and the whole procedure repeated for a number of cycles.

The two applications of PCR to aquatic microbial ecology are:

1 detection of introduced organisms or pathogens against a background of ambient microorganisms;

2 detection of conserved gene sequences in the indigenous microbial community.

The first use of PCR in environmental samples was for detection of *Pseudomonas cepacia* AC1100 in sediments (Steffan *et al.* 1988). Primers were designed to amplify a 1 kb sequence that occurred 15 to 20 times in the *P. cepacia* genome. This procedure resulted in detection of as few as 100 *P. cepacia* cells in 100 g sediment against a background of 10^{11} diverse, nontarget organisms (Steffan *et al.* 1988).

These investigators also designed oligonucleotides to detect parts of the *lacZ*, *lamB*, and *uidA* operons of *E. coli* to enumerate fecal coliforms in aquatic environments (Bej *et al.* 1990; 1991b). The procedure enabled detection of as little as 10 fg DNA and one to five viable *E. coli* cells 100 ml^{-1} water. Similarly, Starnbach *et al.* (1989) used PCR to detect *Legionella* in seeded water samples. As few as 35 colony-forming units could be detected in a water sample and bacteria of other genera were not detected. Bej *et al.* (1991a) extended these technologies for the detection of viable cells of *Legionella pneumophila* in water samples. These adaptations enabled detection of cultivatable and viable but noncultivatable cells, while dead cells were not detected.

The detection of indigenous flora in aquatic environments is an application of this technology with great potential. As mentioned above, Giovannoni *et al.* (1990a,b) and Britschgi & Giovannoni (1991) used primers directed at portions of the 16S rRNA genes to clone genes from marine bacteria of the Sargasso Sea. Zehr & McReynolds (1989) designed degenerate oligonucleotide primers for conserved regions of the *nifH* gene to amplify a 359 base pair fragmented from *Trichodesmium thiebautii*. Degenerate oligonucleotides are primers based upon sequence information that include variable nucleotides for degeneracy in the genetic code, usually in the third codon position. Thus, degenerate oligonucleotide primers have the potential to amplify a family of very closely related genes that vary slightly in nucleotide sequence. The amplified *nifH* gene was purified and cloned for sequence analysis. Sequence analysis of the *Trichodesmium nifH* indicated that this gene was most closely related to those from cyanobacteria, and not heterotrophic nitrogen fixers such as *Azotobacter*. This finding indicates that *Trichodesmium* may be the major nitrogen fixer in *Trichodesmium* tufts and puffs and not associated bacteria.

The gene for the large subunit of Rubisco was amplified in a variety of unicellular algal species in culture and in dissolved and particulate DNA from a Florida reservoir (Paul *et al.* 1990). Unlike other reports of amplification of genes from material embedded in fossils (Golenberg *et al.* 1990), museum specimens, or mummified 4000-year-old Egyptian priests (Paabo *et al.* 1990), amplification of genes from algal cells and environmental DNA required extensive purification of DNA including PVPB treatment and cesium chloride ultracentrifugation. These studies indicate that gene amplification can be used to study the distribution of genes that encode for important ecological functions in aquatic environments.

There are several limitations to the use of gene amplification in environmental samples. Before amplification of genes can be accomplished for nonculturable bacteria in the environment, some sequence information is necessary, such that primers can be designed to amplify a target sequence. Therefore, this approach is usually limited to highly conserved genes, such as rRNA sequences or *nifH* and *rbcL* genes.

A second limitation is the need for rigorously purified DNA (as for algal and environmental samples). Rigorous purification will decrease the overall sensitivity of the method because of poor recovery in techniques like density gradient ultracentrifugation.

A third and serious drawback of the method is the potential for production of sequence artifacts produced during the PCR reaction (Liesack *et al.* 1991). Analysis of an apparently pure culture of a strict psychrophilic and barophilic organism indicated that their culture was comprised of two different organisms, as determined by sequence analysis of amplified 16S rRNA genes. In addition to the two 16S rRNA sequences, a third sequence was obtained, that was shown to be a chimera of the former two sequences found. Secondary structure analysis of this third hybrid sequence showed that it could not be a functional rRNA, but rather a molecule assembled during the PCR. This work shows that amplification for cloning and sequencing can result in the manufacture of organisms that in actuality do not exist. These chimera will escape detection unless careful analysis of secondary structure is performed and/or by calculation of separate phylogenetic trees for individual domains (regions) of the sequences (Liesack *et al.* 1991).

2.5.5 Detection of microbial gene expression

The presence of a gene sequence in the environment, detected by either amplification or direct probing, only provides information on the potential (past or present) for a metabolic activity to occur or the presence of a particular target organism. It does not give information on the viability of that organism or whether the gene-encoded enzyme function is operational in that environment. As a result, there has been considerable interest in the detection of gene expression in natural microbial communities (Capone *et al.* 1990; Fleming & Sayler 1990; Tsai & Olson 1990; Pichard & Paul 1991; Tsai *et al.* 1991). One approach to the detection of gene expression is detection of target enzymes by immunological methods (Orellana *et al.* 1988; Capone *et al.* 1990) and is discussed in Section 2.2. The other approach is the detection of messenger RNA (mRNA) transcripts of target gene sequences.

Tsai & Olson (1990) employed mRNA isolation to study expression of *merA* and *merB* transcripts from bacteria isolated from a mercury-contaminated site in Tennessee. Fleming & Sayler (1990) isolated mRNA from *Pseudomonas putida* seeded in soil microcosms to detect expression of catabolic genes. Both groups used probing of slot or dot blotted samples to detect mRNA levels.

Pichard & Paul (1991) combined guanidinium isothiocyanate RNA extraction with sense and antisense probing to detect gene expression in bacteria added to sea water and natural phytoplankton populations. Antisense probes were used to hybridize to the mRNA targets, and sense probes of equal specific activity were used as controls for nonspecific hybridization and/or contamination of the RNA with target DNA. This methodology enabled detection of mRNA of the target gene *nptII* (neomycin phosphotransferase) in 10^4 bacteria per milliliter (100 ml sample) sea water. *rbcL* gene expression in indigenous phytoplankton populations in oligotrophic surface waters near the Dry Tortugas was detected using the *Synechococcus rbcL* antisense and sense RNA probes. Greatest levels of *rbcL* mRNA were detected during daylight hours with minimal expression at night.

Detection of gene expression by mRNA isolation and analysis is another technology of tremendous potential for application to aquatic microbiology. Additionally, developments in mRNA amplification (Kwoh *et al.* 1989; Guatelli *et al.* 1990) may increase sensitivity to be comparable to that of PCR.

2.6 Transmission electron microscopy

TEM has not been widely applied to problems in microbial ecology. However, when done so, the results have usually been striking (Sieburth 1979; Chisholm *et al.* 1988). TEM has been used primarily for gathering descriptive information, and has not been generally viewed as a technique that could provide quantitative data, except for morphological measurements.

The existence of viruses and bacteriophages in the aquatic environments have been known for some time (Zobell 1946; Spencer 1955; Wiebe & Liston 1968; Baross *et al.* 1978), and attempts to quantitate these were based upon plaque titers on cultivated hosts or some variation of that method. Early reports of direct visualization of virus particles from aquatic samples by TEM (Johnson & Sieburth 1978; Sieburth 1979; Torrella & Morita 1979; Ewert & Paynter 1980) suggested these forms were abundant, yet quantitative estimates were not made.

The capability to directly count viruses akin to bacterial direct counts by epifluorescence microscopy indicated that there were many more viruses present ($\sim 10^3 - 10^7$ times more) in aquatic environments than previously thought (Proctor *et al.* 1988; Sieburth *et al.* 1988; Bergh *et al.* 1989). These observations have been corroborated by further studies (Figure 2.6) (Bratbak *et al.* 1990; Proctor & Fuhrman 1990; Suttle *et al.* 1990; Jiang *et al.* 1991; Paul *et al.* 1991), even though a variety of methods have been employed to concentrate viruses, e.g., tangential flow filtration, ultracentrifugation, and vortex flow filtration, and enumerate them, e.g., DAPI epifluorescence microscopy and TEM. Table 2.1 shows the distribution of viral abundances from Tampa Bay, the Gulf of Mexico, and the Bahamas Bank, performed by a combination of vortex flow filtration, ultracentrifugation, and TEM (Paul *et al.* 1991). The relationship between viral and bacterial abundance for data from several investigators appears in Figure 2.7. Borsheim *et al.* (1990) observed that virus abundance was 10 times that of bacterial abundance, based upon observations made during a spring bloom in a Norwegian fjord. However, bacterial numbers during this study were never greater than $4 \times 10^5\,ml^{-1}$, which seems unusually low for a coastal environment. Our data (see Figure 2.7 and Table 2.1) suggest that the abundance of viruses is usually on the same order of magnitude as bacterial direct counts, and slightly higher in estuarine samples. However, differences in viral abundances caused by differences in viral concentration and enumeration methodologies are not known.

The observation of the abundance of viruses in aquatic environments raises a myriad of ecological and genetic questions, some of which are currently being answered. Bergh *et al.* (1989) estimated that one-third of the bacteria in the water column might experience a

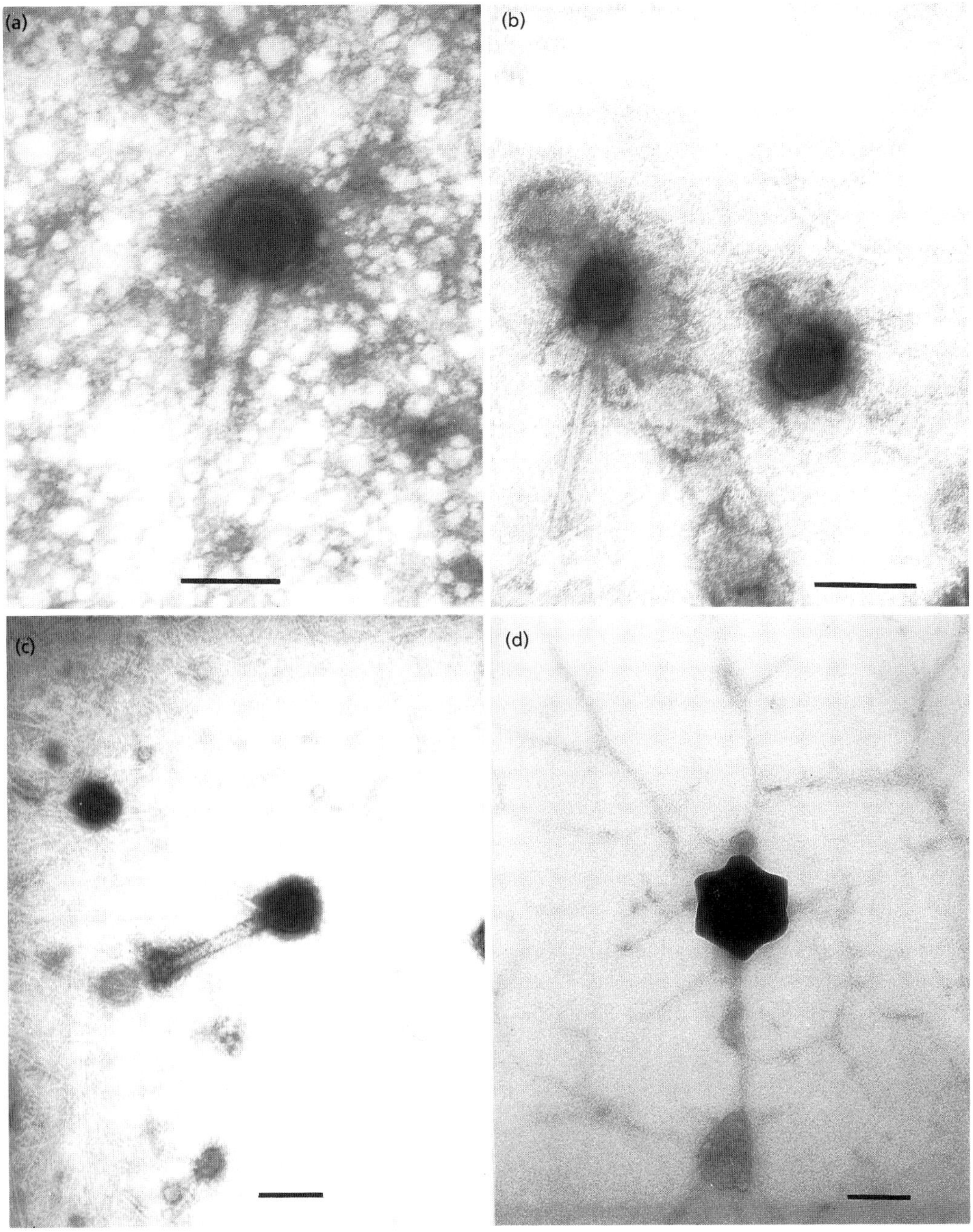

(a) (b) (c) (d)

Figure 2.6 Bacteriophage and virus-like particles collected by vortex flow filtration of surface waters from (a) the Medard Reservoir, Valrico, Florida, (b) Bayboro Harbor, Tampa Bay, (c) the mouth of Tampa Bay, and (d) near Loggerhead Key, Dry Tortugas. All samples were glutaraldehyde fixed and stained with uranyl sulfate. Bar in each figure is 100 nm. From J.H. Paul & S.C. Jiang (unpublished).

phage attack per day, and that the production rate was 1.8×10^7 phages per milliliter per day. Bratbak *et al.* (1990) followed the dynamics of a phytoplankton bloom, and correlated increases in viruses and bacterioplankton with the onset of the bloom.

Proctor & Fuhrman (1990; 1991) viewed thin sections of natural bacterial populations by TEM and found that 3.2% and 1.5% of the heterotrophic bacteria and cyanobacteria, respectively, contained near complete virus particles intracellularly. Those workers estimated that perhaps as high as 70% of the heterotrophic bacteria could be infected, and that phage lysis could contribute to anywhere from 10 to 100% of the bacterial mortality. Heldal & Bratbak (1991) found the number of viruses to change on a diel basis along with bacterial number and bacterial activity in a coastal marine environment. The fraction of bacteria containing mature viral particles ranged from 2 to 16%, and the number of viruses released from these bacteria averaged 50 per cell (range 10 to 300). These data led to an estimate of phage lysis of the bacterial population of 2 to 24% hr^{-1} (Heldal & Bratbak 1991).

Table 2.1 Viral abundance in Tampa Bay, the Eastern Gulf of Mexico, and the Bahamas Bank as determined by TEM direct counts on vortex flow filtration[1] (VFF)-concentrated water samples

Sample	Initial volume of sample (liters)	Volume after VFF concentration (ml)	10^6 viruses ml^{-1} (initial volume)
Bayboro Harbor, Tampa Bay	3	25	34 ± 14
Mouth of Tampa Bay	3	21.3	17.4 ± 2.5
Hookers Point, Tampa Bay	20	57.0	13.3 ± 5.5
St Pete Pier, Tampa Bay	5	46.0	6.04 ± 1.49
Florida Bay (24°39.00N, 81°54.25W)	20	18.9	2.0 ± 0.1
Loggerhead Key, Dry Tortugas	15	8.4	1.5 ± 1.0
Joulter's Cay, Bahamas	10	50.0	2.3 ± 1.1
Gulf of Mexico (26°38.96N, 85°20.00W)	20	17.4	0.24 ± 0.06
Gulf of Mexico (24°49.94N, 85°20.00W)	30	16.0	0.44 ± 0.15
NW Providence Channel, Bahamas	25	36.4	0.42 ± 0.06

[1] VFF is a type of ultrafiltration using Taylor vortices. The system used was manufactured by Membrex Inc., Garfield, New Jersey.

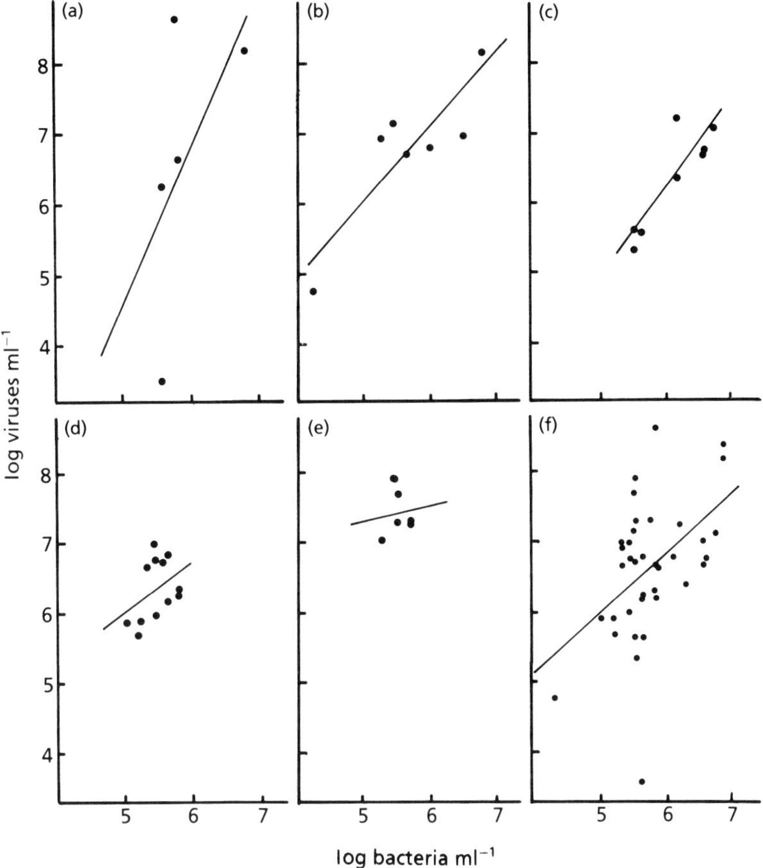

Figure 2.7 Viral direct counts plotted as a function of bacterial direct counts. The lines are the result of least sequences regression analysis performed on the data. Significant correlations were found for data in (b) ($0.01 < P < 0.02$), (c) ($0.002 < P < 0.005$), and (f) ($0.002 < P < 0.005$). Data from (a) Proctor & Fuhrman (1990), (b) Bergh *et al.* (1989), (c) Paul *et al.* (unpublished), (d) Bratbak *et al.* (1990), (e) Borsheim *et al.* (1990), and (f) data from all of these investigators.

The role of viruses in the control of autotrophic populations has come to light largely through the work of Suttle *et al.* (1990). These workers isolated viruses from sea water that could infect a variety of primary producers, including diatoms, cryptophytes, prasinophytes, and chroocoid cyanobacteria. These investigators also demonstrated that natural viral populations concentrated by ultrafiltration could inhibit primary production of the ambient phytoplankton population.

The recent observation of the abundance of viruses raises questions on their quantitative effect on microbial ecosystems. If protozoan grazers are the dominant consumers of bacterial production, then bacterial production has the potential to be a link to higher trophic levels. If viral infection and lysis is the major mechanism of bacterial mortality, then bacterial production is lost to extracellular organic matter, and bacteria are a sink or mere intermediate in the cycling of dissolved organic matter. And, if the latter is true, what happens to the "viral" production?

2.7 Summary and future outlook

The development of new methodologies to understand microorganisms in their environments has become a science unto itself. The need for these methods comes from the fact that classic methodologies of medical and industrial microbiology developed over the last hundred years produce artifacts when applied to natural populations in the environment.

The technologies presented above all possess their respective limitations. Perhaps the immunological approach is best for enumeration of specific microbial serotypes, but it requires that the serotype has previously been isolated in culture. Flow cytometers are ideal for characterizing autofluorescent populations, but are often criticized for being "black boxes" because the data obtained are not often verifiable by other methods. 16S phylogenetic analysis has revolutionized our understanding of the types of organisms in aquatic environments, but only yields phylogenetic data. The use of 16S technology to enumerate specific microbial types is in its infancy, and these methods provide no information on rates of ecological processes. Image analysis technology is perhaps best suited to enumeration and biomass estimation, but again provides no information on the rates of processes. Gene expression technology has the potential to yield process rate information, but may be limited only to a few conserved gene loci. PCR can enable detection of genes at extremely low abundances, but quantitation of original gene abundance is difficult. Each of these technologies is best suited to address specific and often limited questions in aquatic microbial ecology.

Most technologic breakthroughs in environmental microbiology have had their roots in medical microbiology or related fields. For example, the flow cytometer had its infancy in the study of pathogenic bacteria. Although necessity has been described as the mother of invention, certainly the market size drives technologic development (particularly for new instrumentation) and researchers in microbial ecology are a small market by any standards.

The rapid identification of bacteria in clinical samples may be the technology next adapted to aquatic microbial ecology. Most automated systems are simply small-scale cultivation techniques that identify bacteria by standard biochemical phenotypes. As such, these are of limited utility to the environmental microbiologist. PCR has been used in rapid clinical bacterial identification, and perhaps a combination of automated PCR of 16S rRNA genes coupled with automated sequencing may be the future of phylogenetic characterization of microbial communities. Ideally such an instrument could determine the relative abundance and taxonomic identity of any bacterium in a liquid or sediment sample. A second instrument might measure the metabolic rates of the bacterial components, and the rates of the ecological and geochemical functions they catalyze. Eventually, we may see a compact field model that need only be pointed at the body of water to be analyzed, an aquatic microbial ecologist's "tricorder." A satellite version of this instrumentation would be extremely useful. A coupling of remote sensing technology with modern molecular technology just may yield the breakthroughs needed to answer global ecological questions of the present and future.

Acknowledgments

I am grateful to Bess Ward, Jed Fuhrman, Ed DeLong, Sallie Chisholm, Bob Olson, Michael Sieracki, and Norm Pace for providing reprints, preprints, submitted manuscripts, and data of their work for use in this review. I thank Sunny Jiang, Marc Frisher, Scott Pichard, David Williams, and Denise Bombard for help in production of this manuscript. Support was provided by NSF grants OCE 8817172, OCE 9022036, and the Florida High Technology and Industry Council.

References

Amann R.I., Krumholz L. & Stahl D.A. (1990a) Fluorescent-oligonucleotide probing of whole cells for determinative, phylogenetic, and environmental studies in microbiology. *J. Bacteriol.* **172**, 762–770.

Amann R.I., Binder B.J., Olson R.J., Chisholm S.W., Devereux R. & Stahl D.A. (1990b) Combination of 16S rRNA-targeted oligonucleotide probes with flow cytometry for analyzing mixed microbial populations. *Appl. Environ. Microbiol.* **56**, 1919–1925.

Amann R.I., Springer N., Ludwig W., Görtz H.-D. & Scheifer K.-H. (1991) Identification *in situ* and phylogeny of uncultured bacterial endosymbionts. *Nature* **351**, 161–164.

Armbrust E.V., Bowen J.D., Olson R.J. & Chisholm S.W. (1989) Effect of light on the cell cycle of a marine *Synechococcus* strain. *Appl. Environ. Microbiol.* **55**, 425–432.

Armbrust E.V., Chisholm S.W. & Olson R.J. (1990) Role of light and the cell cycle on the induction of spermatogenesis in a centric diatom. *J. Phycol.* **26**, 470–478.

Azam F., Fenchel T., Field J.G., Gray J.S., Meyer-Reil L.A. & Thingstad F. (1983) The ecological role of water-column microbes in the sea. *Mar. Ecol. Prog. Ser.* **10**, 257–263.

Baden D.G., Mende T.J. & Szmant A.M. (1988) Brevetoxins and binding: sodium channels vs. antibodies. In Yentsch C.M., Mague F.C. & Horan P.K. (eds) *Immunochemical Approaches to Coastal, Estuarine, and Oceanographic Questions. Lecture Notes on Coastal and Estuarine Studies*, No. 25, pp. 134–144. Springer Verlag, New York.

Barkay G.M. & Liebert C. (1990) Genes encoding mercuric reductases from selected Gram-negative aquatic bacteria have a low degree of homology with *merA* of transposon Tn501. *Appl. Environ. Microbiol.* **56**, 1695–1701.

Barkay T., Fouts D.L. & Olson B.H. (1985) Preparation of a DNA gene probe for detection of mercury resistance in Gram-negative bacterial communities. *Appl.*

Environ. Microbiol. **49**, 1196–1202.
Barkay T., Liebert C. & Gillman M. (1989a) Hybridization of DNA probes with whole-community genome for detection of genes that encode microbial responses to pollutants: *mer* genes and Hg^{2+} resistance. *Appl. Environ. Microbiol.* **55**, 1574–1577.
Barkay T., Liebert C. & Gillman M. (1989b) Environmental significance of the potential for *mer* (Tn21)-mediated reduction of Hg^{2+} to Hg^{o} in natural water. *Appl. Environ. Microbiol.* **55**, 1196–1202.
Baross J., Liston J. & Morita R.Y. (1978) Incidence of *Vibrio parahaemolyticus* bacteriophages and other *Vibrio* bacteriophages in marine samples. *Appl. Environ. Microbiol.* **36**, 492–499.
Bej A.K., Steffan R.J., DiCesare J., Huff L. & Atlas R.M. (1990) Detection of coliform bacteria in water by polymerase chain reaction and gene probes. *Appl. Environ. Microbiol.* **56**, 307–314.
Bej A.K., Mahubani M.H. & Atlas R.M. (1991a) Detection of viable *Legionella pneumophila* in water by polymerase chain reaction and gene probe methods. *Appl. Environ. Microbiol.* **57**, 597–600.
Bej A.K., McCarty S.C. & Atlas R.M. (1991b) Detection of coliform bacteria and *Escherichia coli* by multiplex polymerase chain reaction: comparison with defined substrate and plating methods for water quality monitoring. *Appl. Environ. Microbiol.* **57**, 2429–2432.
Bergh O., Borsheim K.Y., Bratbak G. & Heldal M. (1989) High abundance of viruses found in aquatic environments. *Nature* **340**, 467–468.
Bermudez M. & Hazen T.C. (1988) Phenotypic and genotypic comparison of *Escherichia coli* from pristine tropical waters. *Appl. Environ. Microbiol.* **54**, 979–983.
Binder B.J. & Chisholm S.W. (1990) Relationship between DNA cycle and growth rate in *Synechococcus* strain PCC 6301. *J. Bacteriol.* **172**, 2313–2319.
Bjornsen P.K. (1986) Automatic determination of bacterioplankton biomass by image analysis. *Appl. Environ. Microbiol.* **51**, 1199–1204.
Bohlool B.B. & Schmidt E.L. (1980) The immunofluorescent approach in microbial ecology. *Adv. Microbiol. Ecol.* **4**, 203–241.
Borsheim K.Y., Bratbak G. & Heldal M. (1990) Environmental biomass estimation of planktonic bacteria and viruses by transmission electron microscopy. *Appl. Environ. Microbiol.* **56**, 352–356.
Bratbak G., Heldal M., Norland S. & Thingstad T.F. (1990) Viruses as partners in spring bloom microbial trophodynamics. *Appl. Environ. Microbiol.* **56**, 1400–1405.
Britschgi T.B. & Giovannoni S.J. (1991) Phylogenetic analysis of a natural marine bacterioplankton population by rRNA gene cloning and sequencing. *Appl. Environ. Microbiol.* **57**, 1707–1713.
Campbell L. (1988) Identification of marine chroococcoid cyanobacteria by immunofluorescence. In Yentsch C.M., Mague F.C. & Horan P.K. (eds) *Immunochemical Approaches to Coastal, Estuarine, and Oceanographic Questions. Lecture Notes on Coastal and Estuarine Studies*, No. 25, pp. 208–229. Springer Verlag, New York.
Campbell L. & Carpenter E.J. (1987) Characterization of phycoerythrin-containing *Synechococcus* spp. populations by immunofluorescence. *J. Plank. Res.* **9**, 1167–1181.
Campbell L. & Iturriaga R. (1988) Identification of *Synechococcus* sp. in the Sargasso Sea by immunofluorescence and fluorescence excitation spectroscopy performed on individual cells. *Limnol. Oceanogr.* **33**, 1196–1201.
Capone D.G., O'Neil J.M., Zehr J. & Carpenter E.J. (1990) Basis for diel variation in nitrogenase activity in the marine planktonic cyanobacterium *Trichodesmium thiebautii. Appl. Environ. Microbiol.* **56**, 3532–3536.
Carpenter E.J. & Chang J. (1988) Species-specific phytoplankton growth rates via diel

DNA synthesis cycles. I. Concept of the method. *Mar. Ecol. Progr. Ser.* **43**, 105–111.

Cavanaugh C.M., Veenhuis M. & Abbott M.S. (1988) Immunochemical localization of ribulose-1,5-bisphosphate carboxylase (RuBPCase) in the symbiont-containing gills of Atlantic coast bivalve *Solemya velum. EOS* **69**, 1099.

Chang J. & Carpenter E.J. (1988) Species specific growth rates via diel DNA synthesis. II. DNA quantification and model verification in the dinoflagellate *Heterocapsa triquetra. Mar. Ecol. Prog. Ser.* **44**, 287–296.

Chisholm S.W., Olson R.J., Zettler E.R., Goericke R., Waterburg J.B. & Welschmeyer N.A. (1988) A novel free-living prochlorophyte abundant in the oceanic euphotic zone. *Nature* **334**, 340–343.

Currin C.A., Paerl H.W., Suba G.K. & Alberte R.S. (1990) Immunofluorescence detection and characterization of N_2-fixing microorganisms from aquatic environments. *Limnol. Oceanogr.* **35**, 59–71.

David A.W. & Paul J.H. (1989) Enumeration and sizing of aquatic bacteria by use of a silicon-intensified target camera-linked image analysis system. *J. Microbiol. Meth.* **9**, 257–266.

DeLong E.F. (1991) Molecular systematics, microbial ecology and single cell analysis. *Nato Advanced Study Institute Symposium, Single Cell and Particle Analysis in Oceanography*, Italy (in press).

DeLong E.F., Schmidt T.M. & Pace N.R. (1989a) Analysis of single cells and oligotrophic picoplankton populations using 16S rRNA sequences. In Hattori T., Ishida Y., Maruyama Y., Morita R. & Uchida A. (eds) *Recent Advances in Microbial Ecology*, pp. 697–701. Japan Scientific Societies Press, Tokyo.

DeLong E.F., Wickham G.S. & Pace N.R. (1989b) Phylogenetic stains: ribosomal RNA-based probes for the identification of single cells. *Science* **243**, 1360–1363.

Distel D.L., DeLong E.F. & Waterbury J.B. (1991) Phylogenetic characterization and *in situ* localization of the bacterial symbiont of shipworms (Teredinidaer: Bivalvia) by using 16S rRNA sequence analysis and oligodeoxynucleotide probe hybridization. *Appl. Environ. Microbiol.* **57**, 2376–2382.

Ewert D.L. & Paynter M.J.B. (1980) Technique for determining total bacterial virus counts in complex aqueous systems. *Appl. Environ. Microbiol.* **39**, 253–260.

Fleming J.T. & Sayler G.S. (1990) Isolation and characterization of mRNAs transcribed by catabolic genes from soil sediment microorganisms, p. 320. *Abst. Ann. Meet. American Society for Microbiology*, Washington DC.

Francisco R.L., Mah R.A. & Rabin A.C. (1973) Acridine orange-epifluorescence technique for counting bacteria in natural waters. *Trans. Am. Microsc. Soc.* **92**, 416–421.

Fredrickson J.K., Bezdicek D.F., Brockman F.J. & Li S.W. (1988) Enumeration of Tn5 mutant bacteria in soil using a most-probable number-DNA hybridization procedure. *Appl. Environ. Microbiol.* **54**, 446–453.

Fuhrman J.A. & Azam F.A. (1982) Thymidine incorporation as a measure of heterotrophic bacterioplankton production in marine surface waters: evaluation and field results. *Mar. Biol.* **66**, 109–120.

Fuhrman J.A., Comeau D.E., Hagstrom A. & Chan A.M. (1988) Extraction from natural planktonic microorganisms of DNA suitable for molecular biological studies. *Appl. Environ. Microbiol.* **54**, 1426–1429.

Giovannoni S.J., DeLong E.F., Olsen G.J. & Pace N.R. (1988) Phylogenetic group-specific oligonucleotide probes for identification of single microbial cells. *J. Bacteriol.* **170**, 720–726.

Giovannoni S.J., Britschgi T.B., Moyer C.L. & Field K.G. (1990a) Genetic diversity in Sargasso Sea bacterioplankton. *Nature* **345**, 60–65.

Giovannoni S.J., DeLong E.F., Schmidt T.M. & Pace N.R. (1990b) Tangential flow filtration and preliminary phylogenetic analysis of marine picoplankton. *Appl. Environ. Microbiol.* **56**, 2572–2575.

Golenberg E.M., Giannosi D.E., Clegg M.T., Smiley C.J., Durbin M., Henderson D. &

Zurawski G. (1990) Chloroplast DNA sequence from a Miocene *Magnolia* species. *Nature* **344**, 656–658.

Guatelli J.C., Whitfield K.M., Kwoh D.Y., Barringer K.J., Richman D.D. & Gingeras T.R. (1990) Isothermal *in vitro* amplification of nucleic acids by a multienzyme reaction modelled after retroviral replication. *Proc. Natl Acad. Sci. USA.* **87**, 1874–1878.

Guire P.E., Duquette P.H., Amos R.A., Behrens J.C., Josephson M.W. & Chambers R.P. (1988) Preparation of model haptens to express common epitopes of PSP toxins. In Yentsch C.M., Mague F.C. & Horan P.K. (eds) *Immunochemical Approaches to Coastal, Estuarine, and Oceanographic Questions. Lecture Notes on Coastal and Estuarine Studies*, No. 25, pp. 145–154. Springer Verlag, New York.

Gunderson K., Mountain C.W., Taylor D., Ohye R. & Shen J. (1972) Some chemical and microbiological observations in the Pacific Ocean off the Hawaiian Islands. *Limnol. Oceanogr.* **17**, 524–531.

Hazen T.C. & Jimenez L. (1988) Enumeration and identification of bacteria from environmental samples using nucleic acid probes. *Microbiol. Sci.* **5**, 340–343.

Heldal M. & Bratbak G. (1991) Production and decay of viruses in aquatic environments. *Mar. Ecol. Prog. Ser.* **72**, 205–212.

Hiraoka Y., Sedat J.W. & Agard D.A. (1987) The use of a charge coupled device for quantitative optical microscopy of biological structures. *Science* **238**, 36–41.

Hobbie J.E., Daley R.J. & Jasper S. (1977) Use of Nuclepore filters for counting bacteria by fluorescence microscopy. *Appl. Environ. Microbiol.* **33**, 1225–1228.

Hokama Y., Hinda S.A.A., Kobayashi M.N., Nakagawa L.K., Shirai L.K. & Miyaharo J.T. (1988) Monoclonal antibodies (monabs) to ciguatoxin and related polyethers. In Yentsch C.M., Mague F.C. & Horan P.K. (eds) *Immunochemical Approaches to Coastal, Estuarine, and Oceanographic Questions. Lecture Notes on Estuarine and Coastal Studies*, No. 25, pp. 155–166. Springer Verlag, New York.

Holben W.E., Jansson J.K., Chelm B.K. & Tiedje J.M. (1988) DNA probe method for the detection of specific microorganisms in the soil bacterial community. *Appl. Environ. Microbiol.* **54**, 703–711.

Ingraham L.J., Maaloe O. & Neidhardt F.C. (1983) *Growth of the bacterial cell.* Sinauer Associates, Sunderland, Massachusetts.

Jiang S.C., Thurmond J.M., Rose J.B. & Paul J.H. (1991) The abundance of viruses and viral DNA contribution to dissolved DNA in seawater. *Abst. N47, American Society for Microbiology*, Washington DC.

Jiang X., Estes M.K., Metcalf T.G. & Melnick J.L. (1986) Detection of hepatitis A virus in seeded estuarine samples by hybridization with cDNA probes. *Appl. Environ. Microbiol.* **52**, 711–717.

Jiang X., Graham D.Y., Wang K. & Estes M.K. (1990) Norwalk virus genome cloning and characterization. *Science* **250**, 1580–1583.

Johnson P.W. & Sieburth J. McN. (1978) Morphology of non-cultured bacterioplankton from estuarine, shelf, and open water, p. 178. *Abst. Ann. Meet. American Society for Microbiology*, Washington DC.

Karl D.M. (1981) Simultaneous rates of ribonucleic acid and deoxyribonucleic acid synthesis for estimating growth and cell division of aquatic microbial communities. *Appl. Environ. Microbiol.* **42**, 802–810.

Korner H. & Zumft W.G. (1989) Expression of denitrification enzymes in response to the dissolved oxygen level and respiratory substrate in continuous culture of *Pseudomonas stutzeri*. *Appl. Environ. Microbiol.* **55**, 1670–1676.

Krambeck C., Krambeck H.-J. & Overbeck J. (1981) Microcomputer-assisted biomass determination of plankton bacteria on scanning electron micrographs. *Appl. Environ. Microbiol.* **42**, 142–149.

Kwoh D.Y., Davis G.R., Whitfield K.M., Chappelle H.L., DiMichello L.J. & Gringeras T.R. (1989) Transcription based amplification system and detection of amplified human immunodeficiency virus type 1 with a bead-based sandwich hybridization format. *Proc. Natl Acad. Sci. USA.* **86**, 1173–1177.

Lee S. & Fuhrman J.A. (1990) DNA hybridization to compare species compositions of natural bacterioplankton assemblages. *Appl. Environ. Microbiol.* **56**, 739–746.

Lee S. & Fuhrman J.A. (1991a) Spatial and temporal variations of natural bacterioplankton assemblages studied by total genomic DNA cross hybridization. *Limnol. Oceanogr.* **36**, 1277–1287.

Lee S. & Fuhrman J.A. (1991b) Species composition shift of confined bacterioplankton studied at the level of community DNA. *Mar. Ecol. Prog. Ser.* **79**, 195–201.

Li W.K.W. & Wood A.M. (1988) Vertical distribution of North Atlantic ultraplankton: analysis by flow cytometry and epifluorescence microscopy. *Deep-Sea Res.* **35**, 1615–1638.

Liesack W., Weyland H. & Stackebrandt E. (1991) Potential risks of gene amplification by PCR as determined by 16S rRNA analysis of a mixed-culture of strict barophilic bacteria. *Microb. Ecol.* **21**, 191–198.

Morris J.G. Jr, Wright A.C., Roberts D.M., Wood P.K., Simpson L.M. & Oliver J.D. (1987) Identification of environmental *Vibrio vulnificus* isolates with a DNA probe for the cytotoxin-hemolysin gene. *Appl. Environ. Microbiol.* **53**, 193–195.

Ogram A., Sayler G.S. & Barkay T. (1987) The exraction and purification of microbial DNA from sediments. *J. Microbiol. Meth.* **7**, 57–66.

Olsen G.J., Lane D.J., Giovannoni S.J., Pace N.R. & Stahl D.A. (1986) Microbial ecology and evolution: a ribosomal RNA approach. *Annu. Rev. Microbiol.* **40**, 337–365.

Olson R.J., Frankel S.L., Chisholm S.W. & Shapiro H.M. (1983) An inexpensive flow cytometer for the analysis of fluorescence signals in phytoplankton: chlorophyll and DNA distributions. *J. Exp. Mar. Biol. Ecol.* **68**, 129–144.

Olson R.J., Vaulot D. & Chisholm S.W. (1985) Marine phytoplankton distributions measured using shipboard flow cytometry. *Deep-Sea Res.* **32**, 1273–1280.

Olson R.J., Chisholm S.W., Zettler E.R. & Armbrust E.V. (1988) Analysis of *Synechococcus* pigment types in the sea using single and dual beam flow cytometry. *Deep-Sea Res.* **35**, 425–440.

Olson R.J., Chisholm S.W., Zettler E.R., Atabet M.A. & Dusenberry J.A. (1990) Spatial and temporal distributions of prochlorophyte picoplankton in the North Atlantic ocean. *Deep-Sea Res.* **37**, 1033–1051.

Orellana M.V., Perry M.J. & Watson B.A. (1988) Probes for assessing single cell primary production: antibodies against ribulose 1,5 bisphosphate carboxylase (RuBPCASE) and peridinin/chlorophyll *a* (PCP). In Yentsch C.M., Mague F.C. & Horan P.K. (eds) *Immunochemical Approaches to Coastal, Estuarine, and Oceanographic Questions. Lecture Notes on Coastal and Estuarine Studies*, No. 25, pp. 243–262. Springer Verlag, New York.

Paabo S., Irwin D.M. & Wilson A.C. (1990) DNA damage promotes jumping between templates during enzymatic amplification. *J. Biol. Chem.* **265**, 4718–4721.

Paerl H.W., Priscu J.C. & Brawner D.L. (1989) Immunochemical localization of nitrogenase in marine *Trichodesmium* aggregates: relationship to N_2 fixation potential. *Appl. Environ. Microbiol.* **55**, 2965–2975.

Paul J.H., Cazares L. & Thurmond J. (1990) Amplification of the *rbcL* gene from dissolved and particulate DNA from aquatic environments. *Appl. Environ. Microbiol.* **56**, 1963–1966.

Paul, J.H., Jiang, S.C. & Rose J.B. (1991) Concentration of viruses and dissolved DNA from aquatic environments by vortex flow filtration. *Appl. Environ. Microbiol.* **57**, 2197–2204.

Pichard S.L. & Paul J.H. (1991) Detection of gene expression in genetically engineered microorganisms and natural phytoplankton populations in the marine environment by mRNA analysis. *Appl. Environ. Microbiol.* **57**, 1721–1727.

Proctor L.M. & Fuhrman J.A. (1990) Viral mortality of marine bacteria and cyanobacteria. *Nature* **343**, 60–62.

Proctor L.M. & Fuhrman J.A. (1991) Roles of viral infection in organic particle flux.

Mar. Ecol. Prog. Ser. (in press).

Proctor L.M., Fuhrman J.A. & Ledbetter M.C. (1988) Marine bacteriophages and bacterial mortality. *EOS* **69**, 1111–1112.

Robertson B.R. & Button D.K. (1989) Characterizing aquatic bacteria according to population, cell size, and apparent DNA content by flow cytometry. *Cytometry* **10**, 70–76.

Saiki R.K., Scharf S., Faloona F., Mullis K.B., Horn G.T., Erlich H.A. & Arnheim N. (1985) Enzymatic amplification of β-globin genomic sequences and restriction site analysis for diagnosis of sickle cell anemia. *Science* **230**, 1350–1354.

Saiki R.K., Gelfand D.H., Steffel S., Scharf S.J., Higuchi R., Horn G.T., Mullis K.B. & Erlich H.A. (1988) Primer directed enzymatic amplification of DNA with a thermostable DNA polymerase. *Science* **239**, 487–491.

Sanders C.A., Yajko D.M., Hyun W., Langlois R.G., Nassos P.S., Fulwyler M.J. & Hadley W.K. (1990) Determination of guanine-plus-cytosine content of bacterial DNA by dual-laser flow cytometry. *J. Gen. Microbiol.* **136**, 359–365.

Sayler G.S., Shields M.S., Tedford E., Breen A., Hooper S., Sirotkin K. & Davis J. (1985) Application of DNA–DNA colony hybridization to the detection of catabolic genotypes in the environmental samples. *Appl. Environ. Microbiol.* **49**, 1295–1303.

Schmidt T.M., DeLong E.F. & Pace N.R. (1991) Analysis of marine picoplankton community by 16S rRNA gene cloning and sequencing. *J. Bacteriol.* **173**, 4371–4378.

Shapiro L. (1988) Immunofluorescent characterization of the eukaryote ultraplankton. In Yentsch C.M., Mague F.C. & Horan P.K. (eds) *Immunochemical Approaches to Coastal, Estuarine, and Oceanographic Questions. Lecture Notes on Coastal and Estuarine Studies*, No. 25, p. 242. Springer Verlag, New York.

Shapiro L.P., Campbell L. & Haugin E. (1989a) Immunochemical recognition of phytoplankton species. *Mar. Ecol. Prog. Ser.* **57**, 219–224.

Shapiro L.P., Haugin E.M., Keller M.D., Bidigare R.R., Campbell L. & Guillard R.R.L. (1989b) Taxonomic affinities of marine coccoid ultraphytoplankton. A comparison of immunochemical surface antigen cross-reactions and HPLC chloroplast signatures. *J. Phycol.* **25**, 794–797.

Sieburth J.McN. (1971) Distribution and activity of oceanic bacteria. *Deep-Sea Res.* **18**, 1111–1121.

Sieburth J.McN. (1979) *Sea Microbes*. Oxford University Press, London.

Sieburth J.McN., Johnson P.W. & Hargraves P.E. (1988) Ultrastructure and ecology of *Aurecoccus anophagefferens* gen. *et* sp. Nov. (Chrysophyceae): the dominant picoplankter during a bloom in Narragansett Bay, Rhode Island, Summer 1985. *J. Phycol.* **24**, 416–425.

Sieracki M.E. & Viles C.L. (1990) Color image-analyzed fluorescence microscopy: a new tool for marine microbial ecology. *Oceanography* **3**, 30–36.

Sieracki M.E. & Webb K.L. (1991) Application of image analyzed fluorescence microscopy for characterizing planktonic bacteria and protists. In Reid P.C., Turly C.M. & Burkill P.H. (eds) *Protozoa and their Role in Marine Processes.* Springer Verlag, Berlin, Heidelberg, New York.

Sieracki M.E., Johnson P.W. & Sieburth J.McN. (1985) Detection, enumeration, and sizing of planktonic bacteria by image-averaged epifluorescence microscopy. *Appl. Environ. Microbiol.* **49**, 799–810.

Sieracki M.E., Reichenbach S.E. & Webb K.L. (1989a) Evaluation of automated threshold selection method for accurately sizing microscopic fluorescent cells by image analysis. *Appl. Environ. Microbiol.* **55**, 2762–2773.

Sieracki M.E., Viles C.L. & Webb K.L. (1989b) Algorithm to estimate cell biovolume using image analyzed microscopy. *Cytometry* **10**, 551–557.

Simonet P., Le N.T., DuCross E.T. & Barden R. (1988) Identification of *Frankia* strain by direct DNA hybridization of crushed nodules. *Appl. Environ. Microbiol.* **54**, 2500–2503.

Sommerville C.C., Knight I.T., Straube W.L. & Colwell R.R. (1989) Simple, rapid method for direct isolation of nucleic acids from aquatic environments. *Appl. Environ. Microbiol.* 55, 548–554.

Spencer R. (1955) A marine bacteriophage. *Nature* **175**, 690.

Stahl D.A., Flesher B., Mansfield H.R. & Montgomery L. (1988) Use of phylogenetically based hybridization probes for studies of ruminal microbial ecology. *Appl. Environ. Microbiol.* **54**, 1079–1084.

Starnbach M.N., Falkow S. & Tompkins L.S. (1989) Species-specific detection of *Legionella pneumophila* in water by DNA amplification and hybridization. *J. Clin. Microbiol.* **27**, 1257–1261.

Steffan R.J., Goksoyr J., Bej A.K. & Atlas R.M. (1988) Recovery of DNA from soils and sediments. *Appl. Environ. Microbiol.* **54**, 2908–2915.

Suttle C.A., Chan A.M. & Cottrell M.T. (1990) Infection of phytoplankton by viruses and reduction of primary productivity. *Nature* **387**, 467–469.

Torrella F. & Morita R.Y. (1979) Evidence by electron micrographs for a high incidence of bacteriophage particles in the waters of Yaquina Bay, Oregon. Ecological and taxonomic implications. *Appl. Environ. Microbiol.* **37**, 774–778.

Tsai Y.-L. & Olson B.H. (1990) Effects of Hg^{2+}, CH_3–Hg^+, and temperature on the expression of mercury resistance genes in environmental bacteria. *Appl. Environ. Microbiol.* **56**, 3260–3272.

Tsai Y.-L., Park M.J. & Olson B.H. (1991) Rapid method for direct extraction of mRNA from seeded soils. *Appl. Environ. Microbiol.* **57**, 765–768.

Van Dilla M.A., Langlois R.G., Pinkel D., Yajko D. & Hadley K. (1983) Bacterial characterization by flow cytometry. *Science* **220**, 620–622.

Vaulot D. & Chisholm S.W. (1987) Flow cytometric analysis of spermatogenesis in the diatom *Thalassiosira weisflogii* (Bacillariophyceae). *J. Phycol.* **23**, 132–137.

Walia S., Khan A. & Rosenthal N. (1990) Construction and application of DNA probes for detection of polychlorinated biphenyl-degrading genotypes in toxic organic-contaminated soil environments. *Appl. Environ. Microbiol.* **56**, 254–259.

Ward B.B. (1982) Oceanic distribution of ammonium-oxidizing bacteria determined by immunofluorescence assay. *J. Mar. Res.* **40**, 1155–1172.

Ward B.B. (1984) Combined autoradiography and immunofluorescence for estimation of single cell activity by ammonium-oxidizing bacteria. *Limnol. Oceanogr.* **29**, 402–410.

Ward B.B. & Carlucci A.F. (1985) Marine ammonia and nitrite oxidizing bacteria: serological diversity determined by immunofluorescence in culture and in the environment. *Appl. Environ. Microbiol.* **50**, 194–201.

Ward B.B., Olson R.J. & Perry M.J. (1982) Microbial nitrification rates in the primary nitrite maximum off southern California. *Deep-Sea Res.* **29**, 247–255.

Ward B.B., Talbot M.C. & Perry M.J. (1984) Contributions of phytoplankton and nitrifying bacteria to ammonium and nitrite dynamics in coastal waters. *Cont. Shelf Res.* **3**, 383–398.

Ward D.M., Weller R. & Bateson M.M. (1990a) 16S rRNA sequences reveal uncultured inhabitants of a well-studied thermal community. *FEMS Microbiol. Rev.* **75**, 105–116.

Ward D.M., Weller R. & Bateson M.M. (1990b) 16S rRNA sequences reveal numerous uncultured microorganisms in a natural community. *Nature* **345**, 63–65.

Weller R. & Ward D.M. (1989) Selective recovery of 16S rRNA sequences from natural microbial communities in the form of cDNA. *Appl. Environ. Microbiol.* **55**, 1818–1822.

Wiebe W.J. & Liston J. (1968) Isolation and characterization of a marine bacteriophage. *Mar. Biol.* **1**, 244–249.

Woese C.R. (1987) Bacterial evolution. *Microbiol. Rev.* **51**, 221–271.

Wood A.M., Horan P.K., Muirhead K., Phinney D.A., Yentsch C.M. & Waterbury J.B. (1985) Discrimination between types of pigments in marine *Synechococcus* spp. by scanning spectroscopy, epifluorescence microscopy, and flow cytometry. *Limnol.*

Oceanogr. **30**, 1303–1315.

Yentsch C.M., Horan P.K., Muirhead K., Dortch Q., Haugen E., Legendre L., Murphy L.S., Perry M.J., Phinney D.A., Pomponi S.A., Spinrod R.W., Wood M., Yentsch C.S. & Zahuranec B.J. (1983) Flow cytometry and cell sorting: a technique for analysis and sorting aquatic particles. *Limnol. Oceanogr.* **28**, 1275–1280.

Zehr J.P. & McReynolds L.A. (1989) Use of degenerate oligonucleotides for amplification of the *nifH* gene from the marine cyanobacterium *Trichodesmium thiebautii. Appl. Environ. Microbiol.* **55**, 2522–2526.

Zehr J.P. Limberger R.J., Ohki K. & Fujita Y. (1990) Antiserum to nitrogenase generated from an amplified DNA fragment from natural populations of *Trichodesmium* spp. *Appl. Environ. Microbiol.* **56**, 3527–3531.

Zobell C. (1946) *Marine Microbiology*. Chronica Botanica, Waltham.

3

Allan E. Konopka

Distribution and activity of microorganisms in lakes: effects of physical processes

3.1 Introduction

Less than 1% of the water on earth is contained in freshwater systems (lakes, rivers, and ground water). However, the impact of microbial processes in these systems can be much greater than the volumetric importance of fresh water in the hydrosphere, due to the magnitude and diversity of nutrient inputs from the surrounding watershed. Although this statement pertains primarily to lakes and rivers, recent work on subsurface environments (see Chapter 8) has shown that there are also substantial microbial populations in the subsurface.

In this introductory chapter on freshwater systems, examples from lakes will be used to illustrate how physical, chemical, and biological processes interact to affect aquatic ecosystems. The general principle that such interactions are important in microbial ecology holds true for lotic and groundwater systems, although the relative importance of specific processes may differ.

The central thesis of this chapter is that lakes do not operate as homogeneously mixed reaction vessels. There are vertical and temporal heterogeneities in the water column caused by physical processes such as heating and turbulent mixing. These heterogeneities affect the distribution and consequences of microbially mediated and purely chemical reactions which effect the productivity of the aquatic ecosystem. In addition, microbes have evolved physiological adaptations which allow them to average out the vertical and temporal heterogeneities in nutrient availability which result from the physical processes operating in lakes.

3.2 Spatial heterogeneities in lakes

Two important physical processes in lakes are solar heating and turbulent mixing of the water column due to wind action. In some lakes, these processes can lead to thermal stratification of the water column during the summer (Figure 3.1). The upper layer (the

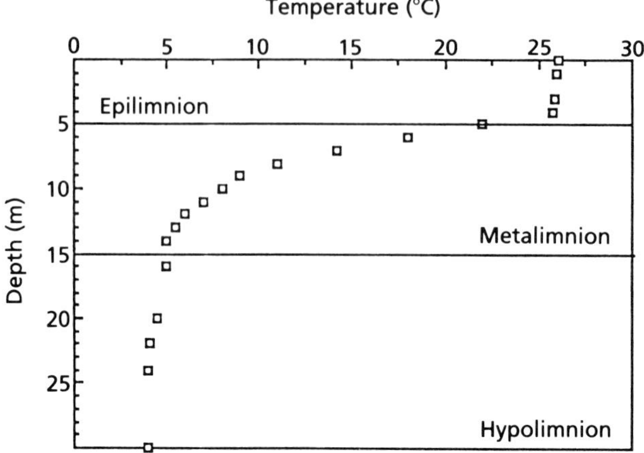

Figure 3.1 The temperature profile in Crooked Lake, Indiana on August 9, 1982. The approximate boundaries of the epi-, meta-, and hypolimnion are indicated.

epilimnion) comprises water which is uniformly warm and mixed freely by wind action. The metalimnion is a transition zone in which there is a steep thermal gradient. Within the metalimnion is the thermocline, defined as the plane in which temperature falls most rapidly. The bottom zone of cold and relatively undisturbed water is called the hypolimnion.

The significance of these thermal discontinuities is that the density of water is a function of its temperature. Water heated at the surface is less dense than deeper, colder water, and, therefore, work must be done to mix these waters. If the density gradient is large enough to reduce vertical turbulence, turbulent mixing will be confined to the epilimnion. That is, whereas vertical mixing over a scale of meters may take place in much less than a day in the epilimnion, it may take more than a month in the thermocline (Denman & Gargett 1983). During the period of thermal stratification, gradients of chemical and microbial species can become established in the stable strata of the water column.

Density gradients in the deeper strata of lakes can also arise from differences in salinity. These lakes are termed meromictic; the strata below the salinity gradient (the chemocline) do not mix with the upper strata. The chemocline may arise from the intrusion of saline water, or from the accumulation of salts derived from the microbial decomposition of organic matter.

The depth of thermal stratification in a lake is a function of morphometric factors. If the lake is too shallow, wind power will be capable of turbulently mixing the water column to the bottom, and thermal stratification will not occur. The depth of wind mixing is best predicted from the maximum effective length, i.e., the two most distant points on the shoreline over which wind and waves may act without interruptions from land or islands (Hanna 1990). Geographic location affects the annual cycle of thermal stratification. Stratified lakes in temperate latitudes are usually dimictic, i.e., they undergo two periods of complete mixing (in the spring and fall) and have periods of density stratification in summer and winter. In winter, ice or water with temperatures of between 0 and 4°C overlie a layer of denser water at 4°C. The thermal properties of lakes are treated in detail by Hutchinson (1957) and Wetzel (1983), and general overviews of the physical processes of turbulent mixing are given by Harris (1986) and Reynolds (1989).

The density gradient produced by thermal stratification will lead to chemical gradients in the metalimnion and hypolimnion. These chemical gradients are often a product of microbial metabolism. The vertical distribution of oxygen is an example of these processes. After the onset of thermal stratification, phytoplankton may settle out of the epilimnion, and their organic matter is then mineralized by oxygen-consuming aerobic chemoheterotrophic bacteria in the hypolimnion or more importantly, at the sediment–water interface (Fallon & Brock 1980).

If net photosynthesis does not occur in the hypolimnion, the consumed oxygen cannot be replenished, because the hypolimnion is physically isolated from the atmosphere and therefore transport occurs primarily by molecular diffusion. If the mass of organic matter delivered to the bottom strata is large enough, anaerobiosis will develop. The rate of hypolimnetic oxygen consumption is a function of not only the flux of organic matter to the hypolimnion, i.e., the productivity in the epilimnion, but also the hypolimnetic temperature (Cornett & Rigler 1980). In addition, the morphometry of the lake is important. For example, Crooked Lake and Little Crooked Lake are two lakes in north-eastern Indiana which have roughly similar productivity and water chemistry. The volume of hypolimnetic water is much smaller in shallower Little Crooked Lake, and as a result, anaerobic strata appear earlier in the year, and span depths from the bottom up to 8 m (Figure 3.2).

Several other vertical gradients are relevant to microbial processes in lakes (Figure 3.3). First among these is the light gradient. Light intensity decreases exponentially with depth. In addition to the absorption of light by water, additional light attenuation is due to:

1 coloration of the water (Bowling *et al.* 1986);
2 abiogenic particles (such as clay);
3 phytoplankton (Talling *et al.* 1973).

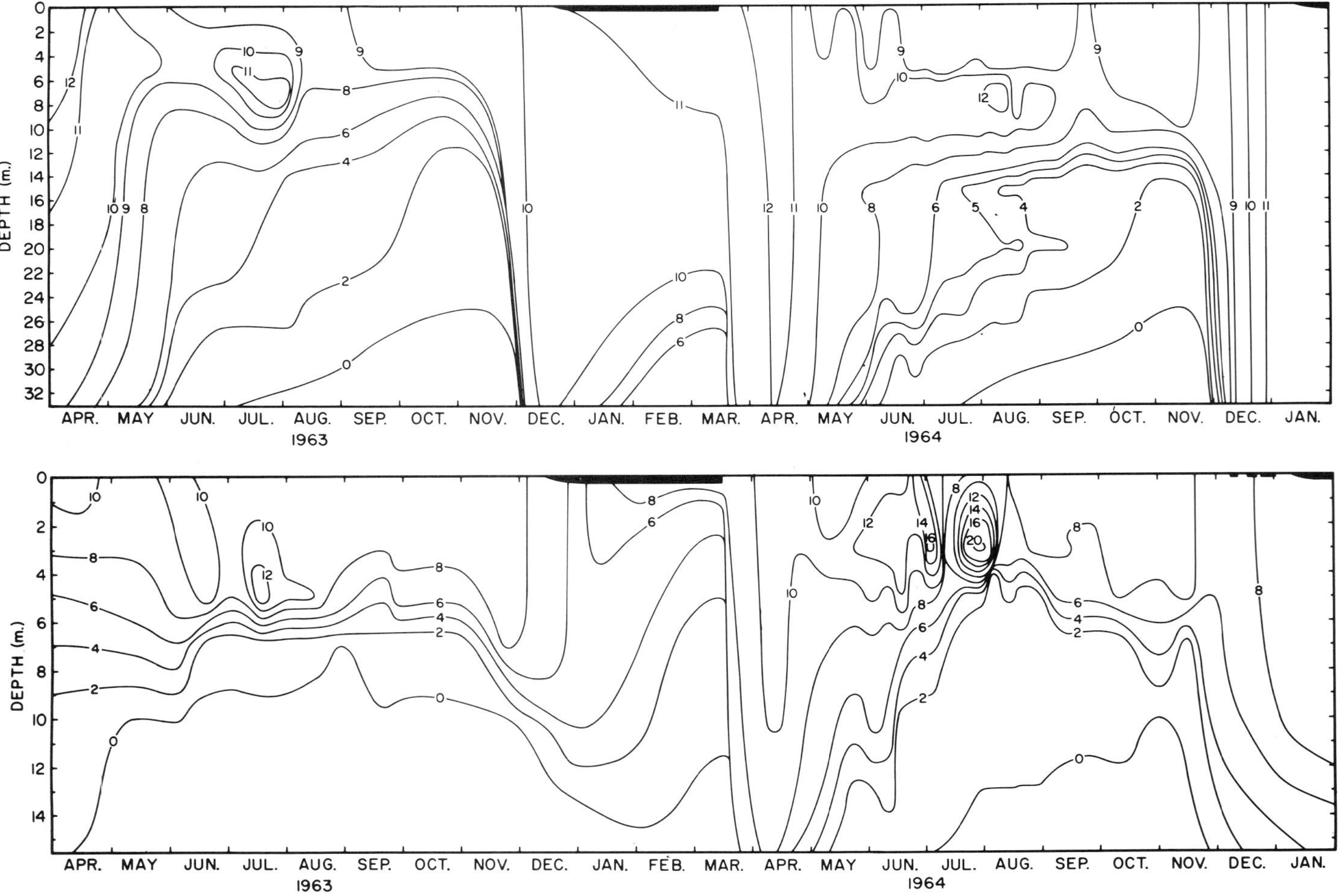

Figure 3.2 Dissolved oxygen concentration ($mg\,l^{-1}$) as a function of depth and time of year for Crooked Lake (upper) and Little Crooked Lake (lower). The lines (isopleths) connect points of equal oxygen concentration. From Wetzel (1983).

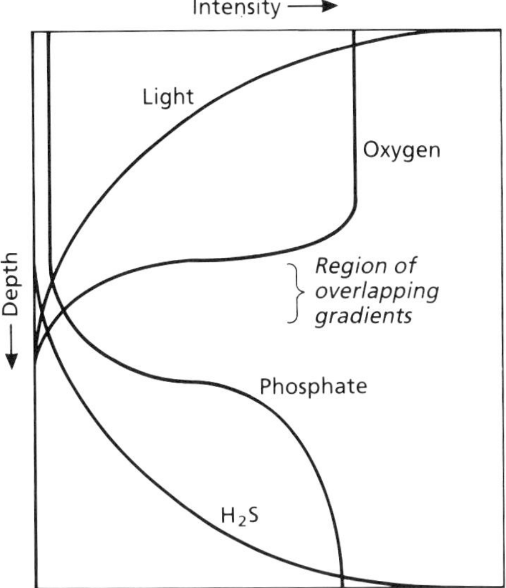

Figure 3.3 An idealized schematic of vertical gradients in lakes. Stable strata where these gradients overlap are regions where specialized physiological types of microbes can form layers. The existence of overlapping gradients is determined by lake morphometry, physical factors, and biological processes in the lake (see Figure 3.4).

As a rule of thumb, the depth of the euphotic zone (z_{eu}), the region where net photosynthesis can occur, is the depth at which the irradiance is 1% of that at the lake surface (I_0). Of special note are lakes in which z_{eu} is greater than the depth of the mixed zone (z_m). Stratified layers of phototrophic microorganisms (cyanobacteria, motile eukaryotic algae, such as dinoflagellates or chrysophytes, or photosynthetic sulfur bacteria), may develop in these layers, occasionally at depths where the irradiance is $<1\%$ I_0. For example, photosynthetic sulfur bacteria were first detected at depths where irradiance was 0.015–0.3% of I_0 in a set of Wisconsin lakes (Parkin & Brock 1980). Although the phototrophs in these layers are exposed to rather low irradiances, they may have access to higher concentrations of inorganic macronutrients or essential electron donors (hydrogen sulfide in the case of photosynthetic sulfur bacteria), in the metalimnion or hypolimnion than do phototrophs in the epilimnion.

Vertical gradients of inorganic forms of nitrogen and phosphorus occur in stratified lakes. In the epilimnion, these nutrients are depleted by photoassimilation. Although the major fraction of these nutrients may be recycled in the epilimnion (Fallon & Brock 1980), the remainder sediments to the stable regions of the water column and is remineralized. Therefore, a vertical gradient of macronutrient availability develops during summer stratification.

Once the sediment–water interface has become anaerobic, the reduced products of anaerobic microbial metabolism in the sediments (hydrogen sulfide and methane) will diffuse from the sediments, and form gradients in the water column.

3.3 Microbial processes in the metalimnion and hypolimnion

The depths at which these vertical gradients intersect provide niches which can be exploited by specific physiological groups of microbes. When the overlying water is oxic, these intersections occur at the sediment–water interface. In lakes with anaerobic hypolimnia, the interface between the metalimnion and hypolimnion is of most importance. The intersecting gradients provide a narrow zone in which certain microbial processes can occur; as a consequence, discrete layers of physiologically specialized microbes may arise in stratified lakes. Gorlenko *et al.* (1983) have compiled a number of studies of the vertical distribution of microbial activities in lakes.

3.3.1 The light–macronutrient gradient

For oxygenic phototrophs (cyanobacteria and eukaryotic algae), the major determinants of the production rate and amount of biomass in a lake are light and macronutrients in the form of phosphorus and nitrogen. If z_{eu} is greater than z_m, then the stable regions are environments in which the availability of nitrogen and/or phosphorus may be much greater than in the epilimnion, and in which irradiance levels, although low, are adequate for net growth. The problem for these organisms is to maintain vertical position. These phototrophs are much larger in size than most heterotrophic bacteria – cell diameters are usually >3 μm, and some form filaments which give them maximum dimensions >100 μm (Reynolds 1984). To a first approximation, the terminal velocity of sinking ($m\,s^{-1}$) is described by Stokes' Law:

$$v_s = 2gr^2(\rho' - \rho)/9\eta, \tag{3.1}$$

in which g is gravitational acceleration ($m\,s^{-2}$), r is the radius of the particle (m), ρ' and ρ are the densities ($kg\,m^{-3}$) of the particle and the medium, and η is the viscosity of the medium ($kg\,m^{-1}\,s^{-1}$). The density of organisms varies, both phylogenetically, for example, diatoms which have silicon frustules are denser than other phytoplanktons and physiologically, due to the accumulation of dense polymers (Guerrero *et al.* 1984), but the most important factor in Stokes' Law is cell size, because velocity increases as the square of the cell's radius. Thus, whereas single-celled bacteria with a radius of 1 μm may sink about 1 mm d^{-1}, and the sinking rate of a spherical algal cell of 10 μm diameter would be on the order of 10 cm d^{-1}, filamentous or colonial phototrophs could have sinking velocities measured in meters per day.

Therefore, larger organisms which form stratified layers must have a means of maintaining their vertical position in the water column. The eukaryotic algae which have been reported to form stratified layers include dinoflagellates, such as *Ceratium hirundinella* (Heaney & Talling 1980), chrysophytes (Pick *et al.* 1984a), and cryptomonads. All of these organisms are capable of flagellar motility, and therefore can undergo controlled movements to maintain their position within discrete layers. The nonmotile species of eukaryotic algae which stratify in stable layers have low sinking velocities because they are either small in size or produce large, hydrated mucilaginous sheaths which reduce the colony density to a value close to that of water.

Filamentous cyanobacteria, such as planktonic *Oscillatoria*, which contain intracellular gas vesicles can achieve neutral buoyancy, and thereby form discrete layers in lakes (Konopka 1982). The environmental factors that influence buoyancy regulation and its physiological basis will be discussed in Section 3.5.

Do these metalimnetic layers represent zones of active production and growth? The irradiance at the top of these layers is usually less than 5–10% of that at the lake surface. Metalimnetic layers have been reported to contribute up to 20–40% of primary production in the water column in lakes with layers of chrysophytes (Fee 1976) or cyanobacteria (Konopka 1983). The presumption is that the flux of inorganic macronutrients in the metalimnion is adequate to support growth of phototrophs in this zone. Pick *et al.* (1984a) have presented a contrasting view for a metalimnetic chrysophyte peak in Jack's Lake, Ontario. They concluded that the metalimnetic layer arose from the vertical migration of algae which had grown in the epilimnion. They detected no subsequent net increase in metalimnetic biomass. Furthermore, they suggested that these organisms were not supplied with nutrients diffusing from the hypolimnion, but had cellular reserves when they migrated from the epilimnion (Pick *et al.* 1984b). Thus, not all metalimnetic layers may represent cases where vertical discontinuities in nutrient availability make the metalimnion an optimal environment for growth; it may also be a refuge in which loss factors for phytoplankton are minimized.

3.3.2 *The light–hydrogen sulfide gradient*

The requirements for growth of anoxygenic phototrophs (the purple and green sulfur bacteria) are light, the absence of oxygen, and a suitable electron donor (hydrogen sulfide) for photosynthesis. Net growth of anoxygenic phototrophs can occur at irradiances much less than 1% of the surface irradiance (Parkin & Brock 1980; van Gemerden 1980). Stratified layers of phototrophic bacteria are most often found in small deep lakes protected from wind mixing; how-

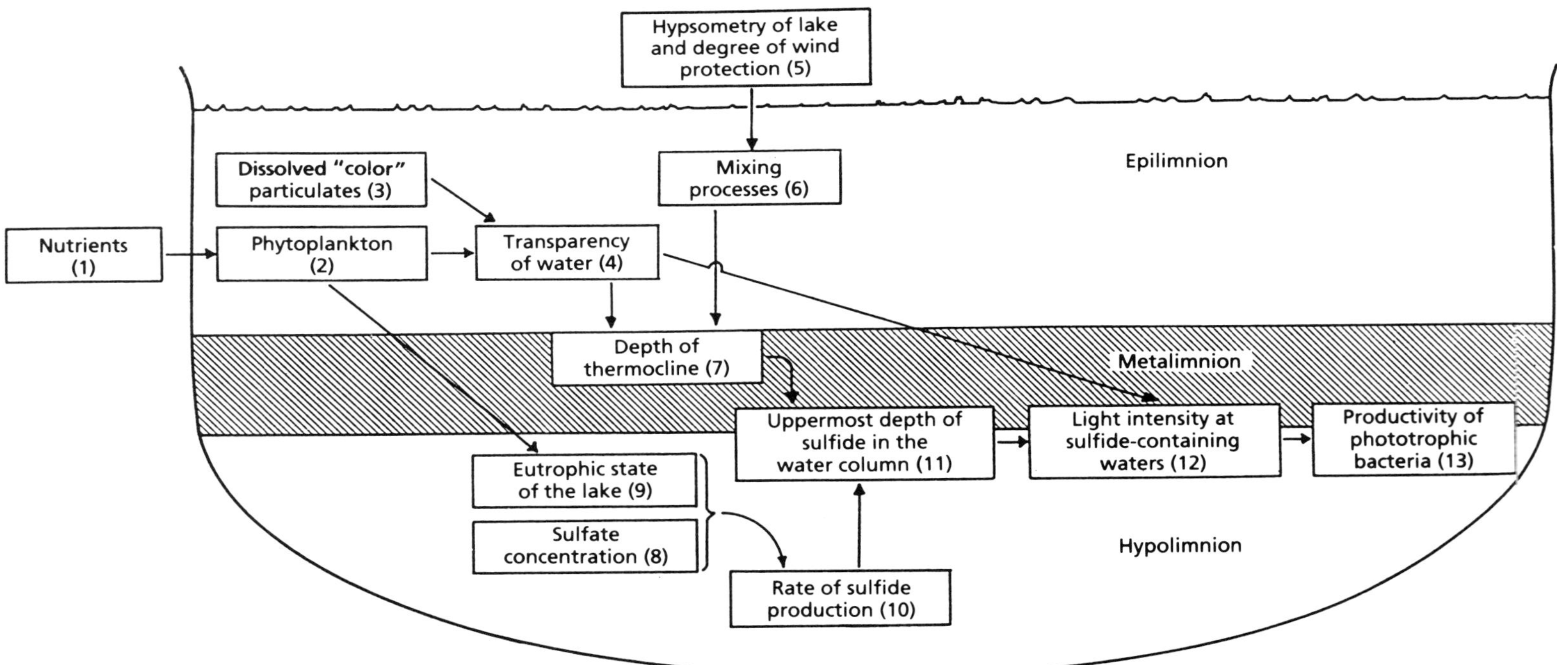

Figure 3.4 Factors which affect the light intensity which penetrates to sulfide-containing waters. Inorganic nutrient concentrations (1) affect phytoplankton growth (2) which, with other particulates and dissolved "color" (3) determine transparency (4). Morphometry of basin and degree of wind protection (5) determine degree and intensity of mixing (6). Mixing processes together with transparency determine thermocline depth (7). Rate of sulfide production (10) in hypolimnion is influenced by sulfate concentration (8) and eutrophic state (9). The latter effect occurs because sulfate reduction is driven by catabolism of organic matter. Net rate of sulfide formation determines upper depth at which sulfide is present, with thermocline defining the upper limit (11). The upper depth of sulfide and water transparency demarcate the light intensity at sulfide-containing water (12), which is the major factor which determines the primary production of phototrophic sulfur bacteria (13). From Parkin & Brock (1980).

ever, a set of physical, chemical, and biological factors interact to determine whether light penetrates to the sulfide-containing anaerobic zone in a lake (Figure 3.4). Dense layers containing more than 100 μg bacteriochlorophyll per liter can occur. Although visually dramatic, the contribution of bacterial photosynthesis to total primary production in the water column is usually small due to the low irradiances at which they stratify. A compilation of data from 16 lakes indicated that photosynthetic bacterial production only contributed more than 10% of total primary production if the phototrophs stratified at a depth where the irradiance exceeded 1% of the surface irradiance (Parkin & Brock 1980).

There can be several groups of anoxygenic phototrophs in these strata, and they stratify at different depths. The primary environmental factors which influence what species occur appear to be light intensity, the spectral quality of the light, and sulfide concentration (Abella & Garcia-Gil 1988). If the light intensity reaching the sulfide-containing layer is relatively high, a layer of purple sulfur bacteria occurs nearest the surface (Guerrero *et al.* 1985); purple bacteria have been shown to require higher irradiances to grow at the same rate as green bacteria (Biebl & Pfennig 1978). The spectral quality of the light which passes through the purple *Chromatium* layer will only support the growth of green *Chlorobium* species. In lakes where the irradiance which reaches the sulfide-containing strata is <0.01% of the surface irradiance, *Pelodictyon* sp. or consortia of *Chlorobium* and motile heterotrophic bacteria dominate if the sulfide concentration is in the micromole range, whereas *Chlorobium* sp. dominate habitats in which the hydrogen sulfide concentration exceeds 1 mmol.

3.3.3 Gradients of oxygen and reduced substances

Just as the overlapping gradients of light and hydrogen sulfide create a niche for anaerobic sulfide oxidation by anoxygenic phototrophs, an overlap of gradients of oxygen diffusing from the epilimnion, and reduced inorganic compounds, such as hydrogen sulfide or methane diffusing from the sediment–water interface results in layers of aerobic chemoautotrophic bacteria within the stable region of the water column. The factors controlling the vertical and temporal distribution of methane oxidation have been investigated in Lake Mendota (Fallon *et al.* 1980; Harrits & Hanson 1980) and in experimentally fertilized Lake 227 (Rudd & Hamilton 1975; Rudd *et al.* 1976). These two studies illustrate the importance of lake morphometry and physical and chemical processes upon microbial metabolism. In Lake 227, methane oxidation during summer stratification was restricted to a zone 1–2 m in depth, whereas in Lake Mendota, methane oxidation rates were higher and extended over a 5 m zone in the thermocline. Lake Mendota has a much larger surface area than

Lake 227, and therefore is more susceptible to wind turbulence. The eddy diffusivity in the thermocline of Lake Mendota is 100-fold greater than in Lake 227 (Fallon *et al.* 1980), hence the diffusion of methane in the thermocline will be greater. The narrow zone of methane oxidation in Lake 227 may have been due to the absence of dissolved inorganic nitrogen in the thermocline. As a result, the methane-oxidizing bacteria were dependent upon nitrogen fixation, and this oxygen-sensitive process restricted their distribution to strata where the oxygen concentration was $<1\,mg\,l^{-1}$ (Rudd *et al.* 1976).

The vertical stratification of bacterial species in the metalimnion and hypolimnion is in fact much greater than indicated by the above considerations. Several microscopic studies of anaerobic hypolimnia in lakes (Caldwell & Tiedje 1975a,b; Fliermans *et al.* 1975) have found a wide morphological diversity of microbes. Specific bacteria are found in discrete layers 10–20 cm in breadth. Many of these bacteria contain gas vesicles (Walsby 1974), which could provide neutral buoyancy and permit microstratification in the hypolimnion. Many of the morphological forms observed in these samples have never been grown in culture, and their specific contributions to the chemical cycles in lakes are unknown.

3.4 Temporal heterogeneities in lakes

Temporal changes in environmental factors occur on a broad range of time scales, ranging from decades (long-term changes in climate) to seasons (annual changes in thermal stratification) to days (local meteorological events) to minutes (consequences of turbulent vertical mixing in the water column). The biological processes affected by temporal changes on these various scales may differ. For surface waters, Harris (1986) has interrelated the scales of temporal and spatial variability in surface waters to the time over which specific biological processes operate. Whereas physiological processes, such as nutrient uptake, may respond to environmental changes that occur on time scales of minutes to days, changes in community composition are expected to reflect environmental changes which occur over periods of weeks to months. In the sections below, consideration is given to a few temporal heterogeneities that occur on different time scales and their impact upon microbial processes in lakes.

3.4.1 Seasonal changes in thermal stratification in temperate, dimictic lakes

In lake ecosystems, the seasonal changes in the species composition of the phytoplankton have been scrutinized most intensively. A mechanistic model (Sommer *et al.* 1986) of plankton seasonal succession in an idealized lake has been presented by the Plankton Ecology Group. Their model identifies seasonal changes in light and nutrient availability, as well as the population dynamics of herbivores and carnivores, as important factors in the seasonal changes in primary productivity and algal species composition. Thermal stratification plays an important role in these processes, because it alters light availability, and creates a sink for inorganic macronutrients, such as NH_4^+ or PO_4^{3-}, which may become limiting in the epilimnion. The discussion of the seasonal succession of phytoplankton presented here will be brief. More extensive treatments can be found in recent monographs on phytoplankton ecology (Reynolds 1984; Harris 1986), and in the classic work of Hutchinson (1967).

As an example, a eutrophic lake found at temperate latitudes will be examined. During early spring, concentrations of dissolved nutrients have reached their annual maximum. The water column is isothermal, so it can be completely mixed by turbulent wind action. The intensity of solar radiation is low and the length of the day is short. As a consequence, primary production is limited by the average irradiance to which phytoplankton are exposed. The average irradiance is related to the ratio $z_{eu}:z_m$. Therefore, it will be less in deeper lakes than in shallow ones. As day length and insolation increase during the spring, primary production is stimulated. Diatoms, such as *Asterionella* or *Fragilaria*, often dominate because of their capacities to grow efficiently under low $z_{eu}:z_m$ ratios (Harris 1978), and because sinking of these nonmotile forms is not a selective disadvantage in the turbulently mixed water column.

The spring diatom bloom may end due to depletion of silicon, an essential element for diatom growth, or because of the onset of thermal stratification. If there is a period of low wind turbulence after the onset of thermal stratification, diatoms can be lost from the productive zone of the lake due to sedimentation. The sinking of the spring diatom bloom can represent the major flux of phosphorus to the sediments in a lake during the year (Edmondson & Lehman 1981).

The onset of lake stratification may stimulate the growth of small algae which have rapid growth rates – "*r*-strategists," as defined by Kilham & Kilham (1980). Members of the Chlorophyceae or Cryptophyceae often dominate the phytoplankton during these periods. The concentrations of dissolved macronutrients are still relatively high, and the average irradiance to which cells are exposed increases substantially, because z_m decreases by a factor of two or more (Hanna 1990). However, these species are readily grazed by zooplankton, and a large increase in the zooplankton population leads to the "clear water" phase. There is now a selection for phytoplankton species which are resistant to grazing, e.g., due to their large size, but also not susceptible to losses due to sinking out of the epilimnion on calm days. Adaptations which could accomplish these goals include the production of mucilaginous sheaths (which decrease sinking velocity due to the reduction in particle density), flagellar motility, or buoyancy conferred by gas vesicles in cyanobacteria.

In the fall, the thermal structure of the lake is destabilized, as convectional cooling leads to a deepening of the epilimnion. Phytoplankton production is stimulated by the reintroduction of limiting macronutrients (nitrogen or phosphorus) from the metalimnion to the epilimnion. Average irradiance decreases as daily insolation and day length decrease during the fall and z_{eu} decreases. This eventually limits phytoplankton production. Eventually, wind power is sufficient to overcome the weakened density gradient, and the lake is completely mixed (fall overturn). In winter, under ice cover, the lake can again stratify, but it is the colder water (<4°C) which is nearer the surface. Primary production is low, especially if light penetration is attenuated by snow on the lake ice (Ragotzkie 1978).

The annual changes in thermal structure also affect microbial groups other than the phytoplankton. As discussed in Section 3.3.3, chemoautotrophic methane oxidizers form a metalimnetic layer during summer stratification. However, after fall turnover the methane which had accumulated in the anaerobic hypolimnion during the summer is distributed throughout the aerobic water column. In Lake 227, >90% of the total annual oxidation of methane occurred during and just after fall overturn (Rudd & Hamilton 1978). The chemoautotrophic oxidation of ammonia to nitrate may follow a pattern similar to methane oxidation, as ammonia is converted to nitrate after fall overturn and during the winter in Lake Mendota (Brock 1985). The subsequent fate of this nitrate depends upon thermal structure the following summer. Whereas nitrate in the epilimnion is photoassimilated (if ammonia supplies are limiting), nitrate in the anaerobic hypolimnion is used as a terminal electron acceptor by heterotrophic bacteria (Brezonik & Lee 1968).

3.4.2 Changes in irradiance on time scales of 1 day or less

Photosynthetic production of organic matter is dependent upon solar radiation, and the irradiance to which phototrophs are exposed changes over time scales from hours to seconds due to several physical processes. The most obvious one is the diurnal change in irradiance at the lake surface due to the earth's rotation on its axis. The irradiance at the lake surface is also altered by cloud cover. Irradiance on cloudy days may be one-tenth the value on a bright sunny day.

The turbulent mixing of the mixed zone generates a periodicity in the irradiance to which individual phytoplankton cells are exposed. There are two components to this periodicity which must be considered. The first is the time it takes for vertical movement around an

Table 3.1 Vertical movement of microbial cells around the mixed zone as a function of the depth of the mixed zone and wind speed. Calculations made using Lagrangian vertical displacement model from Denman & Gargett (1983)

Wind speed ($m\,s^{-1}$)	Time (hr) for microbe to completely circulate around mixed zone if mixed depth equals:				
	2 m	5 m	10 m	15 m	20 m
2	2.9	45	36	1200	2900
5	0.2	3	23	78	190
7.5	0.06	1	7	23	55
10	0.02	0.4	3	10	23
12	0.01	0.2	2	6	13
15	0.007	0.1	1	3	7

eddy. Denman & Gargett (1983) have analyzed the time scales of vertical mixing in stratified and unstratified water columns with a formulation that involved observable quantities, such as the vertical density gradient (obtained from vertical profiles of temperature) and wind speed across the water surface. The time scale around the largest eddies is inversely proportional to the wind speed. In Lake Constance, the velocity of eddies was calculated to be large enough to sustain mixing around a 10 m column of water in less than 15 min (Reynolds 1989). If the lifetime of individual eddies is considered in the analysis (Denman & Gargett 1983), the cycle time is increased, but at mixing depths and wind speeds found in many freshwater lakes particles would be transported through the mixed zone in times less than 1 day (Table 3.1).

The second significant component is $z_{eu}:z_m$ ratio. If z_m is greater than z_{eu}, a portion of the cycle time during turbulent mixing is spent in "dark" water, where photosynthesis is not possible. This may occur in the spring when the lake is turbulently mixed to the bottom, or in eutrophic lakes after thermal stratification. Lake Mendota is an example of the latter case. Whereas z_{eu} is only 4 m, z_m is 12 m in summer. Therefore, if the epilimnion is turbulently mixed, the phytoplankton are exposed to light only one-third of the cycle time.

3.4.3 Temporal changes in nutrient inputs

Unpredictable climatic events can cause pulses of limiting nutrients to enter the aquatic system. Two of the ways in which this might occur are as follows. Rainstorms may leach nutrients from the surrounding watershed, and deliver these as a pulse to the lake via surface runoff. In thermally stratified lakes, episodes of strong winds may generate enough turbulent kinetic energy to deepen the epilimnion, and entrain nutrient-enriched water from the metalimnion. This has the effect of increasing nutrient availability in the mixed layer.

3.5 Biological responses to temporal heterogeneities in lakes

The type of response microorganisms make to the heterogeneities which occur in lakes depends upon the time scales over which these heterogeneities occur (Harris 1986). Thus, changes which occur over seasons or weeks will result in a change in the species composition in the habitat, by altering the selection pressures which operate in that environment. One such change is thermal stratification in the spring, which increases the average irradiance in the water column and makes losses due to sinking an important factor in phytoplankton competition in the epilimnion. Another example is the development of overlapping gradients of methane and oxygen in the metalimnion, which creates the appropriate conditions for the growth of methane-oxidizing bacteria.

Individual species can alter their patterns of gene expression and macromolecular composition in response to environmental conditions. In brief, nutrient-limited organisms increase their capacity for transport of the limiting nutrient. In addition, they regulate the uptake rates of nonlimiting nutrients and rates of synthesis of functional macromolecules, such as ribosomes, so that these are not present in great excess above the capacity to utilize them (Tempest *et al.* 1983). Although the types of physiological adaptations which occur have been identified (Harder *et al.* 1984; Poindexter 1987), the kinetics with which these changes occur have rarely been analyzed. Post *et al.* (1985) and Post (1986) found that after shifts in growth irradiance, changes in the photosynthetic apparatus occurred with first-order kinetics. Roughly one to two generation times elapsed before the adaptations were completed. The generation times of not only phytoplankton but also bacteria (Lovell & Konopka 1985) in lakes generally exceed one day. Thus, changes in macromolecular composition will occur in response to heterogeneities which persist over time scales of a few days.

The physiological responses made by microbes over the time scale of one to two generation times may allow metabolic responses to environmental fluctuations which occur over short time scales. As mentioned above, cells whose growth rate is limited by a specific nutrient have a higher nutrient-saturated transport rate (V_{max}) for the limiting nutrient than do cells limited by another factor. If such cells encounter a pulse of limiting nutrient, the transport rate greatly exceeds the rate at which the element can be used for biosynthesis of functional molecules, such as RNA and protein. Under these transient conditions, storage forms of the nutrient become important. They permit microorganisms to accumulate the limiting nutrient when it is available, and to maintain growth for a period of time after the nutrient pulse has been exhausted. Thus, the storage of

Table 3.2 Growth rates of *Oscillatoria aghardii* grown on 24-hour light–dark cycles. Growth rates (μ) are presented for the light period and the dark period, the observed "average" growth rate over a 24-hour cycle, and a calculated "average" growth rate which assumes there is no polysaccharide storage in the light or growth in the dark. Data from Post *et al.* (1986)

Photoperiod (hr)	Rate of protein synthesis (h^{-1})		Observed μ (h^{-1})	Calculated μ (h^{-1})[1]
	Light	Dark		
2	0.021	0.007	0.007	0.003
4	0.023	0.010	0.012	0.006
8	0.027	0.014	0.018	0.011
12	0.028	0.031	0.029	0.017
16	0.033	0.033	0.033	0.023
24	0.034	–	0.034	–

[1] Assumes that $\mu = 0.034\ hr^{-1}$ in the light period and $\mu = 0$ during the dark period.

limiting nutrients which occurs in pulses allows microbes to buffer the biosynthetic machinery from external fluctuations.

Two specific examples of this phenomenon can be found in the response of cyanobacteria to temporal fluctuations in the supply of light energy or phosphate. In cyanobacteria, glycogen serves as a storage form of carbon and energy (Smith 1982). When *Oscillatoria aghardii* was grown in a set of 24-hour light–dark cycles, the rate of polysaccharide accumulation in the light was inversely related to the length of the photoperiod (Post *et al.* 1986). Protein synthesis continued during the dark period via chemoheterotrophic metabolism of intracellular polysaccharide (Table 3.2). The storage of polysaccharide in the light, and its utilization in the dark resulted in a higher overall growth rate, than if the organism had only grown at the maximum possible rate ($0.034\ hr^{-1}$) during the light period (last column of Table 3.2).

Phosphate-limited microorganisms exhibit a high V_{max} for phosphate uptake (Rosenberg *et al.* 1977; Gotham & Rhee 1981). When phosphorus-limited *Synechococcus* was given a pulse of phosphate, polyphosphates accounted for more than 60% of transported phosphate from 3–7 hours following the pulse. There was a subsequent decrease in polyphosphate, as the phosphorus was used for synthesis of functional macromolecules, such as RNA (Grillo & Gibson 1979). In this way, species can maximize their capacity to accumulate pulses of limiting nutrients when in competition with other species, and then subsequently use cellular reserves to drive biomass production in the absence of exogenous nutrient supplies.

3.6 Biological responses to vertical heterogeneities in lakes

Vertical gradients of factors essential for growth occur in freshwater lakes. If the gradients overlap, then accurate vertical positioning is essential to exploit the zone of overlap. If the gradients do not overlap, then controlled vertical movements in the water column are necessary to exploit these resources.

The factors which determine the types of vertical heterogeneities that microbial species can exploit are their size and capacity for movement. Nonmotile species can exploit overlapping gradients if the cells are small so that the sinking rate is less than their growth rate. However, organisms which are either motile or possess gas vesicles can accurately position themselves in the gradients, and respond to temporal changes in the gradients.

The photosynthetic sulfur bacteria which form stratified layers in lakes usually possess either flagella or gas vesicles. The vertical gradient of hydrogen sulfide may change during the day in these lakes, due to light-driven hydrogen sulfide oxidation during the day, and hydrogen sulfide accumulation during the night. Thus, the phototrophs might be expected to be higher in the water column in the morning, to take advantage of higher irradiances, and move to deeper strata during the day as they deplete hydrogen sulfide. This has been observed in Lake Belovod, where motile *Chromatium* migrated 2 m (Sorokin 1970) and in Rotsee, in which gas vacuolate *Thiopedia* decreased their vertical position by 1 m during the day (Kohler *et al.* 1984).

If the vertical gradients of essential factors for growth do not overlap, large scale vertical movements are necessary. The gradients of light and inorganic nutrients (nitrogen and phosphorus) in thermally stratified eutrophic lakes provide an example where vertical migrations on the scale of several meters per day can occur. The speed of movement by either flagellar motility or buoyancy changes in gas vacuolate microbes is a function of cell size. Among the flagellates, the dinoflagellates, such as *Ceratium*, have a swimming speed high enough to carry out vertical migrations up to $10\,\mathrm{m\,d^{-1}}$ (Eppley *et al.* 1968), although chrysophytes (Pick & Lean 1984) and cryptomonads (Salonen *et al.* 1984) have also been demonstrated to carry out diurnal migrations of 1–2 m.

The floating and sinking speeds of gas vacuolate microbes are described by Stokes' Law (see Equation 3.1). Therefore, vertical excursions on the order of meters are only possible by colony-forming cyanobacteria (Ganf 1974; Konopka *et al.* 1978). Gas vacuolate cyanobacteria which occur as single filaments (with a length of 50 μm and a radius of 5 μm) can move about $0.5\,\mathrm{m\,d^{-1}}$ (Walsby 1988). Kromkamp & Walsby (1990) have developed a mathematical model which simulates the diurnal changes in buoyancy experienced by cyanobacteria in lakes. Their simulations illustrated that gas vacuolate cyanobacteria which form large colonies can carry out large-scale vertical excursions, but were incapable of forming stratified layers. In contrast, single cyanobacterial filaments did form subsurface stratified layers.

A microbial cell which possesses enough low-density gas vesicles to counterbalance the cellular content of molecules denser than water will be buoyant. However, to effectively exploit the vertical gradients found in lakes, the organism must be able to regulate its buoyancy in response to environmental factors. Cyanobacteria which form stratified layers and those which carry out large vertical excursions use similar mechanisms of regulation, in which buoyancy responds to the relative availability of light and inorganic nutrients. Three mechanisms of buoyancy regulation have been described, and are summarized below. The first two operate over time periods much less than a generation time, whereas the third mechanism requires time periods equal to or greater than a generation time.

Firstly, collapse of gas vesicles by turgor pressure. The weakest gas vesicles in some strains of *Anabaena* and *Aphanizomenon* can be collapsed by pressures of 300–400 kPa. After growth under long photoperiods, a shift from low to high irradiance can raise the intracellular turgor pressure to more than 400 kPa, thereby collapsing some of the gas vesicles (Dinsdale & Walsby 1972). However, under the photoperiods which cyanobacteria experience in nature, this mechanism is likely to be less important than the accumulation of polysaccharide ballast (Kromkamp *et al.* 1986).

Secondly, ballast accumulation. As noted in Section 3.5, cyanobacteria respond to the diurnal periodicity in irradiance by storing polysaccharide during the light period. In several cyanobacteria, the amount of polysaccharide stored during the light period during growth on a light–dark cycle, or after a shift-up in irradiance, was sufficient to counter the excess lift provided by gas vesicles (Gibson 1978; Kromkamp & Mur 1984; Thomas & Walsby 1985; Utkilen *et al.* 1985; Kromkamp *et al.* 1986). Konopka (1984) hypothesized that gas vacuolate cyanobacteria would regulate their buoyancy in response to the availability of light and nutrients (Figure 3.5). Cells possess neutral buoyancy when the rate of photosynthetic energy generation balances the rate at which this energy is used to synthesize catalytically active biomass (proteins and nucleic acids). If the rate of carbon dioxide fixation (determined in the short term by irradiance and carbon dioxide availability) is less than the capacity for growth (determined by inorganic nutrient concentration), the organisms will be buoyant and float towards the surface. This exposes the cell to higher irradiances, and may increase the rate of energy generation

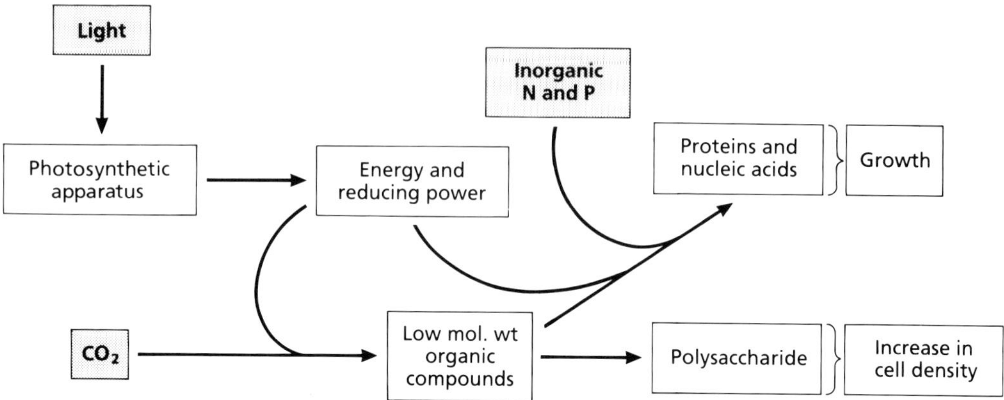

Figure 3.5 Relationships among environmental factors important to short-term buoyancy regulation in gas vacuolate cyanobacteria.

and carbon dioxide fixation. The fate of this additional fixed carbon determines whether the organism will lose buoyancy at a particular irradiance. If inorganic nutrients are not limiting growth, the photosynthate can be used to accelerate the synthesis of proteins and nucleic acids. The organism remains buoyant, and grows faster at the higher irradiance. However, if inorganic nutrient availability limits growth rate to a value below that possible at the incident light intensity, the photosynthate is stored as polysaccharide (Kromkamp *et al.* 1986) and within a few hours, enough may accumulate that the cells begin to sink.

The interaction between light and nutrient availability has been demonstrated in laboratory studies (Konopka *et al.* 1987a,b). Low-buoyancy, phosphorus-limited cells increased buoyancy when irradiance was reduced, whereas buoyant, light-limited cells lost buoyancy if a phosphorus limitation was imposed, even though the irradiance was constant. In field studies, Klemer (1976) and Klemer *et al.* (1985) found that gas vacuolate *Oscillatoria* which stratified at depths of low irradiance in the metalimnion, increased their buoyancy and formed epilimnetic blooms when they were enriched with limiting nutrients. These results illustrate that neither light nor nutrient availability is of overriding importance for buoyancy regulation. Rather, it is the relationship between them that is critical.

Thirdly, regulation of gas vesicle formation. Differences in rates of gas vesicle synthesis in response to irradiance have been found in steady-state cultures of *Aphanizomenon flos-aquae* (Kromkamp *et al.* 1986), and after transitions in irradiance in *Oscillatoria aghardii* (Utkilen *et al.* 1985) and *Microcystis* BC84/1 (Deacon & Walsby 1990). In continuous cultures of *Oscillatoria*, gas vacuolation increased after nitrogen limitation was relieved (Klemer *et al.* 1982).

The specific regulatory mechanism that controls the rate of gas vesicle synthesis in these organisms is unknown, but in *Calothrix*, where gas vesicle formation is developmentally regulated, the expression of the gas vesicle genes is controlled at the transcriptional level (Csiszar *et al.* 1987). In contrast, the cellular gas vesicle content of *Microcystis aeruginosa* AK1 was similar across a spectrum of growth rates in light- or phosphate-limited continuous cultures (Konopka *et al.* 1987b; Kromkamp *et al.* 1988).

These examples illustrate that a dynamic set of processes operate in lakes. Physical processes, such as solar insolation and wind-driven turbulent mixing, are important environmental factors to which microorganisms must respond. Some consequences of these physical processes and microbial activity in lakes are spatial and temporal heterogeneities in the chemical environment. The resulting chemical heterogeneities lead to physiological adaptations in microorganisms as they attempt to maximize growth or survival in the aquatic environment, and to a broad diversity of niches in space and time, in which different microbial species may have a selective advantage.

References

Abella C.A. & Garcia-Gil J. (1988) Diel migration as a mechanism for enrichment of natural populations of branching species of *Pelodictyon*. In Olson J.M., Ormerod J.G., Amesz J., Stackebrandt E. & Truper H.G. (eds) *Green Photosynthetic Bacteria*, pp. 269–285. Plenum, New York.

Biebl H. & Pfennig N. (1978) Growth yields of green sulfur bacteria in mixed cultures with sulfur and sulfate-reducing bacteria. *Arch. Microbiol.* **117**, 9–16.

Bowling L.C., Steane M.S. & Tyler P.A. (1986) The spectral distribution and attenuation of underwater irradiance in Tasmanian inland waters. *Freshwat. Biol.* **16**, 313–335.

Brezonik P.L. & Lee G.F. (1968) Denitrification as a nitrogen sink in Lake Mendota, Wisconsin. *Environ. Sci. Technol.* **2**, 120–125.

Brock T.D. (1985) *A Eutrophic Lake*. Springer Verlag, New York.

Caldwell D.E. & Tiedje J.M. (1975a) A morphological study of anaerobic bacteria from the hypolimnia of two Michigan lakes. *Can. J. Microbiol.* **21**, 362–376.

Caldwell D.E. & Tiedje J.M. (1975b) The structure of anaerobic bacterial communities in the hypolimnia of several Michigan lakes. *Can. J. Microbiol.* **21**, 377–385.

Cornett R.J. & Rigler F.H. (1980) The areal hypolimnetic oxygen deficit: an empirical test of the model. *Limnol. Oceanogr.* **25**, 672–679.

Csiszar K., Houmard J., Damerval T. & Tandeau de Marsac N. (1987) Transcriptional analysis of the cyanobacterial *gvpABC* operon in differentiated cells: occurrence of an antisense RNA complementary to three overlapping transcripts. *Gene* **60**, 29–37.

Deacon C. & Walsby A.E. (1990) Gas vesicle formation in the dark, and in light of different irradiances by the cyanobacterium *Microcystis* sp. *Br. Phycol. J.* **25**, 133–139.

Denman K.L. & Gargett A.E. (1983) Time and space scales of vertical mixing and advection of phytoplankton in the upper ocean. *Limnol. Oceanogr.* **28**, 801–815.

Dinsdale M.T. & Walsby A.E. (1972) The interrelations of cell turgor pressure, gas-vacuolation, and buoyancy in a blue-green alga. *J. Exp. Botany* **28**, 409–415.

Edmondson W.T. & Lehman J.T. (1981) The effect of changes in the nutrient income on the condition of Lake Washington. *Limnol. Oceanogr.* **26**, 1–29.

Eppley R.W., Holm-Hansen O. & Strickland J.D.H. (1968) Some observations on the vertical migration of dinoflagellates. *J. Phycol.* **10**, 15–23.

Fallon R.D. & Brock T.D. (1980) Planktonic blue-green algae: production, sedimentation, and decomposition in Lake Mendota, Wisconsin. *Limnol. Oceanogr.* **25**, 72–78.

Fallon R.D., Harrits S., Hanson R.S. & Brock T.D. (1980) The role of methane in internal carbon cycling in Lake Mendota during summer stratification. *Limnol. Oceanogr.* **25**, 357–360.

Fee E. (1976) The vertical and seasonal distribution of chlorophyll in lakes of the Experimental Lakes Area, northwestern Ontario: implications for lake productivity. *Limnol. Oceanogr.* **21**, 767–783.

Fliermans C.B., Schneider C.P. & Schmidt E.L. (1975) Direct measurement of bacterial stratification in Minnesota lakes. *Arch. Hydrobiol.* **76**, 248–255.

Ganf G.G. (1974) Diurnal mixing and the vertical distribution of phytoplankton in a shallow equatorial lake (Lake George, Uganda). *J. Ecol.* **62**, 611–629.

Gibson C.E. (1978) Carbohydrate content as an ecological tool in the study of planktonic blue-green algae. *Verh. Int. Verein. Theor. Angew. Limnol.* **20**, 630–635.

Gorlenko V.M., Dubinina G.A. & Kuznetsov S.I. (1983) *The Ecology of Aquatic Micro-organisms.* E. Schweizerbart'sche Verlagsbuchhandlung, Stuttgart.

Gotham I.J. & Rhee G.-Y. (1981) Comparative kinetic studies of phosphate-limited growth and phosphate uptake in phytoplankton in continuous culture. *J. Phycol.* **17**, 257–265.

Grillo J. & Gibson J. (1979) Regulation of phosphate accumulation in the unicellular cyanobacterium *Synechococcus*. *J. Bacteriol.* **140**, 508–517.

Guerrero R., Mas J. & Pedros-Alio C. (1984) Buoyant density changes due to intracellular content of sulfur in *Chromatium warmingii* and *Chromatium vinosum*. *Arch. Microbiol.* **137**, 350–356.

Guerrero R., Montesinos E., Pedros-Alio C., Esteve I., Mas J., van Gemerden H., Hofman P.A.G. & Bakker J.F. (1985) Phototrophic sulfur bacteria in two Spanish lakes: vertical distribution and limiting factors. *Limnol. Oceanogr.* **30**, 919–931.

Hanna M. (1990) Evaluation of models predicting mixing depth. *Can. J. Fish. Aquat. Sci.* **47**, 940–947.

Harder W., Dijkhuizen L. & Veldkamp H. (1984) Environmental regulation of microbial metabolism. *Symp. Soc. Gen. Microbiol.* **36**, 51–94.

Harris G.P. (1978) Photosynthesis, productivity and growth: the physiological ecology of phytoplankton. *Arch. Hydrobiol. Beih. Ergebn. Limnol.* **10**, 1–171.

Harris G.P. (1986) *Phytoplankton Ecology*. Chapman and Hall, London.

Harrits S.M. & Hanson R.S. (1980) Stratification of aerobic methane-oxidizing organisms in Lake Mendota, Madison, Wisconsin. *Limnol. Oceanogr.* **25**, 412–421.

Heaney S.I. & Talling J.F. (1980) Dynamic aspects of dinoflagellate distribution patterns in a small, productive lake. *J. Ecol.* **68**, 75–94.

Hutchinson G.E. (1957) *A Treatise on Limnology*. John Wiley & Sons, New York.

Hutchinson G.E. (1967) *A Treatise on Limnology*, vol. 2. John Wiley & Sons, New York.

Kilham P. & Kilham S.S. (1980) The evolutionary ecology of phytoplankton. In Morris I. (ed.) *The Physiological Ecology of Phytoplankton*, pp. 571–597. University of California Press, Berkeley.

Klemer A. (1976) The vertical distribution of *Oscillatoria aghardii* var *isothrix*. *Arch. Hydrobiol.* **78**, 343–362.

Klemer A.R., Feuillade J. & Feuillade M. (1982) Cyanobacterial blooms: carbon and nitrogen limitation have opposite effects on the buoyancy of *Oscillatoria*. *Science* **215**, 1629–1631.

Klemer A.R., Pierson D.C. & Whiteside M.C. (1985) Blue-green algal (cyanobacterial) nutrition, buoyancy, and bloom formation. *Verh. Int. Verein. Theor. Angew.*

Limnol. **22**, 2791–2798.

Kohler H.-P., Ahring B., Abella C., Ingvorsen K., Kewoleh H., Laczko E., Stupperich E. & Tomei F. (1984) Bacteriological studies on the sulfur cycle in the anaerobic part of the hypolimnion and in the surface sediments of Rotsee in Switzerland. *FEMS Microbiol. Lett.* **21**, 279–289.

Konopka A. (1982) Buoyancy regulation and vertical migration by *Oscillatoria rubescens* in Crooked Lake, Indiana. *Br. Phycol. J.* **17**, 427–442.

Konopka A. (1983) Epilimnetic and metalimnetic primary production in an Indiana hardwater lake. *Can. J. Fish. Aquat. Sci.* **40**, 792–798.

Konopka A. (1984) Effect of light-nutrient interactions on buoyancy regulation by planktonic cyanobacteria. In Klug M.J. & Reddy C. (eds) *Current Perspectives in Microbial Ecology*, pp. 41–48. American Society for Microbiology, Washington DC.

Konopka A., Brock T.D. & Walsby A.E. (1978) Buoyancy regulation by planktonic blue-green algae in Lake Mendota, Wisconsin. *Arch. Hydrobiol.* **83**, 524–537.

Konopka A., Kromkamp J. & Mur L.R. (1987a) Regulation of gas vesicle content and buoyancy in light- or phosphate-limited cultures of *Aphanizomenon flos-aquae* (Cyanophyta). *J. Phycol.* **23**, 70–78.

Konopka A., Kromkamp J. & Mur L.R. (1987b) Buoyancy regulation in phosphate-limited cultures of *Microcystis aeruginosa*. *FEMS Microbiol. Ecol.* **45**, 135–142.

Kromkamp J.C. & Mur L.R. (1984) Buoyant density changes in the cyanobacterium *Microcystis aeruginosa* due to changes in the cellular carbohydrate content. *FEMS Microbiol. Lett.* **25**, 105–109.

Kromkamp J.C. & Walsby A.E. (1990) A computer model of buoyancy and vertical migration in cyanobacteria. *J. Plank. Res.* **12**, 161–184.

Kromkamp J.C., Konopka A. & Mur L.R. (1986) Buoyancy regulation in a strain of *Aphanizomenon flos-aquae* (Cyanophyceae): the importance of carbohydrate accumulation and gas vesicle collapse. *J. Gen. Microbiol.* **132**, 2113–2121.

Kromkamp J.C., Konopka A. & Mur L.R. (1988) Buoyancy regulation in energy-limited cultures of *Microcystis aeruginosa*. *J. Plank. Res.* **10**, 171–183.

Lovell C.R. & Konopka A. (1985) Seasonal bacterial production in a dimictic lake as measured by increases in cell numbers and thymidine incorporation. *Appl. Environ. Microbiol.* **49**, 492–500.

Parkin T.B. & Brock T.D. (1980) Photosynthetic bacterial production in lakes: the effects of light intensity. *Limnol. Oceanogr.* **25**, 711–718.

Pick F.R. & Lean D.R.S. (1984) Diurnal movements of metalimnetic phytoplankton. *J. Phycol.* **20**, 430–436.

Pick F.R., Nalewajko C. & Lean D.R.S. (1984a) The origin of a metalimnetic chrysophyte peak. *Limnol. Oceanogr.* **29**, 125–134.

Pick F.R., Nalewajko C. & Lean D.R.S. (1984b) Nutrient status of metalimnetic phytoplankton peaks. *Limnol. Oceanogr.* **29**, 960–971.

Poindexter J.S. (1987) Bacterial responses to nutrient limitation. *Symp. Soc. Gen. Microbiol.* **41**, 283–317.

Post A.F. (1986) Transient state characteristics of adaptation to changes in light conditions for the cyanobacterium *Oscillatoria aghardii*. I. Pigmentation and photosynthesis. *Arch. Microbiol.* **145**, 353–357.

Post A.F., Dubinsky Z., Wyman K. & Falkowski P.G. (1985) Physiological responses of a marine planktonic diatom to transitions in growth irradiance. *Mar. Ecol. Prog. Ser.* **25**, 141–149.

Post A.F., Loogman J.G. & Mur L.R. (1986) Photosynthesis, carbon flows and growth of *Oscillatoria aghardii* Gomont in environments with a periodic supply of light. *J. Gen. Microbiol.* **132**, 2129–2136.

Ragotzkie R.A. (1978) Heat budgets of lakes. In Lerman A. (ed.) *Lakes: Chemistry, Geology, and Physics*, pp. 1–19. Springer Verlag, New York.

Reynolds C.S. (1984) *The Ecology of Freshwater Phytoplankton*. Cambridge University Press, Cambridge.

Reynolds C.S. (1989) Physical determinants of phytoplankton succession. In Sommer

U. (ed.) *Plankton Ecology*, pp. 9–56. Springer Verlag, Berlin.

Rosenberg H., Gerdes R.G. & Chegwidden K. (1977) Two systems for the uptake of phosphate in *Escherichia coli*. *J. Bacteriol.* **131**, 505–511.

Rudd J.W.M. & Hamilton R.D. (1975) Factors controlling rates of methane oxidation and the distribution of the methane oxidizers in a small stratified lake. *Arch. Hydrobiol.* 75, 522–538.

Rudd J.W.M. & Hamilton R.D. (1978) Methane cycling in a eutrophic shield lake and its effects on whole lake metabolism. *Limnol. Oceanogr.* **23**, 337–348.

Rudd J.W.M., Furutani A., Flett K.J. & Hamilton R.D. (1976) Factors controlling methane oxidation in shield lakes: the role of nitrogen fixation and oxygen concentration. *Limnol. Oceanogr.* **21**, 357–364.

Salonen K., Jones R.I. & Arvola L. (1984) Hypolimnetic phosphorus retrieval by diel vertical migrations of lake phytoplankton. *Freshwat. Biol.* **14**, 431–438.

Smith A.J. (1982) Modes of cyanobacterial carbon metabolism. In Carr N.G. & Whitton B.A. (eds) *The Biology of Cyanobacteria*, pp. 47–85. University of California Press, Berkeley.

Sommer U., Gliwicz Z.M., Lampert W. & Duncan A. (1986) The PEG-model of seasonal succession of planktonic events in freshwaters. *Arch. Hydrobiol.* **106**, 433–471.

Sorokin Yu.I. (1970) Interrelations between sulphur and carbon turnover in meromictic lakes. *Arch. Hydrobiol.* **66**, 391–446.

Talling J.F., Wood R.B., Prosser M.V. & Baxter R.M. (1973) The upper limit of photosynthetic productivity by phytoplankton: evidence from Ethiopian soda lakes. *Freshwat. Biol.* **3**, 53–76.

Tempest D.W., Neijssel O.M. & Zevenboom W. (1983) Properties and performance of microorganisms in laboratory culture; their relevance to growth in natural ecosystems. *Symp. Soc. Gen. Microbiol.* **34**, 119–152.

Thomas R.H. & Walsby A.E. (1985) Buoyancy regulation in a strain of *Microcystis*. *J. Gen. Microbiol.* **131**, 799–809.

Utkilen H.C., Oliver R.L. & Walsby A.E. (1985) Buoyancy regulation in a red *Oscillatoria* unable to collapse gas vacuoles by turgor pressure. *Arch. Hydrobiol.* **102**, 319–329.

van Gemerden H. (1980) Survival of *Chromatium vinosum* at low light intensities. *Arch. Microbiol.* **125**, 115–121.

Walsby A.E. (1974) The identification of gas vacuoles and their abundance in the hypolimnetic bacteria of Arco Lake, Minnesota. *Microb. Ecol.* **1**, 48–58.

Walsby A.E. (1988) Homeostasis in buoyancy regulation by planktonic cyanobacteria. In Whittenbury R., Gould G.W., Banks J.G. & Board R.G. (eds) *Homeostatic Mechanisms in Micro-organisms*, pp. 99–116. Bath University Press, Bath.

Wetzel R.G. (1983) *Limnology*, 2nd edn. W.B. Saunders, Philadelphia, Pennsylvania.

Cornelis L.M. Steenbergen, Jean-Pierre R.A. Sweerts, & Thomas E. Cappenberg

4 Microbial biogeochemical activities in lakes: stratification and eutrophication

4.1 Introduction

Lakes contain many kinds of microorganisms representing an immense variety of metabolic types (Overbeck 1974; Kuznetsov 1977; Jones 1987). Apparently, their recurrent appearance and typical spatial distribution is intimately related to their functional characteristics, matching specific environmental conditions, manifesting the typical stratified state of the deep-lake ecosystem. This concept of their occurrence stems from what we measure as ecological variables, i.e., physical and chemical factors and the properties of the microbial populations themselves, such as their distribution, densities, metabolic requirements, and activities. Obviously, microbial populations whose densities and activities peak in gradient zones in the thermocline and at the sediment–water interface, form a most conspicuous feature in stratifying productive lakes.

The functioning of the pelagic stratified water column may be visualized in a conceptual scheme, in which carbon fulfills the role as carrier of energy and matter (Figure 4.1). Major emphasis is laid on the transport processes, i.e., sedimentation and turbulent diffusion. They constitute the main pathways by which nutrients are distributed within the stratified lake, connecting major sites of production and decomposition of organic material.

The complete oxidation of organic carbon to carbon dioxide can be carried out by microorganisms under oxic conditions in the epilimnion and littoral region of lakes. Supply of oxygen to deeper water strata may become limited upon stratification. The thermocline suppresses the exchange between epi- and hypolimnion. The sedimenting organic particles are decomposed in the anoxic hypolimnion and adjacent sediments to organic compounds of low molecular weight by fermentation processes. These small molecules are utilized by different bacterial groups with subsequently NO_3, MnO_2, FeOH, SO_4, and CO_2 as terminal electron acceptors. This sequence is equal

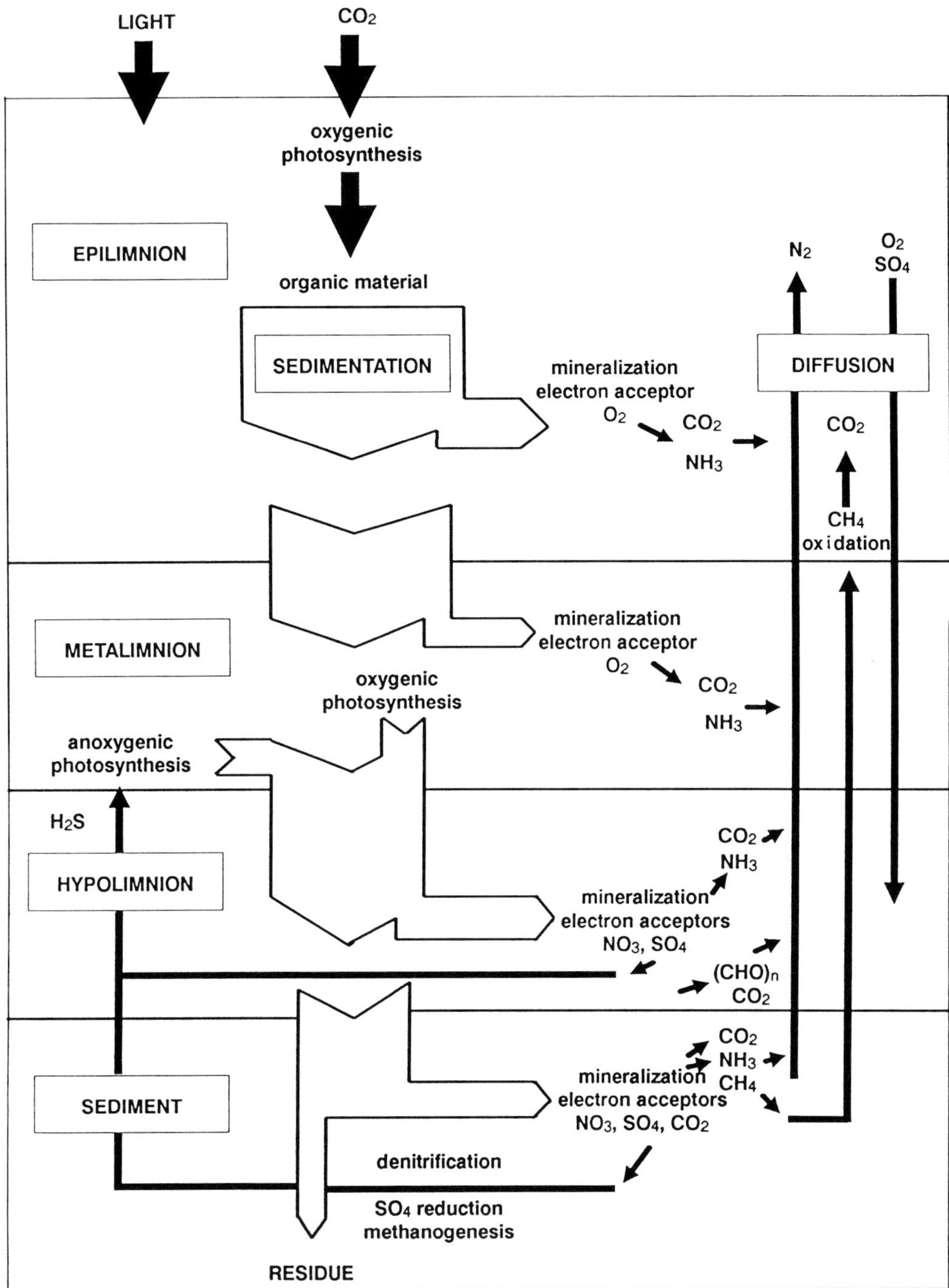

Figure 4.1 Scheme representing the carbon flux through a stratified productive lake showing the zonation of primary production and the series of major consecutive mineralization processes.

as predicted to the thermodynamically predictable energy yield of the individual reactions. Whether this sequence is caused by competition for different bacterial groups is not clear. Spatial overlap of mineralization processes using different electron acceptors also occurs (Canfield & Des Marais 1991). As a result of all these processes, gradient profiles will develop in the stratified lake of the respective electron donors and acceptors: O_2, NO_3, NH_4, SO_4, H_2S, CO_2, and CH_4. In steady-state conditions these gradient profiles can be employed in diagenetic models to estimate the rates of the associated processes, oxygen reduction, denitrification, nitrification, sulfate reduction, and methanogenesis, respectively. In summary, the decomposition of organic carbon produced in primary production is directly associated with the interrelationships among the cycles of carbon, nitrogen, sulfur, and phosphorus, involving bacterial groups that differ with respect to their terminal electron acceptors and competition for electron donors (Cappenberg 1988). Microbial life in lakes can be related to factors such as the lake's morphometry, the quality and characteristics of the catchment area, the climatological conditions, the sediment and water chemistry, management, and, eventually, pollution. These factors govern such variables as the trophic state, the rate of turbulence and turbulent diffusion, anoxic conditions of the hypolimnion, the underwater light climate, and the types of compounds escaping from the sediments.

Clearly our impact may result in degradation of the water quality and ultimately in destruction or restoration of typical microbial populations and processes.

It is the intention in this chapter to highlight structural and functional aspects of microbial populations involved in cycling of nutrients in gradient zones of the lacustrine environment, guided by results of field studies carried out in the small (4.7 ha; mean depth 6 m), warm monomictic, mesoeutrophic Lake Vechten. This lake was formed during 1941 as a result of sand quarrying for road building purposes (Steenbergen & Verdouw 1982). Attention will also be paid to aspects of eutrophication, which have been studied in Lake Vechten and in the Loosdrecht Lakes, a series of shallow interconnected lakes (1800 ha; mean depth 1.8 m) partly man-made as a result of peat extraction in the seventeenth century (Van Liere *et al.* 1984).

4.2 Vertical distribution of microbial assemblages

The study of structure and distribution patterns of microbial assemblages in lakes gained new impetus with the introduction of the analysis of distinct microbial biomarker molecules employed as chemotaxonomic indicators. Microbial biomarkers are chemical components of microorganisms which can be analyzed directly from the environment and be interpreted both quantitatively and qualitatively in terms of *in situ* microbial biomass (Parkes 1987; White 1988).

A fine example of this approach has been the work of Hallegraeff (1977) which concentrated on the analysis of photosynthetic pigments with paper chromatography. Recently, high-performance liquid chromatography (HPLC) techniques for pigment analysis have become available. These permit detection of all types of phototrophs with great sensitivity and high-resolution capabilities (Korthals & Steenbergen 1985; Hurley & Watras 1991). In addition to pigments, polar lipid fatty acids (PLFA) have been used as biomarkers, indicating microbial biomass, community structure, and activity, and providing insight into the occurrence and dynamics of planktonic and benthic species (Fredrickson *et al.* 1986; 1988). In contrast to conventional methods PFLA analysis has many typical attributes which make it very suitable as a chemotaxonomic tool. PLFA are components of the cellular membranes and comprise 1–5% of the cellular dry weight. Their concentration in environmental samples should be proportional to microbial biovolume. Furthermore, their phosphodiester linkage is unstable in most environments, so one may discriminate between live and dead cells. Finally, the chemical structure of PLFA is class- and, in some cases, species-specific.

Reliable techniques for the enumeration of heterotrophic nanoflagellates (HNAN), employing epifluorescence microscopy, have been worked out by Bloem *et al.* 1986. This permits quantitative field studies on these interesting organisms, which peak at distinct sites particularly in the metalimnion of stratified lakes and play a crucial role in the so-called microbial loop (Azam *et al.* 1983).

A recently developed tool for identification and enumeration of microorganisms constitutes flow cytometry (FCM) (Yentsch *et al.* 1983), which is the rapid measurement of particles in a moving fluid; information is obtained through their optical properties, such as morphological features and fluorescence. This powerful technique which can be used on shipboard, enabled Chisholm *et al.* (1988) to identify a new group of phototrophic picoplankton and to detect their depth distribution in the ocean.

4.2.1 *Phototrophic microorganisms, heterotrophic nanoflagellates, and bacteria*

Phototrophic populations have been identified and quantified using class-specific pigments as distinctive biomarker molecules, i.e., signature pigments. Once identified, populations can be recognized and estimated much more easily by, for instance, microscopical counting. In Figure 4.2 and Table 4.1 the vertical distribution of the main signature pigments and associated species from Lake Vechten have been summarized. Eukaryotic algal signature pigments (peridinin, fucoxanthin, alloxanthin), have been found during summer in the

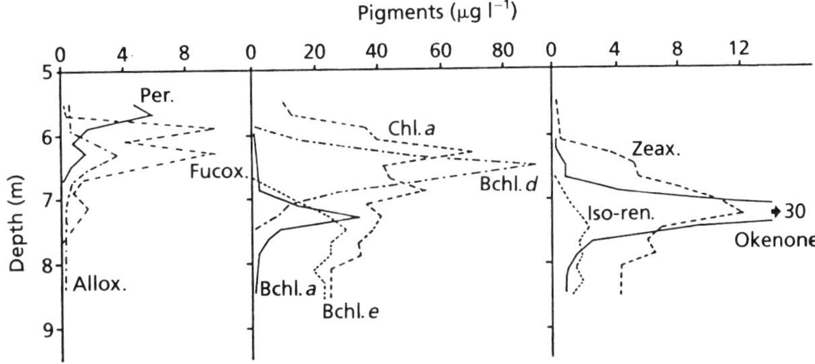

Figure 4.2 Vertical distribution of photosynthetic pigments in meta- and hypolimnion of Lake Vechten on August 29, 1985; associated microbial species are indicated in Table 4.1. From Steenbergen *et al.* (1987).

Table 4.1 Depth distribution and maximum concentration of signature pigments and associated microbial classes and identified species in Lake Vechten. Adapted from Steenbergen & Korthals (1988)

Depth (m)	**Amount** ($\mu g\ l^{-1}$)	**Pigment**	**Classes and species**
0–6	6.0	Peridinin	Dinophyceae *Ceratium hirundinella*
6.3	3.8	Alloxanthin	Cryptophyceae *Cryptomonas*, *Rhodomonas*
6–6.3	8.0	Fucoxanthin	Chrysophyceae *Mallomonas caudata*
6–8	12.0	Zeaxanthin	Cyanobacteria *Synechococcus*
6.5	88.0	Bacteriochl *d*	Green Chlorobiaceae *Chloronema*
7.2	35.0	Bacteriochl *a*	Chromatiaceae *Chromatium*, *Thiopedia*
7.2	30.0	Okenone	*ibid*
7.5	25.0	Bacteriochl *e*	Brown Chlorobiaceae *Chlorobium phaeobacteroides*
7.5	2.0	Iso-renieratene	*ibid*

epilimnion and during winter mixing, which is the nonstratified period from November till April, throughout the water column. During summer stratification prokaryotic pigments were found below eukaryotic pigments, just below the thermocline at the oxygen–sulfide interface (6–7 m). They indicated the presence of various groups of phototrophic bacteria which are rarely exposed to light levels exceeding 1% of that incident at the surface (Steenbergen & Korthals 1982), and were well adapted to the low-ambient light intensity (Steenbergen *et al.* 1987). These bacteria comprise cyanobacterial picoplankton (*Synechococcus* spp.) and green and purple sulfur bacteria (*Chloronema* spp., *Chromatium* spp., *Thiopedia* spp., and *Chlorobium phaeobacteroides*) (see Table 4.1).

Associated with the phototrophic microorganisms in the thermocline are the populations of heterotrophic bacteria, HNAN, and ciliates (Bloem 1989). They form part of the microbial loop and contribute to the flow of energy to higher trophic levels (Stockner 1988). The relationships between these groups of organisms were studied in Lake Vechten by Bloem & Bar-Gilissen (1989). They found distinct peaks of these organisms and their activity during late summer in the thermocline of the lake (see Section 4.3).

To determine the abundance and vertical dispersion of the different groups of phototrophic microorganisms in Lake Vechten, their autofluorescence abilities were measured *in situ* with a field fluorometer and in the laboratory with FCM. The small coccoid cells of *Synechococcus* and the filaments of *Chloronema* could easily be discriminated by FCM. The gas vacuolate *Chloronema* filaments

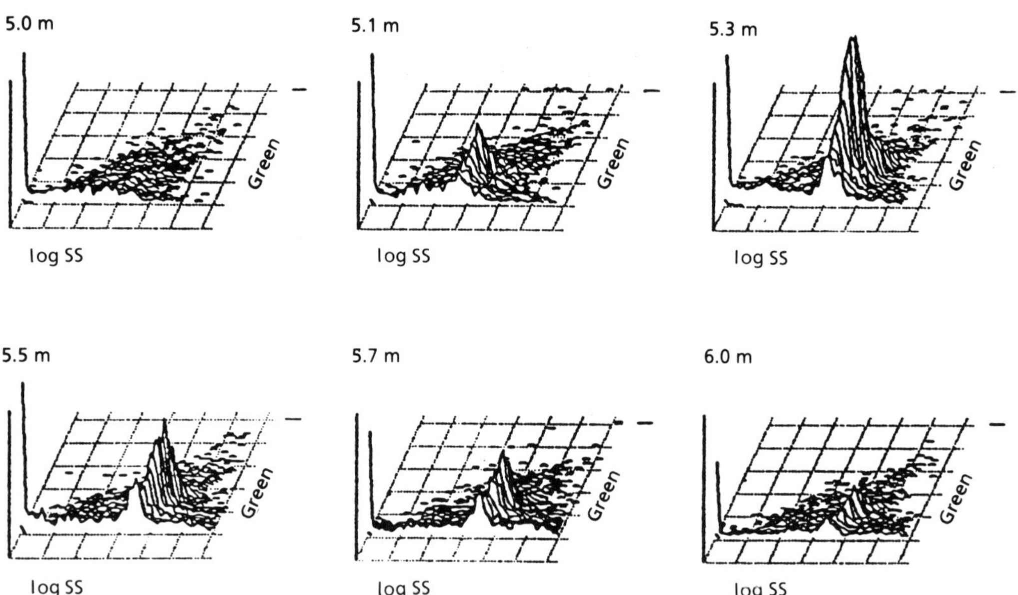

Figure 4.3 Cytograms obtained with FCM (EPICS elite, Coulter Electronics; excitation source 488 nm argon laser) for a population of *Chloronema* sampled at consecutive depths between 5 m and 6 m in the metalimnion of Lake Vechten on August 23, 1990. The *x*, *y*, and *z* axis represent log side scatter (SS), log green fluorescence, and log number of filaments, respectively, counted within an exact time lapse of 180 seconds. Data from H. Verdouw (unpublished).

showed an outstanding sidescatter and a peculiar greenish light emission. The *Synechococcus* cells were characterized by a much smaller sidescatter and by an orange and red fluorescence originating from phycobiliproteins and chlorophyll *a*, respectively. The high sidescatter characteristics and typical green-light emission of the *Chloronema* filaments were used to position their vertical distribution within the 5–6 m depth stratum of the lake (Figure 4.3). Clearly, on August 23, 1990 the *Chloronema* spp. peaked at a depth of approximately 5.3 m, which was confirmed by *in situ* fluorometry.

4.2.2 *Polar lipids as fingerprints*

In order to relate in Lake Vechten the benthic lipid composition to possible sources in the water column, the respective planktonic microbial communities were profiled with respect to their saponifiable PLFA compounds with capillary gas chromatography and mass spectrometry (capillary GC-MS). Epilimnetic samples were easily distinguished from the metalimnetic and sediment samples. The epilimnion, dominated by the dinoflagellate *Ceratium hirundinella*, was characterized by $C_{20:5}$ and $C_{22:6}$ polysaturated fatty acids. The anoxic metalimnion supported a radically different microbial community dominated by phototrophic sulfur bacteria (*Chloronema* spp., *Chromatium* spp., and *Chlorobium phaeobacteroides*) and by *Synechococcus* spp. belonging to the cyanobacteria. Samples analyzed from this water stratum contained high concentrations of C_{16} and C_{18} monounsaturated fatty acids. The fatty acid composition of samples from the hypolimnion and the sediment were qualitatively very similar to those from the metalimnion. Concentrations of methylbranched acids, commonly found in eubacteria, increased with depth in the water column. So, the results suggest that fatty acids derived from the membranes of the dominant epilimnetic phytoplankton were largely metabolized in the oxic part of the water column, while the fatty acids synthesized by prokaryotic microorganisms at and below the metalimnion dominated the sediment.

The concentrations of several unusual fatty acids found in *Desulfovibrio* spp., *Desulfobacter* spp. and *Desulfotomaculum* spp. were inversely related to the sulfate concentration of the hypolimnion (Fredrickson *et al.* 1986). Finally, the composition of PLFA in the sediment showed seasonal changes which could be associated with concentrations of terminal electron acceptors of microbial respiration and thus with physiologically distinct bacterial groups. To facilitate the interpretation of the PLFA data from Lake Vechten samples, a principal component analysis was employed to identify PLFA whose relative concentrations best differentiated planktonic and sedimentary microbial consortia. Results of this analysis were compared with data of the PLFA contents of monocultures of relevant bacteria. Factor

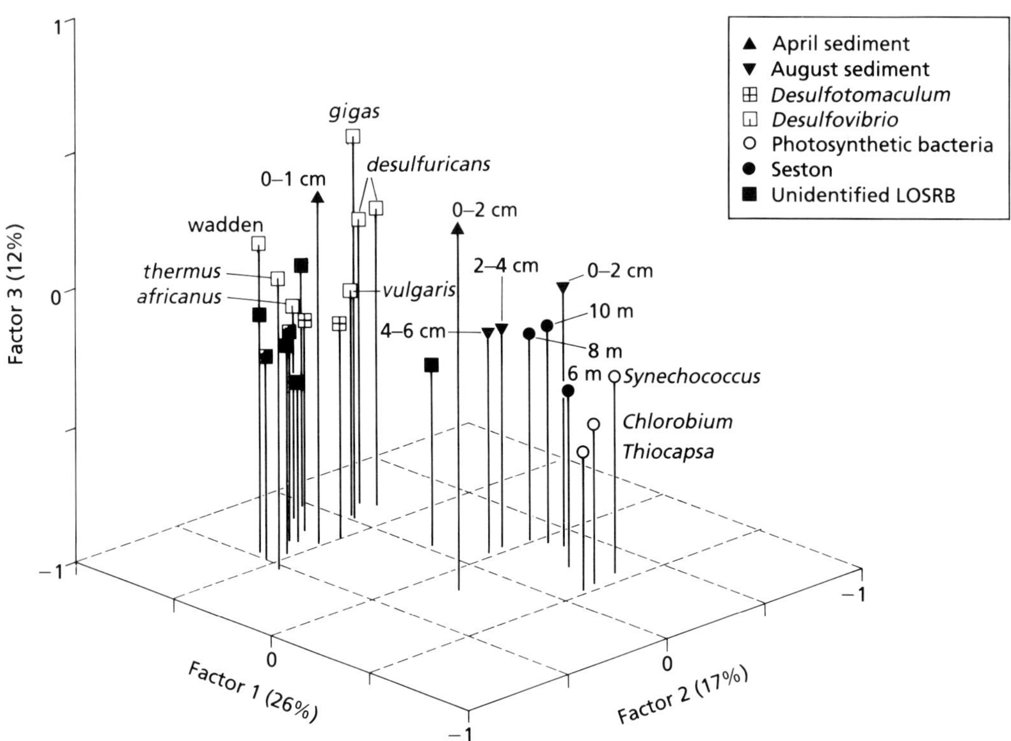

Figure 4.4 Factor analysis of PLFA profiles of selected Lake Vechten seston and sediment samples together with the results of the PLFA analysis of selected monocultures projected into the space defined by factor analysis of PLFA data from 48 bacterial monocultures. LOSRB, lactate-oxidizing sulfate-reducing bacterium. Data from H.L. Fredrickson (unpublished).

analysis showed that 55% of the variance in the total PLFA data set of the bacterial monocultures could be explained by three factors. Points representing PLFA profiles of Lake Vechten microbial assemblages were projected into the space defined by these three factors (Figure 4.4). The orientation of the points in the three-dimensional space could be explained in terms of structure and function of the lake ecosystem. Points representing the PLFA profiles of acetate- and lactate-oxidizing sulfate-reducing bacteria formed a cluster around the point representing the PLFA profile of the microbial assemblage in the superficial sediment (0–1 cm) during winter mixing of the lake. During stratification the PLFA profiles of samples from meta- and hypolimnion and the sediment were clustered close by PLFA profiles originating from monocultures of the metalimnetic phototrophic cyanobacteria and sulfur bacteria. These findings confirmed and supplemented earlier observations on phototrophic bacteria in the metalimnion (see Sections 4.2.1 and 4.3.1) and sulfate reduction in the top layer of the sediment (see Section 4.3.2).

A very useful aspect of the PLFA technique is that labeled isotopes can be incorporated into the procedures. As a result, measurements of microbial activities are obtained. Using ^{14}C-labeled acetate and lactate additions to sediment samples collected from discrete depths, and determining the rate of incorporation of ^{14}C into the lipid fraction it became apparent that potential rates of lipid synthesis were faster in the top sediment layer (0–1 cm depth) than in the deeper (2–3 cm) sediment layers (Fredrickson *et al.* 1988). Molybdate, an inhibitor of sulfate reduction, slowed these rates of lipid synthesis. The degree of inhibition was dependent on the substrate tested and the depth of the sediment layer, indicating that nonsulfate types of respiration were primarily responsible for acetate metabolism at the sediment–water interface, in contrast to the close coupling of lactate–lipid biosynthesis to sulfate reduction.

4.3 The microbial food web

The metalimnion of deep productive lakes constitutes a boundary layer between two completely different environments and is grossly characterized by the simultaneous presence of reduced products diffusing from below and of oxygen penetrating from above, which allows, concomitantly, the growth of aerobic, microaerobic, and anaerobic forms. The supply of essential nutrients for these populations is guaranteed by gravitational force and by diffusion along gradients (see Figure 4.1). The absence of significant turbulence prevents the dispersion of the populations and ensures stability of the chemical gradients.

The metalimnetic microbial assemblage is very complex and contains all kinds of organisms which together may readily comprise a microbial food web. In accepting the food web concept we equip ourselves, at the same time, with useful theories and their techniques for the study of possible relations between the distinct populations, as radioisotopic production measurements and grazing experiments. In addition, realistic estimates of bacterial numbers and bacterial production can be obtained with recently developed techniques. The microbial food web plays a crucial role in the flux of energy and matter and in the recycling of nutrients. Energy produced by phototrophic microorganisms in the form of dissolved organic molecules may be retained and channeled back to the main aquatic food chain via a so-called microbial loop consisting consecutively of bacteria–HNAN–microzooplankton.

4.3.1 Bacterial production and protozoan grazing

In Lake Vechten the bacterial production rates were approximately balanced by the grazing rates of HNAN (Bloem 1989; Bloem & Bar-Gilissen 1989). Cole *et al.* (1988) compared bacterial production estimates obtained in 70 freshwater and marine ecosystems; in the pelagic systems bacterial production ranged between 0.017 and 6.25 μg C $l^{-1} hr^{-1}$ and averaged 20% of the algal primary production. In general, bacterial production showed a significant linear relationship with primary production, which suggests either that bacterial and algal growth are controlled by common limiting factors (nutrients, temperature) or that the bacteria depend on the production of organic substrates by the phototrophs. The latter suggestion seems most likely for the situation in Lake Vechten, where the highest primary and bacterial production and extracellular release by phytoplankton species are located in the metalimnion (Blaauboer *et al.* 1982; Steenbergen & Korthals 1982). This close relationship between primary production and bacterial production has been observed by other authors as well (Gude *et al.* 1985; Lovell & Konopka 1985; Scavia & Laird 1987).

The primary production in the metalimnion of Lake Vechten is carried out mainly by oxygenic and anoxygenic bacteria under light-limiting conditions; algae playing a minor role (cf. Figure 4.2) (Steenbergen & Korthals 1982). Apart from low light intensity the anoxygenic forms suffer from a limited supply of their electron donor sulfide (see Figure 4.1). Using Fick's first law of diffusion:

$$F = -D\, dC/dx \tag{4.1}$$

where F is the flux, D the vertical dispersion coefficient in the lake's thermocline ($0.32 \times 10^{-2}\, cm^2\, s^{-1}$, Verdouw *et al.* 1985) and dC/dx the slope of the concentration gradient of sulfide. Steenbergen *et al.*

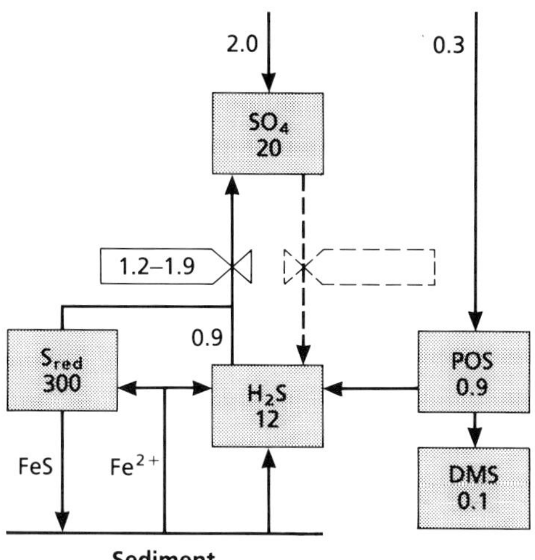

Figure 4.5 Relational diagram of pool sizes ($mmol\, m^{-3}$) and fluxes ($mmol\, m^{-2}\, d^{-1}$) of sulfur compounds at 8 m depth and transformation ($mmol\, m^{-3}\, d^{-1}$) of sulfide to sulfate in the 6.5–7.5 m depth stratum of Lake Vechten on September 24, 1985. From Steenbergen *et al.* (1987).

(1987) calculated the daily sulfide fluxes and compared them with anoxygenic photosynthetic rates. From the simplified scheme constructed from their calculations (Figure 4.5) it can be envisaged that on a daily basis the average vertical flux of sulfide (0.9 mmol $m^{-2} d^{-1}$) and the transformation of sulfide to sulfate by the phototrophic sulfur bacteria (1.2–1.9 mmol $m^{-3} d^{-1}$) were in the same order of magnitude. This strongly suggests that the anoxygenic photosynthesis in the metalimnion of Lake Vechten might be sustained solely by the daily flux of sulfide from below 8 m. In addition, this supply of sulfide to the strata where light penetrates and the oxidation of sulfide in these layers are delicately balanced.

The greatest activity of the microbial populations, which comprise the microbial loop in the metalimnion of Lake Vechten, has been observed during August and September. Based on maximum instantaneous rates obtained from this late summer period, a tentative diagram of the carbon flow through the microbial loop in Lake Vechten was constructed (Figure 4.6) (Bloem 1989). It is crucial

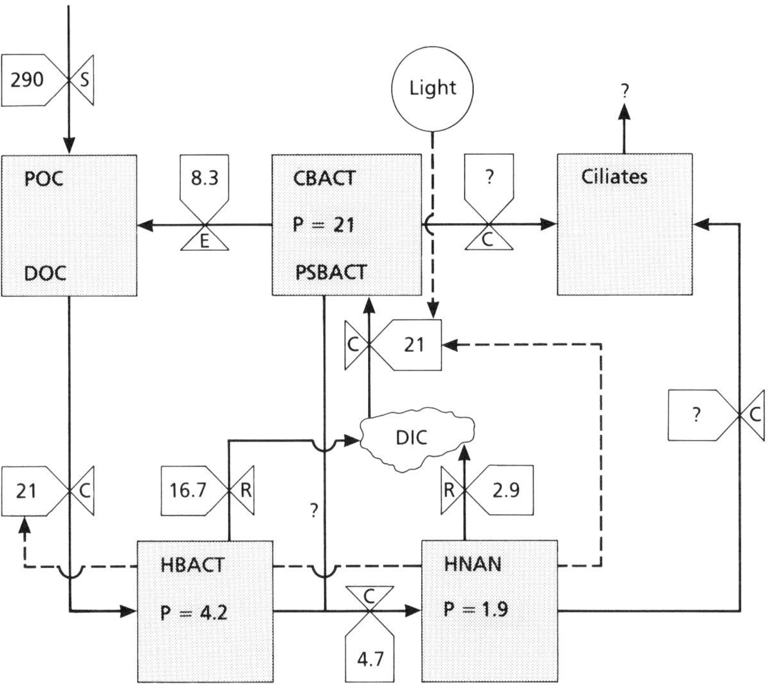

Figure 4.6 Simplified carbon flow diagram of the microbial food web in the metalimnion of Lake Vechten. Estimated maximum rates of sedimentation (S), excretion (E), consumption (C), production (P), and respiration (R) are given in μg C $l^{-1} hr^{-1}$. CBACT, cyanobacteria; DIC, dissolved inorganic carbon; DOC, dissolved organic carbon; HBACT, heterotrophic bacteria (with a yield of 20%); HNAN, heterotrophic nanoflagellates (with a yield of 40%); POC, particulate organic carbon; PSBACT, phototrophic sulfur bacteria. From Bloem (1989).

to note that the maximum consumption of bacteria by HNAN (4.7 μg C $l^{-1} hr^{-1}$) approximates the bacterial production (4.2 μg C $l^{-1} hr^{-1}$). In other words the estimated bacterial production was roughly balanced by the estimated protozoan grazing.

4.3.2 Mineralization

During summer stratification in Lake Vechten high organic-carbon-sedimentation rates (Steenbergen & Verdouw 1984) and maximum primary-production rates were found in the metalimnion (Blaauboer *et al.* 1982; Steenbergen & Korthals 1982). Additionally, comparison of carbon-fixation data, carbon-sedimentation rates and microbial reduction rates of observed fluxes of electron acceptors indicated that approximately 30% of the pelagic primary production was mineralized in these strata (Cappenberg & Verdouw 1982). Thymidine incorporation experiments in this layer indicated a high production rate of heterotrophic bacteria concomitantly with a maximum HNAN density of above 3×10^7 HNAN l^{-1}. So it can be anticipated that mineralization processes are strongly enhanced in this case by HNAN (Bloem 1989).

Of great significance was the observation in experiments with two-stage continuous cultures that addition of HNAN accelerated the mineralization of phosphorus from dissolved organic matter six-fold (Table 4.2) (Bloem *et al.* 1988). Bacteria were grown on yeast extract in the first stage (vessels 1 and 3 in Table 4.2) and fed at a constant rate to the second stage (vessels 2 and 4, without and with HNAN, respectively). The HNAN strongly reduced the amount of bacteria and, obviously, recycled organic phosphorus and nitrogen incorporated in the bacterial biomass. The bulk of the organic carbon was

Table 4.2 Carbon, phosphorus, and nitrogen mineralization rates (average values expressed in μg $l^{-1} h^{-1}$; standard deviation in parentheses; $n = 2$) in two-stage continuous cultures. Presentation of two series of cultures, vessels 1–2 and the vessels 3–4, respectively; in Phase A without HNAN and in Phase B with the addition of HNAN in the second stage (vessel 4). Data from Bloem *et al.* (1988)

	Vessel 1	Vessel 2	Vessel 3	Vessel 4
Phase A				
C	306.1(59.8)	60.7(32.5)	308.0(53.6)	11.2(9.7)
PO_4–P	0.14(1.17)	−0.20(0.22)	−1.38(0.90)	0.14(0.13)
NH_3–N	55.13(6.10)	0.95(2.02)	53.96(3.94)	−0.60(2.15)
NO_3–N	−0.37(0.93)	0.19(0.86)	0.15(0.52)	1.05(0.68)
Phase B				
C	335.1(19.9)	28.9(14.1)	335.1(32.8)	33.6(15.6)
PO_4–P	0.47(0.27)	0.51(0.13)	0.61(0.15)	3.09(0.11)
NH_3–N	62.20(2.36)	2.22(2.98)	60.58(2.13)	20.3(1.90)
NO_3–N	1.38(1.40)	−1.83(1.97)	0.56(0.96)	0.07(0.47)

mineralized in the first stage vessels at rates of 320 μg C l^{-1} hr^{-1}; in the second stage vessels the level was reduced to approximately 30 μg C l^{-1} hr^{-1}, which did not change in the presence of HNAN. In the presence of HNAN the phosphorus and nitrogen mineralization was strongly enhanced from 0.5 to 3.0 μg PO_4–P l^{-1} hr^{-1} and from 2.2 to 20.3 μg NH_3–N l^{-1} hr^{-1}, respectively; nitrate concentrations were not influenced by HNAN (see Table 4.2). However, with regard to nitrogen, much higher mineralization rates were observed in the grazer-free first stage (61 μg NH_3 l^{-1} hr^{-1}). In conclusion, the most important mechanism of phosphate recycling appeared to be direct consumption of heterotrophic bacteria by HNAN. Approximately 70% mineralization of nitrogen was carried out by bacteria. This process was enhanced (30%) in the presence of HNAN. HNAN did not significantly influence the decomposition rate of relatively mineral-rich dissolved organic matter in the cultures, which consisted of yeast extract. Sherr *et al.* (1982) reached the same conclusion, which they attributed to a selective breakdown by HNAN of detritus characterized by high structural carbohydrate and low mineral content.

Most of the carbon mineralization in the metalimnion of Lake Vechten is carried out by the heterotrophic bacteria; their respiration was estimated to be approximately 17 μg C l^{-1} hr^{-1}, while HNAN contributed only 3 μg C l^{-1} hr^{-1} (see Figure 4.6). Phosphorus was mineralized mainly by HNAN, so HNAN play an important role in the recycling of certain nutrients immobilized in bacterial biomass (Bloem & Bar-Gilissen 1989).

4.4 Sediment studies

Since the mid-1980s several detailed studies on the genesis, accretion rate, and stratigraphy of Lake Vechten sediments have become available. They will be outlined below together with an analysis of recent studies on microbially controlled processes in the oxic and anoxic sediment strata.

General characteristics of the profundal sediment of Lake Vechten have been described by Verdouw *et al.* (1985). The sediment consists of a 26–28 cm thick layer of gyttja, i.e., soft, black-colored, sulfide-containing material deposited on a grey-colored mixture of clay and sand. Lake Vechten was formed in 1941 as a result of sand quarrying and the clay originating from the superficial layer was dumped back into the lake after its completion. From the sediment–water interface to the clay boundary the dry weight of the gyttja varied from 15 to 40% of the wet weight. The organic matter content varied from 7 to 15% of dry weight and the organic carbon from 2 to 6% of dry weight.

In Lake Vechten, permanent sediment accretion takes place only below the 8 m isobath and, thus, is restricted to the central region of the deep basins. Above the 8 m depth horizon in the littoriprofundal zone, i.e., the zone immediately below the submerged macroscopic vegetation, and in the littoral region, periods of sediment accumulation are interrupted by periods of resuspension and downward transportation along the bottom, a process closely related to the thermal cycle of the lake. Verdouw *et al.* (1987) used sediment traps to compare this so-called peripheral particle flux with the downward particle flux taking place in the open water. They observed that during stratification the peripheral particle flux amounted to 42% of the total downward flux. This implies that a large part of the permanent profundal sediment originates from material which previously has been deposited on the oxic epilimnetic slopes and has lost a great deal of its organic content before being buried permanently; on a dry weight basis a decrease from 23 to 7% (Verdouw *et al.* 1987).

Recently, sediment cores from the profundal region of Lake Vechten were dated with ^{137}Cs and ^{210}Pb. From the radionuclide profiles it was concluded that the sediment is fairly undisturbed. Vertical sediment accretion rates determined from the clay–sand horizon and from the radionuclide datings varied between 0.60 and 0.74 cm yr^{-1}. Sedimentation rates based on particulate matter collected in sediment traps agreed with results of the ^{137}Cs method with average values of 2.9 and 2.3 kg dry weight $m^{-2} yr^{-1}$, respectively (Steenbergen *et al.* 1989).

Sediments are important sites for mineralization and nutrient recycling; per unit volume sediments contain 10^3–10^4 times more viable bacteria than the overlying water. The heterotrophic activities in the sediment are fueled by the input of organic material from the primary producers in the water column. It is interesting to note that bacteria have a tendency to colonize surfaces, so microbial activity will be highest in surface films and flocs which make up surficial sediment.

4.4.1 *Oxic surface layer*

In sediments of temperate lakes oxygen is generally depleted in the upper few millimeters (Sweerts *et al.* 1986; Carlton *et al.* 1989). Information on processes that consume oxygen in the sediment is scarce. Oxygen is the strongest biologically useful oxidant and is consumed first in the sequence of terminal electron acceptors in sediments. In the anoxic part of the sediment reduced inorganic and organic products of anaerobic mineralization processes are produced; the most important reduced inorganic species are NH_4^+, Fe^{2+}, Mn^{2+}, HS^-, and CH_4. The diffusion of these products to the oxic–anoxic interface facilitates their oxidation. In Lake Erie sediments these inorganic constituents potentially accounted for a third of the sediment oxygen demand (Adams *et al.* 1982). Sediment oxygen consumption is greatly enhanced in the presence of mats of *Beggiatoa*. In addition, these mats play a crucial role in the nitrogen and sulfur cycles at the sediment–water interface (Sweerts *et al.* 1990).

The filamentous gliding bacteria *Beggiatoa* spp. often form conspicuous white mats at the surface of sulfide-rich marine and freshwater sediments and may be responsible for the whole benthic oxygen consumption, converting sulfide to intracellular elemental sulfur or to extracellular sulfate (Jorgensen & Revsbech 1983; Larkin & Strohl 1983). Mats of *Beggiatoa* covered approximately 25% of the profundal bottom of Lake Vechten in the nonstratified period. The population was tentatively identified as *Beggiatoa alba* and contained no significant numbers of other bacteria. Employing microsensors for nitrate, oxygen, and sulfide it was shown by Sweerts *et al.* (1990) from microprofiles that oxygen from the water phase was consumed completely in the upper 50 μm of the mat, whereas nitrate was consumed anaerobically in the next 100 μm (Figure 4.7). So, no oxygen or nitrate penetrated the sediment below the 600 μm-thick mat of *Beggiatoa*. The flux of nitrate was calculated from the linear concentration gradient in the diffusive boundary layer above the mat with Fick's first law of diffusion (see Equation 4.1) and a

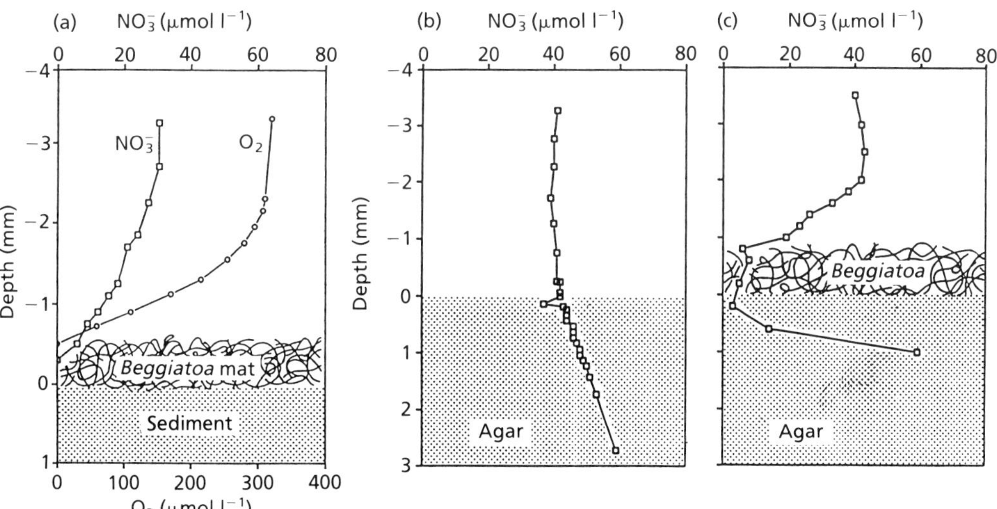

Figure 4.7 (a) Microprofiles of nitrate and oxygen in a mat of *Beggiatoa* spp. on lake sediment (freshly collected lake-sediment cores maintained at 7°C with gentle stirring of the overlying water phase). (b) Microprofiles of nitrate in an agar–water interface without bacteria and (c) with a semipurified tuft of *Beggiatoa*. Agar composition: 0.5% agar, 0.2 mmol l^{-1} acetate, 0.1 mmol l^{-1} nitrate, 0.1 mmol l^{-1} sulfide, and 0.5 mmol l^{-1} ammonium. From Sweerts *et al.* 1990.

molecular diffusion coefficient of nitrate of $1.22 \times 10^{-5}\,cm^2\,s^{-1}$ at 7°C. The uptake of nitrate by *Beggiatoa* covered sediments was $2.5\,mmol\,m^{-2}\,d^{-1}$ and by nearby sediments without *Beggiatoa* the nitrate flux was only $1.3\,mmol\,m^{-2}\,d^{-1}$. The increased nitrate uptake of the mats was explained as denitrification. Addition of $^{15}NO_3^-$ to *Beggiatoa* tufts on agar gave conclusive evidence that denitrification occurred; the $^{15}NO_3^-$ added was recovered as $^{15}N_2$ indicating that nitrate was consumed by denitrification and not by assimilation or dissimilatory reduction to ammonia (Sweerts *et al.* 1990). The results clearly demonstrated the denitrifying capability of this freshwater *Beggiatoa* spp.

In addition to the *Beggiatoa* mat studies in Lake Vechten, research was devoted to other oxygen-consuming processes at the sediment–water interface, both in the profundal gyttja and littoral sands of Lake Vechten. These studies were carried out using pore-water

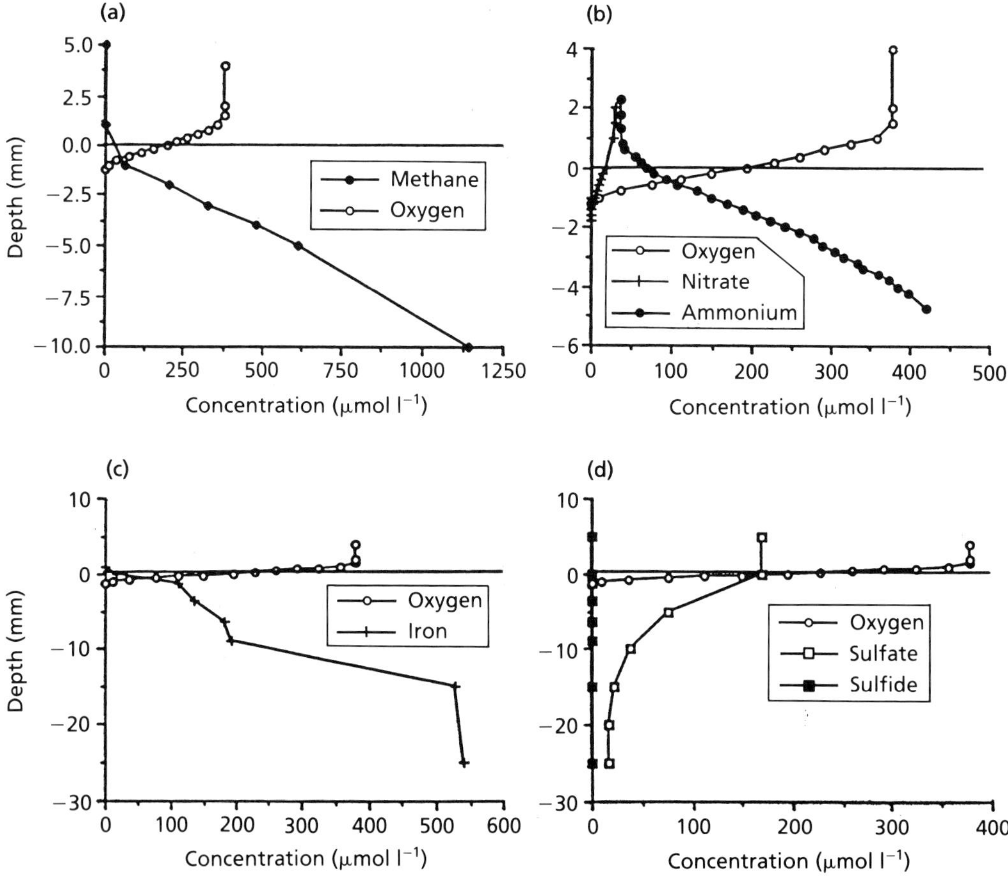

Figure 4.8 Pore-water profiles of: (a) methane and oxygen, (b) ammonium, nitrate, and oxygen, (c) iron and oxygen, and (d) sulfide, sulfate, and oxygen in the profundal sediments of Lake Vechten during overturn. From Sweerts *et al.* (1991).

analysis and core incubation; profiles of oxygen and of the reduced end-products of anaerobic mineralization were measured with micro-electrodes (Figure 4.8). The calculated fluxes based on these profiles and diffusion coefficients (Sweerts *et al.* 1991) were compared with the measured fluxes in the overlying water of incubated, minimally disturbed cores under both oxic and anoxic conditions. Methane (58–64%), ammonium (10–13%), iron (2–6%), and sulfide (<12%) oxidation together accounted for roughly 75% of the oxygen consumed in the profundal sediments. Almost the complete diffusive flux of methane out of the anoxic layer into the oxic surface layer (7.9–9.4 mmol $m^{-2} d^{-1}$) was oxidized, whereas only 27–36% of the ammonium flux (5.1–5.2 mmol $m^{-2} d^{-1}$) was oxidized in the oxic surface layer and 64–73% diffused into the overlying water. These observations compare with those made in Lake Constance where 93% of the produced methane has been found to be oxidized at the sediment–water interface (Frenzel *et al.* 1990). In Lake Washington about half of the methane is oxidized in the upper 7 mm of the sediments and the remainder escapes into the water column (Kuivila *et al.* 1988).

The high oxidation of methane compared to that of ammonium may indicate that the methane-oxidizing bacteria are able to compete successfully with the nitrifying bacteria for oxygen. Since half-saturation constants (K_m) for oxygen are partly in the same range for both bacterial groups, i.e., 8–20 μmol l^{-1}, as are their maximal rate values (V_{max}) (Sharma & Ahlert 1977; Kuivila *et al.* 1988), it seems difficult to predict the outcome of competition for oxygen in a specific environment. There is a need for kinetic studies of the bacterial groups at the sediment–water interface to predict the impact of the different oxidative processes in a specific environment.

In contrast to the profundal sediments, oxidation of reduced end-products of anaerobic respiration was of minor importance (<15%) in littoral sandy sediments. Also, sediment-oxygen consumption rate was higher in the gyttja (29 mmol $m^{-2} d^{-1}$ at 7°C) during winter mixing than in the sandy littoral sediment in both winter (8.7 mmol $m^{-2} d^{-1}$ at 7°C) and summer (18.6 mmol $m^{-2} d^{-1}$ at 22°C).

Finally, considerable bacterial biomass may be produced by the oxidation processes, which in turn becomes available for mineralization or for grazing activities by HNAN at the sediment–water interface. Ranges of 0.56–1.82 and 0.28–1.3 g cells mol N^{-1} have been reported for *Nitrosomonas* and *Nitrobacter*, respectively (Painter 1970). In Lake Vechten sediments this would equal a production of 0.4–1.3 and 0.2–0.7 mg cell $m^{-2} d^{-1}$, respectively. The methane oxidation rate of 9.4 mmol $m^{-2} d^{-1}$ may result in an organic carbon production of 3.8 mmol $m^{-2} d^{-1}$.

4.4.2 Anoxic sediments

Due to its variable origin and susceptibility to ingestion, digestion, and mineralization during settlement in the water column, the total input of organic material in the sediment will vary considerably together with its chemical composition. Many organic compounds can be mineralized completely by microorganisms under oxic conditions, but in productive lakes a large portion of the mineralization is carried out under anoxic conditions, either in the hypolimnion or in the adjacent sediment where supply of oxygen is limited. As outlined in the introduction, anaerobic mineralization proceeds sequentially in space or in time by hydrolysis and depolymerization yielding fermentable monomers. Subsequently, fermentation end-products are consumed by terminal respiration processes, major ones are denitrification, sulfate reduction, and methanogenesis. Clear distinctions exist in anaerobic carbon metabolism between marine and freshwater systems. At the high-sulfate concentration in the sea, electron donor limitation is likely to occur so sulfate reduction exceeds methanogenesis and is responsible for most of the organic carbon oxidation. In fresh waters, sulfate may become limited and methanogenesis is responsible for the bulk of terminal metabolism under anoxic conditions (Jones & Simon 1981; Capone & Kiene 1988).

Recently Cappenberg (1988) addressed the lack of knowledge on the initial breakdown pathways and turnover rates of carbon in the field and evaluated several approaches used in this type of research. Very probably initial exoenzymatic hydrolysis is the rate-limiting step for degradation of organic material.

Respiratory denitrification (Seitzinger 1988) is expected to occur first after oxygen depletion, but due to very low concentrations in overlying water and pore water, nitrate, although thermodynamically a very favorable electron acceptor, does not appear to support a significant fraction of carbon oxidation. In Lake Vechten, maximally 5% of the total electron acceptor consumption could be attributed to denitrification (Steenbergen & Verdouw 1984); Kelly *et al.* (1988) reported values of 0.1–5% in Canadian shield lakes.

Recently, methodological developments in the measurement of *in situ* nitrate concentrations with microsensors (Revsbech *et al.* 1988; Sweerts & de Beer 1989) and sensitive ion chromatographic methods (Hordijk *et al.* 1987) became available. Both techniques have the ability to quantify nitrate reduction directly following nitrate fluxes and depletion rates under *in situ* conditions. It appeared that the penetration depth of nitrate in the profundal gyttja of Lake Vechten was only 1.3 mm, in contrast to 5 mm in littoral sandy sediment. The zone in which denitrification occurred was 500 μm thick and adjacent to the top aerobic sediment layer; denitrification rates

Table 4.3 Flux into the overlying water and consumption rates in the oxic surface layer of profundal sediments of Lake Vechten (7°C) of NH_4^+, Fe^{2+}, H_2S, and CH_4, and the calculated oxygen consumption necessary for their oxidation (calculated using the equations in Sweerts *et al.* 1991). The ranges represent values obtained by core incubations and pore-water profiles

Chemical species	Flux into the overlying water ($mmol\,m^{-2}\,d^{-1}$)	Consumption in oxic surface layer ($mmol\,m^{-2}\,d^{-1}$)	Calculated consumption ($mmol\,m^{-2}\,d^{-1}$)	Oxygen (%)[1]
NH_4^+	3.7–3.8	1.4–1.9	2.8–3.8	10–13
Fe^{2+}	0	2.7–6.8	0.7–1.6	2–6
H_2S	0	<1.8	<3.6	<12
CH_4	<0.1	7.9–9.4	16–19	58–64

[1] Percentage of the total oxygen consumption of 27–29 $mmol\,m^{-2}\,d^{-1}$.

amounted to 0.26–2.5 $mmol\,m^{-2}\,d^{-1}$ at an *in situ* nitrate concentration of 0.030 $mmol\,l^{-1}$ (Sweerts & de Beer 1989).

Bacterial sulfate reduction with the release of the end-product hydrogen sulfide becomes a significant process in organic carbon degradation at low redox potential (−100 mV) (Widdel 1988). As a result, its contribution to total anaerobic carbon mineralization can be considerable (Table 4.3) (Capone & Kiene 1988; Cappenberg 1988). In most lakes, sulfate concentrations in the sediment are much lower than in the overlying water. So, sulfate reduction is restricted to a thin layer near the sediment surface. For the measurement of the very steep sulfate profile and the study of sulfate conversion processes, sophisticated methods have been developed which make use of HPLC and radio-ion chromatography (Hordijk *et al.* 1985). These methods combined with microdistillation offered the possibility to simultaneously measure the depletion of sulfate and the production of sulfide. Clearly this approach proved to be a more reliable estimate of the sulfate reduction rate than that carried out with the $^{35}SO_4^{2-}$ radiotracer technique, which suffers from low recovery of the end-product $^{35}S^{2-}$.

Since sulfate becomes depleted with depth in sediments due to consumption by sulfate reducers, methanogenesis becomes significant in organic carbon mineralization in the deeper sediment strata as exemplified in Figure 4.9 (Cappenberg *et al.* 1984; Ward & Winfrey 1985). The relative contributions of the various degradation processes, as calculated via the oxygen consumption of the end-products, are summarized in Table 4.3.

The basis for the vertical separation of both processes is likely to be competition for common substrates, in particular, acetate, hydrogen, and carbon dioxide. As long as sufficient sulfate is present the sulfate-reducing bacteria will dominate methane-producing

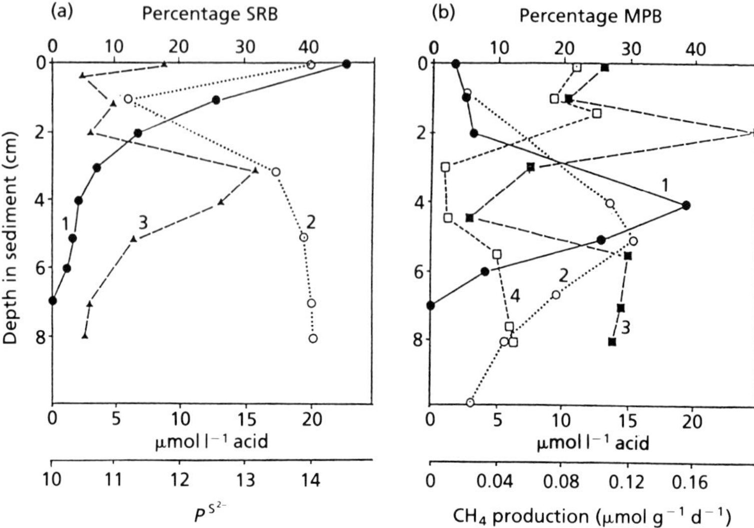

Figure 4.9 Vertical zonation in the gyttja of Lake Vechten of fatty acids and sulfide concentrations, relative bacterial densities of sulfate reducers (SRB) and methane producers (MPB) and methane production rates. (a): 1, SRB; 2, sulfide; 3, lactate, and (b): 1, MPB; 2, methane production; 3, acetate; 4, formate. From Hordijk & Cappenberg (1983).

bacteria, which may be partly explained by the greater affinity of sulfate reducers for acetate (Ward & Winfrey 1985).

It will be clear that the end-products of the fermentative bacteria, including acetate, hydrogen, carbon dioxide, alcohols, and volatile and nonvolatile fatty acids, play a crucial role in the ecology of these terminal bacterial groups. When acetate is regarded as the main precursor for both methanogenesis and sulfate reduction, acetate turnover rates in sediments should provide a good estimate of total mineralization. In Lake Vechten sediments, acetate turnover rates have been measured by Cappenberg & Jongejan (1978) from the depletion of ^{14}C-acetate and recently by Hordijk *et al.* (1990) with capillary gas chromatography employing the stable isotope of deuterated acetate. Acetate depletion could be modeled by first-order kinetics and reached low threshold values ($0.6\,\mu mol\,l^{-1}$), which may indicate that acetogenesis and acetate consumption were in close equilibrium. However, besides acetate, a variety of potential substrates for sulfate-reducing bacterial groups have been reported (Ward & Winfrey 1985; Capone & Kiene 1988). Fredrickson *et al.* (1988) provided evidence for a close relation between sulfate reduction and lactate-coupled lipid biosynthesis, promoting lactate also as a likely substrate for sulfate respiration in lacustrine sediments (see Section 4.2.2).

So far, the estimates of organic carbon decomposition may be accepted with some reservations (see Table 4.3), but there is little doubt that they provide insight into the relative importance of the various mineralization processes occurring in the sediment.

4.5 Aspects of eutrophication

Eutrophication is enrichment by plant nutrients, particularly phosphorus and nitrogen compounds. Here aspects of synthetic or cultural eutrophication are considered (Hutchinson 1973).

The activities of man have in many cases resulted in the undesirable consequence of excess nutrients, along with other pollutants, being introduced into lakes, streams, and estuaries. This causes significant changes in aquatic environments. The excess nutrients greatly accelerate the natural process of eutrophication. The chief sources of enrichment are sewage, artificial fertilizers, and agricultural wastes.

A most conspicuous feature of eutrophication is increased phytoplankton growth and in a number of cases the occurrence of nuisance blooms of cyanobacterial species (Paerl 1988; Steinberg & Hartmann 1988).

In general, solving the eutrophication problem can be achieved by reducing the nutrient loadings of the lake or lake system, particularly the phosphorus concentration. The best method is to reduce the phosphorus at its source by precipitation from the lake's water supply. Examples of lake restoration involving diversion of phosphorus-rich effluent are described by Cullen & Forsberg (1988). Sometimes improvement of water quality may take many years, particularly in shallow lakes which store high amounts of phosphorus in the sediment. Additional measures, such as biomanipulation, dredging, or *in situ* phosphate fixation, may accelerate the restoration (Cooke *et al.* 1986; Hosper 1989).

4.5.1 Long-term changes in a stratified lake

Lake Vechten is situated in a vast rural area and apart from direct precipitation, the lake is fed by agricultural drainage, and surface and groundwater runoff mainly from grassland and scattered crop fields (Steenbergen & Verdouw 1982). The lake has been a study site of the Dutch Limnological Institute for nearly three decades and valuable long-term data sets have been obtained covering routine measurements of phytoplankton and chemical and physical variables. The most complete series of observations begins in 1962 and ends in 1980; less frequent data records exist for the period 1982–1991. Much of the data remains unpublished but scattered details can be found in Golterman (1975) and Blaauboer (1982). Long-term phytoplankton records in relation to nutrient enrichment of English lakes have been published by Lund (1972; 1978).

From 1962 onwards, significant changes have been observed in

Lake Vechten. The most obvious changes were the gradual disappearance of macrophytes (Best 1982), the decrease of average summer Secchi depth from 2.70 m in 1978 to 1.80 m in 1990 and the appearance since 1987 of dense cyanobacterial blooms in spring and early summer. Since 1962 there has been an approximate two-fold increase in total nitrogen level and a four-fold increase in chlorophyll *a* concentrations during summer. Total phosphorus concentrations have not changed. As a result, the total nitrogen/phosphorus ratio (by weight) changed gradually from approximately 20 in the 1960s to 30–40 in the 1980s. This implies a steady increase in phosphorus limitation.

Looking more closely at the chemistry of Lake Vechten reveals increased winter levels of NO_3–N, starting significantly in the early 1970s. Mean yearly values in the 1960s of 153 $\mu g\, NO_3–N\, l^{-1}$ increased to 240 $\mu g\, NO_3–N\, l^{-1}$ in the early 1980s. Mean values in 1989, 1990, and 1991 again showed higher values; 346, 447, and 347 $\mu g\, NO_3–N\, l^{-1}$, respectively. Winter levels of soluble reactive phosphorus (SRP), although very variable (8–20 $\mu g\, SRP\, l^{-1}$), showed no up- or downward trend throughout the years. Sutcliffe *et al.* (1982) have discussed the long-term changes in the chemical composition of surface waters of lakes and tarns of the English Lake District. They showed that both NO_3–N and SRP have increased markedly. Goldman (1988) reported a significant increase of NO_3–N content in Lake Tahoe, but no change in SRP concentration. In both cases the changes in the chemistry of the lakes coincided with an increase in human population and a steep rise in the use of agricultural fertilizers with changes in farming practice.

There exists the possibility to trace the trophic history of a lake back in time beyond the period of first routine observations. This has been done in a number of cases using photosynthetic pigments that persist in lake sediments and may function as paleoecological indicators closely linked to trophic status (Brown 1969; Swain 1985). However, alteration of pigments by diagenesis, particularly by photooxidation in the water column and at the sediment surface, presents a major problem in the interpretation of pigment stratigraphic records (Hurley & Amstrong 1990). Therefore, pigments should be selected carefully on the basis of their susceptibility to degradation and whether they originate in the oxic or in the anoxic zone of the lake (Leavitt & Carpenter 1990). With regard to Lake Vechten, Steenbergen *et al.* (1989) observed a several-fold increase in pigment concentration per unit of organic carbon from the deepest to the superficial sediment layers. This most likely reflected the increase in pelagial primary producers from the early 1960s onwards, as observed during the routine measurements. Pigment stratigraphic records of recent cores taken in the profundal region of Lake Vechten showed characteristic changes in the structure of primary producers since the mid 1980s (Figure 4.10). Since β-carotene belongs to the least labile pigments and is ubiquitous in algae and some phototrophic bacterial groups, it forms a superior indicator of total phototrophic biomass sedimentation. Its profile in the sediment of Lake Vechten (Figure 4.10b) illustrates the stepwise increase of phototrophic microorganisms during the 50-year-long history of the lake (Steenbergen *et al.* 1989). The doubling of the concentration of echinenone, a signature pigment of cyanobacteria, in the top 4 cm layer of the sediment (Figure 4.10d) reflects the appearance, since 1987, of dense cyanobacterial blooms. Apparently due to light

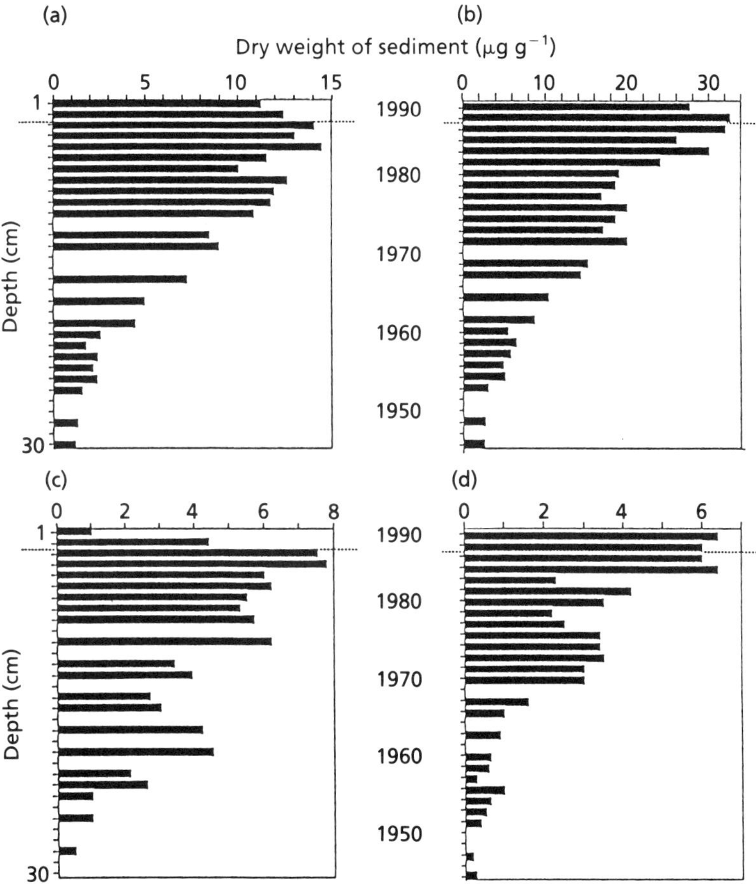

Figure 4.10 Stratigraphy of selected pigments analyzed with HPLC in a core collected from the gyttja in Lake Vechten on December 8, 1990; (a) lutein, (b) β-carotene, (c) bacteriopheophytin *e*, and (d) echinenone. Yearly sediment accretion rates range between 0.60 and 0.74 cm yr^{-1} (see Section 4.4); the dotted line indicates the boundary of the less dense top layer of about 2 cm, which compares approximately to 1988. Data from H.J. Korthals (unpublished); dating from Steenbergen *et al.* (1989).

limitation caused by these blooms it was observed that the familiar metalimnetic population of *Chlorobium phaeobacteroides* (see Figure 4.2 and Table 4.1) has disappeared. This phenomenon is reflected in the dramatic decrease of bacteriopheophytin *e* concentrations in the top layers of the sediment (Figure 4.10c). The same downward trend, but less pronounced, has been observed by the group of green algae, shown in the decreasing lutein concentration in the top layers of the sediment (Figure 4.10a).

4.5.2 Phosphorus metabolism in shallow eutrophic lakes

It is well established that the sediment is an important source of phosphorus in lakes (Bostrom *et al.* 1982; Marsden 1989) and forms a potentially serious threat in lake restoration programs aimed at a substantial reduction of the external phosphorus load. The extent to which phosphate in the sediment contributes to the internal phosphorus cycle of a lake is controlled by the rate of its release into the pore water and transport into the water column. This involves physicochemical and microbial mobilization of dissolved phosphorus and subsequent transportation across the sediment–water interface by, e.g., diffusion, turbulence, or resuspension mechanisms. The processes which play a role in this phosphorus flux have been studied intensively in the Loosdrecht Lakes, a series of shallow interconnected water bodies in the center of The Netherlands in which raw sewage and nutrient-rich water supplies have resulted in eutrophication (Kal *et al.* 1984; Van Liere *et al.* 1984). Since 1984, the phosphorus concentration in the water supply to the lakes was reduced by coagulation with $FeCl_3$. This measure diminished the external phosphorus loading from 1.0 to $0.3\,g\,m^{-2}\,yr^{-1}$.

The surface sediment in the Loosdrecht Lakes consists of so-called "dy" (Wetzel 1983), which is soft, hydrous, and brown-colored and forms a mixture of particulate organic, inorganic, and minerogenic matter and humic substances; the deeper layers are firm peat. The oxygen penetration depth into the sediment is 1.6 mm in summer and increases to 5.4 mm in winter (Sweerts & Cappenberg 1988). Phosphorus content and distribution in the bottom of the lakes was measured by Boers *et al.* (1984). Total phosphorus in the top 0–2 cm zone of the sediment at a representative sampling station was $1.1\,mg\,g^{-1}$ on a dry weight basis. Seven years after the reduction of the external phosphorus load this value has decreased to $0.6\,mg\,g^{-1}$ dry weight. Phosphorus release experiments have been carried out, initially, by Boers *et al.* (1984) and recently by Sinke *et al.* (1990) and Keizer *et al.* (1991) in a continuous flow reactor as described by Boers *et al.* (1984). Sinke & Cappenberg (1988) demonstrated significant differences in phosphorus-release rates by sterilized and nonsterilized sediments which they contributed to microbial processes, eventually influencing sediment phosphorus release in two manners; a direct temperature-dependent reaction and an indirect effect acting on a long-term basis, governed by mineralization. Phosphorus in recent sediments is primarily recycled as phosphate by microbially mediated decomposition of organic matter (Berner 1974). Phosphates may diffuse, adsorb, or complex with inorganic and organic cations and precipitate as authigenic minerals. Particularly in organic-rich sediments, e.g., in the Loosdrecht Lakes, mineralization may play a crucial role in the phosphorus chemistry.

Observations on the influence of mineralization on the phosphorus release from the sediments of the Loosdrecht Lakes were continued year round by Sinke *et al.* (1990), to quantify seasonal variations. They argued that seasonal variations in pore-water chemistry might reflect mineralization processes governed by environmental variables. Chemical characteristics of pore-water profiles were measured together with phosphorus release rates under standardized laboratory conditions. The pore-water composition, particularly the concentrations of NH_4^+, dissolved reactive phosphorus, and methane, varied greatly during the year, being highest in September and October and lowest in January and February, suggesting large seasonal fluctuations of anaerobic mineralization processes. The measured pore-water profiles were used to calculate apparent fluxes towards the sediment–water interface. The calculated fluxes were compared to the initial flux (4 hours) and average flux (21 days) of phosphorus from intact sediment cores measured under standardized conditions. The initial phosphorus fluxes measured in the cores varied between −7.7 and 1330 $\mu mol\, m^{-2} d^{-1}$ and were significantly correlated with the calculated phosphorus fluxes over the sediment–water interface. High correlation between calculated fluxes of ammonia, phosphorus, and methane with the measured initial phosphorus flux, points conclusively to mineralization of organic matter as the main cause of phosphorus release from the sediment. The high values in the fall may be related to increased input of organic material by sedimentation.

Phosphorus-release experiments are usually carried out over several weeks (Boers *et al.* 1984; Keizer *et al.* 1991). Sinke *et al.* (1990) found that the average release over such a period was not related to any of the pore-water characteristics nor to environmental variables. It is therefore very doubtful if measured average fluxes are representative of fluxes occurring *in situ*, mainly because in long-term reactor experiments the sediment is deprived of the input of freshly sedimented organic matter.

Finally, recent research on the phosphorus cycle in the Loosdrecht Lakes' sediments illustrates the crucial role of adsorption of phosphorus under aerobic and anaerobic conditions, the significant removal of dissolved phosphorus by seepage and the importance of wind-induced resuspension which may result in an uptake of phosphates in the food chain (Keizer *et al.* 1991).

4.6 Conclusion

Gradient zones in aquatic ecosystems, particularly the thermocline region in deep lakes and the sediment–water interface, are crucial sites of microbial activities coupled to biogeochemical cycles. Our understanding of structure and function of these sites has progressed in recent years. Improved analysis of microbial biomarkers has helped this development; it forms an important tool for the microbial ecologist to elucidate biological structure and zonation. The technique of PLFA analysis is particularly useful in sediments with their inherent heterogeneity. Calibration of PLFA biomarkers is still in progress and there are difficulties in the interpretation of analyses; however, data on relevant bacterial groups are rapidly becoming available.

During summer, microbial populations may be tightly stratified over depth intervals of a few decimeters in the thermocline of lakes, and constitute highly productive food webs consisting of phototrophic bacteria, heterotrophic bacteria, HNAN, and ciliates. Relatively little is known about microbial food webs in freshwater environments. Recent developments in fluorescence microscopy and the FCM technique provide insights into the distribution and dynamics of microbial assemblages in these thermocline strata, where both thymidine incorporation rates, indicating bacterial production, and HNAN grazing rates estimated with fluorescently labeled bacteria peaked during late summer. There is still considerable controversy over the factors to be used in converting thymidine incorporation into bacterial biomass, and more work is required on the biochemical and physiological basis of this method. Some degree of uncertainty exists on the precise role of HNAN in the recycling of nutrients and there is a great need for detailed research on interrelations between bacteria and HNAN in water–sediment zones.

We have also made substantial progress in understanding the role of sediments. Besides shallow oxic sediments, the anoxic hypolimnetic sediments have been studied most intensively with a great variety of recently introduced methods, such as microsensors, sophisticated chromatography, and radiotracer techniques.

It will be clear that algae and bacteria play key roles in coupling the cycles of phosphorus, nitrogen, sulfur, carbon, and related elements in lakes. However, these groups of organisms may be largely under control of protozoa, and meio- and macrofauna species. The interactions of these latter components of the community may be extremely important in introducing oxygen in deeper sediment layers by bioturbation; however, realistic data from the field are scarce. In addition, fungi certainly play a role in the decomposition of organic carbon, but again more studies are needed in this area.

The greatly increased input of phosphorus and nitrogen to lakes in densely populated areas results in increasing primary productivity. This causes high algal biomass and high settling rates of sestonic particles with concomitant oxygen consumption and creation of anaerobic conditions in the hypolimnion, and even in the upper strata of the metalimnion. In addition, cyanobacterial blooms may appear, and their subsequent decay may consume oxygen in the surface layers and render the lake inhospitable for fish. These types of changes in deep and shallow lakes greatly affect the water quality and make it less suitable for human needs, such as drinking-water supply and for recreation purposes. Therefore, restoration programs for eutrophic and hypereutrophic lakes, based on the results of research, need every support. In addition, agencies involved in eutrophication problems, water pollution control, and management of fish ponds may benefit from the available knowledge of aquatic microbiologists.

Acknowledgments

The work described here was carried out in collaboration with many colleagues and former colleagues at the Netherlands Institute of Ecology – Centre for Limnology, Nieuwersluis, The Netherlands; we are most grateful to all of them. We wish to thank Mrs C.C.C. Janssen-Kroon who prepared the final script and Figure 4.1. Much of the research reported here on the Loosdrecht Lakes' sediments was funded by the Netherlands Integrated Soil Research Programme and by the Department of Public Works of the Province of Utrecht.

References

Adams D.D., Matisoff G. & Snodgrass W.J. (1982) Flux of reduced chemical constituents (Fe^{2+}, Mn^{2+}, NH_4^+ and CH_4) and sediment oxygen demand in Lake Erie. *Hydrobiologia* **92**, 405–414.

Azam F., Fenchel T., Field J.G., Gray J.S., Meyer-Reil L.A. & Thingstad F. (1983) The ecological role of water-column microbes in the sea. *Mar. Ecol. Prog. Ser.* **10**, 257–263.

Berner R.A. (1974) Kinetic models for the early diagenesis of nitrogen, sulfur, phosphorus, and silicon in anoxic marine sediments. In Goldberg E.D. (ed.) *The Sea, vol. 5, Marine Chemistry*, pp. 427–450. John Wiley & Sons, New York.

Best E.P.H. (1982) The aquatic macrophytes of Lake Vechten. Species composition, spatial distribution and production. *Hydrobiologia* **95**, 65–77.

Blaauboer M.C.I. (1982) The phytoplankton species composition and the seasonal periodicity in Lake Vechten from 1956 to 1979. *Hydrobiologia* **95**, 25–36.

Blaauboer M.C.I., Van Keulen R. & Cappenberg Th.E. (1982) Extracellular release of photosynthetic products by freshwater phytoplankton populations, with special reference to the algal species involved. *Freshwat. Biol.* **12**, 559–572.

Bloem J. (1989) *Bacterial Production, Protozoan Grazing, and Mineralization in Stratified Lake Vechten.* PhD thesis, University of Wageningen.

Bloem J. & Bar-Gilissen M.-J.B. (1989) Bacterial activity and protozoan grazing potential in a stratified lake. *Limnol. Oceanogr.* **34**, 297–309.

Bloem J., Bar-Gilissen M.-J.B. & Cappenberg Th.E. (1986) Fixation, counting, and manipulation of heterotrophic nanoflagellates. *Appl. Environ. Microbiol.* 52, 1266–1272.

Bloem J., Starink M., Bar-Gilissen M.-J.B. & Cappenberg Th.E. (1988) Protozoan grazing, bacterial activity, and mineralization in two-stage continuous cultures. *Appl. Environ. Microbiol.* **54**, 3113–3121.

Boers P.C.M., Bongers J.W.Th., Wisselo A.G. & Cappenberg Th.E. (1984) Loosdrecht Lakes restoration project: sediment phosphorus distribution and release from the sediments. *Verh. Int. Verein. Limnol.* 22, 842–847.

Boström B., Jansson M. & Forsberg C. (1982) Phosphorus release from lake sediments. *Arch. Hydrobiol. Beih. Ergebn. Limnol.* **18**, 5–59.

Brown S.R. (1969) Paleolimnological evidence from fossil pigments. *Mitt. Int. Verein. Limnol.* **17**, 95–103.

Canfield D.E. & Des Marais D.J. (1991) Aerobic sulfate reduction in microbial mats. *Science* **251**, 1471–1473.

Capone D.G. & Kiene R.P. (1988) Comparison of microbial dynamics in marine and freshwater sediments: contrasts in anaerobic carbon catabolism. *Limnol. Oceanogr.* 33, 725–749.

Cappenberg Th.E. (1988) Quantification of aerobic and anaerobic carbon mineralization at the sediment–water interface. *Arch. Hydrobiol. Beih. Ergebn. Limnol.* **31**, 307–317.

Cappenberg Th.E. & Jongejan E. (1978) Microenvironments for sulfate reduction and methane production in fresh-water sediments. In Krumbein W.E. (ed.) *Environmental Biogeochemistry and Geobiology*, vol. 1, pp. 129–138. Ann Arbor Science Publishers, Ann Arbor.

Cappenberg Th.E. & Verdouw H. (1982) Sedimentation and breakdown kinetics of organic matter in the anaerobic zone of Lake Vechten. *Hydrobiologia* **95**, 165–179.

Cappenberg Th.E., Hordijk C.A. & Hagenaars C.P.M.M. (1984) A comparison of bacterial sulfate reduction and methanogenesis in the anaerobic sediments of a stratified lake-ecosystem. *Arch. Hydrobiol. Beih. Ergebn. Limnol.* **19**, 191–199.

Carlton W.S., Walker G.S. & Klug M.J. (1989) Relative values of oxygen, nitrate and sulfate to terminal microbial processes in the sediments of Lake Superior. *J. Great Lakes Res.* **15**, 133–140.

Chisholm S.W., Olson R.J., Zettler E.R., Goericke R., Waterbury J.B. & Welschmeyer N.A. (1988) A novel free-living prochlorophyte abundant in the oceanic euphotic zone. *Nature* **334**, 340–343.

Cole J.J., Findlay S. & Pace M.L. (1988) Bacterial production in fresh and saltwater ecosystems: a cross system overview. *Mar. Ecol. Prog. Ser.* **43**, 1–10.

Cooke G.D., Welch E.B., Peterson S.A. & Newroth P.R. (1986) *Lake and Reservoir Restoration*. p. 392, Butterworths Publishers, Boston.

Cullen P. & Forsberg C. (1988) Experience with reducing point sources of phosphorus to lakes. *Hydrobiologia* **170**, 321–336.

Fredrickson H.L., Cappenberg Th.E. & De Leeuw J.W. (1986) Polar lipid fatty acid compositions of Lake Vechten seston, an ecological application of lipid analysis. *FEMS Microbiol. Ecol.* **38**, 381–396.

Fredrickson H.L., Cappenberg Th.E. & De Leeuw J.W. (1988) Relative rates of lipid synthesis from [^{14}C] acetate and [^{14}C] lactate in the surface sediments of a small monomictic lake in relation to seasonal changes in the sedimentary distribution of polar lipid fatty acids. *Arch. Hydrobiol. Beih. Ergebn. Limnol.* **31**, 35–42.

Frenzel P., Thebrath B. & Conrad R. (1990) Oxidation of methane in the oxic surface layer of a deep lake sediment (Lake Constance). *FEMS Microbiol. Ecol.* **73**, 149–158.

Goldman C.R. (1988) Primary productivity, nutrients, and transparency during the early onset of eutrophication in ultra-oligotrophic Lake Tahoe, California–Nevada. *Limnol. Oceanogr.* **33**, 1321–1333.

Golterman H.L. (1975) *Physiological Limnology*. Elsevier, Amsterdam.

Gude H., Haibel B. & Muller H. (1985) Development of planktonic bacterial populations in a water column of Lake Constance (Bodensee-Obersee). *Arch. Hydrobiol.* **105**, 57–77.

Hallegraeff G.M. (1977) Pigment diversity in freshwater phytoplankton. II. Summer-succession in three Dutch lakes with different trophic characteristics. *Int. Revue Ges. Hydrobiol.* **62**, 19–39.

Hordijk C.A. & Cappenberg Th.E. (1983) Quantitative high-pressure liquid chromatography–fluorescence determination of some important lower fatty acids in lake sediments. *Appl. Environ. Microbiol.* **46**, 361–369.

Hordijk C.A., Hagenaars C.P.M.M. & Cappenberg Th.E. (1985) Kinetic studies of bacterial sulfate reduction in freshwater sediments by high-pressure liquid chromatography and microdistillation. *Appl. Environ. Microbiol.* **49**, 434–440.

Hordijk C.A., Snieder M., van Engelen J.J.M. & Cappenberg Th.E. (1987) Estimation of bacterial nitrate reduction rates at *in situ* concentrations in freshwater sediments. *Appl. Environ. Microbiol.* **53**, 217–223.

Hordijk C.A., Burgers I., Phylipsen G.J.M. & Cappenberg Th.E. (1990) Trace determination of lower volatile fatty acids in sediments by gas chromatography with chemically bounded FFAP columns. *J. Chromatogr.* **511**, 317–323.

Hosper S.H. (1989) Biomanipulation, new perspective for restoring shallow, eutrophic lakes in The Netherlands. *Hydrobiol. Bull.* **23**, 5–10.

Hurley J.P. & Armstrong D.E. (1990) Fluxes and transformations of aquatic pigments in Lake Mendota, Wisconsin. *Limnol. Oceanogr.* **35**, 384–398.

Hurley J.P. & Watras C.J. (1991) Identification of bacteriochlorophylls in lakes via reverse-phase HPLC. *Limnol. Oceanogr.* **36**, 307–315.

Hutchinson G.E. (1973) Eutrophication. The scientific background of a contemporary practical problem. *Am. Sci.* **61**, 269–279.

Jones J.G. (1987) Diversity in freshwater microbiology. In Fletcher M., Gray T.R.G. & Jones J.G. (eds) *Ecology of Microbial Communities*, pp. 235–259. Cambridge University Press, Cambridge.

Jones J.G. & Simon B.M. (1981) Differences in microbial decomposition processes in profundal and littoral lake sediments, with particular reference to the nitrogen cycle. *J. Gen. Microbiol.* **123**, 297–312.

Jørgensen B.B. & Revsbech N.P. (1983) Colorless sulfur bacteria, *Beggiatoa* spp. and *Thiovulum* spp., in O_2 and H_2 microgradients. *Appl. Environ. Microbiol.* **45**, 1261–1270.

Kal B.F.M., Engelen G.B. & Cappenberg Th.E. (1984) Loosdrecht Lakes restoration project: hydrology and physico-chemical characteristics of the lakes. *Verh. Int. Verein. Limnol.* **22**, 835–841.

Keizer P., Buysman M.C.N.P. & Cappenberg Th.E. (1991) Sorption and release of phosphorus in a peaty sediment. *Verh. Int. Verein. Limnol.* **24**, 722–725.

Kelly C.A., Rudd J.W.M. & Schindler D.W. (1988) Carbon and electron flow via methanogenesis, SO_4^{2-}, NO_3^-, Fe^{3+}, and Mn^{4+} reduction in the anoxic hypolimnia of three lakes. *Arch. Hydrobiol. Beih. Ergebn. Limnol.* **31**, 333–344.

Korthals H.J. & Steenbergen C.L.M. (1985) Separation and quantification of pigments from natural phototrophic microbial populations. *FEMS Microbiol. Ecol.* **31**, 177–185.

Kuivila K.M., Murray J.W., Devol A.H. & Novelli P.C. (1988) Methane production, sulfate reduction and competition for substrates in the sediments of Lake Washington. *Geochim. Cosmochim. Acta.* **53**, 409–416.

Kuznetsov S.I. (1977) Trends in the development of ecological microbiology. In Droop M.R. & Jannasch H.W. (eds) *Advances in Aquatic Microbiology*, pp. 1–48. Academic Press, London.

Larkin J.M. & Strohl W.R. (1983) *Beggiatoa*, *Thiotrix* and *Tioploca*. *Ann. Rev. Microbiol.* **37**, 341–363.

Leavitt P.R. & Carpenter S.R. (1990) Aphotic pigment degradation in the hypolimnion: implications for sedimentation studies and paleolimnology. *Limnol. Oceanogr.* **35**, 520–534.

Lovell C.R. & Konopka A. (1985) Primary and bacterial production in two dimictic Indiana lakes. *Appl. Environ. Microbiol.* **49**, 485–491.

Lund J.W.G. (1972) Eutrophication. *Proc. R. Soc. Lond. Biol.* **180**, 371–382.

Lund J.W.G. (1978) Changes in the phytoplankton of an English lake, 1945–1977. *Hydrobiol. J.* **14**, 6–21.

Marsden M.W. (1989) Lake restoration by reducing the external phosphorus loading: the influence of sediment phosphorus release. *Freshwat. Biol.* **21**, 139–162.

Overbeck J. (1974) Microbiology and biochemistry. *Mitt. Int. Verein. Limnol.* **20**, 198–228.

Paerl H.W. (1988) Nuisance phytoplankton blooms in coastal, estuarine, and inland waters. *Limnol. Oceanogr.* **33**, 823–847.

Painter H.W. (1970) A review of literature on inorganic nitrogen metabolism in microorganisms. *Wat. Res.* **4**, 393–450.

Parkes R.J. (1987) Analysis of microbial communities within sediments using biomarkers. In Fletcher M., Gray T.R.G. & Jones J.G. (eds) *Ecology of Microbial Communities*, pp. 147–177. Cambridge University Press, Cambridge.

Revsbech N.P., Nielsen L.P., Christensen P.B. & Sørensen J. (1988) Combined oxygen and nitrous oxide microsensor for denitrification studies. *Appl. Environ. Microbiol.* **54**, 2245–2249.

Scavia D. & Laird G.A. (1987) Bacterioplankton in Lake Michigan: dynamics, controls, and significance to carbon flux. *Limnol. Oceanogr.* **32**, 1017–1033.

Seitzinger S.P. (1988) Denitrification in freshwater and coastal marine ecosystems: ecological and geological significance. *Limnol. Oceanogr.* **33**, 702–724.

Sharma B. & Ahlert R.C. (1977) Nitrification and nitrogen removal. *Wat. Res.* **11**, 797–925.

Sherr B.F., Sherr E.B. & Berman T. (1982) Decomposition of organic detritus: a selective role for microflagellate protozoa. *Limnol. Oceanogr.* **27**, 765–769.

Sinke A.J.C. & Cappenberg Th.E. (1988) Influence of bacterial processes on the phosphorus release from sediments in the eutrophic Loosdrecht Lakes, The Netherlands. *Arch. Hydrobiol. Beih. Ergebn. Limnol.* **30**, 5–13.

Sinke A.J.C., Cornelese A.A., Keizer P., Van Tongeren O.F.R. & Cappenberg Th.E. (1990) Mineralization, pore water chemistry and phosphorus release from peaty sediments in the eutrophic Loosdrecht Lakes, The Netherlands. *Freshwat. Biol.* **23**, 587–599.

Steenbergen C.L.M. & Korthals H.J. (1982) Distribution of phototrophic microorganisms in the anaerobic and microaerophilic strata of Lake Vechten (The Netherlands). Pigment analysis and role in primary production. *Limnol. Oceanogr.* **27**, 883–895.

Steenbergen C.L.M. & Verdouw H. (1982) Lake Vechten: aspects of its morphometry, climate, hydrology and physico-chemical characteristics. *Hydrobiologia* **95**, 11–23.

Steenbergen C.L.M. & Verdouw H. (1984) Carbon mineralization in microaerobic and anaerobic strata of Lake Vechten (The Netherlands): diffusion flux calculations and sedimentation measurements. *Arch. Hydrobiol. Beih. Ergebn. Limnol.* **19**, 183–190.

Steenbergen C.L.M. & Korthals H.J. (1988) Vertical distribution of pigments in stratified Lake Vechten (The Netherlands), determined by HPLC. *Arch. Hydrobiol. Beih. Ergebn. Limnol.* **31**, 45–53.

Steenbergen C.L.M., Korthals H.J. & Van Nes M. (1987) Ecological observations on phototrophic sulfur bacteria and the role of these bacteria in the sulfur cycle of monomictic Lake Vechten (The Netherlands). *Acta Academiae Aboensis* **47**, 97–115.

Steenbergen C.L.M., Berger G.W., Korthals H.J. & Verdouw H. (1989) Pigment stratigraphy and trophic status: an evaluation of radionuclide-dated lacustrine sediment. *Geomicrobiol. J.* 7, 207–222.

Steinberg C.E.W. & Hartmann H.M. (1988) Planktonic bloom-forming cyanobacteria and the eutrophication of lakes and rivers. *Freshwat. Biol.* **20**, 279–287.

Stockner J.G. (1988) Phototrophic picoplankton: an overview from marine and freshwater ecosystems. *Limnol. Oceanogr.* **33**, 765–775.

Sutcliffe D.W., Carrick T.R., Heron J., Rigg E., Talling J.F., Woof C. & Lund J.W.G. (1982) Long-term and seasonal changes in the chemical composition of precipitation and surface waters of lakes and tarns in the English Lake District. *Freshwat. Biol.* **12**, 451–506.

Swain E.B. (1985) Measurement and interpretation of sedimentary pigments. *Freshwat. Biol.* **15**, 53–75.

Sweerts J.-P.R.A., Rudd J.W.M. & Kelly C.A. (1986) Metabolic activities in flocculant surface sediments and underlying sandy littoral sediments. *Limnol. Oceanogr.* **31**, 330–338.

Sweerts J.-P.R.A. & Cappenberg Th.E. (1988) Use of microelectrodes for measuring oxygen profiles in lake sediments. *Arch. Hydrobiol. Beih. Ergebn. Limnol.* **31**, 365–371.

Sweerts J.-P.R.A. & de Beer D. (1989) Microelectrode measurements of nitrate gradients in the littoral and profundal sediments of a meso-eutrophic lake (Lake Vechten, The Netherlands). *Appl. Environ. Microbiol.* **55**, 754–757.

Sweerts J.-P.R.A., de Beer D., Nielsen L.P., Verdouw H., van den Heuvel J.C., Cohen Y. & Cappenberg Th.E. (1990) Denitrification by sulphur oxidizing *Beggiatoa* spp. mats of freshwater sediments. *Nature* **344**, 762–763.

Sweerts J.-P.R.A., Bar-Gilissen M.-J.B., Cornelese A.A. & Cappenberg Th.E. (1991) Oxygen-consuming processes at the profundal and littoral sediment–water interface of a small meso-eutrophic lake (Lake Vechten, The Netherlands). *Limnol. Oceanogr.* **36**, 1124–1133.

Van Liere L., Parma S., Mur L.R., Leentvaar P. & Engelen G.B. (1984) Loosdrecht Lakes Restoration Project, an introduction. *Verh. Int. Verein. Limnol.* **22**, 829–834.

Verdouw H., Boers P.C.M. & Dekkers E.M.J. (1985) The dynamics of ammonia in sediments and hypolimnion of Lake Vechten (The Netherlands). *Arch. Hydrobiol.* **105**, 79–92.

Verdouw H., Gons H.J. & Steenbergen C.L.M. (1987) Distribution of particulate matter in relation to the thermal cycle in Lake Vechten (The Netherlands): The significance of transportation along the bottom. *Wat. Res.* **21**, 345–351.

Ward D.M. & Winfrey M.R. (1985) Interactions between methanogenic and sulfate-reducing bacteria in sediments. In Jannasch H.W. & Williams P.J. le B. (eds) *Advances in Microbiology*, vol. 3, pp. 141–179. Academic Press, London.

Wetzel R.G. (1983) *Limnology*. W.B. Saunders, Philadelphia.

White D.C. (1988) Validation of quantitative analysis for microbial biomass, community structure, and metabolic activity. *Arch. Hydrobiol. Beih. Ergebn. Limnol.* **31**, 1–18.

Widdel F. (1988) Microbiology and ecology of sulfate- and sulfur-reducing bacteria. In Zehnder A.J.B. (ed.) *Biology of Anaerobic Microorganisms*, pp. 469–585. John Wiley & Sons, New York.

Yentsch C.M., Horan P.K., Muirhead K., Dortch Q., Haugen E., Legendre L., Murphy L.S., Perry M.J., Phinney D.A., Pomponi S.A., Spinrad R.W., Wood M., Yentsch C.S. & Zahuranec B.J. (1983) Flow cytometry and cell sorting: a technique for analysis and sorting of aquatic particles. *Limnol. Oceanogr.* **28**, 1275–1280.

Jonathan J. Cole & Nina F. Caraco

The pelagic microbial food web of oligotrophic lakes

5.1 Introduction

Pelagic bacteria have been recognized for some time to be an important component of the carbon cycle in lakes. The biomass of bacteria can approach that of the phytoplankton, and the production and respiration of bacteria are major fluxes of carbon (Riemann & Sondergaard 1986a). In oligotrophic lakes, the relative magnitude of these fluxes may be even larger than in eutrophic lakes. Several studies suggest that pelagic bacteria are such a large component of the carbon budget of oligotrophic lakes that it is difficult to reconcile bacterial fluxes with other terms in the food web. Using data from the literature, as well as the author's own experience with an oligotrophic lake in New Hampshire (Mirror Lake), the hypothesis is developed that bacteria themselves might consume a significant fraction of bacterial production in oligotrophic lakes. This could be by direct predation of bacteria upon bacteria or it may be through consumption of bacterial exudates or dissolved compounds lost during lysis.

The hypothesis that bacteria consume bacteria brings into balance a carbon budget that would otherwise not balance, and addresses the three prominent questions in aquatic microbial ecology listed below.

1 Are organic carbon inputs sufficient to support the high rates of bacterial secondary production that we measure? (Scavia 1988; Strayer 1988.)

2 Why do measurements of bacterial production frequently exceed measures of mortality due to grazing? (Pace 1988).

3 How can bacteria metabolize a relatively large fraction of primary production and at the same time not be an important link to higher trophic levels? (Ducklow *et al.* 1986).

This hypothesis is a particular case of the more generalized idea put forth by Strayer (1988) that summed carbon demands of secondary producers may exceed the organic carbon input to a system.

5.2 Bacterial biomass

Two recent reviews suggest that bacterial abundance is strongly related to the abundance of phytoplankton across the trophic gradient (Bird & Kalff 1984; Cole *et al.* 1988; but see Currie 1990), and these reviews are supported by direct experimental evidence in both marine (Hobbie & Cole 1984) and freshwater systems (Riemann & Sondergaard 1986b; Pace & Funke 1991). Intriguing about the empirical relationships is the observation that for a 1000-fold increase in algal biomass, there is about a 100-fold increase in bacterial numbers. While there is still great uncertainty about the controls of bacterial cell size (Fenchel *et al.* 1990), several investigations indicate that cell size is either relatively independent of cell abundance (Van Es & Meyer-Reil 1982) or may actually decrease with increasing cell abundance (Bird & Kalff 1984). Even assuming constant cell size, it is likely, therefore, that the ratio of bacterial to algal biomass is greater in oligotrophic than eutrophic systems (Figure 5.1).

The relatively greater importance of bacterial biomass in oligotrophic lakes has two major implications:

1 as a potential food resource for consumers, bacteria should be relatively more important in oligotrophic systems;

2 as a reservoir of limiting nutrients, bacterial biomass could be critical.

The phosphorus content of pelagic bacteria appears to be enriched

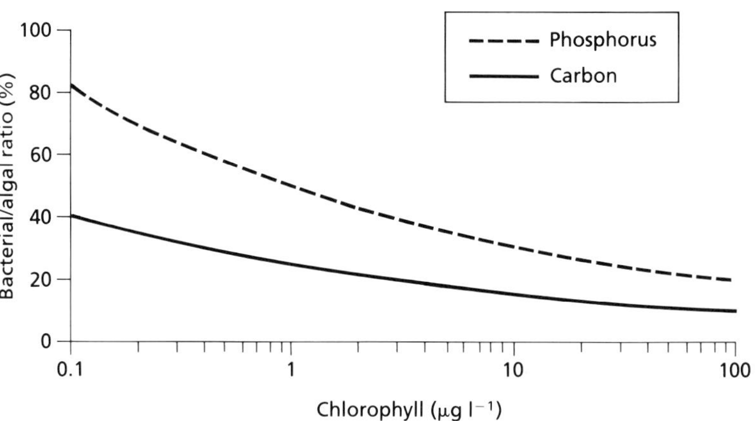

Figure 5.1 Empirical relationship between planktonic bacterial and algal biomass across a trophic gradient. The *x*-axis shows chlorophyll *a* concentrations on a log scale; the *y*-axis shows the ratio of bacterial to phytoplankton biomass (expressed as a percentage) for phosphorus and carbon (indicated on lines). These relationships were computed from the empirical relationship between bacterial numbers and chlorophyll in Bird & Kalff (1984). The following assumptions were made: average bacterial cell size of $0.12\,\mu m^{-3}$; a bacterial cell carbon content of $1.2 \times 10^{-7}\,\mu g\,\mu m^{-3}$; and a phytoplankton carbon/chlorophyll ratio of 50 : 1. For the phosphorus line, algal cells were assumed to have a carbon/phosphorus ratio of 106 : 1 for phytoplankton and 53 : 1 for bacteria (see text).

relative to the carbon/phosphorus ratio of 106:1 commonly observed for phytoplankton (Redfield *et al.* 1963) by some two-fold or more (Bratbak 1985; Nagata 1986; Goldman *et al.* 1987; Vadstein *et al.* 1988). Assuming a molar carbon/phosphorus ratio for bacteria of 53:1, e.g., a phosphorus cell quota of 1.8% or twice the Redfield Ratio, planktonic bacteria could hold about 40–50% as much phosphorus as phytoplankton (assuming a carbon/phosphorus ratio of 106:1) in oligotrophic systems; the importance of this reservoir would decrease dramatically with increasing trophic status (see Figure 5.1). Thus, in phosphorus-limited oligotrophic fresh waters, planktonic bacteria might represent an important sink for this limiting nutrient, and their biomass could limit further increases in the biomass of phytoplankton. It is abundantly clear that bacteria can outcompete phytoplankton for the uptake of phosphorus at low concentrations in both laboratory (Rhee 1972) and natural (Currie & Kalff 1984a,b) systems, an experimental observation which supports the idea that bacteria could be an important phosphorus reservoir particularly in oligotrophic systems. Furthermore, recent data suggest that pelagic bacteria may be limited directly by the supply of phosphorus (Ochs 1991; Toolan *et al.* 1991). Thus, bacteria may remove phosphorus from solution and retain it in their biomass, making it unavailable, or more slowly available to phytoplankton. This effect would be greatest at times of lowest external phosphorus loading, such as during low runoff times, especially for oligotrophic lakes. A similar argument could probably be made for nitrogen in marine settings (Fuhrman *et al.* 1989).

5.3 Bacterial production

Like bacterial abundance, bacterial secondary production is positively related to phytoplankton primary production both across systems and within experimental studies (Hobbie & Cole 1984; Riemann & Sondergaard 1986b; Cole *et al.* 1988). For systems dominated by autochthonous inputs of organic matter from phytoplankton, bacterial production tends to be about 20% of primary production on a volumetric basis for the photic zone and about 30% on an areal basis for the water column as a whole. Based upon the data available at the time, Cole *et al.* (1988) were unable to prove statistically that the relationship between bacterial and algal production was significantly curvilinear, although the form of their empirical equation suggested it was. Data gathered since that review strongly supports a curvilinear relationship (Robarts & Wicks 1990). Nevertheless, the form of the empirical equation relating bacterial to algal production (Equation 5.1) is curvilinear and indicates that, like the situation with biomass, algal production probably increases more rapidly along the trophic gradient than does bacterial production (Figure 5.2):

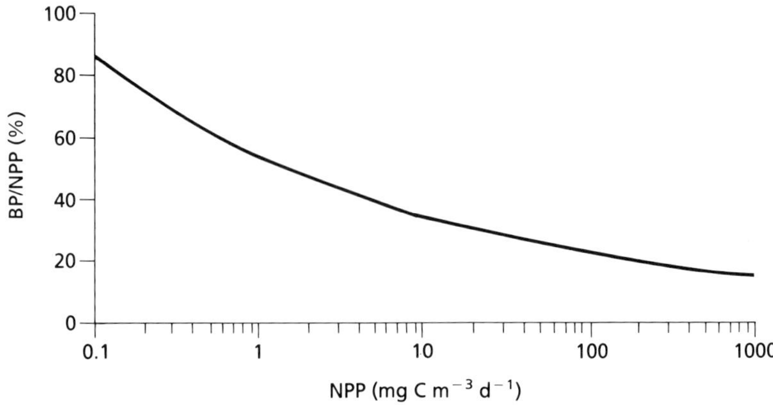

Figure 5.2 Empirical relationship between bacterial and algal production across a trophic gradient. The *x*-axis shows planktonic primary production in the epilimnion on a volumetric basis; the *y*-axis shows the ratio of bacterial to phytoplankton production expressed as a percentage. The equation used here is Equation 5.1 and comes from Cole *et al.* (1988) (see text).

$$BP = CF \times [0.347 \times NPP^{0.8}], \tag{5.1}$$

where BP is bacterial production for surface waters on a volumetric basis ($\mu g\ C\ l^{-1}\ d^{-1}$); NPP is phytoplankton net primary production in the same units. CF is a correction factor equal to 1.56 which is required to convert the original log–log regression to arithmetic units (Cole *et al.* 1988).

Thus, in oligotrophic systems bacterial production could be almost as large as primary production, but would become less important with increasing levels of primary production. The underlying cause of this relationship is not entirely clear. One possibility is that bacterial production is supported, to some extent, by allochthonous carbon sources. If these allochthonous sources are relatively constant along the trophic gradient (which they may not be), these inputs would be relatively more important in oligotrophic lakes simply because primary production was lower. In this vein, Hessen *et al.* (1990) have suggested that humic materials may be the major substrate for aquatic bacteria in highly colored lakes; such lakes have low primary production, and the inputs of allochthonous dissolved organic carbon (DOC) dominates the carbon inputs. The phenomenon of increased relative bacterial production in oligotrophic systems may also result from an increased tendency for algal release of DOC in oligotrophic freshwater (but not marine) systems (Baines & Pace 1991) or, possibly, by limitation of substrate supply to the bacteria in extremely eutrophic conditions (Robarts & Wicks 1990).

Table 5.1 Estimates of epilimnetic bacterial production during summer in Mirror Lake, New Hampshire. A variety of techniques have been employed over a variety of years. $^{35}S[SO_4]$ uptake is the incorporation of ^{35}S into a bacterial-sized fraction; ^{14}C-flux is an estimate based on the decomposition rate of ^{14}C-labeled phytoplankton plus the incorporation of algal extracellular release into bacteria (Cole *et al.* 1982). Thymidine I is the incorporation of thymidine into trichloracetic acid-insoluble material; thymidine II is the incorporation of thymidine into a specific DNA fraction (Ochs 1991)

Year	Method	Epilimnetic bacterial production $\mu g\,C\,l^{-1}d^{-1}$	Reference
1975	$^{35}S[SO_4]$ uptake	5.4	Jordan & Likens (1980)
1979	^{14}C-flux	7.3	Cole *et al.* (1984)
1985	^{3}H-thymidine I	5.6	Cole *et al.* (1989)
1986	^{3}H-thymidine I	8.0	Ochs (1991)
1987	^{3}H-thymidine I	8.3	Ochs (1991)
1988	^{3}H-thymidine II	6.0	Ochs (1991)
Mean ± SD		6.8 ± 1.3	

5.4 A case study

The above analysis would suggest that both in terms of biomass and production pelagic bacteria should be particularly important to carbon and nutrient cycling in oligotrophic lakes. A detailed analysis of one such lake, Mirror Lake, New Hampshire, reveals this to be true but also presents us with some unforeseen problems about how the microbial food web might be structured.

Bacterial production in Mirror Lake has been measured by a variety of techniques and over a number of years (Table 5.1). For the six summer periods for which there are sufficient data, estimates of production range between about 5 and 8 μg C $l^{-1}d^{-1}$ for the epilimnion, with surprisingly little variation between years and across different measurement techniques. Considering the uncertainties associated with all these techniques, the relatively tight clustering gives some confidence in the measurements. Taking the average value and integrating for depth of the epilimnion (through 6.5 m depth) would give a value of about 36 mg C $m^{-2}d^{-1}$. Planktonic primary production has also been measured in Mirror Lake with various minor modifications of the $^{14}C[HCO_3]$ method over a number of years. Annual estimates range from about 21 to 44 g C $m^{-2}yr^{-1}$ from studies from 1973 to 1988 (Ochs 1991); a typical average for summer-time production in the epilimnion (through 6.5 m-depth) is 190 mg C $m^{-2}d^{-1}$. Based on the average of various years and techniques, the secondary production of epilimnetic planktonic bacteria would be about 19% of primary production for the epilimnion. This fraction is slightly lower than predicted in Figure 5.2. If the assumption is made that bacteria grow with growth efficiency of 50%, as they appear to in Mirror Lake on ^{14}C-labeled algal substrates (Cole 1982), then bacteria would metabolize (growth plus respiration) a quantity of organic carbon equal to about 40% of epilimnetic algal primary production. If the growth efficiency of bacteria were lower (10–30%), as some authors argue (Newell *et al.* 1981; Bell & Kuparinen 1984; Schwaerter *et al.* 1988), bacteria would metabolize an even larger fraction of primary production.

Is it possible that nearly 40% (or more) of primary production is metabolized by bacteria in an oligotrophic lake? What consequences does this large flux of carbon have for the food web? Furthermore, Mirror Lake is only moderately oligotrophic (see Figure 5.1). In ultraoligotrophic lakes, if Figure 5.2 is deemed correct, an even larger fraction of algal net primary production (NPP) would presumably flux through the bacteria. During the summer of 1985, an intensive study was made of the organic carbon budget of the epilimnion of Mirror Lake in an attempt to put constraints on the estimates of bacterial production and examine their role in the pelagic food web. The methods and detailed description of that work is published elsewhere (Cole *et al.* 1989; Ochs 1991). Based upon that work, we have assembled a traditional food-web diagram constrained by the values from the epilimnetic carbon budget and other data for Mirror Lake (Figure 5.3). Although the boxes and arrows have been set up in a reasonable way, the scenario in Figure 5.3 is untenable. Since bacterial metabolism is 38% of primary production, we envision a direct assimilation of this much algal carbon by bacteria. Similarly, according to the carbon budget, macrozooplankton would assimilate 77% of phytoplankton primary production of which they respire 54% and use 23% for secondary production (growth, etc.). The carbon to support this demand (for growth plus respiration) by zooplankton comes directly from the phytoplankton (63.2%), bacteria (10.5%), and protozoa (3.3%). The protozoan total demand for organic carbon (3.3% for secondary production and 5% for respiration) could be met entirely by less than half of bacterial production.

The major problem with this scenario is that the demand for algal carbon exceeds the supply (by 1.2%) and leaves no algal material to be exported to the hypolimnion by sedimentation. Sedimentation is a major loss of organic carbon (33%) from the water column in this system (Caraco *et al.* 1988). According to this scenario, we have a quantity equal to 23% of the NPP available for sedimentation after heterotrophic processes; this material would consist entirely of zooplankton remains (see Figure 5.3). When the contents of sediment trap collections from Mirror Lake are examined microscopically, algal cells, many of which appear to be intact, predominate. Furthermore, the carbon/chlorophyll *a* ratio of this sinking material is about

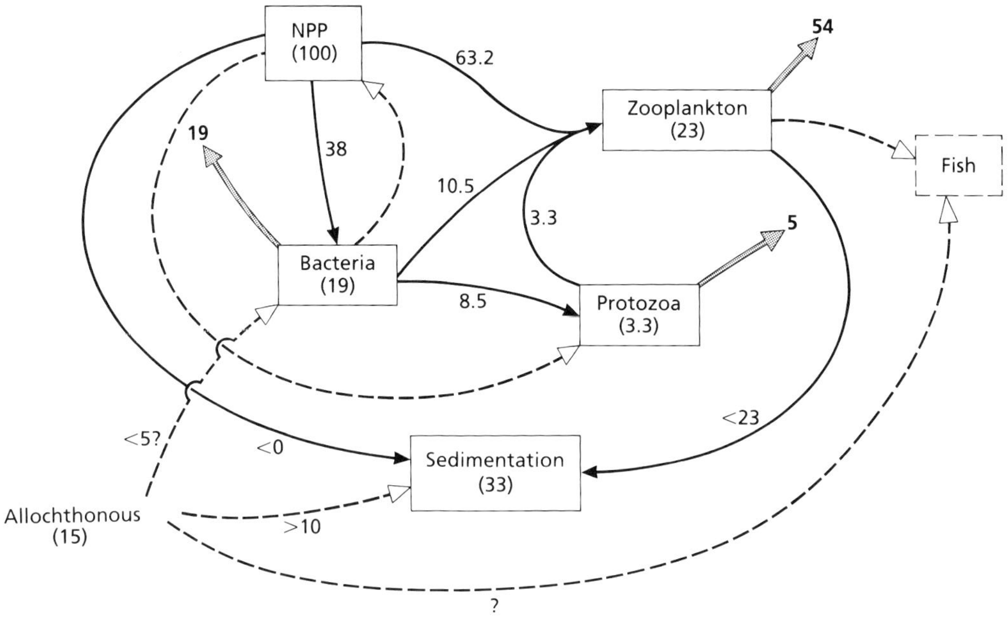

Figure 5.3 Possible food-web structure for an oligotrophic lake. The values are based on the stratified warm-water season for Mirror Lake, New Hampshire. All entries in the boxes are actual measured components and are normalized to 100 units of primary production to make percentage comparison easier; primary production in Mirror Lake would typically be 150–200 mg C m^{-3} d^{-1} during summer. Similarly, the estimates of respiration (indicated by the stipple-filled arrows and bold numbers) are measured.

The interactions between the boxes are hypothetical. The solid lines represent interactions discussed in the text; the dotted lines represent interactions that are less certain. The allochthonous input to fish, for example, is meant to represent the terrestrial insects consumed by fish, which is potentially important in oligotrophic lakes (Carlton & Goldman 1984; Cole *et al.* 1990). The line from bacteria to phytoplankton shows potential consumption of bacteria by phytoplankton (Bird & Kalff 1986; Hessen *et al.* 1990). Factors, such as zooplankton "sloppy feeding," which may make algal carbon available but do not change it, e.g., from algal to zooplankton biomass, are not shown.

In this scenario, although heterotrophic respiration (78 units) is less than algal production, the direct demand on algal carbon by bacteria (38 units) and zooplankton (77 units) exceeds algal production. Furthermore, no algal material is available for sedimentation which is also inconsistent with observations (see text).

80 : 1, consistent with a largely algal source, and inconsistent with a source that would have no algal material in it, as drawn in Figure 5.3.

This food web can be brought into balance, or nearly so, simply by increasing internal recycling (Figure 5.4). If zooplankton are divided into two groups, herbivores (zooplankton 1) that would consume algal material directly (and heterotrophic nanoflagellates as well) and carnivores (zooplankton 2) that consume zooplankton 1, their requirement for algal and bacterial carbon is greatly reduced. That is, although zooplankton production is still 23% of NPP (19.2% for herbivores and 3.8% for carnivores) and zooplankton respiration is still 54% of NPP (45% for herbivores and 9% for carnivores), increasing the number of trophic links lowers the total demand on algal carbon inputs but keeps the respiratory output constant. Increasing the complexity in this way to the bacterial component has a similar but even more pronounced effect due to the presumed greater growth efficiency of bacteria. Suppose, for example, there were three trophic levels of bacteria: those that directly consumed phytoplankton carbon; those that consumed bacteria 1; and those that consumed bacteria 2 (Figure 5.4). This consumption could be envisioned in two major ways. One possibility is that within the microbial food web predatory bacteria, such as *Bdellovibrio* (Starr & Seidler 1971), represent levels 2 and 3. A possibility which is less difficult to imagine is that during growth and death, bacteria, like phytoplankton, release significant quantities of DOC, and that other bacteria metabolize this DOC. This leakage of DOC could be caused by a number of processes including lysis by phage or by bacteria,

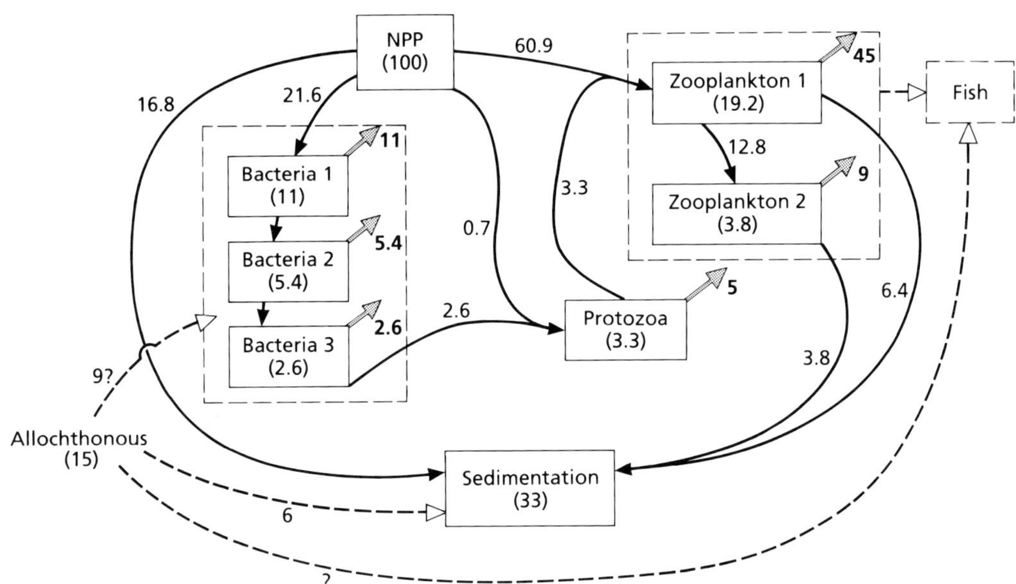

Figure 5.4 A second possible food web using the same data as in Figure 5.3 with a different set of hypothetical interactions. In this case zooplankton have been split into herbivores and carnivores (1 and 2), and bacteria have been split into three levels (1, 2, and 3). Although the production and respiration of zooplankton, bacteria, and protozoa are identical to those in Figure 5.3, allowing bacteria to consume bacteria greatly changes the food web structure and brings the carbon budget into balance (see text). Whereas in Figure 5.3, 19 units of bacterial production plus 19 units of bacterial respiration required 38 units of algal carbon, the same bacterial production and respiration is now met with only 21.6 units.

autolysis, or mechanical damage by grazers. That is, rather than considering the three distinct levels shown in Figure 5.4, the average bacterium may simply receive about 60% of its organic carbon from phytoplankton directly and the remainder from DOC lost by other bacteria. In either case, the effect this increased complexity has within the microbial loop reduces greatly (from 38 to 22; or by 40%) bacterial direct demand on algal carbon. The scenario shown in Figure 5.4 is not meant to represent a unique solution. Rather, it is illustrative of our hypothesis that bacteria could consume significant amounts of bacterial production.

The hypothesis that bacteria consume bacteria brings the food web closer into balance. Now the total direct demand by heterotrophs on algal carbon is 83.2%, leaving a portion (16.8%) directly available for sedimentation. The sedimented material (now accounting for 81% of total sedimentation) would contain an appropriate mix of zooplankton and algal remains in agreement with microscopic examination and measured carbon/chlorophyll ratios. Furthermore, while in the previous food web protozoa consumed only bacteria, in this scenario available bacterial production (after the bacteria have consumed their own portion) is insufficient to support all protozoan demands. Pace (1988) reviewing the literature on bactivory suggested that, in general, protozoan grazing of bacteria is not large enough to consume bacterial production. The idea that bacteria consume bacteria is consistent both with Pace's observation and with the general observation that bacterial numbers do not increase greatly over time. Furthermore, despite the relatively large magnitude of bacterial production, experimental studies indicate that bacteria appear to act more as a respiratory "sink" for carbon than as a "link" to higher trophic levels (Ducklow *et al.* 1986). According to our hypothesis, bacterial production could be both large and a sink. In the food web shown in Figure 5.4, for example, only 2.6% of NPP is returned to higher trophic levels via the microbial loop.

The carbon budget of Mirror Lake is not an extreme case in which bacterial production is especially large. As suggested by Figures 5.1 and 5.2, bacteria would be even more important in more oligotrophic systems, both in terms of biomass and production. The extreme case would probably be the oligotrophic parts of the ocean in which bacterial biomass actually exceeds algal biomass (Fuhrman *et al.* 1989). It is not yet known whether large oligotrophic lakes function in this way, but there is some suggestion that they might (Scavia & Laird 1987; Scavia 1988; Strayer 1988).

Acknowledgments

Financial support was provided in part by the National Science Foundation (BRS-8917962 and BRS-9019873) and by the Mary

Flagler Cary Charitable Trust. We thank R.T. Bell, M.L. Pace, and D.LL. Strayer for comments on the manuscript. This is a contribution to the Institute of Ecosystem Studies and to the Hubbard Brook Ecosystem Study.

References

Baines S.B. & Pace M.L. (1991) The production of dissolved organic matter by phytoplankton and its importance to bacteria: patterns across marine and freshwater systems. *Limnol. Oceanogr.* **36**, 1078–1090.

Bell R.T. & Kuparinen J. (1984) Assessing phytoplankton and bacterioplankton production during early spring in Lake Erken, Sweden. *Appl. Environ. Microbiol.* **48**, 1221–1230.

Bird D.F. & Kalff J. (1984) Empirical relationships between bacterial abundance and chlorophyll concentration in fresh and marine waters. *Can. J. Fish. Aquat. Sci.* **41**, 1015–1023.

Bird D.F. & Kalff J. (1986) Bacterial grazing by planktonic lake algae. *Science* **32**, 277–284.

Bratbak G. (1985) Bacterial volume and biomass estimations. *Appl. Environ. Microbiol.* **49**, 1488–1493.

Caraco N.F., Cole J.J., Likens G.E., Mattson M.D. & Nolan S. (1988) A very imbalanced nutrient budget for Mirror Lake, New Hampshire, USA. *Verh. Int. Verein Theor. Angew. Limnol.* **23**, 170–175.

Carlton R.G. & Goldman C.R. (1984) Effects of a massive swarm of ants on ammonium concentrations in a subalpine lake. *Hydrobiologia* **111**, 113–117.

Cole J.J. (1982) Interactions between bacteria and algae in aquatic ecosystems. *Ann. Rev. Ecol. Syst.* **13**, 291–314.

Cole J.J., Likens G.E. & Strayer D.L. (1982) Photosynthetically produced dissolved organic carbon: an important carbon source for planktonic bacteria. *Limnol. Oceanogr.* **27**, 1080–1090.

Cole J.J., Likens G.E. & Hobbie J.E. (1984) Decomposition of planktonic algae in an oligotrophic lake. *Oikos.* **42**, 257–266.

Cole J.J., Findlay S. & Pace M.L. (1988) Bacterial production in fresh and saltwater ecosystems: a cross-system overview. *Mar. Ecol. Prog. Ser.* **43**, 1–10.

Cole J.J., Caraco N.F., Strayer D.L., Ochs C. & Nolan S. (1989) A detailed organic carbon budget as an ecosystem-level calibration of bacterial respiration in an oligotrophic lake during midsummer. *Limnol. Oceanogr.* **34**, 286–296.

Cole J.J., Caraco N.F. & Likens G.E. (1990) Short-range atmospheric transport: a significant source of phosphorus to an oligotrophic lake. *Limnol. Oceanogr.* **35**, 1230–1237.

Currie D.J. (1990) Large-scale variability and interactions among phytoplankton, bacterioplankton and phosphorus. *Limnol. Oceanogr.* **35**, 1437–1455.

Currie D.J. & Kalff J. (1984a) A comparison of the abilities of freshwater algae and bacteria to acquire and retain phosphorus. *Limnol. Oceanogr.* **29**, 298–310.

Currie D.J. & Kalff J. (1984b) The relative importance of bacterioplankton and phytoplankton in phosphorus uptake in freshwaters. *Limnol. Oceanogr.* **29**, 311–321.

Ducklow H.W., Purdie D.A., Williams P.J. le B. & Davies J.M. (1986) Bacterioplankton: a sink for carbon in a coastal marine plankton community. *Science* **232**, 865–867.

Fenchel T., Kristensen L.D. & Rasmussen L. (1990) Water column anoxia: vertical zonation of planktonic protozoa. *Mar. Ecol. Prog. Ser.* **62**, 1–10.

Fuhrman J.A., Sleeter T.D., Carlson C.A. & Proctor L.M. (1989) Dominance of bacterial biomass in the Sargasso Sea and its ecological implications. *Mar. Ecol. Prog. Ser.* **57**, 207–217.

Goldman J.C., Caron D.A. & Dennet M.R. (1987) Regulation of gross growth efficiency and ammonium regeneration in bacteria by substrate C:N ratio. *Limnol. Oceanogr.* **32**, 1239–1252.

Hessen D.O., Anderson T. & Lyche A. (1990) Carbon metabolism in a humic lake: pool sizes and cycling through zooplankton. *Limnol. Oceanogr.* **35**, 84–99.

Hobbie J.E. & Cole J.J. (1984) Response of a detrital foodweb to eutrophication. *Bull. Mar. Sci.* **35**, 357–363.

Jordan M. & Likens G.E. (1980) Measurement of planktonic bacterial production in an oligotrophic lake. *Limnol. Oceanogr.* **25**, 721–732.

Nagata T. (1986) Carbon and nitrogen content of natural planktonic bacteria. *Appl. Environ. Microbiol.* **52**, 28–32.

Newell R.C., Lucas M.I. & Linley E.A.S. (1981) Rate of degradation and efficiency of conversion of phytoplankton debris by marine microorganisms. *Mar. Ecol. Prog. Ser.* **6**, 123–136.

Ochs C.O. (1991) *Population Dynamics of the Bacterioplankton of Mirror L. New Hampshire*, p. 211. PhD thesis, Cornell University.

Pace M.L. (1988) Bacterial mortality and the fate of bacterial production. *Hydrobiologia* **159**, 41–50.

Pace M.L. & Funke E. (1991) Regulation of planktonic microbial communities by nutrients and herbivores. *Ecology* **72**, 904–914.

Redfield A.C., Ketchum A.C. & Richards F.A. (1963) The influence of organisms on the composition of seawater. In Hill M.N. (ed.) *The Sea*, pp. 26–77. Interscience, New York.

Rhee G.-Y. (1972) Competition between an alga and an aquatic bacterium for phosphate. *Limnol. Oceanogr.* **17**, 505–514.

Riemann B. & Sondergaard M. (eds) (1986a) *Carbon Dynamics in Eutrophic Temperate Lakes*. Elsevier, Amsterdam.

Riemann B. & Sondergaard M. (1986b) Regulation of bacterial secondary production in two eutrophic lakes and in experimental enclosures. *J. Plank. Res.* **8**, 519–536.

Robarts R.D. & Wicks R.J. (1990) Heterotrophic bacterial production and its dependence on autotrophic production in a hypertrophic African reservoir. *Can. J. Fish. Aquat. Sci.* **47**, 1027–1036.

Scavia D. (1988) On the role of bacteria in secondary production. *Limnol. Oceanogr.* **33**, 1220–1224.

Scavia D. & Laird G. (1987) Bacterioplankton in Lake Michigan: dynamics, controls, and significance to carbon flux. *Limnol. Oceanogr.* **32**, 1017–1033.

Scavia D., Laird G.A. & Fahnenstiel G.L. (1986) Production of planktonic bacteria in Lake Michigan. *Limnol. Oceanogr.* **31**, 612–626.

Schwaerter S., Sondergaard M., Riemann B. & Jensen L.M. (1988) Respiration in eutrophic lakes: the contribution of bacterioplankton and bacterial growth yield. *J. Plank. Res.* **10**, 515–531.

Starr M.P. & Seidler R.J. (1971) The Bdellovibrios. *Ann. Rev. Microbiol.* **25**, 649–678.

Strayer D.L. (1988) On the limits to secondary production. *Limnol. Oceanogr.* **33**, 1217–1220.

Toolan T., Wehr J.D. & Findlay S. (1991) Inorganic phosphorus stimulation of bacterioplankton production in a meso-eutrophic lake. *Appl. Environ. Microbiol.* **57**, 2074–2078.

Vadstein O., Jensen A., Olsen Y. & Reinertsen H. (1988) Growth and phosphorus status of limnetic phytoplankton and bacteria. *Limnol. Oceanogr.* **33**, 489–503.

Van Es F.B. & Meyer-Reil L.A. (1982) Biomass and metabolic activity of heterotrophic marine bacteria. *Adv. Microbiol. Ecol.* **6**, 111–170.

Maurice A. Lock

6 Attached microbial communities in rivers

6.1 Introduction

6.1.1 Some definitions

Any wetted surface submerged in a river for more than an hour or so will be coated with microorganisms. These will consist of bacteria, protozoa, and algae (where light levels permit) forming an assemblage on each surface. Names for such site-specific assemblages are given in Table 6.1. However, a generic term for microorganisms attached to surfaces has been slow to materialize. A term with historical precedence (Wetzel 1975) is *aufwuchs* (German for growth upon), but this has not gained significant acceptance. While the word periphyton has gained generic status among algologists as the word to describe algae attached to surfaces, this has not been widely accepted by microbial ecologists. Attached microorganisms of wetted surfaces are generally a mixture of autotrophs and heterotrophs and thus the "phyton" component of the word may have been considered misleading, particularly when using the "epi" system of constructs for specific surfaces. Furthermore, from an etymological standpoint the word periphyton is inappropriate as a generic term even for algal assemblages since it means "around plants." As the "epi" system gained acceptance among aquatic ecologists a generic construct using epi- as the first syllable seemed a logical step. However, the evolution of the scientific vocabulary is a peculiar process and it now seems that aquatic ecologists are settling upon the word biofilm for the generic term for attached microorganisms, both autotrophic and heterotrophic. Consequently, biofilm will be used throughout this chapter and the epi- construct will only be used as an adjective to describe the biofilm, e.g., epilithic biofilm (microorganisms attached to rocks), or in a noun as in epipelon (microorganisms attached to fine sediment surfaces).

6.1.2 Background

In this chapter, I intend to develop an overview of attached microbial communities in rivers in terms of their structure, function, and responses to their environment. It will not be a complete review of the literature because to do so in the context of a chapter would be

Table 6.1 Terms used to describe attached microbial assemblages in aquatic environments. From Wetzel (1975)

Substratum	Noun	Adjective
Rock	Epilithon	Epilithic
Wood[1]	Epixylon	Epixylic
Macrophytes	Epiphyton	Epiphytic
Fine sediment surfaces	Epipelon	Epipelic
Within soft sediments	Epipsammon	Epipsammic

[1] From Sinsabaugh *et al.* (1991).

to lose the broad perspective. The starting point will be the perception of epilithic biofilms as mixed autotrophic and heterotrophic assemblages (Lock *et al.* 1984). The studies on epilithic biofilms will form a basic model for biofilms in general, with the consideration that the greatest shifts in biofilm structure and metabolism are associated with the amount of photosynthetically active radiation. For those readers wishing to trace back the work primarily on the algal components of biofilms I would recommend the review book on this topic by Wetzel (1983).

In Lock (1981) and Lock *et al.* (1984), a structural and functional model of epilithon was proposed. The essential basis of this model was that the biotic elements, bacteria, cyanobacteria, and algae, were fixed to rock surfaces by microbially produced extracellular polysaccharides and that these formed a continuous matrix. Furthermore, it was suggested that this polysaccharide matrix acted as:

1 a site for entrapment of soluble and particulate matter;

2 a retaining and possibly protective mechanism for extracellular enzymes;

3 a retention mechanism for any products of extracellular enzymatic hydrolyses;

4 a framework or "glue" enabling a multistorey of autotrophic and heterotrophic microorganisms to form under suitable conditions.

While this model was specifically developed in the context of microorganisms attached to stone surfaces in streams, it is reasonable to extend this, with some modification, to all wetted surfaces. These would range from fallen trees to surfaces of rooted macrophytes, and from fine surficial/sediments to gravels tens of centimeters below the stream-bed surface. The composition of these biofilms is likely to be very different, with the greatest differences perhaps associated with shifts in the relative importance of autotrophic and heterotrophic microorganisms, as ambient light levels change. As shown in an earlier review, a smaller biomass of algae was found on the undersides of stones compared with the upper surfaces (Lock *et al.* 1984).

In addition, the proportion of autotrophic forms would be expected to diminish with depth in river sediments. In this chapter it is proposed to both confirm and update the epilithon model as a generalized biofilm model and to extend it to other wetted surfaces in rivers. Readers are also referred to the recent review of bacterial biofilms by Costerton *et al.* (1987) for an industrial and biomedical perspective.

6.2 Biofilm structure

Over the last few years a considerable amount of research has emerged on this topic. The agents which hold the living components of the biofilm together are the extracellular polysaccharides or exopolymers (Leppard 1986). The first studies of this material were limited to analytical chemistry (Fazio *et al.* 1982). However, Low & White (1989) described a variety of non-destructive and destructive techniques, some of which were then in an early stage of development. It is now possible to monitor conformational changes in the molecular architecture of exopolymers of living biofilms and to follow the activity of specific genes involved in the adhesion processes in real time. However, detailed descriptions of such techniques are beyond the scope of this chapter and interested readers are directed to that review.

6.2.1 Microscopic architecture

The distribution of microorganisms in a biofilm was originally presented as a generally uniform assemblage (Lock *et al.* 1984). It is now apparent that this was a rather simplistic conception. Several studies have shown the existence of a rather loosely attached assemblage of microorganisms overlying a more firmly attached component. In addition, the firmly attached component is not homogeneous but may have a complex and varied structure (Stock & Ward 1989).

The investigation of biofilm structure has commonly been through the use of scanning and, to a much lesser degree, transmission electron microscopy (TEM). Recently, low-temperature scanning electron microscopy (LTSEM) has been used to demonstrate the existence of an extremely loose outer surface of organisms in an Antarctic lake biofilm (Oppenheim & Paterson 1990). The samples were immediately frozen after collection and then transferred to the specially modified cooled stage of a scanning electron microscope. A comparison of normally freeze-dried SEM material with LTSEM-prepared material indicated that normal freeze-drying resulted in biofilm compaction in three dimensions. The LTSEM technique clearly showed that the upper surface was of a very loose nature, bound together with "copious amounts of mucilage." This in itself posed a problem, making identification of the cells extremely difficult. However, this difficulty was overcome by progressively etching away the mucilage (polysaccharide matrix) by fast-particle etching, permitting a controlled exposure of the underlying material. Even so, the study of this loosely attached layer remains problematic since the very act of sample collection may remove a portion of that component. Oppenheim & Paterson (1990) conclude that in order to fully quantify it, *in situ* freezing techniques are likely to prove necessary. While river biofilms are unlikely to develop as extensive a loose outer layer as those in lakes, it is reasonable to assume some development takes place (Paul & Duthie 1989). From a functional standpoint the loosely attached biofilm components have been shown to be a site of high microbial activity and also to act as an inhibitor of microbial activity in the underlying storeys (Paul & Duthie 1989; Burkholder *et al.* 1990). Thus, this aspect of biofilm structure requires serious consideration and requires a reevaluation of biofilm sampling techniques.

Most recently Lawrence *et al.* (1991) and D.E. Caldwell (personal communication, 1992) have used the technique of laser scanning confocal microscopy to show that "channels" can be present within a fully hydrated biofilm which extend to the surface. If this proves to be a general feature of biofilms, reevaluation of the mode of entry of both organic and inorganic substrates, and the release of products (photosynthate, enzymatic hydrolysates) will be necessary. Instead of the possible modification of solute movement by close proximity to the anionic polysaccharide matrix, their movement in the open "channels" could be more governed by diffusion. In biofilms from fast-flowing turbulent water there is the additional possibility that advective solute movement might take place through a physical pumping action of the biofilm matrix from water repeatedly surging against the biofilm. Such a phenomenon might explain some of the observations of higher metabolic activities associated with biofilms from swiftly flowing water (Lock & John 1979).

Below the loosely attached biofilm component the more adherent microbial population may have a complex, three-dimensional structure at one extreme or be effectively almost a monolayer of microorganisms at the other. The factor most likely to influence such community types is water velocity, with high velocities associated with monolayers and low velocities with complex structures (Luttenton *et al.* 1986; Lamb & Lowe 1987). The development of two such extremes of architecture was diagrammatically represented by Korte & Blinn (1983). The contribution of extracellular polysaccharide to the formation of three-dimensional diatom assemblages was discussed by Roemer *et al.* (1984). Based upon their analysis it would appear that:

1 it allows cell–substratum adhesion;

2 mucilagenous diatom stalks provide increased surface area for

attachment of other autotrophic and heterotrophic microbial cells;
3 dense canopies of stalked diatoms trap suspended microorganisms;
4 binding of nonliving organic or inorganic matter;
5 permits the formation of multitiered microbial assemblages at certain times of the year.

However, the formation of substantial three-demensional assemblages is not restricted to stalked diatoms but also includes bacteria and amorphous detritus (Rounik & Winterbourn 1983). Stock & Ward (1989) reported filamentous green-algae and cyanobacteria forming multitiered assemblages on freshly exposed and weathered rock surfaces. These authors found that the initial colonizers were bacteria, which were rapidly followed by a monolayer of adnate diatoms. Subsequent bacterial growth between the diatoms was thought to be dependent upon formation of refugia between diatom cells and on the extent of mucilage and photosynthate production by the diatoms. Final colonization by the filamentous forms was thought to have been facilitated by the "burying" of the adnate diatom monolayer by bacteria and associated mucilage.

While SEM has been used extensively in examinations of biofilm structure, the transmission electron microscope has had more limited use (Geesey *et al.* 1977; 1978). In these studies biofilms were grown on electron microscope resins which were subsequently sectioned and viewed with TEM. Blenkinsopp & Lock (in press) used this approach in a study of biofilms growing in fast flowing water. Bacteria were found extensively between diatoms, but less frequently above the algal cells. They were, however, consistently found in the attachment pad of the diatoms (Figure 6.1a). Such bacteria would be anticipated to benefit from photosynthate production but could also be of benefit to the diatom in the provision of trace organic substances, e.g., vitamin B12. While the biofilms in this study tended to be ~5 μm thick, at times multi-layered (~25 μm thick) assemblages formed with diatoms secondarily colonizing a layer of cyanobacteria (Figure 6.1b).

The bulk of knowledge concerning biofilm structure has come from biofilms exposed to light (Figure 6.2). However, as demonstrated in Lock *et al.* (1984), biofilms also grow on the undersides of rocks. In addition, it is reasonable to expect to find them attached to the rocks, gravels, and fine sediments extending tens of centimeters into the river and stream beds (Figure 6.2). These subsurface substrata are referred to as the hyporheic zone (Williams 1984). This zone represents an important interface between the terrestrial environment and the aquatic environment. Experimental studies (Wallis *et al.* 1981; Fiebig & Lock 1991) have indicated that terrestrially derived organic matter is discharged through the river bed as the

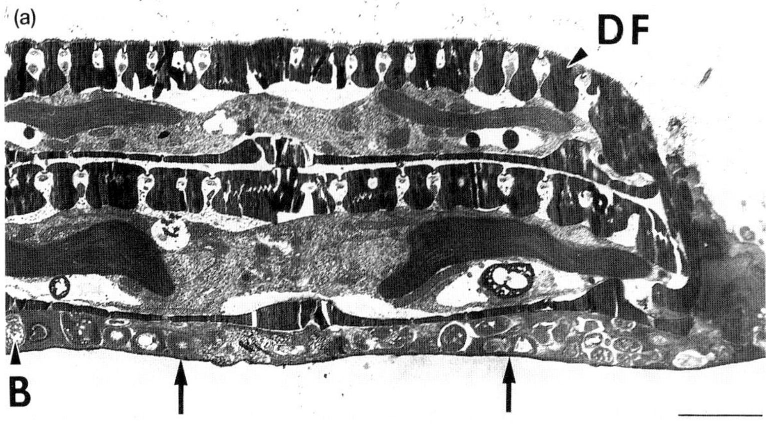

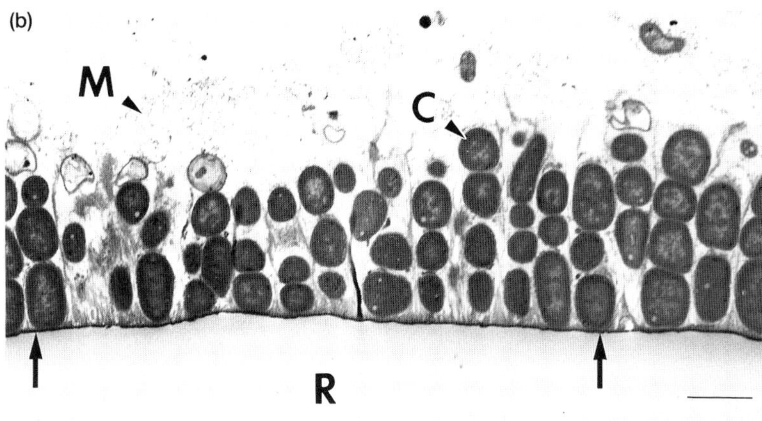

Figure 6.1 (a) A recently divided *Cocconeis* sp. with large numbers of bacteria beneath the diatom frustule. (b) A "palisade" formation of cyanobacteria. Arrows, biofilm–substratum interface; B, bacterium; C, cyanobacterium; DF, diatom frustule; M, exopolysaccharide matrix; R, resin substratum. Scale bar = 2 μm. Micrographs courtesy of S.A. Blenkinsopp.

dissolved organic matter (DOM) in ground water and may become immobilized there. The site for this immobilization is likely to be biofilms coating the subsurface sediments (Smith *et al.* 1986; Baerlöcher & Murdoch 1989) plus microbial flocs sloughed-off from surface and subsurface biofilms, and from particulate organic matter (POC) derived from terrestrial vegetation (Ward *et al.* 1990). To my

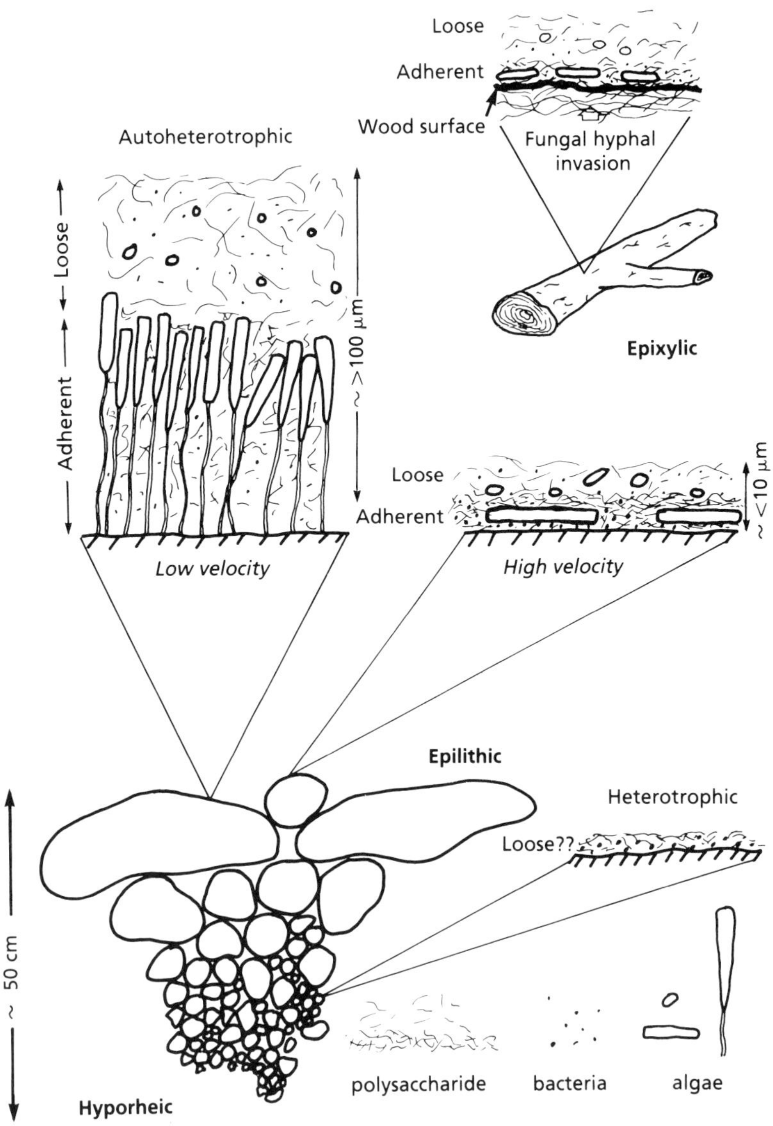

Figure 6.2 Some biofilms from surfaces in rivers: the upper surfaces of stones tend to have algae and bacteria present (autoheterotrophic) while biofilms from within the stream bed (hyporheic) will be predominantly heterotrophic. The thickness of the biofilm is related to current velocity, with thick films being associated with low velocities and thin films with high velocities. The biofilms are also likely to be made up of an adherent layer overlain by a loosely attached layer. The biofilms associated with wood are anticipated to exhibit similar composition and structure to epilithic biofilms with the addition that the products from surface and subsurface fungal hydrolysis of wood may act as an additional source of organic matter/energy. A similar phenomenon may be occurring in biofilms growing on river macrophytes and mosses with photosynthate from the plant being utilised within an epiphytic biofilm. (Allanson 1973).

knowledge, information on the structure of these subsurface microbial assemblages does not exist. However, studies carried out in marine sediments suggest a continuity of microorganisms through sediments, held together with mucopolysaccharides (De Flaun & Mayer 1983; Watling 1988).

If we now turn to the other direction, out into the water column, a further site for attached microorganisms is tree debris. In small pristine streams these may provide large surface areas for colonization, particularly when flow-formed debris dams occur (Bilby & Likens 1980), or from the action of beavers (Naiman & Melillo 1984). It has been suggested that the presence of fallen trees within rivers may have been much more common prior to man's propensity to remove them for navigation and flood management purposes (Triska 1984). In the past, large wood choked rivers may have trapped larger quantities of dissolved nutrients than occurs in the "managed" channels of rivers today. A comparison of biofilms on stones and wood surfaces has recently been made by Sinsabaugh *et al.* (1991), but their methodology of wood sampling was such that microorganisms involved in surface wood decomposition were also included with the surface biofilm. It would now appear that microorganisms are distributed in three dimensions in rivers (see Figure 6.2). These microbes are being exposed to organic and inorganic nutrients flowing down the river or stream (microbes attached to surface sediments and wood and leaf debris), and in addition, may be exposed to those substances flushing up through the bed (subsurface microbes) transported by groundwater discharge. Also, the surfaces of submerged and emergent macrophytes (Allanson 1973) and mosses will act as sites for biofilm formation. Their large surface to volume ratios will result in high levels of exchange of organic and inorganic materials between their associated biofilms and the overlying water.

6.3 Biofilm metabolism

6.3.1 Dissolved organic matter

DOM is one of the major components of the organic energy inputs for streams (Cummins *et al.* 1983) and may represent more than 70% of the organic carbon. Although much of this material is exported out of a study reach, even a small retention of such a large flux is likely to be of major importance to running-water food webs (Meyer 1990). Thus the metabolism of DOM within rivers has been a topic of much interest. An obvious step towards understanding that process would be information on what DOM is. While considerable progress has been made in identifying the simple components, the bulk of the DOM falls into that poorly characterized category of materials known as humic substances. These are complex polyphenolic substances which have complexed with them greater or lesser amounts of more microbially labile materials. It is the extreme chemical heterogeneity of DOM which has made progress on understanding its metabolism somewhat slow. For a recent discussion of DOM, the reader is referred to the excellent book on the topic by Thurman (1985).

Work on DOM metabolism has progressed in two directions. One has been to measure the uptake/respiration of single compounds known to be present in river water and the other has been to conduct studies on the total, or operationally defined subsets of, the DOM pool. The former approach has involved a variety of radiolabeled compounds, such as amino acids, carbohydrates, and monomers of lignin (Ladd *et al.* 1979; Wallis *et al.* 1981). However, the tendency has been to report heterotrophic potential rather than actual uptake rates since the concentration of the substrate was frequently unknown. These are reported as V_{max} specific activity indices which fall within the approximate range of 50–500 µg substrate (per cell × 10^6) hr^{-1}. However, progress with this approach has been restricted by limited knowledge of the composition of the DOM pool (Thurmann 1985) and the availability of radiolabeled substrates. It is clearly not an approach for investigation of the total DOM pool but is of value in answering questions about the roles of specific compounds, particularly in combination with other studies, e.g., tracking the fate of products of exoenzymatic hydrolyses.

By far the greatest attention has been given to examinations of the metabolism of all or parts of the DOM pool. The earliest of these involved monitoring the disappearance of poorly characterized DOM, such as water extracts of soils or autumn-shed leaves in the presence of stream sediments. These laboratory-based studies usually involved adding the substrates at high concentrations (several milligrams per liter) to laboratory streams (Cummins *et al.* 1972) or to river sediment cores (Lock & Hynes 1976). DOM was observed to disappear at rates ranging from <5 to $>1000\,mg\,C\,m^{-2}\,hr^{-1}$. Kaplan & Bott (1983; 1985) used this approach with a diverse assemblage of DOM sources; forest floor leaf litter, bovine manure, a green alga (*Ulothrix*), and riparian vegetation *Impatiens capensis* and *Lindera benzoin*. They found that the microflora of stream beds exhibited an acclimation to DOM sources, e.g., a minimum of 48 hours exposure to bovine manure DOM (~45% acetic acid) was required before measurable changes in sediment microbial biomass or activity could be detected. Of equal interest was their finding that this acclimation response was retained for up to 72 hours in the absence of the stimulus, i.e., unamended stream water. Furthermore, they also demonstrated that specific DOM molecular size classes and functional chemical groups were selectively removed by stream sediments. They found preferential removal of low molecular weight compounds (<10 000 apparent molecular weight (AMW)) and that carbohydrates played a dominant role in sediment microbial metabolism.

Further evidence of an acclimation of stream-bed microorganisms to the DOM source was provided by McArthur & Marzolf (1986). They examined the uptake of riparian grass and tree-leaf leachates in prairie streams draining grassland (head water streams) and forest (lower reaches). They found that grass leachates were readily taken up in the grassland or forest sites but the leaf leachates were only readily taken up in the forest sites (naturally exposed to leaf and grass leachates) with much less uptake in the grassland sites (naturally exposed to only grass leachates).

In all these studies, rates of uptake are dependent upon specific chemical composition (usually unknown) and substrate concentration (under nonsaturating conditions). Since widely varying compositions and concentrations were used in different experiments comparison is difficult. In addition, it was apparent from these studies that it was very difficult to ascribe uptake of DOM to biotic or abiotic activity. The partitioning of these two processes has been attempted through the use of biocides, e.g., mercury (Dahm 1981), or sterilization techniques, e.g., autoclaving or tyndallization (Lock & Hynes 1976). However, both approaches resulted in greater or lesser releases of DOM from the stream sediments. Saturation of putative adsorption sites may occur under these conditions, resulting in an underestimation of abiotic uptake. Thus the extent of abiotic uptake of DOM remains to be resolved (McDowell 1985).

The *in situ* uptake of DOM by stream beds has also been monitored. Substrates have ranged from the poorly characterized materials, such as water extracts of leaves (Lush & Hynes 1978), to input–output determinations of compounds or classes of compounds along a section of stream (Telang & Hodgson 1983), or the uptake of specific compounds, such as individual amino acids (Fiebig 1988). In this last study, the largest removal was recorded for total fatty acids which decreased by 82% over the 1 km study reach while total amino acids (the least changes) decreased by only 12%. Fiebig (1988) added amino acids with an inert tracer (sodium iodide) to monitor dilution rate. He found considerable variation in removal rates of amino acids over a 200 m section (from 283 nmol to 19 mmol $m^{-2} hr^{-1}$). The above studies have also shown an apparent release of DOM from stream sediments. In the study of Telang & Hodgson (1983) releases of dissolved organic carbon (DOC) and hydrocarbons were observed while Fiebig (1988) found substantial releases of one particular amino acid, glutamic acid, with release rates reaching 4.8 mmol $m^{-2} hr^{-1}$; however, uptake/release of glutamic acid was erratic. Undefined DOC releases have also been noted by Kaplan & Bott (1983).

Studies on operationally defined components of the DOM pool have been conducted extensively in my own laboratory. In essence, the metabolic response of biofilms – as heat output – (Lock & Ford 1983) to perfusion with river water with organic matter >1000 AMW was compared with a whole river water control. Two major points came out from this work: firstly, biofilms apparently react very slowly to major reductions in the DOM supply (Lock & Ford 1985), and secondly, in certain rivers (usually brown water), removal of the >1000 AMW fraction results in substantial increases (often several-fold) in metabolic heat output (Ford & Lock 1987; Freeman *et al.* 1990). This increase in heat output led to the suggestion that there were substances in the removed fraction which were in some way inhibiting biofilm metabolism. This inhibitory phenomenon was substantiated by Freeman & Lock (1992) using both natural and "model" (polyvinylpyrrolidone) recalcitrant >1000 AMW compounds. However, further studies on the mechanisms of this inhibition are needed. The resilience of biofilms to massive changes in the DOM supply was also confirmed by Freeman (1990). Perfusion of a biofilm with an organic-matter-free inorganic medium for 3 weeks resulted in a steady rise in microbial activity (^{14}C-acetate incorporation into lipids). This observation suggests the presence of a substantial organic matter reservoir within the biofilm.

Thus, studies which had originally been aimed at determining the relative utility of different molecular size fractions of organic matter to biofilm metabolism revealed some unexpected metabolic behavior of natural river biofilms.

6.3.2 Fine particulate organic matter (FPOM)

While much effort has been expended upon the role of DOM as a carbon/energy resource for attached microorganisms, little is known about the role of suspended POM. The presence of POM is well documented in rivers where it represents a major food resource for filter-feeding animals as FPOM (Wotton 1990). Its origins are potentially diverse; aggregation of DOM, microbial maceration of leaf, wood, and aquatic macrophyte debris, sloughed biofilm, and disaggregation of fecal pellets of river animals (Ward *et al.* 1990). With such origins it is reasonable to expect it would also serve as a carbon/energy resource for attached microorganisms. However, I am unaware of any major evaluation of its potential in biofilm metabolism. Certainly from an anecdotal standpoint, studies on biofilm structure have noted the presence of POM obviously not of *in situ* origin (Madsen 1972; Karlstrom 1978). Given the aggregative properties of microbial consortia it is likely that FPOM in suspension will tend to adhere to biofilms or other attached microbial communities in rivers. Once adhered to the biofilm, FPOM is likely to be incorporated into the metabolism of the microbial communities and to influence them on a local scale.

6.3.3 Enzymology

Since the majority of organic matter synthesized by autotrophs is polymeric or macromolecular, microorganisms require a process, or processes, to reduce these compounds to monomeric units capable of being transported into the cell. Two possibilities come to mind, a physicochemical process, such as photolysis (Lock *et al.* under review), or biologically mediated enzymatic degradation. It is probably worth stressing that in the absence of one or other of these processes, organic matter would accumulate within all ecosystems with a consequent removal of elements from the influence of the biosphere. Much work on extracellular enzymes has been conducted in soil ecosystems (Burns 1978; 1989) and only in the mid 1980s was attention directed in any significant way to the enzymology of organic matter decomposition in aquatic ecosystems (Lock 1990).

The detection of biofilm enzyme activity has most frequently been through the use of chromogenic (Sinsabaugh & Linkins 1988) or fluorogenic substrates (Sinsabaugh & Linkins 1988; Jones & Lock 1989; Marxsen & Witzel 1990). Endocellulases have been measured viscometrically, exocellulases have been estimated by the generation of reducing sugar and phenol oxidase activity has been estimated using changes in absorbance of L-DOPA (Sinsabaugh *et al.* 1991).

Sinsabaugh & Linkins (1988) is the earliest paper known to me on this topic in rivers. They conducted 11 assays for exoenzyme activity in river biofilms over an 8-month period. The biofilms were from slow ($1-10\,cm\,s^{-1}$) and fast ($\geqslant 80\,cm\,s^{-1}$) parts of a fourth-order river in the Adirondack Mountains, USA, grown in the presence and absence of light. The most readily detected enzymes were phenol oxidase; phosphatase; alpha-1,4-glucosidase; beta-1,4-*N*-acetyl glucosamidase; and beta-1,4-glucosidase. In general terms they found that as biofilm biomass increased, so also did the sum total of all enzyme activities. However, a somewhat surprising finding was that there was little difference between the enzymology of either light-grown or dark-grown biofilms despite structural divergence between the two types in summer and the fall. While one might have anticipated that the internally derived substrates from algal exudation might have altered enzyme expression and/or activity, this clearly was not the case. This suggests that there may be a degree of segregation of enzymes and substrates within biofilms.

Jones & Lock (1989) showed in a study of two contrasted lotic systems that differences in exoenzymatic activities do not necessarily reflect differences in their respective microbial populations. In the biofilms from the oligotrophic Nant Waen stream, while absolute enzyme activity was lower, there was a higher cell-specific esterase activity than in biofilms from the more eutrophic Clywedog River. This may have been due to a greater amount of enzyme production per cell or to the enhanced retention of extracellular enzymes within the polysaccharide matrix of oligotrophic biofilms. Evidence of the latter phenomenon comes from work on soil enzymes (Burns 1978) where extracellular enzyme longevity has been attributed to their stabilization with mucopolysaccharides and humic moieties.

Most recently, Sinsabaugh *et al.* (1991) have shown that the amount of enzyme activity in a biofilm can be related to the underlying substratum. Out of a suite of seven enzyme assays, with the exception of β-xylosidase, increases in enzyme activity of an epixylic (wood) biofilm ranged from two to 50 times that of a corresponding epilithic biofilm from the same river. Given the similarity in the hydrodynamic regimes under which each biofilm developed, the substantial differences observed in biofilm structure and function were attributed to the underlying substratum. In particular they noted a much greater prevalence of fungi. However, they originally hypothesized that total biomass of biofilm (irrespective of composition) would probably converge due to a common suite of limiting factors at the water–biofilm interface. However, this was not found to be so. Much greater biomass was supported on the wood substratum. This led them to suggest that the epilithic biofilms may have been carbon limited. However their data, and hence conclusions, may be influenced by their biofilm sampling methodology which included decomposing surface wood along with the biofilm.

We now have a picture of river biofilms bathed in solutions of DOC, in concentrations sometimes reaching tens of milligrams per liter; yet, they may still be limited by organic carbon because of its enzymatic recalcitrance. Clearly the degree of enzymatic recalcitrance of DOM is likely to vary according to the river and its associated catchment. However, the hydrolysis of DOM may also be influenced by component(s) in that organic matter pool. Freeman *et al.* (1990) showed that removal of organic matter >1000 AMW from river water perfusing river biofilms, frequently resulted in an increase in enzyme activity. This suggested that there may be substances within the DOM pool which act as enzyme inhibitors. Variations in the efficacy of such inhibition between rivers and through time could provide additional insights into the control of river ecosystem metabolism.

6.4 Influence of water movement

6.4.1 Nutrient uptake and metabolism

Early studies on this subject were conducted by Whitford and Schumacher working on filamentous algae (Whitford 1960; Whitford & Schumacher 1961; 1964; Schumacher & Whitford 1965). They found that increases in water velocities (0–18 cm s^{-1}) resulted in 43-fold increases in phosphorus uptake and two- to three-fold increases in respiration. Lock & John (1979) empirically determined whether such a phenomenon occurred with biofilms. They found under phosphorus-limiting conditions that a small increase in water velocity (0–5 cm s^{-1}) was sufficient to maximize the uptake of phosphorus (2–2.5 times the control). While a substantial increase had been noted, this was over an order of magnitude less than that observed by Whitford and Schumacher for filamentous algae. Lock & John (1979) concluded that this was probably due to differences in the flow regime in and around an algal tuft compared with those associated with the comparatively smooth surface of a biofilm. Thus increased water velocities served to maintain steep diffusion gradients to the outer filaments but also rendered more filament surface area available through stimulating greater water exchange within the tuft.

Riber & Wetzel (1987) conducted a study of boundary layer and internal diffusion effects on phosphorus fluxes into a lake biofilm. They altered water velocities between 0.01 and 1.0 cm s^{-1} under laminar flow conditions. In most cases they found that the phosphorus flux into the biofilm (measured using a tracer approach) was limited by boundary layer mass transfer. However, deviations from this pattern were noted, and attributed to biofilm patchiness and internal water flows within the biofilm matrix. This was demonstrated graphically in an associated dye visualization study (Figure 6.3). These findings parallel those of Whitford & Schumacher's (1961) concerning movement of water into versus across filamentous tufts. Furthermore, the recent findings of Lawrence *et al.* (1991) and Caldwell (personal communication, 1992) concerning "channels" within biofilms, highlights the need for critical studies of the comparative roles of diffusive versus advective transport of moieties into and out of river biofilms.

While studies are as yet few, there is little doubt that the chemical milieu within biofilms is very different from the bulk medium. Understanding is likely to come from the combination of fluid dynamic/diffusion studies, such as Riber & Wetzel (1987), and the waste-water engineering approaches of Harremoës (1978) and Ritman & McCarty (1978; 1981) concerning diffusion within biofilms and microbial flocs. Readers are referred to Cunningham (1989) for a general review of hydrodynamics and solute transport at the fluid–biofilm interface.

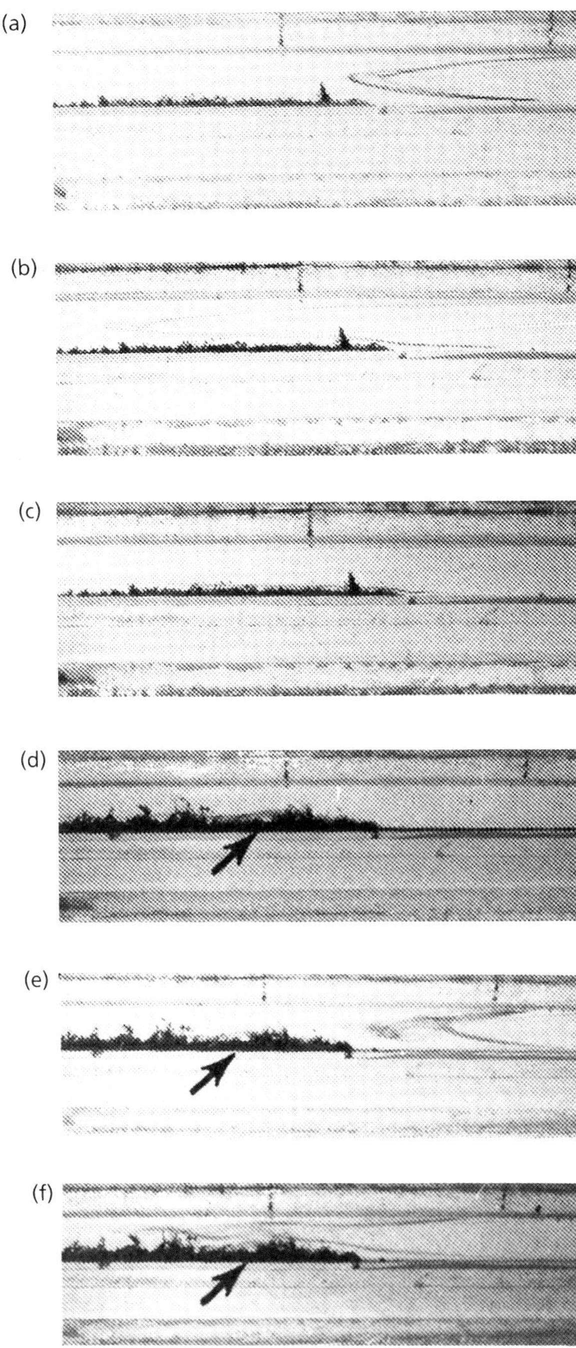

Figure 6.3 Visualization of laminar flow in laboratory flow chamber. (a–c) Three stages in the passage of a dye plume over a medium-biomass biofilm. (d) Dye creeping slowly into a mat after application at the bottom of the chamber, the arrow indicates dye trapped in a zone of almost stagnant water. (e,f) A plume passing over the same preparation while the "dead-zone" remains unaffected. Scale bar = 1 cm. From Riber & Wetzel (1987).

6.4.2 Taxonomic composition, biomass, and activity

Algal composition of stream biofilms has frequently been observed to exhibit a great deal of spatial variability. Experimental investigation of this phenomenon has been relatively recent. Stevenson (1990) showed that changes in current patterns around objects in flowing water can cause great changes in the composition of diatom assemblages. Those in sheltered areas were of higher abundance, more species, and lower evenness than assemblages exposed to direct current. It was also shown that the immigration rates of diatoms were greater in areas protected from the direct effect of stream flow. These general findings were supported by the study of Lamb & Lowe (1987) who observed that diatom densities were more than three times as dense in a slow current (15 cm s^{-1}) as in a fast current (40 cm s^{-1}). They also noted that the three-dimensional structures of the biofilm from the two flow regimes were different. The assemblages in the slow currents exhibited greater vertical height development (demonstrated with SEM). This growth habit had the effect of increasing the amount of substratum available for subsequent colonization and growth, resulting in an increase in community diversity.

The development of three-dimensional biofilms can be on such a scale that it can locally modify flow patterns. Reiter & Carlson (1986) used a laboratory flume to distinguish between what they described as "free water" and "local" velocities, the local velocity being that closely associated with the algal mat (0.7–3.7 mm above). In their study they set up three flumes with differing free-water velocities (71, 42, and 20 cm s^{-1}) and followed the colonization of the bed over ~6 weeks. At the start of the experiments, the "local velocities" were close to the free-water velocities, however, during the course of the experiment they diverged widely and by the end of the study the local velocities were all uniformly low (~10 cm s^{-1}). This was attributed to the friction associated with the algal mats. They conclude that mats growing in widely differing free-water velocity regimes may be experiencing similar velocity regimes at the mat surface. Although the mats under study were particularly thick (0.5–1.5 cm), the general observation may also hold with thinner biofilm assemblages.

While algae have been the major object of attention in flow studies, one recent paper has examined the behavior of bacteria in the boundary layer (Lawrence & Caldwell 1987). In a surface colonization study, they showed that four discrete types of cell movement occurred. These were designated as packing, spreading, shedding, and rolling. Each of these maneuvers was seemingly associated with a specific-species population. The packing maneuver resulted in a monolayer of continuous cells while spreading maneuvers resulted in a monolayer of adjacent cells (separated by 5–20 μm). In the shedding maneuver, mother cells attached perpendicular to the surface and shed daughter cells. In the rolling maneuver, cells were loosely attached and continuously somersaulted across the surface as they grew and divided. When flow was eliminated in the culture chamber, the packing colonies dispersed (a response to diffusion limitation?) while the other maneuvers were unaffected. They also noted that the highest growth rates were associated with the solitary maneuvers of rolling and shedding. The study provided a fascinating insight into a poorly understood aspect of biofilm dynamics and an approach that would be worth exploiting further.

6.4.3 *Effect of storm flows*

Although it is well documented that biomass removal occurs during storms (Tett *et al.* 1978; Rounik & Gregory 1981; Biggs & Close 1989) there is little information on the subsequent structure and activity of biofilms after such events. Blenkinsopp (1989) and Blenkinsopp & Lock (in press) artificially simulated storm flow using a high-velocity flume based upon a design by Vogel & La Barbera (1978). The response of biofilms to storm-flow velocities (237 $cm\,s^{-1}$) was compared with ambient-flow conditions (60 $cm\,s^{-1}$) for biofilm structure and metabolic activity.

A TEM study of the biofilms showed that exposure to storm flow conditions for 12 hours resulted in substantial changes in biofilm structure. Some of the diatoms of the biofilm (*Cocconeis* sp.) were damaged and the density of bacteria between the algal cells was greatly reduced. However, the storm-flow simulation had little impact upon the cyanobacteria. The corresponding analysis of metabolic activity using an electron transport system assay (Blenkinsopp & Lock 1990), showed that activity was substantially reduced (~50%) immediately after exposure to flood conditions in the flume. Recovery of these impacted biofilms was monitored over 24 days. Within 2 days the control and impacted biofilms showed no significant difference and remained so until conclusion of the study. While this was only a single study, it does suggest that river biofilms may be resilient to flooding events. This finding is supported by earlier studies on the artificial disruption of biofilms which showed a similar initial effect (Ladd *et al.* 1979).

In a study by Stevenson (1990), the impact of a natural storm event upon benthic algae was followed. Algae were monitored the day before the storm and for 11 days afterwards. An increase in density of live cells and a decrease in the proportions of sexually reproducing cells was noted. The benthic diatoms grew rapidly from 1.1×10^7 cells per centimeter to a perceived carrying capacity for the rocks of 5.2×10^7 cells per centimeter within only 8 days. This led Stevenson to suggest that flood scour has a positive effect upon diatoms, providing the stones of the river are not overturned. This provides further support for the potentially positive effect of flood scour on biofilm production.

6.5 Internal nutrient fluxes and cycling

6.5.1 Autotrophy

Autotrophic production can be chemosynthetic or photosynthetic. To my knowledge, information on potential chemosynthetic activity has not been directly reported. In production studies using $^{14}C\text{-}HCO_3$, the amount of carbon fixation in the dark is subtracted from the amount of carbon fixation in the light as a matter of course, but those former values do not get reported. As such, it is generally accepted that the amount of chemosynthetic production (dark carbon dioxide fixation) is a very small fraction of photosynthetic production. This contention is supported by studies from an arctic river (J.R. Vestal personal communication, 1987) and from northern temperate streams (C.M. Wood personal communication, 1992). If the dark fixation is indeed chemosynthesis and not an abiotic inorganic transformation, then an exploration of dark carbon dioxide fixation throughout the depth of the stream bed (hyporheic zone) may still be worthwhile.

Rates of photosynthetic primary production have been found to span five orders of magnitude (Lock 1981). In that brief survey, the carbon dioxide fixation ranged from 0.06 to $47\,\mu g\,C\,cm^{-2}\,hr^{-1}$. More recent studies have tended to measure primary production in the context of assessing the relative proportion of autotrophic versus heterotrophic metabolism. These studies have generally employed light–dark oxygen techniques (Naiman 1983; Bott *et al.* 1985; Edwards & Meyer 1987; King & Cummins 1989a,b) and showed that shifts between autotrophic/heterotrophic metabolism changed according to the extent of light occlusion by riparian trees. While such information is of great importance in the understanding of ecosystem level phenomena, a finer scale of study is necessary to explore cell–cell microbial interactions.

6.5.2 Algae–bacterial linkage

Haack & McFeters (1982) showed what appeared to be a direct link between the decline of a large epilithic phototrophic population and the increase in catabolic activity of the associated sessile bacteria. Here it was assumed that the bacteria responded to the lytic products of a declining algal population. Kaplan & Bott (1982) demonstrated the presence of diel fluctuations in DOC which they attributed to benthic algal excretion. Concentrations could increase during the course of a single day by as much as 40% of the daily minimum. Laboratory studies suggested that bacterial uptake could change both the concentration and composition, with the latter shifting toward high molecular weight substances through the proposed preferential consumption of low molecular weight products. More recently, Kaplan & Bott (1989) demonstrated a strong link between living epilithic algae and epilithic bacteria. They showed that bacteria associated with algae (light-grown) grew twice as fast as those bacteria grown in the dark. However, dark-grown bacteria were also stimulated by presumed algal DOC excretion changes as they modified the river-water DOC pool. The latter phenomenon was presumed to be a consequence of carbon recycling during transport downstream. They also made the interesting observation that bacteria associated with sediments were growing so slowly that they did not respond to either algal excretion from direct neighbors or from algal excretion products transported in the water. Further support for the linkage between epilithic algal production and bacterial production comes from Stock & Ward (1989). These authors studied biofilm development on old and new rock surfaces, finding similar activities over time with bacterial production up to 50% of algal production.

Thus there is a small but growing body of evidence which supports the hypothesis that algae and bacteria in biofilms are indeed closely linked (Lock 1981). Studies using some form of a pulse-chase procedure would be most useful in establishing an unequivocal metabolic link between these two biofilm components.

6.5.3 *Nutrient cycling*

While the potential for nutrient recycling in biofilms has long been recognized it is only recently that this topic has been addressed, albeit indirectly. In a study on metabolic attributes of biofilms, Paul & Duthie (1989) determined the spiraling length for several released products of metabolism. The spiral length is the distance traveled by a substance in a stream before assimilation by attached organisms of the stream bed (Newbold *et al.* 1981). Thus if a nutrient was at limiting concentrations one would expect these to be rapidly recycled, i.e., having a short spiral length, while nutrients in plentiful supply would be recycled less rapidly and hence have a long spiral length. In this study, biofilms grown on glass slides were incubated with radiolabeled bicarbonate, glucose, and phosphorus. The slides were then rinsed free of unincorporated substrate, placed at the top of an artificial stream channel and the amount of incorporation of released products by biofilms at 1, 2, and 3 m downstream was monitored. From these data the spiraling lengths were calculated. They observed that the spiraling length for phosphorus reduced from 30.4 m for an immature biofilm to 14.1 m for one that was well developed. Both of these spiraling lengths were noted as being short in comparison with literature values and hence were indicative of a vigorous recycling of phosphorus within the biofilm.

In a subsequent study Paul *et al.* (1991) examined spiraling lengths in two rivers with differing levels of phosphorus. They anticipated that a longer spiraling length would be associated with the river with the lower phosphate concentrations and associated lower biofilm biomass. However, spiraling lengths were found to be the same. These authors speculated that this was a consequence of increased efficiency of phosphorus uptake in the low biomass biofilm. Such an observation had also been made by McColl (1974) at the whole-stream level where a nutrient-poor stream assimilated added phosphate rapidly while, conversely, a nutrient-rich stream did not.

Paul *et al.* (1991) also showed that organic carbon released from biofilms was reassimilated by downstream biofilms. However, they did caution against the absolute extrapolation of the data, but it seems reasonable to conclude that in principle, an organic carbon recycling process had been demonstrated.

6.6 Conclusion

This overview of structure and functioning of attached microorganisms in rivers has been largely based upon epilithic biofilms. However, as pointed out in the introduction, all wetted surfaces are likely to support a biofilm. So, are the biofilms from other surfaces likely to be very different from epilithic ones? Most certainly the surface could influence the type of microorganisms present. For example, the epiphytic biofilms are likely to utilize the exudates from the underlying plant (Allanson 1973) or in the case of decaying leaves in streams, those biofilms may utilize the soluble products of leaf decomposition. Similarly biofilms overlying deep organic sediments could potentially assimilate the products of anaerobic decomposition. However, in broad structural and functional terms, I suspect that the biofilms from different surfaces will be convergent rather than divergent in their properties.

Thus, there is now a substantial body of evidence supporting the concept of a river biofilm as a potential assemblage of algae, bacteria, cyanobacteria, and, to a lesser extent, fungi. These organisms are held together within a matrix of polysaccharide, with the presence or absence of photosynthetic forms governed by light levels (Figure 6.4). The possibility has also recently emerged that the biofilm matrix may be criss-crossed with channels which may be exchanging water and associated solutes with the external milieu. However, there would also seem to be a need to now distinguish between a strongly adherent component and a loosely attached surface layer. The latter aspect may be particularly important to our understanding of biofilm metabolism, in that this layer may be very metabolically active. Clearly biofilms have the ability to take up and metabolize DOM but unequivocal evidence is still required to distinguish the biotic from abiotic uptake mechanisms. In addition, the role of POM has yet to

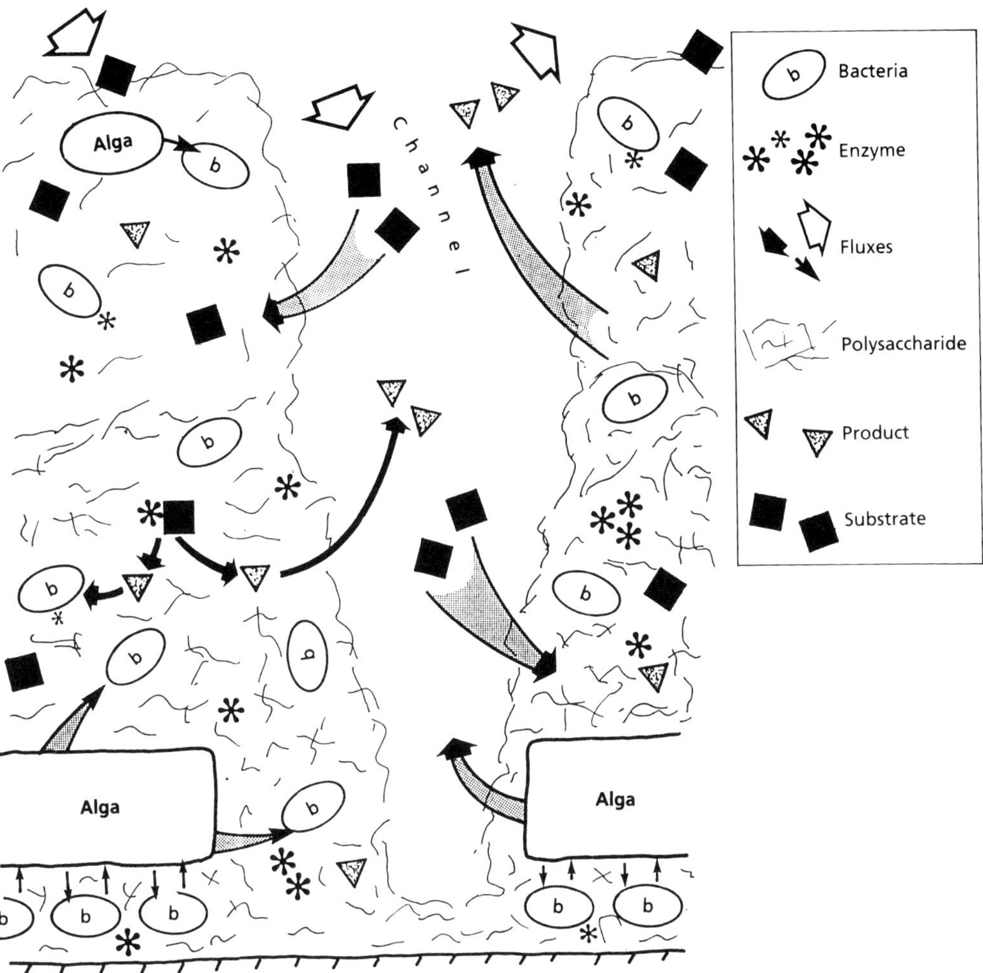

Figure 6.4 A synthesis of biofilm structure and functioning. Biofilms may now be seen to consist of microbes supported by a polysaccharide matrix potentially permeated by channels. These channels could act as conduits for organic and inorganic nutrients with diffusion processes being augmented by possible pumping of the biofilm through stream-water surging. Algae release photosynthate which may be taken up directly by bacteria or may leak from the biofilm into the overlying water. Bacteria secrete enzymes which may persist in the biofilms for an as yet undetermined period after the death of the secreting cell. The enzymes may hydrolyze substrates released from the algae or from DOM and POM trapped from the overlying water. The products from an enzymatic hydrolysis may then be taken up by microbial cells, sequestered by the polysaccharide matrix or diffuse out into the overlying stream water.

be seriously explored. Within the biofilm there is clear evidence of hydrolytic enzymes but nothing as yet is known of their half-lives. If enzymes are protected within a biofilm, and could thus "serve" many generations of microorganisms, then this would be a most interesting facet of biofilm ecology. Most recently, evidence has been obtained of the potential for biofilm exchanges of soluble, organic, and inorganic substances. These may be derived from *in situ* primary production, enzymatic hydrolysates or directly from the water column. It would seem that it is the quantification and understanding of the within-biofilm transfers and transductions which represent the challenge for the future.

Acknowledgments

I wish to thank Felix Baerlöcher and Louis Kaplan for valuable comments on a first draft of the manuscript. I wish to express my sincere thanks to Nerys Sanpher, Pat Kellett, and Joan Macrae for word processing.

References

Allanson B.R. (1973) The fine structure of the periphyton of *Chara* sp. and *Potamogeton natans* from Wytham Pond, Oxford and its significance to the macrophyte–periphyton metabolic model of R.G. Wetzel and H.C. Allen. *Freshwat. Biol.* **3**, 535–542.

Baerlöcher F. & Murdoch J.H. (1989) Hyporheic biofilms – a potential food source for interstitial animals. *Hydrobiologia* **184**, 61–67.

Biggs B.J.F. & Close M.E. (1989) Periphyton biomass dynamics in gravel bed rivers: the relative effects of flows and nutrients. *Freshwat. Biol.* **22**, 209–231.

Bilby R.E. & Likens G.E. (1980) Importance of organic debris dams in the structure and function of stream ecosystems. *Ecology* **61**, 1107–1115.

Blenkinsopp S.A. (1989) *The Ultra-Structure, Storage Products and Electron Transport System Activity of River Biofilms*. Unpublished PhD thesis, University of Wales.

Blenkinsopp S.A. & Lock M.A. (1990) The measurement of electron transport system activity in river biofilms. *Wat. Res.* **24**, 441–445.

Blenkinsopp S.A. & Lock M.A. (In press) The impact of stormflow on river biofilm architecture. *J. Phycol.*

Bott T.L., Brock J.T., Dunn C.S., Naiman R.J., Ovink R.W. & Petersen R.C. (1985) Benthic community metabolism in four temperate stream systems: an inter-biome comparison and evaluation of the river continuum concept. *Hydrobiologia* **123**, 3–45.

Burkholder J.M., Wetzel R.G. & Klomparens K.L. (1990) Direct comparison of phosphate uptake by adnate and loosely attached microalgae within an intact biofilm matrix. *Appl. Environ. Microbiol.* **56**, 2882–2890.

Burns R.G. (1978) *Soil Enzymes*. Academic Press, New York.

Burns R.G. (1989) Microbial and enzymic activities in soil biofilms. In Characklis W.G. & Wilderer P.A. (eds) *Structure and Function of Biofilms*. John Wiley & Sons, New York.

Costerton J.W., Cheng K.J., Geesey G.G., Ladd T.I., Nickel J.C., Dasgupta M. & Marrie T.J. (1987) Bacterial biofilms in nature and disease. *Ann. Rev. Microbiol.* **41**, 435–464.

Cummins K.W., Klug J.J., Wetzel R.Ge, Petersen R.C., Suberkropp K.F., Manny B.A., Wuychechk J.C. & Howard F.O. (1972) Organic enrichment with leaf leachate in experimental lotic ecosystems. *BioScience* **22**, 719–722.

Cummins K.W., Sedell J.R., Swanson F.J., Minshall G.W., Fisher S.J., Cushing C.E., Petersen R.C. & Vannote R.L. (1983) Organic matter budgets for stream ecosystems: problems in their evaluation. In Barnes J.R. & Minshall G.W. (eds) *Stream Ecology, Application and Testing of General Ecological Theory*. Plenum, New York.

Cunningham A.B. (1989) Hydrodynamics and solute transport at the fluid–biofilm interface. In Characklis W.G. & Wilderer P.A. (eds) *Structure and Function of Biofilms*, pp. 19–31. John Wiley & Sons, New York.

Dahm C.N. (1981) Pathways and mechanisms for removal of dissolved organic carbon from leaf leachates in streams. *Can. J. Fish. Aquat. Sci.* **38**, 68–76.

De Flaun M.F. & Mayer L.M. (1983) Relationships between bacteria and grain surfaces in intertidal sediments. *Limnol. Oceanogr.* **28**, 873–886.

Edwards R.T. & Meyer J.L. (1987) Metabolism of a sub-tropical low gradient blackwater river. *Freshwat. Biol.* **17**, 251–263.

Fazio S.A., Uhlinger D.J., Parker J.H. & White D.C. (1982) Estimations of uronic acids as quantitative measures of extracellular and cell wall polysaccharide polymers from environmental samples. *Appl. Environ. Microbiol.* **43**, 1151–1159.

Fiebig D.M. (1988) *Riparian Zone and Stream Water Chemistries and Organic Matter Immobilisation at the Stream-Bed Interface.* Unpublished PhD thesis, University of Wales.

Fiebig D.M. & Lock M.A. (1991) Immobilisation of dissolved organic matter from groundwater discharging through the stream bed. *Freshwat. Biol.* **26**, 45–55.

Ford T.E. & Lock M.A. (1987) Epilithic metabolism of dissolved organic carbon in boreal forest rivers. *FEMS Microb. Ecol.* **45**, 89–97.

Freeman C. (1990) *The Role of High and Low Molecular Weight Organic Substances in Biofilm Metabolism*, p. 174. Unpublished PhD thesis, University of Wales.

Freeman C. & Lock M.A. (1992) Recalcitrant high molecular weight material, an inhibitor of microbial metabolism in river biofilms. *Appl. Environ. Microbiol.* **58**, 2030–2033.

Freeman C., Lock M.A., Marxsen J. & Jones S.E. (1990) Inhibitory effects of high molecular weight dissolved organic matter upon metabolic processes in biofilms from contrasting rivers and streams. *Freshwat. Biol.* **24**, 159–166.

Geesey G.G., Richardson W.T., Yeomans H.G., Irvin R.T. & Costerton J.W. (1977) Microscopic examination of natural sessile bacterial populations from an alpine stream. *Can. J. Microbiol.* **23**, 1733–1736.

Geesey G.G., Mutch R., Costerton J.W. & Green R.B. (1978) Sessile bacteria: an important component of the microbial population in small mountain streams. *Limnol. Oceanogr.* **23**, 1214–1223.

Haack T.K. & McFeters G.A. (1982) Microbial dynamics of an epilithic mat community in a high alpine stream. *Appl. Environ. Microbiol.* **43**, 702–707.

Harremoës P. (1978) Biofilm kinetics. In Mitchell R. (ed.) *Water Pollution Microbiology II.* John Wiley & Sons, New York.

Jones S.E. & Lock M.A. (1989) Hydrolytic enzyme activity in the biofilms from two contrasted streams. *Freshwat. Biol.* **22**, 289–296.

Kaplan L.A. & Bott T.L. (1982) Diel fluctuations of DOC generated by algae in a piedmont stream. *Limnol. Oceanogr.* **27**, 1091–1100.

Kaplan L.A. & Bott T.L. (1983) Microbial heterotrophic utilisation of dissolved organic matter in a piedmont stream. *Freshwat. Biol.* **13**, 363–377.

Kaplan L.A. & Bott T.L. (1985) Acclimation of stream-bed heterotrophic microflora: metabolic responses to dissolved organic matter. *Freshwat. Biol.* **15**, 479–492.

Kaplan L.A. & Bott T.L. (1989) Diel fluctuations in bacterial activity on stream-bed substrata during vernal algal blooms: effects of temperature, water chemistry and habitat. *Limnol. Oceanogr.* **34**, 718–733.

Karlstrom V. (1978) Role of the organic layer on stones in detrital metabolism in streams. *Verh. Int. Verein. Limnol.* **20**, 1463–1470.

King D.K. & Cummins K.W. (1989a) Autotrophic–heterotrophic community metabolism relationships of a woodland stream. *J. Freshwat. Ecol.* **5**, 205–218.

King D.K. & Cummins K.W. (1989b) Factors affecting autotrophic–heterotrophic relationships of a woodland stream. *J. Freshwat. Ecol.* **5**, 219–230.

Korte V.L. & Blinn D.W. (1983) Diatom colonisation on artificial substrata in pool and riffle zones studied by light and scanning electron microscopy. *J. Phycol.* **19**, 332–341.

Ladd T.I., Costerton J.W. & Geesey G.G. (1979) Determination of the heterotrophic activity of epilithic microbial populations. In Costerton J.W. & Colwell R.R. (eds) *Native Aquatic Bacteria: Enumeration, Activity and Ecology*, pp. 180–195. American Society for Testing and Materials STP 695, Philadelphia.

Lamb M.A. & Lowe R.L. (1987) Effects of current velocity on the physical structuring of diatom (Bacillariophyceae) communities. *Ohio J. Sci.* **87**, 72–78.

Lawrence J.R. & Caldwell D.E. (1987) Behaviour of bacterial stream populations within the hydrodynamic boundary layers of surface microenvironments. *Microb. Ecol.* **14**, 15–27.

Lawrence J.R., Korber D.R., Hoyle B.D., Costerton J.W. & Caldwell D.E. (1991) Optical sectioning of microbial biofilms. *J. Bact.* **173**, 6558–6567.

Leppard G.G. (1986) The fibrillar matrix component of lacustrine biofilms. *Wat. Res.* **20**, 697–702.

Lock M.A. (1981) River epilithon (biofilm) – a light and organic energy transducer. In Lock M.A. & Williams D.D. (eds) *Perspectives in Running Water Ecology*, pp. 3–40. Plenum, New York.

Lock M.A. (1990) The dynamics of dissolved and particulate organic material over the substratum of water bodies. In Wotton R.S. (ed.) *The Biology of Particles in Aquatic Ecosystems*. CRC Press, Boca Raton.

Lock M.A. & Hynes H.B.N. (1976) The fate of "dissolved" organic carbon derived from autumn-shed maple leaves (*Acer saccharum*) in a temperate hard-water stream. *Limnol. Oceanogr.* **21**, 436–443.

Lock M.A. & John P.H. (1979) The effect of flow patterns on uptake of phosphorus by river periphyton. *Limnol. Oceanogr.* **24**, 376–383.

Lock M.A. & Ford T.E. (1983) An inexpensive flow microcalorimeter for measuring the heat production of attached and sedimentary aquatic microorganisms. *Appl. Environ. Microbiol.* **46**, 463–467.

Lock M.A. & Ford T.E. (1985) Microcalorimetric approach to determine relationships between energy supply and metabolism in river epilithon. *Appl. Environ. Microbiol.* **49**, 408–412.

Lock M., Wallace R.R., Costerton J.W., Ventullo R.M. & Charlton S.E. (1984) River epilithon (biofilm): toward a structural functional model. *Oikos.* **42**, 10–22.

Lock M.A., Ford T.E., Fiebig D.M. & Miller M.R. (under review) The role of photolysis in an arctic river.

Low C.F.S. & White D.C. (1989) Regulation of external polymer production in benthic microbial communities. In Cohen Y. & Rosenberg E. (eds) *Microbial Mats – Physiological Ecology of Benthic Microbial Communities*. American Society of Microbiology, Washington DC.

Lush D. & Hynes H.B.N. (1978) The uptake by dissolved organic matter by a small spring stream. *Hydrobiologia* **60**, 271–275.

Luttenton M.L., Vansteenberg J.B. & Rada R.G. (1986) Phycoperiphyton in selected reaches of the upper Mississippi River: community composition, architecture and productivity. *Hydrobiologia* **136**, 31–46.

Madsen B.L. (1972) Detritus on stones in small streams. *Mem. Inst. Ital. Idrobiol.* **29** (Suppl.), 385–403.

Marxsen J. & Witzel K.P. (1990) Measurement of exoenzymatic activity in streambed sediments using methylumbelliferyl-substrates. *Arch. Hydrobiol. Beih. Ergebn. Limnol.* **34**, 21–28.

McArthur J.V. & Marzolf G.R. (1986) Interactions of the bacterial assemblage of a prairie stream with dissolved organic carbon from riparian vegetation. *Hydrobiologia* **134**, 192–199.

McColl R.H.S. (1974) Self-purification of small freshwater streams: phosphate, nitrate and ammonia removal. *N.Z. J. Mar. Freshwat. Res.* **8**, 375–388.

McDowell W.H. (1985) Kinetics and mechanisms of dissolved organic carbon retention in a head-water stream. *Biogeochemistry* **1**, 329–352.

Meyer J.L. (1990) A black water perspective on riverine ecosystems. *BioScience* **40**, 643–651.

Naiman R.J. (1983) The annual pattern and spatial distribution of aquatic oxygen metabolism in boreal forest watersheds. *Ecol. Monogr.* **53**, 73–94.

Naiman R.J. & Melillo J.M. (1984) Nitrogen budget of a sub-arctic stream altered by beaver (*Castor canadensis*). *Oecologia* **62**, 150–155.

Newbold J.D., Elwood J.W., O'Neil R.V. & Van Winkle W. (1981) Measuring nutrient spiralling in streams. *Can. J. Fish. Aquat. Sci.* **38**, 860–863.

Oppenheim D.R. & Paterson D.M. (1990) The fine structure of an algal mat from a freshwater maritime Antarctic lake. *Can. J. Bot.* **68**, 174–183.

Paul B.J. & Duthie H.C. (1989) Nutrient cycling in the epilithon of running waters. *Can. J. Bot.* **67**, 2302–2309.

Paul B.J., Duthie H.C. & Taylor W.D. (1991) Nutrient cycling by biofilms in running waters of differing nutrient status. *J. N. Am. Benthol. Soc.* **10**, 31–41.

Reiter M.A. & Carlson R.E. (1986) Current velocity in streams and the composition of benthic algal mats. *Can. J. Fish. Aquat. Sci.* **43**, 1156–1162.

Riber H.H. & Wetzel R.G. (1987) Boundary layer and internal diffusion effects on phosphorus fluxes in lake periphyton. *Limnol. Oceanogr.* **32**, 1181–1194.

Ritman B.E. & McCarty P.L. (1978) Variable order-model of bacterial film kinetics. *ASCE J. Environ. Eng. Div.* **104**, 889–900.

Ritman B.E. & McCarty P.L. (1981) Substrate flux into biofilms of any thickness. *ASCE J. Environ. Eng. Div.* **107**, 831–849.

Roemer S.C., Hoagland K.D. & Rosowski J.R. (1984) Development of a freshwater periphyton community as influenced by diatom mucilages. *Can. J. Bot.* **62**, 1779–1813.

Rounik J.S. & Gregory S.V. (1981) Temporal changes in periphyton standing crop during an unusually dry winter in the streams of the Western Cascades, Oregon. *Hydrobiologia* **83**, 197–205.

Rounik J. & Winterbourn M.J. (1983) Formation structure and utilisation of stone surface organic layers in 2 New Zealand streams. *Freshwat. Biol.* **13**, 57–72.

Schumacher G.J. & Whitford L.A. (1965) Respiration and ^{32}P uptake in various species of freshwater algae as affected by a current. *J. Phycol.* **1**, 78–80.

Sinsabaugh R.L. & Linkins A.E. (1988) Exoenzyme activity associated with lotic epilithon. *Freshwat. Biol.* **20**, 249–261.

Sinsabaugh R.L., Golladay S.W. & Linkins A.E. (1991) Comparison of epilithic and epixylic biofilm development in a boreal river. *Freshwat. Biol.* **25**, 179–187.

Smith G.A., Nickels J.S., Kerger B.D., Davis J.D., Collins S.P., Wilson J.T., McNabb J.F. & White D.C. (1986) Quantitative characterization of microbial biomass and community structure in subsurface material: a prokaryotic consortium responsive to organic contamination. *Can. J. Microbiol.* **32**, 104–111.

Stevenson R.J. (1990) Benthic algal community dynamics in a stream during and after a spate. *J. N. Am. Benthol. Soc.* **9**, 277–288.

Stock M.S. & Ward A.K. (1989) Establishment of a bed-rock epilithic community in a small stream: microbial (algal and bacterial) metabolism and physical structure. *Can. J. Fish. Aquat. Sci.* **46**, 1874–1883.

Telang S.A. & Hodgson G.W. (1983) Fate of organic compounds in mountain streams of the Marmot Basin: a process reactor model. *Hydrobiologia* **98**, 185–198.

Tett P., Gallegos C., Kelly M.G., Hornberger G.M. & Cosby G.J. (1978) Relationships among substrate, flow and benthic microalgal pigment density in the Mechums River, Virginia. *Limnol. Oceanogr.* **3**, 785–797.

Thurman E.M. (1985) *Organic Geochemistry of Natural Waters.* Nijhoff/Junk, Dordrecht.

Triska F.J. (1984) Role of wood debris in modifying channel geomorphology and riparian areas of a large lowland river under pristine conditions: An historical case study. *Verh. Int. Verein Theor. Angew. Limnol.* **22**, 1876–1892.

Vogel S. & La Barbera (1978) Simple flow tanks for research and teaching. *BioScience* **28**, 638–643.

Wallis P.M., Hynes H.B.N. & Telang S.A. (1981) The importance of groundwater in the transportation of allochthonous dissolved organic matter to the streams draining a small mountain basin. *Hydrobiologia* 79, 77–84.

Ward G.M., Ward A.K., Dahm C.N. & Aumen N.G. (1990) Origin and formation of organic and inorganic particles in aquatic systems. In Wotton R.S. (ed.) *The Biology of Particles in Aquatic Systems*, pp. 37–56. CRC Press, Boca Raton.

Watling L. (1988) Small-scale features of marine sediments and their importance to the study of deposit feeding. *Mar. Ecol. Prog. Ser.* **47**, 135–144.

Wetzel R.G. (1975) *Limnology*. W.B. Saunders, New York.

Wetzel R.G. (1983) *Periphyton of Freshwater Ecosystems*. Nijhoff/Junk, New York.

Whitford L.A. (1960) The current effect and growth of freshwater algae. *Trans. Am. Microsc. Soc.* **79**, 302–309.

Whitford L.A. & Schumacher G.J. (1961) Effect of current on mineral uptake and respiration by freshwater algae. *Limnol. Oceanogr.* **6**, 423–425.

Whitford L.A. & Schumacher G.J. (1964) Effect of a current on respiration and mineral uptake in *Spirogyra* and *Oedogonium*. *Ecology* **45**, 168–170.

Williams D.D. (1984) The hyporheic zone as a habitat for aquatic insects and associated arthropods. In Resh V.H. & Rosenberg D.M. (eds) *The Ecology of Aquatic Insects*. Praeger, New York.

Wotton R.S. (1990) Particulate and dissolved organic materials as food. In Wotton R.S. (ed.) *The Biology of Particles in Aquatic Systems*, pp. 213–262. CRC Press, Boca Raton.

7

Louis A. Kaplan & J. Denis Newbold

Biogeochemistry of dissolved organic carbon entering streams

The slope of the valley, the depth and permeability of the soil, and the patterns of precipitation greatly affect the pattern of flow in a stream, which is a prime ecological factor; and more subtly, they effect the dissolved content of water. (H.B.N. Hynes, 1975)

7.1 Introduction

Hynes (1975) suggested that advances in forest hydrology and soil science should change the way that stream ecologists view streams. He argued for expanding the context of stream ecology to encompass processes that occur below, beside, and upslope of the stream channel. He pointed to the influence of these processes on water chemistry and proposed that the hydrological pathways through which water reaches streams have a major influence on stream-water chemistry. His remarks also emphasized the dominance of heterotrophy in stream metabolic processes, and in this context he again focused attention on terrestrial processes, observing that "most of the dissolved organic matter goes into solution in or on the soil, which is where most of the litter falls rather than into the stream itself."

Advances in stream ecosystem theory and research within the last decade have underscored the insightful nature of Hynes' comments. The temporal and spatial boundaries of stream ecosystems have been expanded (Minshall 1988; Ward 1989), with spatial boundaries pushed laterally upslope through the riparian zone and floodplain into the uplands, and vertically down and laterally out into the hyporheos, the interface between the stream channel and the ground water (Triska *et al.* 1989). This broader view of stream ecosystems, emphasizing dynamic terrestrial/aquatic linkages, has focused attention on the quantitative and qualitative importance of watershed sources of dissolved organic carbon (DOC) to metabolism in streams (Dahm *et al.* 1987; Meyer *et al.* 1987; Naiman *et al.* 1987; Marxsen 1988; Ford & Naiman 1989; Meyer 1990; Meyer & Edwards 1990), the influence of soil processes (Dawson *et al.* 1981; McDowell & Wood 1984), and hydrological pathways (McDowell & Likens 1988; Triska *et al.* 1989; Fiebig *et al.* 1990) on water chemistry, and the contribution of hyporheic heterotrophy to ecosystem metabolism (Grimm & Fisher 1984) and nutrient cycling (Duff & Triska 1990; Valett *et al.* 1990).

Organic carbon dynamics are central to stream ecosystem studies because they link energy flow and nutrient cycling. DOC may contribute 30–75% of the total energy input into stream ecosystems (Fisher & Likens 1973; Fisher 1977; Mulholland 1981). Significant strides have been made towards understanding DOC sources and processing in streams:

1 bacteria readily metabolize leaf leachates with uptake rates which can be ascribed to slow and fast components (Cummins *et al.* 1972; Wetzel & Manny 1972; McDowell & Fisher 1976);

2 stream beds are the primary site of DOC processing in small streams (Lock & Hynes 1976; Dahm 1981);

3 specific molecular components of the DOC pool are selectively utilized (Kaplan & Bott 1983; Moran & Hodson 1989);

4 DOC in transport can support a large component of benthic bacterial metabolism (Bott *et al.* 1984);

5 bacterial uptake and growth efficiency are dependent upon the source of DOC (Kaplan & Bott 1983; Edwards & Meyer 1987; Findlay *et al.* 1987);

6 stream-bed heterotrophic bacteria can respond rapidly to algal generated diel changes in DOC quantity and quality (Kaplan & Bott 1989);

7 high molecular weight fractions of DOC from black waters can support growth of suspended bacteria in some rivers (Meyer *et al.* 1987), but inhibit bacterial activity on biofilms in others (Ford & Lock 1987; Freeman *et al.* 1990);

8 anaerobic processes and wetlands can contribute high concentrations of DOC to streams (Dahm *et al.* 1987; Naiman *et al.* 1987; Ford & Naiman 1988; Eckhardt & Moore 1990);

9 abiotic and biotic mechanisms are important in controlling the movement of DOC from the watershed to streams (McDowell & Likens 1988).

Despite these advances, there is still only a very poor understanding of how much DOC actually enters streams and rivers and, concomitantly, how much DOC is utilized within these ecosystems. Stream energy budget studies (Fisher & Likens 1973; Naiman 1976; Fisher 1977; Mulholland 1981; Richey *et al.* 1990) have not rigorously quantified subsurface inputs or directly estimated DOC utilization. A recent energy budget of the Amazon River (Richey *et al.* 1990) found that inputs of total organic carbon (TOC) – dissolved plus particulate – including autochthonous production, were only about half that required to account for respiration, indicating that inputs of either DOC or particulate organic carbon (POC), but probably both, were seriously underestimated. These difficulties in budgeting DOC reflect unanswered questions in three major areas:

1 the sources of terrestrially derived DOC entering streams under base flow and storm flow condition;

2 the biological lability and molecular composition of DOC from terrestrial sources (Hynes 1983; Meyer *et al.* 1988);

3 the influence of heterotrophy within the riparian zone and the hyporheos on DOC dynamics in streams (Rutherford & Hynes 1987; Meyer 1988; Fiebig *et al.* 1990).

The implications of these questions extend well beyond the energy dynamics of the stream itself. Uncertainties regarding inputs to streams are simultaneously uncertainties regarding processes in, and losses from, the terrestrial environment. Downstream implications may be even more significant. Deuser (1988) pointed out the delivery of DOC to the oceans by rivers is sufficient to account for much of the oceans' DOC and may account for substantial heterotrophy within the oceans. The riparian–hyporheic interface, therefore, might be thought of as a controlling gateway of potentially major significance.

In this chapter, we review the literature on DOC concentrations in ground water and stream water and describe a conceptual model of the mechanisms controlling the quantity and quality of DOC moving from the landscape through the riparian–hyporheic interface. Our initial focus is the presentation of a simple mass-balance model to explain the concentrations of DOC in stream waters. We then develop a more general model that recognizes that the mix of terrestrial inputs involves not only variation in concentration but variation in quality.

7.1.1 *Hypotheses*

Conflicting evidence available on concentrations of DOC in ground waters and stream waters has been segregated into two categories dubbed the "dilution" and "stripping" hypotheses by Rutherford & Hynes (1987). The stripping hypothesis, suggested by Wallis (1981) and Wallis *et al.* (1981), and elaborated by Hynes (1983), views the hyporheos as the primary site of processing for terrestrially derived DOC, and considers the processing to be primarily biological. Under this hypothesis, streams are supplied by ground water that is high in DOC, but much of the DOC is removed in the hyporheos, yielding stream water that is low in DOC. The primary data supporting the stripping hypothesis are derived from mass-balance calculations.

The dilution hypothesis derives from a series of observations that groundwater DOC concentrations are lower than stream-water DOC concentrations and thus "dilute" DOC concentrations in the stream (Rutherford & Hynes 1987). Our perspective is that the dilution hypothesis is only a statement of differences between groundwater and stream-water concentrations; it is not a mechanistic alternative to the stripping hypothesis and does not address the question of sources of stream-water DOC. Below we present a more general model which is derived from the dilution hypothesis, extending beyond the observations to a mechanism for generating the differences between groundwater and stream-water DOC concentrations. The terrestrial source/riparian–hyporheic interface model consists of three basic components:

1 stream-water DOC is largely determined as a mix of DOC from terrestrial sources with high and low concentrations;

2 the hydrological, geochemical, and biochemical factors that determine the concentration of DOC in terrestrial sources, also determines the quality of DOC, especially its biological lability;

3 stream-water DOC concentrations may be further altered by passage of terrestrial sources through the riparian and hyporheic zones.

7.2 Dissolved organic carbon sources and sinks

7.2.1 *Terrestrial sources*

Ground water derived from flow in bedrock fractures is generally low in DOC. A survey of ground water from five aquifer types throughout the continental USA, found that the median concentration for all aquifers ranged from 0.5 to 0.7 $mg\,l^{-1}$ (Leenheer *et al.* 1974). These concentrations are less than most base flow concentrations for surface waters (Malcolm & Durum 1976), and in fact, most studies of DOC in ground water and stream water have shown that groundwater DOC concentrations, usually sampled from surface springs, are equal to or less than stream-water concentrations (Fisher & Likens 1973; Baker *et al.* 1974; McDowell & Fisher 1976; Comiskey 1978; Kaplan *et al.* 1980; Wallace *et al.* 1982; Meyer & Tate 1983).

Soil water, in contrast to ground water, is often high in DOC. Studies of water in saturated and unsaturated soil zones have been conducted with wells and lysimeters. The investigations of Comiskey (1978), Meyer & Tate (1983), and Fiebig *et al.* (1990) indicated that concentrations of DOC in soil water exceeded stream-water concentrations. However, the concentrations decline with increasing depth in the soil column (Dawson *et al.* 1981; Wallis *et al.* 1981; McDowell & Wood 1984; Cronan & Aiken 1985; Hornung *et al.* 1990). McDowell & Wood (1984) observed that DOC concentrations declined sharply through the A and B horizons of a well-drained podzol, and that concentrations at the bottom of the B horizon closely approximated those of the stream water. They (as did McDowell & Likens 1988) argued that the terrestrial ecosystem is an efficient scavenger of carbon, and that abiotic adsorption of DOC–iron complexes onto mineral soils (podzolization) was the primary mechanism controlling the concentration of DOC in water reaching the stream channel. McDowell & Likens (1988) noted that terrestrial processing can be "short circuited," such as when storms alter flow pathways and deliver high concentrations of DOC to the stream, but they considered these events to be transient pulses which are rapidly adsorbed by stream-bed sediments, and subsequently metabolized by the stream microflora (McDowell 1985).

Where stream-water DOC exceeds that of ground water it is likely that the additional DOC comes from soil water that has not been extensively processed. This source includes DOC draining directly into the stream from near-stream zones of saturated soils that typically make up the "partial contributing areas" or "variable source areas" that generate overland flow during storms (Dunne & Black 1970; Gburek 1990), and DOC that moves to the stream by translatory downslope flow from unsaturated soils when soil moisture exceeds field capacity (Hewlett & Hibbert 1967; Harr 1977; Wilson *et al.* 1990). Hemond (1990), citing evidence from seasonal changes of DOC concentrations in streams, the existence of low-DOC surface waters, and an inverse relationship of DOC and sulfate concentrations, has suggested that in small glaciated catchments, riparian wetlands are the dominant source of DOC to streams rather than upland sources which would require much longer flow paths and have increased opportunities for sorption. In watersheds with extensive riparian wetlands, it is fairly clear that soil-water DOC is important. Stream water in these watersheds is higher in DOC than in other watersheds (Brinson 1976; Mulholland & Kuenzler 1979), and the levels correlate with the extent of wetlands (Eckhardt & Moore 1990). On the other hand, high stream-water concentrations under base flow conditions have been reported for a subcatchment of Harp Lake, a boreal drainage basin without wetlands (Schiff *et al.* 1990). Schiff *et al.* (1990) speculated that the DOC came from either an in-stream source or a riparian organic-rich layer through which the ground water percolated. In the Panola Mountain drainage, an unglaciated Georgia Piedmont catchment, Hooper *et al.* (1990) used end-member mixing analysis to calculate that upland soil water contributed at least 10%, and often as much as 40%, to streamflow during nonstorm periods. Both the Harp Lake and the Panola Mountain studies emphasized the influence of seasonally wet or dry periods on inputs of terrestrially derived DOC to streams.

Another line of evidence for direct inputs of either wetland or upland soil-water DOC comes from the behavior of stream-water DOC concentrations during and after storms. There are numerous reports of DOC concentration increases during storms, including Grieve (1984), McDowell & Likens (1988), and Sedell & Dahm (1990). Soil-water DOC probably contributes to this increase (Hynes 1983; McDowell & Likens 1988; Fiebig *et al.* 1990; Mulholland *et al.* 1990), but this has not been directly measured or demonstrated. Evidence that soil water is important comes from the pattern of DOC decline after storms: in the White Clay Creek, stream-water DOC concentration declines more slowly than discharge and continues to decline gradually until the next storm (Larson 1978). Similar patterns have been observed in other streams (Comiskey 1978; McDowell & Likens 1988). The less-than-proportionate decline would be predicted if most of the storm flow were to originate from soil water with higher concentrations than the ground water, which dominates at base flow. The extended period of decline indicates that soil water may contribute substantially to the DOC even as its contribution to flow becomes small. Mixing of soil-water DOC with ground water may occur in the near-stream shallow ground water, rather than in the stream itself.

With the inclusion of soil-water sources, stream-water DOC concentrations might be explained as a simple mixing of groundwater sources from the soils and bedrock. Under this conceptual terrestrial source model, stream-water DOC is terrestrially controlled via two primary factors:

1 the relative DOC concentrations of the source pools;

2 the relative hydrological contribution of the respective pools to stream flow.

This model differs from the terrestrial processing model presented above (McDowell & Wood 1984; McDowell & Likens 1988) in that the input of soil-water DOC to streams is viewed as a continuous, though modulating, phenomenon, rather than a transient pulse associated only with storm flows. In other words, expanding and contracting soil-water inputs during both base flow and storm flow periods are layered on top of more constant bedrock groundwater inputs.

7.2.2 *In-stream sources*

Some stream-water DOC is generated within the stream itself. One in-stream source is extracellular release from primary producers. In White Clay Creek in the Pennsylvania Piedmont, DOC concentrations in spring seeps at base flow are generally $<700\,\mu g\,l^{-1}$, and can be as low as $300\,\mu g\,l^{-1}$ (Kaplan *et al.* 1980). Recently we have measured similarly low-DOC concentrations in ground water taken from shallow wells in well-drained soils within the riparian zone (Kaplan 1992). Stream-water DOC at base flow typically exceeds these groundwater concentrations by 1.0–1.5 mg $C\,l^{-1}$ (Kaplan *et al.* 1980). Approximately 0.5 mg $C\,l^{-1}$ of this difference can be accounted for by release of DOC from primary production in the *aufwuchs*, but only at times of peak primary production that last for periods of a few days to at most several weeks in the intervals between scouring storm flows (Kaplan & Bott 1982). Another in-stream source is biotic and abiotic degradation of stored POM (Meyer *et al.* 1988; Moran & Hodson 1989), which could involve both autochthonous or allochthonous forms. In the White Clay Creek, a production of 0.1% d^{-1} from the 126 $g\,m^{-2}$ pool of benthic POC (Minshall *et al.* 1983) would account for 46 $g\,C\,m^{-2}\,yr^{-1}$, a value in the same range as the 25 $g\,C\,m^{-2}\,yr^{-1}$ estimated for Creeping Swamp, North Carolina (Mulholland 1981). Although intended to be conservatively high, this DOC release would account for only 0.4 $mg\,l^{-1}$ of DOC in White Clay Creek, leaving 0.6–1.1 $mg\,l^{-1}$ unaccounted for, except during peak production periods. In addition, a portion of these releases may be degraded within the stream, further reducing the net effect on concentrations.

7.2.3 Soils as dissolved organic carbon sinks

The podzolization hypothesis is a mechanistic alternative to the stripping hypothesis, claiming that DOC from terrestrial sources is processed within the terrestrial environment rather than in the stream. High densities of microorganisms in the A soil horizon are expected to degrade highly labile DOC (McDowell & Wood 1984), but geochemical immobilization in the B horizon extends the residence time of less-labile DOC within the soil column, perhaps by several decades, permitting biochemical oxidation to occur (McDowell & Likens 1988). This oxidation, in turn would regenerate adsorption sites on the soils leading to a steady state between inputs and removal. From the isotherms of McDowell & Wood (1984), we estimate that sorption capacity is sufficient to maintain efficient removal of DOC, provided that oxidation occurs at ~1% yr^{-1}. Rates of oxygen uptake by different horizons in a podzol under *Pinus sylvestris* suggest that oxidation exceeds 1% yr^{-1} (Swift *et al.* 1979).

7.2.4 Hyporheos as a dissolved organic carbon sink

A very different conception of the fate of terrestrially derived DOC arises from studies of watersheds where groundwater DOC exceeds that of the stream. Wallis *et al.* (1981) measured high concentrations of DOC not only in the unsaturated soil zone, but also in the saturated zone. Other reports of DOC concentrations in ground waters exceeding those in stream waters include watersheds in the boreal forest (Ford & Naiman 1989), old growth redwood forests

(Duff & Triska 1990), spruce forests in Wales (Fiebig *et al.* 1990), and the New Jersey Pinelands (Lord *et al.* 1990). This pattern is the basis for the view that ground water is stripped of its high DOC load as it passes up through the stream bed (Wallis 1981; Wallis *et al.* 1981; Hynes 1983). According to the stripping hypothesis, groundwater DOC is an important source of organic carbon for the stream ecosystem, and stream-water DOC concentrations are controlled by biological processes in the hyporheos rather than biogeochemical processes in the terrestrial environment.

We are skeptical regarding the stripping hypothesis, both because the mechanism lacks plausibility and because the evidence supporting it is weak. Wallis (1981) explained stripping in terms of differences between the terrestrial and stream subsurface environment in the Marmot Basin, Alberta. Low temperature, redox potential (Eh), oxygen, and possibly low nutrient levels would have slowed microbial degradation in the ground water, whereas the well-aerated, warmer, and more nutrient-rich environment of the hyporheos would have stimulated DOC utilization. The temperature argument would not hold during the relatively long winters in Alberta, Canada, but that point aside, the notion that the hyporheos is a metabolic dynamo fueled by ground water DOC must be examined more closely. Flow paths of ground water in a watershed converge at the stream. This focusing of watershed ground water into a relatively small zone, even if the stream bed were expanded several-fold by the inclusion of a large hyporheos, means that exceedingly efficient geochemical immobilization and high rates of biochemical processing would be necessary to remove the DOC from ground water during its transit through the stream bed. This also implies either that a substantial fraction of the DOC reaching the stream is sufficiently labile to be degraded within a relatively short residence time in the hyporheos, i.e., a few hours to a few weeks, or that the hyporheos is capable of adsorbing and immobilizing DOC for long enough periods, e.g., months to decades, to degrade refractory forms. In the Marmot Basin 98% of the carbon identified in the ground water consisted of humic and fulvic acids, tannins, and lignins (Wallis *et al.* 1981), leading Wallis (1981) to conclude that "despite the high uptake rates found during field studies, dissolved organic matter in ground water must still be considered to be refractory in nature." Yet it remains unexplained how these refractory forms could be held within the hyporheos long enough for significant degradation to occur.

Evidence presented for DOC stripping comes primarily from mass-balance calculations. From stream-water and groundwater concentrations, and the increase in discharge through a 150 m reach, Wallis (1981) estimated an uptake rate of $210\,mg\,C\,m^{-2}\,h^{-1}$ in the Marmot

Basin. Using a similar approach, Fiebig *et al.* (1990) calculated an uptake rate of 285 $mg\,C\,m^{-2}\,hr^{-1}$ in a Welsh stream. These are very high values, even considering that uptake in the hyporheos involves a column of sediment and not just an area of stream-bed surface. These values exceed the range of most reported biotic uptake rates for stream-water DOC by a factor of 10 to 30 as well as exceeding uptake rates of fresh leaf-leachate DOC (Kaplan & Bott 1983). Moreover, Wallis' estimate was nearly one order of magnitude higher than rates he measured in laboratory perfusion experiments (Wallis 1981). One explanation for the high uptake rate estimates reported for the hyporheos (Wallis 1981; Fiebig *et al.* 1990) is that the sampled ground water is not representative of that feeding the streams, and that much of the stream flow comes from a source with lower DOC concentration, perhaps deeper ground water which upwells from directly below the stream. This underscores a general problem of the mass-balance approach; unless all the sources are known, the calculations may be erroneous. Invoking a deep groundwater source which is low in DOC would resolve the inconsistencies surrounding the data from the Marmot Basin. In fact, all the reports of groundwater DOC exceeding stream-water concentrations derive from shallow groundwater samples – the very samples most susceptible to soil-water influence.

We do not argue that stripping never occurs, but rather that stripping seems unlikely to occur at rates sufficient to effect major differences between terrestrial and stream-water DOC concentrations, or to support very high rates of heterotrophy in stream beds. Some degree of stripping is likely to occur whenever labile DOC enters a stream. The entry of labile DOC, in turn, involves inputs of relatively high-concentration DOC from sources, such as soil water, that have not undergone extensive degradation. These are the same sources that we have suggested contribute significantly to DOC concentrations observed in streams. Thus stripping may, to some degree, reduce the impact of these high-DOC sources on the final stream-water concentrations. Whether this reduction is large depends on whether the additional DOC is sufficiently labile to be removed at significant rates in the hyporheos, and on the ability of hyporheic sediments to adsorb and immobilize a large pool of DOC. We suspect there are streams, particularly black-water streams, where stripping does not occur. High concentrations of DOC move from the watershed, through the hyporheos, and downstream without a marked reduction in DOC concentration; these may be a special case. We also suspect that there are streams where stripping of DOC from ground water and soil water occur simultaneously with the release of DOC from the large pool of slowly degrading benthic POC. Clearly, there exists a need to measure the sources and biological lability of terrestrially derived DOC.

7.3 Processes common across biomes

Export of organic carbon by rivers in different biomes has been estimated to equal 1% of net primary productivity in a watershed (Thurman 1985a). Export expressed on an areal basis, e.g., as $g\,C\,m^{-2}\,yr^{-1}$, represents the product of the TOC concentration and the "specific discharge," or annual runoff, e.g., as $m^3\,m^{-2}\,yr^{-1}$ or simply $m\,yr^{-1}$). Much of the variation among biomes reflects variation in runoff (Brinson 1976; Meybeck 1981; 1982; Mulholland & Watts 1982), which itself is a function of precipitation, is correlated with primary production, and ranges from $<0.1\,m\,yr^{-1}$ in arid areas and grasslands to $1-2\,m\,yr^{-1}$ or more in rainforests. Concentrations of TOC and DOC, on the other hand, show no consistent trend with primary productivity across biomes (Table 7.1). The highest average DOC concentrations, for example, appear in the taiga or boreal forest, a region of relatively low productivity. As Brinson (1976) suggested, it is probably a complex of interactions among factors such as hydrology, productivity, and decomposer activity, that actually underlies the seemingly simple associations of carbon export with runoff and primary productivity.

One potentially important component of these interactions undoubtedly involves the influence of decomposition rate on the accumulation of soil organic matter. For a given rate of organic matter input (or litter production), the standing stock of organic matter varies inversely with decomposition rate (k_r), as illustrated in Figure 7.1. The decomposition rate may vary over two orders of magnitude, and in the sequence of biomes shown in the diagram, its influence on the organic matter standing stock is larger than that of inputs. If the rate of leaching losses of DOC from the accumulated

Table 7.1 DOC levels in rivers and the range of TOC export rates compared to terrestrial primary production. From Meybeck 1982

Environment	DOC[1] $mg\,l^{-1}$	Average TOC export rate $g\,C\,m^{-2}\,yr^{-1}$	Mean net primary production[2] $g\,C\,m^{-2}\,yr^{-1}$
Tundra	2	0.6	65
Taiga	10	2.5	360
Temperate	3	4.0	225–585
Tropical[3]	6	6.5	315–900
Semiarid	3	0.3	32

[1] From Meybeck (1981).
[2] From Whittaker & Likens (1973).
[3] Savanna to tropical rain forest.

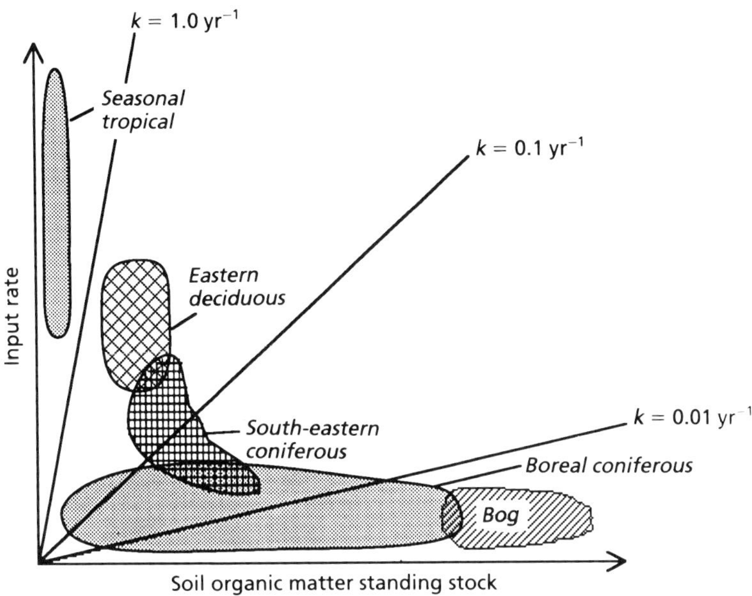

Figure 7.1 Relationship of inputs (net terrestrial primary production) to soil organic matter across biomes. From Swift *et al.* (1979).

organic matter (k_L) were directly proportional to k_r, export would be proportional to inputs and independent of accumulation, as in the constant 1% loss rate noted above, i.e., export = $k_L \times$ organic matter = $k_L \times \text{input}/k_r \propto$ input. However, leaching of plant litter is an abiotic process that is not constrained by ambient temperatures (Nykvist 1963). Thus leaching could proceed independently of decomposition, but some interaction is more likely. Leaching probably supplies soluble substrates for decomposers and, as the cascade structure of decomposition processes suggests (Swift *et al.* 1979), decomposition will produce leachable substances over time. The transport of leaching losses from an ecosystem competes with decomposition within the ecosystem, and tends to increase with increasing organic matter accumulation. Thus, export of DOC would become more important relative to losses via respiratory degradation. This cascade model therefore explains why carbon exports might show far less variability among biomes than would be indicated by carbon standing stocks alone. Yet at the same time, it suggests that in environments where plant litter decomposition rates are limited by environmental conditions – such as in the boreal forest, wetlands, and pineland sections of south-eastern temperate forests – DOC export and stream-water concentrations may be relatively high.

Temperature clearly has a major influence on decomposition and, hence, standing stock of organic matter. However, high concentrations of DOC in rivers draining temperate and tropical wetlands

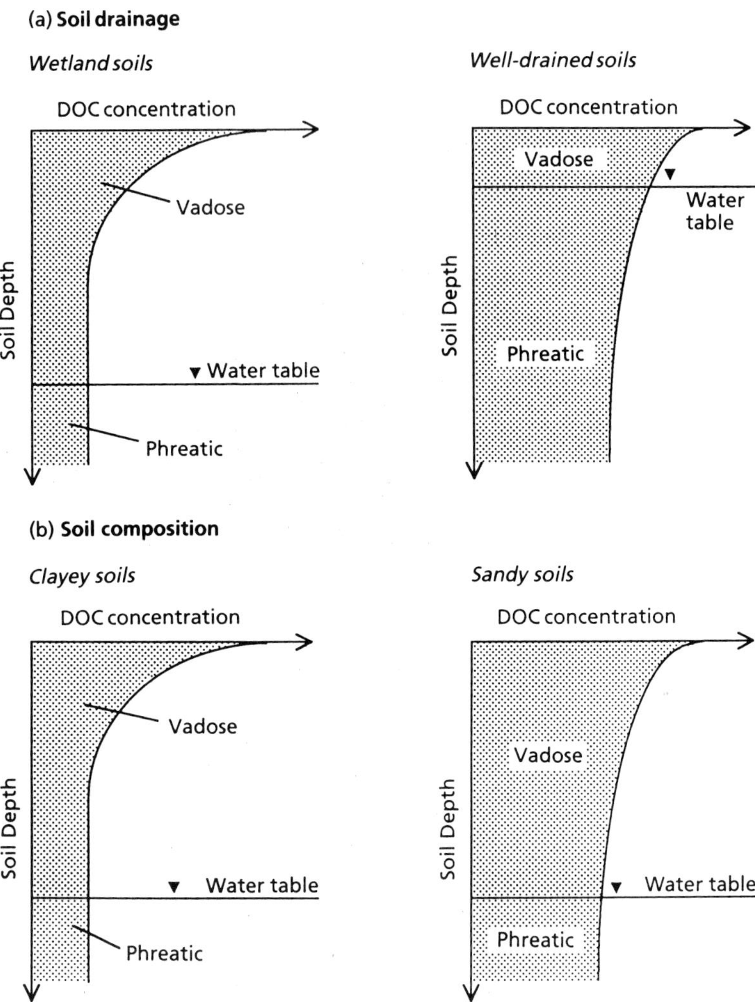

Figure 7.2 Influence of (a) soil drainage and (b) soil composition on the concentration of DOC in vadose and phreatic zones.

(Brinson 1976; Mulholland & Kuenzler 1979; Ittekkot *et al.* 1986; Richey *et al.* 1990) suggest that hydrological factors, such as precipitation and drainage patterns, may also control rates of decomposition. Some moisture is required for decomposition, but saturated and anoxic conditions can greatly reduce rates of decomposition even under warm conditions (Kilham & Alexander 1984). For this reason, the contribution of soil-water DOC to ground water and streams is expected to vary with soil drainage (Figure 7.2a). Soil composition also is important to the geochemical processing of DOC, and the contribution of DOC from soil water to streams is expected to increase along the spectrum from reactive clayey soils (alumina) to less reactive sandy soils (silica) (Figure 7.2b). These factors account

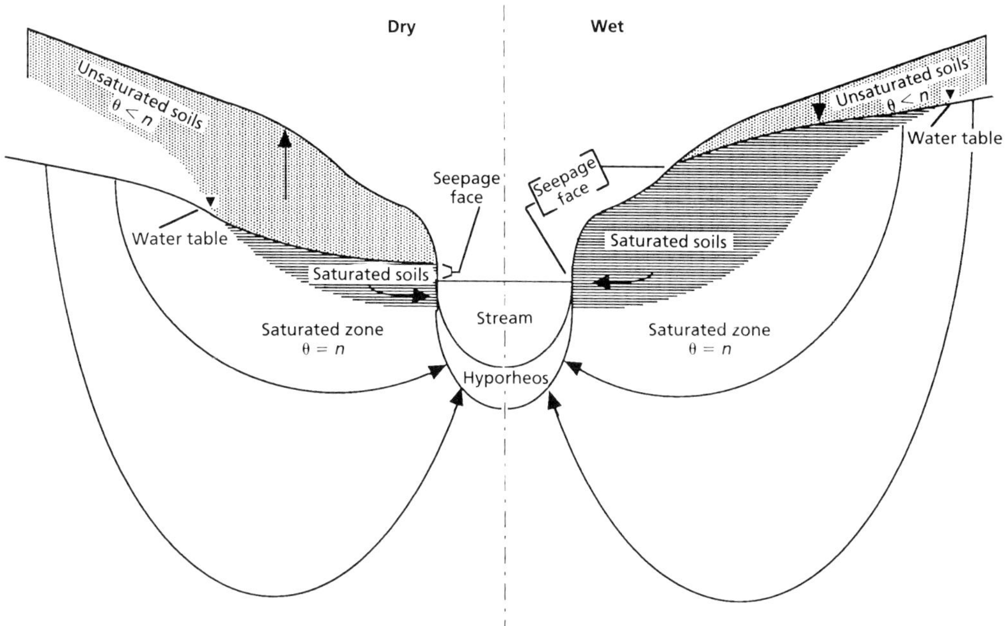

Figure 7.3 Expansion and contraction of DOC inputs from saturated soils under wet and dry conditions. Arrows represent water movement.

for the relative placement in Figure 7.1 of North American eastern deciduous forest, and the south-eastern coniferous forest – two habitats with very different hydrology and soil types within the temperate zone.

Within a watershed, we expect that areas of saturated soils, and their influence on stream-water DOC, will change seasonally, contracting between storms, constricting during periods of extended freezing or drought/heat wave, and expanding greatly during storms, possibly to contribute the majority of storm runoff (Figure 7.3). In the aftermath of storms, contributions of DOC from unsaturated soils to ground water should depend on the degree of saturation, and gradually decline as the watershed dries (Figure 7.4).

We have up to this point, presented a simple mass-balance model in which DOC concentrations in stream water represent a mix of terrestrial sources of differing concentrations, modified by in-stream additions and removal of DOC. We have also discussed some aspects of the processing and control of DOC concentrations in the terrestrial source pools. We now attempt to tie these ideas together in a general model that recognizes that the mix of terrestrial inputs involves not only variation in concentration but variation in lability of DOC from terrestrial sources arriving at the riparian and hyporheic zones. This model involves alternative scenarios which we describe below and formulate as specific predictions.

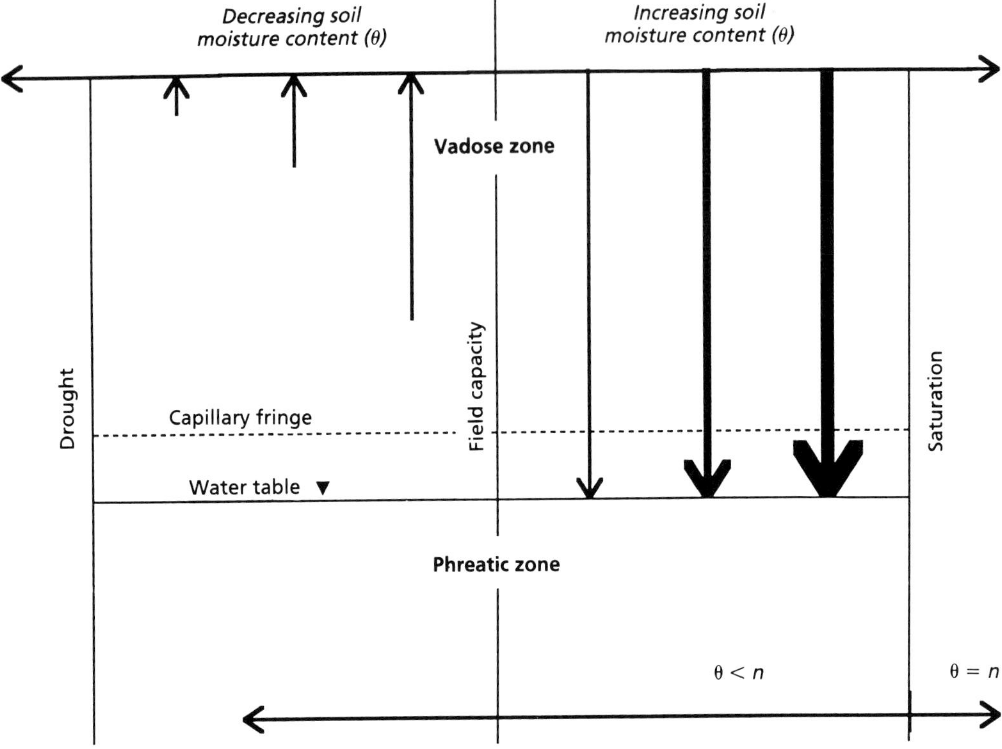

Figure 7.4 Changes in the flux of DOC from unsaturated soils under varying conditions of soil moisture. Arrow size is related to flux. Coupling of vadose and phreatic zones occurs when soil moisture exceeds field capacity; n is the porosity.

DOC leached from decomposing litter has a broad spectrum of lability, ranging from compounds that may be mineralized within minutes, e.g., simple sugars, to those taking perhaps decades, even under optimal conditions (Cummins *et al.* 1972; Moran & Hodson 1989). The bulk of this DOC probably falls into a category of "intermediate" lability, which under optimal conditions may mineralize on a time scale of a few weeks to several years. Many of these compounds undergo a series of transformations (notably formation of humic substances) during degradation. This is represented by the biopolymer degradation model of humic substance formation (Hedges 1988), and DOC undergoing this degradation reaches the stream through a variety of hydrological pathways, involving differing residence times as well as differing conditions for degradation. Assuming, provisionally, that the degradative sequence is the same under all environmental conditions, we would expect a consistent positive relationship between the concentration of DOC at a particular point in the subsurface environment and the lability of the remaining mixture, i.e., labile compounds are metabolized first. DOC

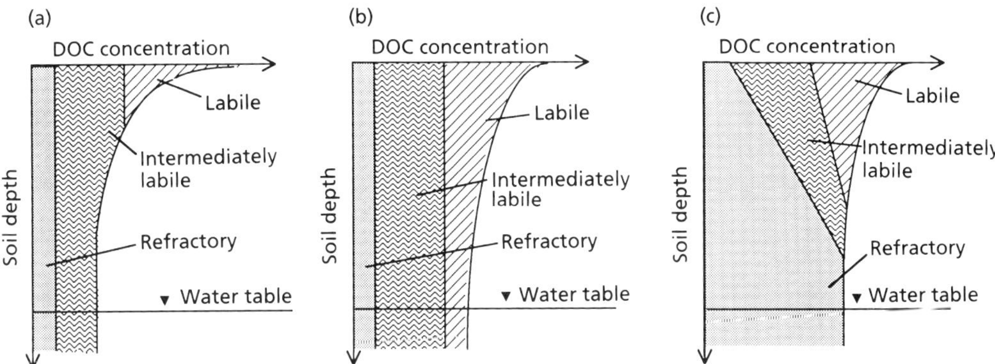

Figure 7.5 Profiles of DOC concentration and lability generated under alternative degradative pathways: (a) labile components are completely degraded within the soil horizon; (b) labile components are "protected" from degradation and are preserved as labile compounds until transported to locations where environmental conditions are conducive to degradation; (c) labile components are "protected" from degradation and converted to refractory forms.

reaching the stream represents the residual of this process and therefore consists of a mix of highly refractory, low-DOC water that has undergone extensive degradation, and varying quantities of higher DOC water that is less degraded and more labile (Figure 7.5a). The extensively degraded DOC may have an average age of several decades, while the age of less-degraded DOC may range from a few days to a year or so. Thus even the latter labile category would consist only of compounds of intermediate lability or that are refractory. Once this mixture reaches the stream, its residence time in the hyporheos and within the drainage basin may be only a few hours to a few days. From the perspective of the stream ecosystem, it would appear as entirely refractory, supporting very little heterotrophic activity.

The degradation of plant litter is not the same under all environmental conditions. There are major differences between aerobic and anaerobic degradation rates (Benner *et al.* 1986), pathways and end-products (Kilham & Alexander 1984; Blackburn 1987). Nutrient limitation (nitrogen and phosphorus) can alter rates of microbial organic carbon oxidation as well as end-product chemistry (Nickels *et al.* 1979), and inhibition of degradation by secondary plant compounds (Stout 1989) probably has an impact on end-product lability as well. These alternative pathways for degradative processes would predict, for example, that DOC entering streams from saturated and anaerobic soils, soils which are poor in inorganic nutrients, or habitats vegetated by plants containing high concentrations of inhibitory substances, contains some highly labile components. These components have been protected from terrestrial decomposition, but may be susceptible to decomposition within the stream, i.e., stripping would occur (Figure 7.5b). Alternatively, high concentrations of labile DOC in the terrestrial environment, protected from decomposition by whatever mechanism, may be converted to humic substances and thus become refractory (Figure 7.5c). This represents the abiotic condensation model of humic substance formation (Hedges 1988). In this case, stripping would be small and exports would be high.

7.4 Quantifying dissolved organic carbon sources using organic carbon tracers and end-member analysis

Major challenges in the study of stream ecosystem metabolism include:

1 quantifying the flux of terrestrial DOC sources entering streams from soil water and ground water;
2 measuring the biological lability of these sources and assessing how they change seasonally and across biomes under different temperature and hydrological conditions;
3 assessing the role of the hyporheos in altering DOC concentrations in stream water.

At this time we limit our focus to a discussion of analytical approaches and a mathematical solution that have the potential to quantify the flux of terrestrial DOC sources to streams. In practice, this is a two step process; first, markers for terrestrial DOC sources are developed, and then a mixing model is used to quantify the proportion of each source present in the stream water.

The DOC in stream water may be considered as a mix of DOC from terrestrial sources modified by processes that occur within the stream. The DOC in potential terrestrial sources – phreatic zone water (deep ground water and water from saturated soils) and vadose zone water (water from unsaturated soils) – can be sampled and characterized, but the proportions of stream water supplied from each are not known. Cronan (1990) assumed hydrological transfer fluxes from DOC source compartments in order to simulate stream-water DOC concentrations in forested watersheds under a variety of climatic and hydrological conditions. That exercise demonstrates that if the proportionate water-volume contributions of different terrestrial sources to the stream water were known, then from these proportions, and concentrations of the respective sources, a theoretical DOC concentration for the stream could be calculated on the assumption of simple mixing. Departures of calculated from observed DOC concentrations would indicate the influence of in-stream processes.

Inorganic analytes have been used to obtain estimates of water sources for streams in acid deposition research (Christophersen *et al.* 1982; Bergstrom *et al.* 1985; Gherini *et al.* 1985; Nikolaidis *et al.* 1988; Hooper *et al.* 1990). Analytes have included pH, alkalinity, Ca^{2+}, Mg^{2+}, Na^{+}, K^{+}, NH_4^{+}, Cl^{-}, NO_3^{-}, PO_4^{3-}, SO_4^{2-}, Br^{-}, and

SiO_2. Most recently, Hooper *et al.* (1990) successfully used these analytes in an end-member mixing model analysis to separate three water sources in a piedmont watershed: deep ground water, saturated soil water draining a hill slope, and organic layer water, collected 10 cm below the surface. These three sources provided distinct end-members with respect to several analytes. Water from the organic source contributed to stream chemistry only during storm flows. Hill (1990) and Mulholland *et al.* (1990) have similarly identified distinct signatures of ground water and soil water from inorganic analytes.

In the simplest form of a mixing model, the proportions of two sources can be calculated from the concentration of the mixture, if the concentrations of the sources are known. More generally, the contributions from n sources can be determined if the concentrations of $n - 1$ analytes are known from each source and the relative concentrations in each is independent of the others. This is done by constructing n simultaneous equations, one for each of the $n - 1$ analytes plus a mass-balance constraint (Christophersen *et al.* 1990). If the actual number of distinct sources is limited, e.g., two or three, then additional analytes provide redundancy necessary to verify the basic assumptions of a limited number of sources and conservative behavior of the analytes. One problem with the mixing models developed to date is that they are without provision for in-stream sources or sinks. Below we suggest an approach which uses organic rather than inorganic markers for DOC sources and modifies the mixing model to account for the possible influence of in-stream removal or additions of the constituents.

In suggesting an alternative approach to the development of chemical fingerprints of DOC sources through the identification of analytes that are constituents of terrestrial DOC, we begin from two premises:

1 DOC concentrations in streams represent a mixture of terrestrial sources (ground water, soil water from saturated and unsaturated soils) with minor in-stream influence attributable to in-stream production and consumption of DOC;

2 humic substances in soil water and ground water have distinct chemical signatures that reflect the diagenetic history of the DOC and these chemical differences can be used to separate and quantify sources of stream-water DOC.

We believe that the DOC-marker approach is fundamentally stronger than one based upon inorganic analytes because these markers reflect the DOC diagenesis directly, whereas processes influencing concentrations of inorganic analytes may be influenced by geochemical factors that vary independently of DOC processing.

DOC for an average stream consists of 40% fulvic acids, 10%

humic acids, 30% hydrophilic acids, and 20% simple compounds including carbohydrates, carboxylic acids, amino acids, and hydrocarbons (Thurman 1985a). Much of the DOC consisting of small molecules is biologically labile (Kaplan & Bott 1983), and thus would not make good tracers of sources and transport. The more refractory humic substances make better tracers of DOC but are difficult to analyze on a molecular level. Two potential approaches that use analyses on different levels are:

1 molecular analysis of humic and fulvic acid CuO oxidation products, which will produce a suite of molecular markers traceable to terrestrial sources;

2 a "bulk" chemical analysis, fluorescence.

The utility of lignin degradation products as molecular markers of vascular plant contributions to humic materials in different biomes has been demonstrated with soils, sediments, and DOC from the Columbia River, Washington, and the Amazon River (Hedges & Ertel 1982; Hedges *et al.* 1982; Ertel & Hedges 1984; Ertel *et al.* 1984; Hedges *et al.* 1986). Because lignin is only produced by vascular plants, its building blocks (substituted phenols) are unequivocal tracers of terrestrial processes. Furthermore, lignin is relatively refractory, with traceable structural alterations upon diagenesis (Ertel *et al.* 1984). Moran & Hodson (1989) have suggested that the polysaccharide components of DOC derived from lignocellulose are selectively degraded by bacteria, resulting in a DOC pool enriched in lignin-derived components. Aromatic ring hydroxylation and demethylation of arylmethyl ethers does occur during degradation of lignin by fungi, and results in catechol formation (Kirk *et al.* 1978; Larson & Rockwell 1980). To avoid contamination of the markers by free substituted phenols, humic and fulvic acids would first need to be separated from the rest of the DOC.

The molecular markers produced by CuO oxidation of lignin-type material are substituted phenols which fall into three structural categories as follows: (1) vanillyl, (2) syringyl, and (3) cinnamyl (Hedges & Mann 1979). The relative concentrations of vanillyl, syringyl, and cinnamyl phenols vary with the source of vegetation (Hedges & Mann 1979), but more importantly for our purposes, the ratios of the acidic phenol to the aldehyde reflect the diagenetic history of the humic matter (Ertel *et al.* 1984). Ratios of the components within phenolic compound classes are conservative in plant lignins until diagenetic oxidation, then acidic moieties increase in concentration relative to the aldehydes. Humics and fulvics analyzed from two water sources by Ertel *et al.* (1984) demonstrated a significant increase in the vanillic acid/vanillin (*Va*/*Vh*) and syringic acid/syringaldehyde (*Sa*/*Sh*) ratios, more so in the fulvic acid fractions than in the humic acid fractions and total humic substances. These patterns strongly suggest that the humic substance signatures will differ between waters with different diagenetic histories. Deep ground water, water from saturated soils, and water from unsaturated soils should have very different lignin signatures because the bulk DOC pool has undergone differing levels of diagenesis, and these signatures will differ from those of humics generated from within stream sediments. Thurman (1985b) characterized humic substances in deep ground waters and concluded that they differ from humics in surface waters in color, concentration, and oxygen content. The use of lignin oxidation products as humic substance signatures extends the work of Ertel *et al.* (1986) and Hedges *et al.* (1986) in using definitive end-members measured from the diagenesis of terrestrial sources. In previous uses, plant material was used to identify end-members, but this was found to be complicated by the alteration of the phenols with diagenesis; we suggest the exploitation of this complication.

"Bulk" chemical analysis of fluorescence spectra (Larson & Rockwell 1980; Cabaniss & Shuman 1987; Ewald *et al.* 1988) is also potentially useful as a DOC pool fingerprint. Fluorescence spectra of heterogeneous macromolecular material, such as fulvic and humic acids, generally demonstrate broad spectra with few structural features due to the complexity and multitude of functional groups present. However, certain features provide information which can be used to determine relative differences. For example, Ewald *et al.* (1988) noticed a red shift in emission maxima with an increase in molecular weight of fulvic acids produced by microbes. Larson & Rockwell (1980) also noted a red shift in both emission and excitation spectra as DOC was transported from spring seeps through a stream in the White Clay Creek basin. They hypothesized that this resulted from the biochemical and photolytic oxidation of the phenolic moieties from vanillin-type to coumarin-type structural units because of the similarities between the fluorescent spectra of the seep DOC and vanillin, and stream DOC and coumarins. Specialized techniques of fluorescence depolarization spectroscopy and phase-shift fluorometry have been used to characterize aquatic humic fractions (Goldberg & Negomir 1989). Another technique, synchronous fluorescence scanning, which links emission and excitation scanning, has been used as a way of fingerprinting DOC from different sources, because unlike simple excitation and emission spectra, synchronized spectra contain more structural features (Cabaniss & Shuman 1987). Once spectra were obtained for four river sources, spectra of various mixtures of those sources could be decomposed into the contributing sources from linear regression of a set of simultaneous equations. There are chemical interferences of fluorescence, such as metal ion concentration and pH; these effects would need to be reduced by cleanup with a cation exchange resin and pH adjustment.

The measures just discussed could be used to calculate the proportions of stream flow emanating from terrestrial sources by means of a mixing model approach, also described above. Although the goal would be to identify analytes that are unaffected by in-stream sources and sinks, this may not be possible to achieve completely. We expect microbial removal of humics to be negligible, but some removal may occur through sorption of humics onto sediments. Additionally, humics may be formed and added to the stream through degradation of particulate organic matter stored in the sediments. If in-stream processing were to have an unknown and independent effect on each analyte, it would be necessary to introduce an additional unknown variable into each equation of the mixing model, and no solution would be possible. However, we suggest that a net loss or gain can be estimated and accounted for, provided that the "signature" of the loss or gain is known. That is, it is not necessary to have an independent estimate of the magnitude of loss or gain, but it is necessary to know the ratios among constituents of interest of any removal or addition process.

We present the modified mixing model for the case of two analytes, *a* and *b*, and two water sources, *g*, e.g., deep ground water, and *r*, e.g., riparian soil water. Mass balance for analyte *a* in the stream is given by:

$$C_{ag}Q_g + C_{ar}Q_r + F_a = C_{as}Q_s,$$

in which C represents concentrations, e.g., $\mathrm{g\,m^{-3}}$, in the ground water (C_{ag}), riparian water (C_{ar}), and stream water (C_{as}); Q represents water flow, e.g., $\mathrm{m^3\,s^{-1}}$, subscripted as above; and F_a represents an input flux to the stream, e.g., $\mathrm{g\,s^{-1}}$, from in-stream sources (negative if there is net removal). The stream is bounded by its head waters and some point at which C_{as} and Q_s are measured and includes all tributaries.

The second analyte, *b*, is described analogously by:

$$C_{bg}Q_g + C_{br}Q_r + \alpha F_a = C_{bs}Q_s,$$

in which $\alpha = F_b/F_a$, where F_b is the input flux of *b* from in-stream processing. α is experimentally determined as described below.

The final equation is simply the water mass balance:

$$Q_g + Q_r = Q_s.$$

This system of three equations can be written as a matrix:

$$\begin{pmatrix} C_{ag} & C_{ar} & 1 \\ C_{bg} & C_{br} & \alpha \\ 1 & 1 & 0 \end{pmatrix} \begin{pmatrix} Q_g \\ Q_r \\ F_a \end{pmatrix} = \begin{pmatrix} C_{as}Q_s \\ C_{bs}Q_s \\ Q_s \end{pmatrix}$$

or:

$$A\mathbf{X} = \mathbf{Y}$$

in which **X** is the vector of unknowns to be solved for including the two source flows and the in-stream processing term F_a. A solution for **X** can be obtained provided that:

1 the concentration of at least one analyte differs between the two sources;

2 the ratio of concentrations between the two sources, for at least one analyte, is not equal to α.

For example, in using lignin oxidation products we expect:

1 concentrations of all analytes to differ between soil water and ground water;

2 that more extensive oxidation of the deep ground water will produce higher ratios of acidic moieties to aldehydes, than that found in DOC released from the *in situ* processing of POC in stream beds.

An approach to the experimental determination of the "signatures" of in-stream source and sink processes would involve perfusion experiments in which groundwater DOC is passed through stream sediments under laboratory conditions and inputs and outputs are measured (Wallis 1981). Experimentally obtaining reliable estimates of α (the relative effect of sources and sinks on the two analytes) and the use of these signatures to estimate and compensate for nonconservative behavior is a simpler task than estimating actual fluxes, e.g., F_a, because the former should be relatively unaffected by spatial variability in sediment carbon standing stocks and flow rates.

If a large number of analytes were measured, the effect would be to add more rows to the matrix A and the vector **Y**. If the additional rows in A are linearly independent, i.e., the additional analytes do not vary in direct proportion to other analytes among all source waters, this provides a means for verifying the consistency of the model and/or for resolving DOC sources into more than two terrestrial pools (in effect solving for more rows in vector **X**). The problem can be defined in terms of multiple linear regression (Cabaniss & Shuman 1987; Christopherson *et al.* 1990), in which the dependent variables are the measurements in the stream (the **Y** vector above), the independent variables are the columns in the A matrix, and the estimated coefficients are given by the **X** vector. Additional analytes would allow estimation of error terms for the source pools and additional source pools would be represented by more columns in the A matrix and rows in the **X** vector.

7.5 Conclusion

The current interest in the role of the hyporheos and the riparian zones in stream ecosystem metabolism is evidence that advances in forest hydrology, soil science, and geochemistry have changed the ways in which stream ecologists view streams. Much of the progress in understanding how water gets to streams and especially the attention focused on soil processes can be traced back to interdisciplinary research on acid deposition (Christopherson & Neal 1990). There have been an increased number of linked measurements of groundwater, soil-water, and stream-water chemistry by stream ecologists, and use of mixing models. It seems clear that further progress in quantifying the contributions of specific sources of stream-water DOC will require improved chemical fingerprinting of the sources or end-members and application of hydrological mixing models.

To go beyond quantitative estimates of flux, to evaluating the contribution of terrestrially derived DOC to stream ecosystem metabolism, we must consider the quality of the DOC, with particular emphasis on biological lability. Linking the watershed processes that determine the concentration of DOC in terrestrial sources to the quality of DOC entering streams, involves a shift in emphasis from hydrology, soil science, and geochemistry to biogeochemistry and microbial ecology. The challenges to the microbial ecologist include the measurements of biological lability of the sources and biogeochemical processes in the riparian zone and the hyporheos. Perfusion experiments with stream-bed substrata or stream-side sediments are one approach. These require realistic flow rates, residence times, and physical and chemical gradients. Analysis of the data will require appropriate scaling of the perfusion chambers to the riparian zone or the hyporheos (Stream Solute Workshop 1990). Recent advances in microbial ecology using microelectrodes to measure oxygen, nitrous oxide, temperature, sulfide, and pH (Revsbech & Jorgensen 1986; Nielsen *et al.* 1990), and immobilized biofilms to measure concentrations of organic carbon susceptible to metabolism by communities of heterotrophic bacteria (Lucena *et al.* 1991) should prove to be valuable techniques for the generation of these data.

References

Baker C.D., Bartlett P.D., Farr I.S. & Williams G.I. (1974) Improved methods for the measurement of dissolved and particulate organic carbon in fresh water and their application to chalk streams. *Freshwat. Biol.* 4, 467–481.

Benner R., Peele E.R. & Hodson R.E. (1986) Microbial utilization of dissolved organic matter from leaves of the red mangrove, *Rhizophora mangle*, in the Fresh Creek Estuary, Bahamas. *Estuarine, Coastal and Shelf Science* 23, 607–619.

Bergstrom S., Carlsson B., Sandberg G. & Maxe L. (1985) Integrated modelling of runoff, alkalinity and pH on a daily basis. *Nordic Hydrology* **16**, 89–104.

Blackburn T.H. (1987) Role and impact of anaerobic microbial processes in aquatic systems. In Moriarty D.J.W. & Pullin R.S.V. (eds) *Detritus and Microbial Ecology in Aquaculture. ICLARM Conference Proceedings 14*, pp. 32–53. International Center for Living Aquatic Resources Management, Manila.

Bott T.L., Kaplan L.A. & Kuserk F.T. (1984) Benthic bacterial biomass supported by streamwater dissolved organic matter. *Microb. Ecol.* **10**, 335–344.

Brinson M.M. (1976) Organic matter losses from four watersheds in the humid tropics. *Limnol. Oceanogr.* **21**, 572–582.

Cabaniss S.E. & Shuman M.S. (1987) Synchronous fluorescence spectra of natural waters: tracing sources of dissolved organic matter. *Mar. Chem.* **21**, 37–50.

Christopherson N. & Neal C. (1990) Linking hydrological, geochemical, and soil chemistry processes on the catchment scale: an interplay between modeling and field work. *Wat. Resources Res.* **26**, 3077–3086.

Christopherson N., Seip H.M. & Wright R.F. (1982) A model for streamwater chemistry at Birkenes, Norway. *Wat. Resources Res.* **18**, 977–996.

Christopherson N., Neal C., Hooper R.P. & Vogt R.D. (1990) Modelling streamwater chemistry as a mixture of soilwater end-members: a step towards second-generation acidification models. *J. Hydrol.* **116**, 307–320.

Comiskey C.E. (1978) *Aspects of the Organic Carbon Cycle on Walker Branch Watershed: a Study of Land/Water Interaction.* PhD thesis, The University of Tennessee.

Cronan C.S. (1990) Patterns of organic acid transport from forested watersheds to aquatic ecosystems. In Perdue E.M. & Gjessing E.T. (eds) *Organic Acids in Aquatic Ecosystems*, pp. 245–260. John Wiley & Sons, New York.

Cronan C.S. & Aiken G.R. (1985) Chemistry and transport of soluble humic substances in forested watersheds of the Adirondack Park, New York. *Geochim. Cosmochim. Acta* **49**, 1697–1705.

Cummins K.W., Klug J.J., Wetzel R.G., Petersen R.C., Suberkropp K.F., Manny B.A., Wuycheck J.C. & Howard F.O. (1972) Organic enrichment with leaf leachate in experimental lotic ecosystems. *BioScience* **22**, 719–722.

Dahm C.N. (1981) Pathways and mechanisms for removal of dissolved organic carbon from leaf leachate in streams. *Can. J. Fish. Aquat. Sci.* **38**, 68–76.

Dahm C.N., Trotter E.H. & Sedell J.R. (1987) Role of anaerobic zones and processes in stream ecosystem productivity. In Averett R.C. & McKnight D.M. (eds) *Chemical Quality of Water and the Hydrologic Cycle.* Lewis Publishers, Chelsea, Michigan.

Dawson H.J., Hrutfiord B.F., Zasoski R.J. & Ugolini F.C. (1981) The molecular weight and origin of yellow organic acids. *Soil Science* **182**, 191–199.

Deuser W.G. (1988) Whither organic carbon? *Nature* **332**, 396–397.

Duff J.H. & Triska F.J. (1990) Denitrification in sediments from the hyporheic zone adjacent to a small forested stream. *Can. J. Fish. Aquat. Sci.* **47**, 1140–1147.

Dunne T. & Black R.D. (1970) Partial area contributions to storm runoff in a small New England Watershed. *Wat. Resources Res.* **6**, 1296–1311.

Eckhardt B.W. & Moore T.R. (1990) Controls on dissolved organic carbon concentrations in streams, Southern Quebec. *Can. J. Fish. Aquat. Sci.* **47**, 1537–1544.

Edwards R.T. & Meyer J.L. (1987) Metabolism of a sub-tropical low gradient blackwater river. *Freshwat. Biol.* **17**, 251–264.

Ertel J.R. & Hedges J.I. (1984) The lignin component of humic substances: distribution among soil and sedimentary humic, fulvic, and base-insoluble fractions. *Geochim. Cosmochim. Acta* **48**, 2065–2074.

Ertel J.R., Hedges J.I. & Perdue E.M. (1984) Lignin signature of aquatic humic substances. *Science* **223**, 485–487.

Ertel J.R., Hedges J.I., Devol A.H., Richey J.E. & Ribeiro M. de N.G. (1986)

Dissolved humic substances of the Amazon River system. *Limnol. Oceanogr.* **31**, 739–754.

Ewald M., Berger P. & Visser S.A. (1988) UV-Visible absorption and fluorescence properties of fulvic acids of microbial origin as functions of their molecular weights. *Geoderma* **43**, 11–20.

Fiebig D.M., Lock M.A. & Neal C. (1990) Soil water in the riparian zone as a source of carbon for a headwater stream. *J. Hydrol.* **116**, 217–237.

Findlay S., Carlough L., Crocker M.T., Gill H.K., Meyer J.L. & Smith P.J. (1987) Bacterial growth on macrophyte leachate and fate of bacterial production. *Limnol. Oceanogr.* **31**, 1355–1341.

Fisher S.G. (1977) Organic matter processing by a stream-segment ecosystem: Fort River, Massachusetts, U.S.A. *Int. Rev. Ges. Hydrobiol.* **32**, 701–727.

Fisher S.G. & Likens G.E. (1973) Energy flow in Bear Brook, New Hampshire: an integrative approach to stream ecosystem metabolism. *Ecol. Monographs* **48**, 421–439.

Ford T.E. & Lock M.A. (1987) Epilithic metabolism of dissolved organic carbon in boreal forest rivers. *FEMS Microbial. Ecol.* **45**, 89–97.

Ford T.E. & Naiman R.J. (1988) Alteration of carbon cycling by beaver: methane evasion rates from boreal forest streams and rivers. *Can. J. Zoology* **66**, 529–539.

Ford T.E. & Naiman R.J. (1989) Groundwater–surface water relationships in boreal forests watersheds – dissolved organic carbon and inorganic nutrient dynamics. *Can. J. Fish. Aquat. Sci.* **46**, 41–49.

Freeman C., Lock M.A., Marxsen J. & Jones S.E. (1990) Inhibitory effects of high molecular weight dissolved organic matter upon metabolic processes in biofilms from contrasting rivers and streams. *Freshwat. Biol.* **24**, 159–166.

Gburek W.J. (1990) Initial contributing area of a small watershed. *J. Hydrol.* **118**, 387–403.

Gherini S., Mok L., Hudson R.J.M., Davis G.F., Chen C. & Goldstein R. (1985) The ILWAS model: formulation and application. *Water, Air, and Soil Pollution* **26**, 95–113.

Goldberg M.C. & Negomir P.M. (1989) Characterization of aquatic humic acid fractions by fluorescence depolarization spectroscopy. In Goldberg M.C. (ed.) *Luminescence Applications in Biological, Chemical, Environmental, and Hydrological Sciences, ACS Symposium Series No. 383*, pp. 180–205.

Grieve I.C. (1984) Relationships among dissolved organic matter, iron and discharge in a moorland stream. *Earth Surface Processes and Landforms* **9**, 35–41.

Grimm N.B. & Fisher S.G. (1984) Exchange between interstitial and surface water: implications for stream metabolism and nutrient cycling. *Hydrobiologia* **111**, 219–228.

Harr R.D. (1977) Water flux in soil and subsoil on a steep forested slope. *J. Hydrol.* **33**, 37–58.

Hedges J.I. (1988) Polymerization of humic substances in natural environments. In Frimmel F.H. & Christman R.F. (eds) *Humic Substances and Their Role in the Environment*, pp. 45–58. John Wiley & Sons, New York.

Hedges J.I. & Mann D.C. (1979) The characterization of plant tissues by their lignin oxidation products. *Geochim. Cosmochim. Acta* **43**, 1803–1807.

Hedges J.I. & Ertel J.R. (1982) Characterization of lignin by gas capillary chromatography of cupric oxide oxidation products. *Anal. Chem.* **54**, 174–178.

Hedges J.I., Ertel J.R. & Leopold E.B. (1982) Lignin geochemistry of a Late Quaternary sediment core from Lake Washington. *Geochim. Cosmochim. Acta* **46**, 1869–1877.

Hedges J.I., Ertel J.R., Quay P.D., Grootes P.M., Richey J.E., Devol A.H., Farwell G.W., Schmidt F.W. & Salati E. (1986) Organic carbon-14 in the Amazon River system. *Science* **231**, 1129–1131.

Hemond H. (1990) Wetlands as a source of dissolved organic carbon to surface waters. In Perdue E.M. & Gjessing E.T. (eds) *Organic Acids in Aquatic Ecosystems,* pp. 301–313. John Wiley & Sons, New York.

Hewlett J.D. & Hibbert A.R. (1967) Factors affecting the response of small watersheds to precipitation in humid areas. In Sopper W.E. & Lull H.W. (eds) *International Symposium on Forest Hydrology*, pp. 275–290. Pergamon, New York.

Hill A.R. (1990) Groundwater cation concentrations in the riparian zone of a forested headwater stream. *Hydrological Processes* **4**, 121–130.

Hooper R.P., Christophersen N. & Peters N.E. (1990) Modelling streamwater chemistry as a mixture of soilwater end-members – an application to the Panola Mountain Catchment, Georgia, U.S.A. *J. Hydrol.* **116**, 321–343.

Hornung M., Reynolds B., Stevens P.A. & Hughes S. (1990) Water quality changes from input to stream. In Edwards R.W., Gee A.S. & Stoner J.H. (eds) *Acid Waters in Wales*, pp. 223–240. Kluwer Academic Publishers, Dordrecht.

Hynes H.B.N. (1975) Edgardo Baldi Memorial Lecture. The stream and its valley. *Verh. Int. Verein. Theor. Angew. Limnol.* **19**, 1–15.

Hynes H.B.N. (1983) Groundwater and stream ecology. *Hydrobiologia* **100**, 93–99.

Ittekkot V., Safiullad S. & Arain R. (1986) Nature of organic matter in rivers with deep sea connections: The Ganges-Brahmaputra and Indus. *The Science of the Total Environment* **58**, 93–107.

Kaplan L.A. (1992) Comparison of high-temperature and persulfate oxidation methods for the determination of dissolved organic carbon in freshwaters. *Limnol. Oceanogr.* **37**, 1119–1125.

Kaplan L.A. & Bott T.L. (1982) Diel fluctuations of DOC generated by algae in a piedmont stream. *Limnol. Oceanogr.* **27**, 1091–1100.

Kaplan L.A. & Bott T.L. (1983) Microbial heterotrophic utilization of dissolved organic matter in a piedmont stream. *Freshwat. Biol.* **13**, 363–377.

Kaplan L.A. & Bott T.L. (1989) Diel fluctuations in bacterial activity on streambed substrata during vernal algal blooms: effects of temperature, water chemistry, and habitat. *Limnol. Oceanogr.* **34**, 718–733.

Kaplan L.A., Larson R.A. & Bott T.L. (1980) Patterns of dissolved organic carbon in transport. *Limnol. Oceanogr.* **15**, 1034–1043.

Kilham O.W. & Alexander M. (1984) A basis for organic matter accumulation in soils under anaerobiosis. *Soil Science* **137**, 419–427.

Kirk T.K., Yang H.H. & Keyser P. (1978) The chemistry and physiology of the fungal degradation of lignin. *Developments in Industrial Microbiology* **19**, 51–61.

Larson R.A. (1978) Dissolved organic matter of a low-coloured stream. *Freshwat. Biol.* **8**, 91–104.

Larson R.A. & Rockwell A.L. (1980) Fluorescence spectra of water-soluble humic materials and some potential precursors. *Arch. Hydrobiol.* **89**, 416–425.

Leenheer J.A., Malcolm R.L., McKinley P.W. & Eccles L.A. (1974) Occurrence of dissolved organic carbon in selected groundwater samples in the United States. *Journal of Research U.S. Geological Survey* **2**, 361–369.

Lock M.A. & Hynes H.B.N. (1976) The fate of "dissolved" organic carbon derived from autumn-shed maple leaves (*Acer saccharum*) in a temperate hard-water stream. *Limnol. Oceanogr.* **21**, 436–443.

Lord D.G., Barringer J.L., Johnsson P.A., Schuster P.F., Walker R.L., Fairchild J.E., Sroka B.N. & Jacobsen E. (1990) Hydrogeochemical data from an acidic deposition study at McDonalds Branch basin in the New Jersey Pinelands. *US Geological Survey Open-File Report 88-500.*

Lucena F., Frias J. & Ribas F. (1991) A new dynamic approach to the determination of biodegradable dissolved organic carbon in water. *Environ. Tech.* **12**, 343–347.

Malcolm R.L. & Durum W.H. (1976) Organic carbon and nitrogen concentrations and annual organic carbon load of six selected rivers of the United States. *US Geological Survey Water-Supply Paper 1817-F.*

Marxsen J. (1988) Evaluation of the importance of bacteria in the carbon flow of a small open grassland stream, the Breitenbach. *Arch. Hydrobiol.* **111**, 339–350.

McDowell W.H. (1985) Kinetics and mechanisms of dissolved organic carbon retention in a headwater stream. *Biogeochemistry* **1**, 329–352.

McDowell W.H. & Fisher S.G. (1976) Autumnal processing of dissolved organic matter in a small woodland stream ecosystem. *Ecology* **57**, 561–569.

McDowell W.H. & Wood T. (1984) Podzolization: soil processes control dissolved organic carbon concentrations in stream water. *Soil Science* **137**, 23–32.

McDowell W.H. & Likens G.E. (1988) Origin, composition, and flux of dissolved organic carbon in the Hubbard Brook Valley. *Ecol. Monogr.* **58**, 177–195.

Meybeck M. (1981) River transport of organic carbon to the ocean. In Likens G.E. (ed.) *Flux of Organic Carbon by Rivers to the Oceans*, pp. 219–269. U.S. Department of Energy, NTIS Report CONE 8009140.

Meybeck M. (1982) Carbon, nitrogen, and phosphorus transport by world rivers. *Am. J. Sci.* **282**, 401–450.

Meyer J.L. (1988) Benthic bacterial biomass and production in a blackwater river. *Proceedings of the International Association of Theoretical and Applied Limnology* **23**, 1832–1838.

Meyer J.L. (1990) A blackwater perspective on riverine ecosystems. *BioScience* **40**, 643–651.

Meyer J.L. & Tate C.M. (1983) The effects of watershed disturbance on dissolved organic carbon dynamics of a stream. *Ecology* **64**, 33–44.

Meyer J.L. & Edwards R.T. (1990) Ecosystem metabolism and turnover of organic carbon along a blackwater river continuum. *Ecology* **71**, 668–677.

Meyer J.L., Edwards R.T. & Risley R. (1987) Bacterial growth on dissolved organic carbon from a blackwater river. *Microb. Ecol.* **13**, 13–30.

Meyer J.L., McDowell W.H., Bott T.L., Elwood J.W., Ishizaki C., Melack J.M., Peckarsky B.L., Peterson B.J. & Rublee P.A. (1988) Elemental dynamics in streams. *J. N. Am. Benthol. Soc.* 7, 410–432.

Minshall G.W. (1988) Stream ecosystem theory: a global perspective. *J. N. Am. Benthol. Soc.* 7, 263–288.

Minshall G.W., Petersen R.C., Cummins K.W., Bott T.L., Sedell J.R., Cushing C.E. & Vannote R.L. (1983) Interbiome comparison of stream ecosystem dynamics. *Ecol. Monogr.* **53**, 1–25.

Moran M.A. & Hodson R.E. (1989) Formation and bacterial utilization of dissolved organic carbon derived from detrital lignocellulose. *Limnol. Oceanogr.* **34**, 1034–1047.

Mulholland P.J. (1981) Organic carbon flow in a swamp-stream ecosystem. *Ecol. Monogr.* **51**, 307–322.

Mulholland P.J. & Kuenzler E.J. (1979) Organic carbon export from upland and forested wetland watersheds. *Limnol. Oceanogr.* **24**, 960–966.

Mulholland P.J. & Watts J.A. (1982) Transport of organic carbon to the oceans by rivers of North America: a synthesis of existing data. *Tellus* **34**, 176–186.

Mulholland P.J., Wilson G.V. & Jardine P.M. (1990) Hydrogeochemical response of a forested watershed to storms: effect of preferential flow along shallow and deep pathways. *Wat. Resources Res.* **26**, 3021–3036.

Naiman R.J. (1976) Primary production, standing stock, and export of organic matter in a Mohave Desert thermal stream. *Limnol. Oceanogr.* **21**, 60–73.

Naiman R.J., Melillo J.M., Lock M.A., Ford T.E. & Reice S.R. (1987) Longitudinal patterns of ecosystem processes and community structure in a subarctic river continuum. *Ecology* **68**, 1139–1156.

Nickels J.S., King J.D. & White D.C. (1979) Poly-β-Hydroxybutyrate accumulation as a measure of unbalanced growth of the estuarine detrital microbiota. *Appl. Environ. Microbiol.* **37**, 459–465.

Nielsen L.P., Christensen P.B., Revsbech N.P. & Sorensen J. (1990) Denitrification and photosynthesis in stream sediment studied with microsensor and wholecore techniques. *Limnol. Oceanogr.* **35**, 1135–1144.

Nikolaidis N.P., Rajaram H., Schnoor J.L. & Georgakakos K.L. (1988) A generalized soft water acidification model. *Wat. Resources Res.* **24**, 1983–1996.

Nykvist N. (1963) Leaching and decomposition of water soluble organic substances from different types of leaf and needle litter. *Studia Forestalia Suecica* **3**, 1–31.

Revsbech N.P. & Jorgensen B.B. (1986) Microelectrodes: their use in microbial ecology. In Marshall K.C. (ed.) *Advances in Microbial Ecology*, vol. 9, pp. 293–352. Plenum, New York.

Richey J.E., Hedges J.I., Devol A.H., Quay P.D., Victoria R., Martinelli L. & Forsberg B.R. (1990) Biogeochemistry of carbon in the Amazon River. *Limnol. Oceanogr.* **35**, 352–371.

Rutherford J.E. & Hynes H.B.N. (1987) Dissolved organic carbon in streams and groundwater. *Hydrobiologia* **154**, 33–48.

Schiff S.L., Aravena R., Trumbore S.E. & Dillon P.J. (1990) Dissolved organic carbon cycling in forested watersheds: a carbon isotope approach. *Wat. Resources Res.* **26**, 2949–2957.

Sedell J.R. & Dahm C.N. (1990) Spatial and temporal scales of dissolved organic carbon in riverine ecosystems. In Perdue E.M. & Gjessing E.T. (eds) *Organic Acids in Aquatic Ecosystems*, pp. 261–279. John Wiley & Sons, New York.

Stream Solute Workshop (1990) Concepts and methods for assessing solute dynamics in stream ecosystems. *J. N. Am. Benthol. Soc.* **9**, 95–119.

Stout R.J. (1989) Effects of condensed tannins on leaf processing in mid-latitude and tropical streams: a theoretical approach. *Can. J. Fish. Aquat. Sci.* **46**, 1097–1106.

Swift M.J., Heal O.W. & Anderson J.M. (1979) *Decomposition in Terrestrial Ecosystems. Studies in Ecology Volume 5.* University of California Press, Berkeley.

Thurman E.M. (1985a) *Organic Geochemistry of Natural Waters.* Nijhoff/Junk, Dordrecht.

Thurman E.M. (1985b) Humic substances in groundwater. In McKnight D.M. (ed.) *Humic Substances in Soil, Sediment, and Water: Geochemistry, Isolation, and Characterization*, pp. 87–103. John Wiley & Sons, New York.

Triska F.J., Kennedy V.C., Avanzino R.J., Zellweger G.W. & Bencala K.E. (1989) Retention and transport of nutrients in a third-order stream in northwestern California: hyporheic processes. *Ecology* **70**, 1893–1905.

Valett H.M., Fisher S.G. & Stanley E.H. (1990) Physical and chemical characteristics of the hyporheic zone of a Sonoran Desert stream. *J. N. Am. Benthol. Soc.* **9**, 201–215.

Wallace J.B., Ross D.H. & Meyer J.L. (1982) Seston and dissolved organic carbon dynamics in a southern Appalachian stream. *Ecology* **63**, 824–838.

Wallis P.M. (1981) The uptake of dissolved organic matter in groundwater by stream sediments – a case study. In Lock M.A. & Williams D.D. (eds) *Perspectives in Running Water Ecology*, pp. 97–111. Plenum, New York.

Wallis P.M., Hynes H.B.N. & Telang S.A. (1981) The importance of groundwater in the transportation of allochthonous dissolved organic matter to the streams draining a small mountain basin. *Hydrobiologia* **79**, 77–90.

Ward J.V. (1989) The four-dimensional nature of lotic ecosystems. *J. N. Am. Benthol. Soc.* **8**, 2–8.

Wetzel R.G. & Manny B.A. (1972) Decomposition of dissolved organic carbon and nitrogen compounds from leaves in an experimental hardwater stream. *Limnol. Oceanogr.* **17**, 927–931.

Whittaker R.H. & Likens G.E. (1973) Carbon in the biota. In Woodwell G.M. & Pecan E.V. (eds) *Carbon and the Biosphere*, pp. 281–302. US Atomic Energy Comm., NTIS Report CONF 720510.

Wilson G.V., Jardine P.M., Luxmoore R.J. & Jones J.R. (1990) Hydrology of a forested hillslope during storm events. *Geoderma* **46**, 119–138.

Eugene L. Madsen & William C. Ghiorse

8 Groundwater microbiology: subsurface ecosystem processes

8.1 Introduction

8.1.1 Groundwater habitat: definition

Continental areas of the earth are typically composed of materials in the following vertical sequence: the A and B soil horizons; the C soil horizon (from which the other soil horizons may have been derived) (Brady 1984); an unsaturated (or vadose) zone (that begins with the C soil horizon and ends at the water table); and a capillary fringe zone residing directly above a saturated zone which may extend through many different geological strata. Where does the soil habitat end and the groundwater habitat begin? For the purposes of this chapter, the groundwater habitat begins immediately below the B soil horizon where soil scientists traditionally have felt that major biological activity ceases (Alexander 1977; Brady 1984). It is important to acknowledge, however, that the transition between soil and groundwater habitats is not delineated by soil horizons *per se*. Indeed, plant roots may penetrate the C soil horizon, thereby supplementing ground water with organic carbon compounds that may stimulate microbial activity.

In a hydrogeological sense, "ground water" refers to water that is easily extractable from saturated, highly permeable geological strata known as aquifers (Davis & DeWiest 1966; Freeze & Cherry 1979). However, to microorganisms living in microhabitats, all available forms of water may be important. Therefore, we have chosen a broader definition of ground water which includes capillary water, water vapor, and water within aquifer formations. The groundwater habitat includes the unsaturated zone because it may contain significant amounts of biologically available water. Also, unsaturated zones may be transiently saturated during recharge events and they may influence both the chemistry and microbiology of the saturated zone. Thus throughout this chapter, "ground water" refers to all subsurface water found beneath the soil A and B horizons that is available to sustain and influence microbial life in the terrestrial subsurface. Implicit in this definition is the synonymity of the phrases "groundwater" habitat and "terrestrial subsurface" (Ghiorse & Wilson 1988).

8.1.2 Groundwater microbiology: background

The study of life in groundwater habitats was of relatively minor scientific interest prior to the 1970s (see Section 8.2.3) (Ghiorse & Wilson 1988). Limited, though insightful, information on subsurface microorganisms had been assembled from geological drilling efforts associated with exploration for petroleum and sulfur deposits (Zobell 1952; Kuznetsov *et al.* 1963; McNabb & Dunlap 1975). Much of the data were equivocal, however, because waters and sediments were not obtained in a manner which prevented microbiological contamination from surface sources. In the 1970s, development of aseptic sampling techniques (McNabb & Mallard 1984; Phelps *et al.* 1989a) was prompted by rising concerns about groundwater protection and the need to characterize both subsurface microbial communities and their potential impact upon groundwater quality (Ghiorse & Wilson 1988). The impetus for developing aseptic coring methods and their implementation triggered a surge of subsurface microbiological research whose momentum, scope, and variety has continued to grow into the 1990s. Subsurface microbiology research has progressed from the initial phases of documenting the presence of microorganisms in aquifer samples to current interests in basic

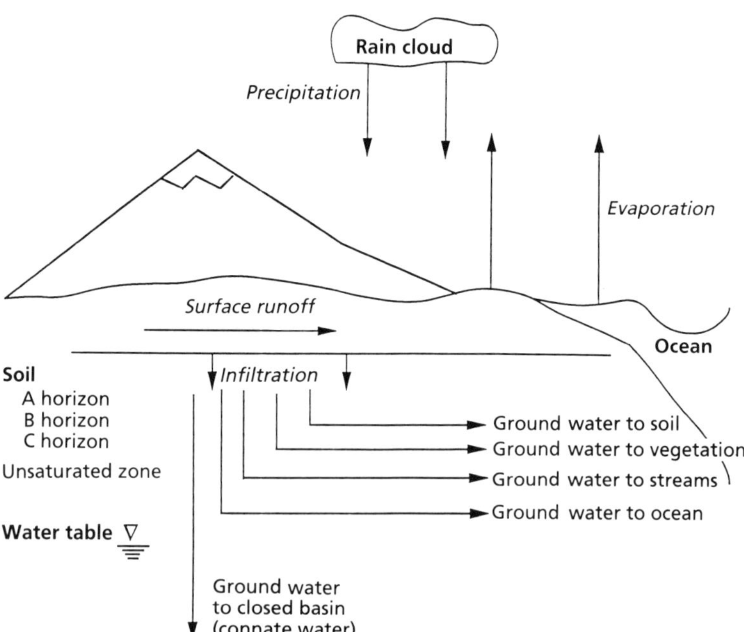

Figure 8.1 The hydrological cycle. Adapted from Davis & DeWiest (1966) and van der Leeden *et al.* (1990).

research on deep aquifer ecology and biogeochemistry and in stimulating *in situ* metabolic activity for aquifer biorestoration.[1]

Several significant reviews pertinent to subsurface microbiology were published during the 1980s (Bitton & Gerba 1984; Ward *et al.* 1985; Ghiorse & Wilson 1988; Lee *et al.* 1988; Balkwill 1990; Hirsch *et al.* 1992). Since then there has been an explosion of publications in this area. Thus, a comprehensive review of all scientific information on subsurface microbiology is now beyond the scope of a chapter such as this one. In keeping with the themes of the present volume, we emphasize relationships of the groundwater habitat to other aquatic habitats and endeavor to provide the reader with a conceptual understanding of groundwater microbiology and, where possible, subsurface ecosystem processes. Our review of the literature is based principally on the substantial number of papers on these subjects published since 1985.

8.2 The groundwater habitat

8.2.1 Role of ground water in the hydrological cycle

The terrestrial subsurface is an important component of the landscape through which water passes as it cycles between the atmosphere, soil, lakes, streams, and oceans (Figure 8.1). Once water has infiltrated below the surface layer of soil, it has several possible fates. It may:

1 return to soil via capillary, gaseous, or saturated transport;
2 be intercepted by plant roots;
3 reach streams, lakes, or ponds via saturated flow;
4 directly reach the ocean via saturated flow;
5 enter a closed deep continental basin.

Regardless of the flow path taken through the subsurface, ground water remains in the biosphere. However, the residence time before water exits the subsurface is highly variable. Return of subsurface water to the soil may occur within a few days or weeks, although return from a deep continental basin may require thousands of years (Table 8.1, Section I).

8.2.2 Physical and biological relationships to other aquatic habitats

Table 8.1 provides a concise means of appreciating the similarities and differences between the subsurface and other aquatic environments. Ground water can be found beneath all continental surfaces (which represent 29% of the Earth's surface area); therefore, on a global scale, groundwater habitats are vast. Assuming that microbial life is possible to a temperature-limited depth of about 4000 m

[1] While this chapter was being printed, an elaborate *in situ* bioremediation project was in progress at the US Department of Energy's Savannah River Site. The project utilizes horizontal wells, injection of gases, methanotrophic microorganisms and both chemical and microbiological monitoring to document the elimination of trichloroethylene from contaminated ground water (Hazen 1992).

Table 8.1 Contrasts between the groundwater and other aquatic habitats

Characteristics	Marine habitat	Freshwater habitat	Groundwater habitat	References[1]
I Physical				
Global location	Ocean basins	Continental depressions, valleys, basins	Beneath continental subsoils	Gross (1987); Wetzel (1983); Brady (1984), Freeze & Cherry (1979)
Per cent of global surface area	70.8	0.2	29.0	Ehrlich *et al.* (1977), Freeze & Cherry (1979)
Percentage of global[2] water volume	94	<0.01	4	Freeze & Cherry (1979)
Residence time for water	~4000 yr	2 w–~10 yr	2 w–10 000 yr	Freeze & Cherry (1979)
Hydrological regime	Relatively deep water; very low percentage solids; little if any unsaturated zone; hydrological stratification	Relatively shallow water; low percentage solids; little if any unsaturated zone, though some streams are ephemeral; hydrological stratification	Interstitial water in solid matrix of variable porosity and variable degree of saturation; unsaturated zone may be substantial; hydrological and geological stratification	Gross (1987); Wetzel (1983); Davis & DeWiest (1966)
II Biological				
Biota	Higher plants, animals, protists, fungi, prokaryotes	Higher plants, animals, protists, fungi, prokaryotes	Primarily prokaryotes; protists, rare algae and cave dwelling animals	Raven & Johnson (1989), Nybakken (1988); Wetzel (1983); Ghiorse & Wilson (1988), this chapter
Food chain	Photosynthesis, rare chemosynthesis, heterotrophy	Photosynthesis, rare chemosynthesis, heterotrophy	Heterotrophy, chemosynthesis possible	Nybakken (1988); Wetzel (1983); Ghiorse & Wilson (1988)
Nutrient status	Nutrient-poor regions widespread; high productivity in upwelling zones	Broad range of oligotrophic, mesotrophic, and eutrophic conditions	Low levels of DOC and other nutrients common but many nutrient-rich waters, i.e., beneath landfills, reported	Nybakken (1988); Wetzel (1983); Freeze & Cherry (1979), Miller *et al.* (1990), Thorstenson *et al.* (1979), Baedecker & Back (1979)
Water flow	Flow paths well defined	Flow paths well defined	Flow paths difficult to define	Gross (1987); Wetzel (1983); Freeze & Cherry (1979)

continued

Table 8.1 *Continued*

Characteristics	Marine habitat	Freshwater habitat	Groundwater habitat	References[1]
III Factors affecting scientific investigation				
Accessibility of habitat	Coastal waters accessible; deep waters may require specialized equipment	Many streams and lakes easily accessed, deep waters may require specialized equipment	Largely inaccessible; excavation and/or drilling equipment essential	Nybakken (1988); Wetzel (1983); Freeze & Cherry (1979)
Motivation for exploration	Transportation, fisheries, mining, wildlife protection, biosphere maintenance	Transportation, fisheries, drinking water, wildlife protection, ecosystem maintenance	Mining, drinking water	Nybakken (1988); Wetzel (1983); Freeze & Cherry (1979)
Maturity of microbiological studies	Relatively mature	Relatively mature	Immature, aseptic sampling 1970s	Zobell (1946); Kuznetsov (1970); Ghiorse & Wilson (1988)

[1] Where appropriate, references for marine, freshwater, and subsurface habitats appear consecutively and are separated by semicolons.
[2] The remainder (2%) of water is in ice-caps and glaciers.

(Kuznetsov *et al.* 1963; McNabb & Dunlap 1975; Ghiorse & Wilson 1988) and further assuming that the biomass per kilogram of subsurface sediments is 1000-fold less on average than that found in the top 1 m of agricultural soil, then the total biomass in subsurface habitats may be 40 times greater than that in the top 1 m of soil. The global volume of ground water in the subsurface is several hundredfold greater than that of rivers and lakes but less than one-twentieth that of the oceans. The water in the lakes and oceans may be influenced by suspended particles and their attached biota to varying degrees depending on the concentration of particles. Contrastingly, most ground water is interstitial, i.e., it resides within a matrix of minerals with variable chemistry, porosity, and degree of saturation. Thus in the subsurface, particulate materials may have a much more profound influence on microbiology than they do in other aquatic environments. For example, with the exception for cavernous limestone aquifers (see below), all life forms larger than microorganisms are effectively excluded from most groundwater habitats. Also the large surface area provided by the sediments may strongly influence the physical and chemical conditions of the groundwater habitat by sorbing microbial cells and by altering concentrations of dissolved aqueous constituents at the surfaces (Ghiorse & Wilson 1988; van

Loosdrecht *et al.* 1990). In arid climates, the unsaturated zone may be hundreds of meters deep. Rainfall in such desert climates may be insufficient to allow saturated infiltration to reach to the water table, except in restricted low-lying areas (Davis & DeWiest 1966). Therefore, rather large areal portions of deserts may have unsaturated zones beneath them with little or no saturated water flux. Such zones of desiccation are not found in marine and freshwater habitats. In groundwater habitats, the absence of water flux through unsaturated zones may have major influence on the survival and activity of microorganisms. The microbiology of these zones has begun to be investigated[2] (Brockman *et al.* 1992), but this is as yet incomplete. Unsaturated zone microbiology is an important area for future research.

Another prominent characteristic of groundwater habitats is their geological and chemical stratification (see Table 8.1, Section I). At a given location, the vertical profile beneath the Earth's surface reflects a unique and complex history of geological and chemical events, e.g., sedimentation, erosion, volcanism, tectonic activity, dissolution, precipitation, and biogeochemical activity. The result often is a heterogeneous geological profile whose complexity may be compounded by variations in pore-water chemistry that may stem from localized aberrations in mineral phases or inorganic or organic solute concentrations.

Besides physical and chemical contrasts, there are also key biological distinctions between the groundwater habitats and other aquatic milieus (see Table 8.1, Section II). The subsurface biological community is unique in that it consists primarily of unicellular bacteria, fungi, and protozoa; though algae may be present under some circumstances (Sinclair & Ghiorse 1987; 1989; Sinclair *et al.* 1990). Larger organisms occur only rarely in subterranean caves or cavernous aquifers that are connected to the surface (Longley 1981; Hynes 1983; Ghiorse & Wilson 1988). Other aquatic fauna, such as amphipods, may inhabit ecotone habitats, such as river-bank or lake sediments that are transitional to the subsurface (Gilbert *et al.* 1990). Because sunlight is precluded from groundwater environments, photosynthesis is impossible. Thus, the food chain in aquifers is primarily heterotrophic, reliant either upon an influx of dissolved

[2] At the time this chapter was being written, the US Department of Energy, Office for Health and Environmental Research, Deep Subsurface Microbiology Program was conducting exploratory research to characterize microbial populations in deep unsaturated zones at sites in the arid Intermountain Basin areas of the western United States. The preliminary data showed that these zones contained very low numbers of culturable microorganisms (D.L. Balkwill, F.S. Colwell, J.K. Fredrickson, W.C. Ghiorse, T. Kieft, E.L. Madsen, T.J. Phelps, D.C. White unpublished data).

organic carbon (DOC) along a given hydrological flow path or organic materials of sedimentary origin that subsequently may have been metamorphosed (Barker 1979; van Krevelen 1984). The potential for microbial oxidation of reduced inorganic energy sources has been demonstrated in some subsurface sediments, e.g., sulfide, hydrogen, ammonium (Fredrickson *et al.* 1989), and iron (F.S. Colwell personal communication, 1991). These energy sources may be utilized for autotrophic carbon dioxide fixation or other microbial metabolic functions (Kelly 1981; Kuenen & Bos 1989) as has been demonstrated in deep-sea thermal vent regions (Jannasch & Mottl 1985). However, to our knowledge, aerobic chemosynthetic food chains have not been described in the terrestrial subsurface. The possibility for such food chains does exist, especially in areas of geothermal activity and in ultradeep boreholes where reduced inorganic energy sources encounter oxygenated water.

A characteristic that the subsurface shares with other aquatic habitats is a wide variation in nutrient levels (see Sections 8.3.7 and 8.3.8 and Table 8.1). Beneath localized sites, such as landfills (Baedecker & Back 1979; Beeman & Suflita 1990), ground water may contain high concentrations of DOC and therefore be dominated by anaerobic processes. But subsurface habitats can also be nutrient-poor, as reflected by analyses of carbon, nitrogen, and phosphorus and by measures of the physiological status of microorganisms (see Section 8.3.4). Unlike other aquatic habitats, the inaccessibility of the subsurface is a major barrier to thoroughly defining the factors (spatial heterogeneity of sediments, clay lenses, fissures, and other geological characteristics) that govern water flow. Thus, for the most part, flow paths and their influence on subsurface ecology are obscure. Spatial and temporal relationships taken for granted in ecological studies of marine and freshwater habitats are problematic in groundwater ecology.

8.2.3 Contrasting histories of scientific investigation among aquatic habitats

An historical awareness of scientific investigations of aquatic habitats also aids in their comparison (see Table 8.1, Section III). Human curiosity, ingenuity, and material gain are among the primary motivations for scientific investigation of the biosphere. Given the inaccessibility of the terrestrial subsurface and the absence of plants and animals that have stimulated inquiry into other aquatic habitats, it is not surprising that investigations of subsurface ecosystems were delayed for centuries. Clearly, oceans, lakes, ponds, rivers, and streams and their flora and fauna have been critical for the development of civilizations (Odum 1971; Wetzel 1983; Nybakken 1988). When the fragile nature of fisheries and other facets of the biosphere became apparent in the twentieth century, wildlife protection joined resource utilization and management as major concerns in marine and freshwater habitats. As mentioned in Section 8.1.2, economic geology and groundwater contamination, resulting from modern agricultural practices, deep well injection, and improper management of industrial waste (Keswick 1984), have greatly influenced the development of subsurface microbiology in recent years. Unlike marine and freshwater environments, the primary motivation for studying subsurface ecology has had little to do with protecting wildlife or maintaining ecosystem function. After all, other than microorganisms, there is no wildlife to speak of in aquifers. Furthermore, the function of subsurface ecosystems may be extremely difficult and costly to study (see Sections 8.3.5–8.3.8). But the study of aquifer ecosystems for the purpose of restoring contaminated drinking water does have a direct impact upon human health and financial well-being. Thus, a major motivation for advancing the science of subsurface microbiology has been the need to characterize microbial distribution, abundance, and activities in aquifer ecosystems, and especially the potential of the resident microflora to eliminate toxic organic compounds. This potential can then be compared to that in other aquatic and terrestrial environments where microbial activity is well studied (Odum 1971; Alexander 1977; 1981; Dagley 1977; Leahy & Colwell 1990). The potential impact of subsurface microbial activity upon a broad range of geochemical processes, including nutrient cycling, petroleum recovery, and transport or immobilization of heavy metals and radioisotopes, provides additional impetus for advancing the study of groundwater microbiology and subsurface ecosystem processes.

8.3 Microorganisms and their activities in the groundwater habitat

8.3.1 Source of groundwater microflora

We envision two possible sources for microorganisms in extant aquifer sediments:

1 the organisms and their descendants may have remained with the sediments ever since initial colonization during surface deposition;

2 the organisms may have been transported to the buried strata via hydrologic flow.[3]

Though microbial colonization may seem a trivial issue for habitats at the surface, the colonization of subsurface habitats has generated interest and speculation among hydrologists, geologists, and environmental microbiologists (see discussion in Sargent & Fliermans 1989).

Comparisons of microbiological communities within vertical profiles extending from the surface soil to subsurface sediments have

[3] If one is examining water pumped from aquifers (well water) a third possible source is colonization from surface sources. For this reason, sediment samples are preferable to well-water samples in performing subsurface microbiological investigations.

shown that the bacteria found in different geological strata can be physiologically (Balkwill 1989; Balkwill *et al.* 1989; Fredrickson *et al.* 1991b) and morphologically (Bone & Balkwill 1988; Beloin *et al.* 1988) distinct. Although such comparisons reveal that subsurface microbial communities may be different from the surface soil communities above, they do not reveal the developmental history of subsurface microflora. It is feasible that surface sediments were colonized by microorganisms prior to their burial (see above). Microbial communities in deep confined sediments, especially the more consolidated ones, may have existed in the sediments since the time of deposition. Such communities would be cut off from surface influences, perhaps for millions of years. Thus, they could have evolved unique subsurface phenotypes. Indeed, the contrasting physiological characteristics of aerobic heterotrophic bacteria from different deep geological strata of the Savannah River Site (SRS) in the Atlantic coastal plain (Balkwill 1989; Balkwill *et al.* 1989; Fredrickson *et al.* 1991b), may reflect this source for groundwater microorganisms. However, it is equally feasible that the stratified distribution of subsurface microorganisms could have been caused by vertical and horizontal immigration patterns by water-borne soil microorganisms (see above). Subsequent to dispersal to particular sediments, distinct communities could further develop if changing environmental conditions produced shifts in the populations based on natural selection of fitness traits (Alexander 1977; Atlas & Bartha 1987). Virtually all of the experimental evidence for explaining the origins of subsurface microorganisms addresses their transport from surface environments. Recent laboratory studies suggest that cell surface properties affect transport of microorganisms through soil (Gannon *et al.* 1991a) and that mineralogy and/or solution chemistry influence attachment of microorganisms to aquifer material (Scholl *et al.* 1990; Gannon *et al.* 1991b). Also, Reynolds *et al.* (1989) found that motility, chemotaxis, and metabolism governed the penetration of *Escherichia coli* cells in columns packed with sand. Furthermore, Lindqvist & Bengtsson (1991), after developing and experimentally testing mathematical dispersal models for subsurface bacteria, concluded that sorption kinetics, associations between cells and fine particles, and cell surface hydrophobicity, as governed by low-nutrient conditions, were major influences for the dispersal of subsurface bacteria.

While the above-mentioned laboratory studies illustrate basic principles governing movement of bacterial cells in porous media, field observations are relatively rare. Field evidence for colonization of actual aquifer sediments by microorganisms from surface waters was obtained by Beloin *et al.* (1988) who detected filamentous green algae and cyanobacteria in a semiconfined high-yield aquifer at a depth of 7.5 m in the subsurface at a site in Lula, Oklahoma, but not elsewhere in the profile. Sinclair & Ghiorse (1989) also found algae in aquifer sediments from as deep as 214 m in two of four deep boreholes in the Atlantic coastal plain at the SRS. The two SRS boreholes containing algae were the closest to recharge areas and they were in regions with a history of drilling activity. Colwell (1989) speculated that flooding events would be required to transport bacteria in a semiarid high desert soil to the 70 m depth where they were detected.

Studies have traced the movement of viruses and bacteria through subsoil of septic filtration systems and other shallow subsurface field sites (McCoy & Hagedorn 1979; Gerba 1983; Powelson *et al.* 1990). These studies focused primarily on the movement of pathogens which are alien to the subsurface. They give little information on the problem of colonization by the long-term residents of groundwater habitats. Using filtration theory, Harvey & Garabedian (1991) developed a model for the movement of subsurface bacteria in a Cape Cod, Massachusetts, aquifer (Harvey *et al.* 1984). The predictions of the filtration theory model were largely verified for movement of bacteria over a distance of 7 m at the field site. These results notwithstanding, it is clear that much more work is needed to answer fundamental questions about the sources of subsurface microorganisms. Colonization factors, such as the depth and age of buried sediment, its mode and time of deposition, its geochemical and biological history, and its stratigraphic relationship to recharge areas upgradient in the hydrological flow path, must be considered in attempts to explain sources of microorganisms in the sediment. Experiments addressing the influences of these factors have yet to be designed and implemented.

8.3.2 Metabolic diversity of microorganisms

The accumulating evidence indicates that most of the known physiological types of bacteria that we have come to expect in moderate (as opposed to extreme) marine and freshwater habitats on the Earth's surface are also present in moderate subsurface habitats.[4] Ghiorse & Wilson (1988) compiled an extensive list of reports examining microorganisms and their potential metabolic activities in relatively shallow drinking-water aquifer sediments. A variety of experimental

[4] Of course, we also would expect the subsurface to contain microorganisms with specialized physiological traits in regions dominated by environmental extremes. Up to now such regions, e.g., ultradeep sediments and sediments in volcanic regions, have not been investigated thoroughly. Current drilling in very-dry sediments of the western United States may provide results on subsurface microorganisms from regions of very-low water potential.

Table 8.2 Summary of microbiological groups detected in subsurface habitats

Microbiological group	Sample type	Method of detection	References
Aerobic chemoheterotrophic bacteria	Well water	Microscopy, ATP, loss of various hydrocarbons in bottles incubated with water from unpolluted and polluted sites	Aamand *et al.* (1989)
	Shallow aquifer contaminated with jet fuel	$^{14}CO_2$ evolution and ^{14}C incorporation into biomass from ^{14}C-glucose, but not ^{14}C-benzene or toluene	Aelion & Bradley (1991)
	Well water from shallow aquifer at hazardous waste site	^{14}C-glucose and toluene mineralization, ^{14}C-amino acid incorporation, MPN and direct microscopic counts	Armstrong *et al.* (1991)
	Shallow aquifer	Microscopy; membrane lipid, cell wall components, ATP	Balkwill *et al.* (1988)
also actinomycetes and fungi	Deep coastal plain aquifer	Microscopy, agar media	Balkwill (1989)
	Deep coastal plain aquifer	21 physiological tests applied to 1112 isolated bacteria	Balkwill *et al.* (1989)
	Shallow sediment	Microscopy, analysis of cell constituents applied to bacterial isolates grown on glass beads or gravel	Bengtsson (1989)
	Shallow aquifer	Isolation and characterization of *Pseudomonas geomorphus*	Breen *et al.* (1989)
	Sandy aquifer	Field study using 42 monitoring wells, 10 sampling periods over 3 years. Water analysis and computer modeling showed loss of benzene, toluene, xylene	Chiang *et al.* (1989)
also manganese solubilizers	Shallow sediments	Screening of 105 isolates for Mn-reducing activity, protein assays, growth with and without added MnO_2	Di-Ruggiero & Gounot (1990)
	Pond sediments	$^{14}CO_2$ evolution from radiolabeled surfactants, microscopy, fluorescein diacetate hydrolysis	Federle & Pastwa (1988)
	Deep coastal plain aquifer	Enrichment cultures able to mineralize trichloroethylene	Fliermans *et al.* (1988)
	Deep coastal plain aquifer	Growth on agar media, agarose gel electrophoresis for detection of plasmids	Fredrickson *et al.* (1988)
	Deep coastal plain aquifer	Microscopy, growth on various compounds, $^{14}CO_2$ evolution from toluene, naphthalene; hybridization with pWWO and NAH7 plasmids	Fredrickson *et al.* (1991a)

continued on p. 178

Table 8.2 *Continued*

Microbiological group	Sample type	Method of detection	References
	Deep coastal plain aquifer	Agar media, 108 physiological tests applied to 198 isolated bacteria	Fredrickson *et al.* (1991b)
	Deep coastal plain aquifer	$^{14}CO_2$ evolution from acetate, 4-methoxybenzoate, phenol, growth on agar media	Hicks & Fredrickson (1989)
	Well water	Colonization of sterile sediments placed into a groundwater well, subsequent microscopy and plating on agar media	Hirsch & Rades-Rohkohl (1990)
	Well waters	Microscopy, ATP, uptake of ^{14}C-glucose, acetate, and naphthalene. ^{3}H-thymidine incorporation	Jensen (1989)
	Deep coastal plain aquifer	DNA homology, moles guanine + cytosine, physiological tests	Jiménez (1990)
	Shallow aquifer	Analysis of water from a field site, computer modeling, $^{14}CO_2$ evolution from ^{14}C-cresol, -phenol and -naphthalene	Klecka *et al.* (1990)
	Water and sediment from a shallow aquifer	Microscopy, agar media, 155 morphological, and physiological tests applied to 2700 isolated bacteria	Kölbel-Boelke *et al.* (1988a,b)
	Shallow unsaturated zone	$^{14}CO_2$ evolution and ^{14}C incorporation into biomass from ^{14}C-glucose, -phenol, and -aniline; microscopy, agar media, ^{32}P incorporation into phospholipid	Konopka & Turco (1991)
	Shallow infiltration basin	Growth on agar and liquid media	Lindqvist & Bengtsson (1991)
	Deep coastal plain aquifer	Incubated core material, no added organic carbon	McMahon *et al.* (1990)
	Shallow aquifer	Growth response to hydrocarbons, protein dendrogram analysis and physiological tests performed on 300 isolated bacteria	Ridgway *et al.* (1990)
	Shallow aquifer	$^{14}CO_2$ evolution from phenol, paranitrophenol, ethylene dibromide, glucose	Swindoll *et al.* (1988a)
	Shallow aquifer	$^{14}CO_2$ evolution and ^{14}C incorporation into biomass from a variety of natural and xenobiotic organic compounds	Swindoll *et al.* (1988b)
	Anoxic shallow aquifer	ATP, disappearance of alkylbenzenes and halogenated aliphatic hydrocarbons from laboratory cultures of aquifer slurries	Wilson *et al.* (1986)

continued

Table 8.2 *Continued*

Microbiological group	Sample type	Method of detection	References
Iron precipitating and Mn-oxidizing bacteria, aerobic chemoheterotrophic bacteria	Well waters, aquifers, sediments	Microscopy, agar media	Hirsch & Rades-Rohkohl (1988)
Lithotrophic and heterotrophic bacteria	Deep coastal plain aquifer	Agar media, MPN enrichments	Fredrickson *et al.* (1989)
Methanotrophic bacteria	Shallow aquifer	Enrichment cultures able to metabolize both methane and trichloroethylene	Alvarez-Cohen & McCarty (1991)
	Shallow aquifer	Enrichment cultures able to metabolize both methane and trichloroethylene	Henry & Grbic-Galic (1991)
	Shallow aquifer	Laboratory incubation of aquifer material and pure cultures, demonstrating cell growth, oxidation of both methane and trichloroethylene	Henry & Grbic-Galic (1990)
Aerobic and anaerobic chemoheterotrophic bacteria	Shallow aquifer	Microscopy and agar media	Bone & Balkwill (1988)
	Deep coastal plain aquifers	Agar media, metabolism of quinoline	Brockman *et al.* (1989)
	Shallow aquifers	Laboratory column breakthrough curves for chlorophenols, nitrophenols, and organo-phosphorus pesticides	Kjeldsen *et al.* (1990)
	Deep coastal plain aquifer	$^{14}CO_2$ evolution from glucose and indole. $^{35}SO_4$ reduction	Madsen & Bollag (1989)
	Deep coastal plain aquifer	Incorporation of radiolabeled acetate and thymidine into biomass. $^{14}CO_2$, $^{14}CH_4$ evolution from acetate and glucose; reduction of $^{35}SO_4$	Phelps *et al.* (1989b); Phelps & White (1991)
	Sandy subsoils	Microscopy, low and high nutrient agar media	van Beelen & Fleuren-Kemilä (1989)
	Shallow aquifer, peat, subsoils	$^{14}CO_2$ evolution from ^{14}C-glucose and -acetate, microscopy, toxicity assays using pentachlorophenol, Zn, Cd, chlorite	van Beelen *et al.* (1991)
Protozoa and aerobic chemoheterotrophic bacteria	Shallow aquifer	Microscopy, agar media, MPN enrichments, ATP	Beloin *et al.* (1988)
	Four subsurface sediments	Phospholipid analysis, enzymatic hydrolysis of fluorescein	Federle *et al.* (1986)

continued on p. 180

Table 8.2 *Continued*

Microbiological group	Sample type	Method of detection	References
also actinomycetes and fungi	Shallow sandy aquifer	Phospholipid-derived fatty acids, agar media, $^{14}CO_2$ evolution from radiolabeled alkylbenzene sulfonate and alcohol ethoxylate, ^{3}H incorporation into biomass	Federle *et al.* (1990)
	Various shallow aquifers	Microscopy, agar media, MPN enrichment	Sinclair (1991)
also actinomycetes, fungi, and algae	Deep coastal plain aquifer	Microscopy, agar media, MPN enrichment, ATP	Sinclair & Ghiorse (1989)
also fungi	Aquifer sediments	Microscopy, agar media, MPN enrichment	Sinclair *et al.* (1990)
also fungi	Shallow aquifer	Microscopy, agar media, MPN enrichment	Sinclair & Ghiorse (1987)
Other anaerobic microorganisms			
Fermenters and sulfate reducers	Deep coastal plain aquifers and aquitards	Pore-water analysis, production of CO_2, formate, and acetate from radiolabeled glucose	McMahon & Chapelle (1991)
Nitrate and nitrous oxide reducers	Shallow aquifer	Loss of NO_3^- and N_2O coupled to disappearance of benzene, toluene, ethylbenzene, xylenes, and trimethylbenzene	Hutchins (1991)
Denitrifiers	Deep coastal plain aquifer	N_2O production	Francis *et al.* (1989)
Denitrifiers	Shallow aquifer	Disappearance of toluene, ethylbenzene, xylenes and 1,2,4-trimethylbenzene from aquifer slurries	Hutchins *et al.* (1991)
Denitrifiers	Shallow aquifer	Loss of benzene, toluene, and xylenes from aquifer slurries; N_2O production	Major *et al.* (1988)
Denitrifiers	Shallow aquifer	ATP, MPN enumeration of denitrifiers; nitrate loss, and N_2O production	Obenhuber & Lowrance (1991)
Denitrifiers	Shallow aquifer	N_2O production, NO_3^- loss in sediments	Slater & Capone (1987)
Denitrifiers	Shallow coastal aquifer	Production of N_2O by aquifer slurries	Smith & Duff (1988)
Dissimilatory and assimilatory nitrate reducers	Shallow aquifer	Laboratory column microcosms ^{15}N gas chromatography/mass spectrometry, ^{3}H-thymidine incorporation	Bengtsson & Annadotter (1989)
Denitrifying and aerobic chemoheterotrophs	Shallow septic tank sediments	$^{14}CO_2$ evolution from NTA	Ward (1986)

continued

Table 8.2 *Continued*

Microbiological group	Sample type	Method of detection	References
Iron-reducing bacteria	Deep coastal plain aquifer	Laboratory incubations of sediments and a pure culture, growth on acetate, conversion of Fe(III) to Fe(II)	Lovley *et al.* (1990)
Methanogens	Shallow aquifer	HPLC and MS analysis show reductive debromination of a herbicide in aquifer slurries	Adrian & Suflita (1990)
Methanogens, sulfate reducers, denitrifiers	Subsoils, aquifers (several sites)	Agar plates, disappearance of methanol and tert-butylalcohol from aquifer slurries	Hickman *et al.* (1989)
Methanogens, other anaerobes	Shallow aquifer	Disappearance of tetrachloroethylene and toluene added to aquifer slurries	Sewell & Gibson (1991)
	Shallow aquifer	Disappearance of hydroxybenzoate isomers added to aquifer slurries	Kuhn *et al.* (1989)
Sulfate reducers	Shallow aquifer	Disappearance of cresol isomers added to aquifer slurries, $^{14}CO_2$ evolution from ^{14}C-*p*-cresol	Suflita *et al.* (1989)
Sulfate reducers, nitrate reducers	Shallow aquifer	Disappearance of cresol isomers and hydroxybenzoic acid and sulfate added to aquifer slurries	Ramanand & Suflita (1991)
Methanogens, sulfate reducers, acetogens, coliforms, fecal streptococci	Deep coastal plain aquifer	MPN enrichments, HPLC analysis of media	Jones *et al.* (1989)
Methanogens, sulfate reducers	Shallow anoxic aquifer	Disappearance of chloroanilines added to aquifer slurries	Kuhn *et al.* (1990)
Facultative anaerobic nonfermenting heterotrophs, anaerobic chemoheterotrophs, methanogenic bacteria, sulfate-reducing bacteria	Deep granitic water	Microscopy, agar media, liquid enrichment media	Pederson & Ekendahl (1990)
Metabolically active microorganisms	Three shallow aquifers	Tritiated thymidine incorporation	Thorn & Ventullo (1988)
	Well water	Microscopy, filtration, staining for viable cells	King & Parker (1988)
	Well water	Microscopy, filtration, staining for viable cells	Marxsen (1988)

ATP, adenosine triphosphate; HPLC, high-performance liquid chromatography; MPN, most probable number enumeration; MS, mass spectrometry; NTA, nitrilotriacetic acid.

techniques performed on samples from many sites revealed wide-ranging metabolic capabilities of both aerobic and anaerobic microorganisms. Kölbel-Boelke *et al.* (1988b) listed physiological properties of 2700 aerobic heterotrophic bacteria isolated from a Pleistocene sand aquifer in northern Germany. More recently, Balkwill (1990) tabulated 27 metabolic groups or physiological types of aerobic and anaerobic bacteria isolated from a single deep aquifer drilling site in South Carolina. To further illustrate the breadth of metabolic diversity found among subsurface microorganisms in shallow and deep aquifer systems we have summarized results of recent studies (published since 1985) in which the physiological types or activities of subsurface microorganisms were determined (Table 8.2). In evaluating the meaning of these results, it is important to recognize the distinction between metabolic potential, determined by incubating field samples in the laboratory, and *in situ* microbial activity (see Section 8.3.6). Despite this caveat, it is clear from Table 8.2 that the metabolic diversity of subsurface microbial communities, like those

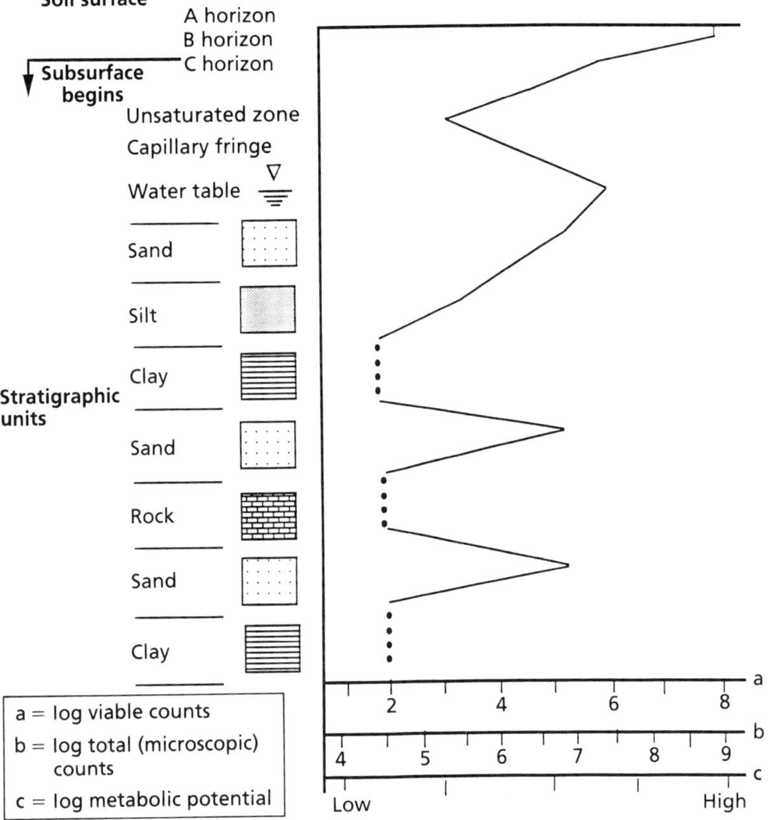

Figure 8.2 Generalized vertical profile showing relationships between microbiological parameters and geologic stratigraphy. (See text for discussion and references.)

of all other moderate habitats in the biosphere, is substantial. This diversity has major implications for all aspects of groundwater microbiology, especially for biodegradation of organic contaminants and the biogeochemical cycling of carbon, nitrogen, and other nutrient elements in groundwater ecosystems (see Section 8.3.8).

8.3.3 Major microbiological groups and their spatial distribution

The distribution patterns of microorganisms on large (meter), small (centimeter), and micro (micrometer) scales are major considerations in developing an understanding of the groundwater habitat. Large-scale spatial distribution patterns of microorganisms, especially bacteria, and their relationships to the sedimentology, geochemistry, hydrology, and stratigraphy of the geological profile have emerged as a common theme in many recent investigations of groundwater systems (Table 8.2). The small and microscale distribution patterns are much more difficult to study. They are only now being investigated (J.K. Fredrickson personal communication, 1992). Therefore, in the following pages we review the existing information on large-scale distribution patterns of bacteria and other microorganisms in subsurface sediments.

8.3.3.1 Bacteria

A generalized model of the relationships between typical hydrogeological zones in the terrestrial subsurface and the abundance of unicellular bacteria and their metabolic potential is presented in Figure 8.2. Surface soils in temperate regions typically contain 10^8–10^9 bacteria per gram. The major chemical and physical influences which govern bacterial abundances in soil, available organic carbon, nitrogen, phosphorus, sulfur, moisture, pH, electron acceptors, grazing by predators, immigration of microorganisms from other habitats, etc. (Alexander 1977), are modified in the subsurface along the hydrological flow paths. The nature of the geological stratum, mineral type, particle size distribution, texture, hydraulic conductivity, etc., also may determine the abundance and distribution of bacteria in a given subsurface zone.

In descending from the A to B soil horizons into the C horizon, a decline in nutrient levels is accompanied by a proportional decline in bacterial abundance (see Figure 8.2). This decline may be very drastic. Indeed, many reports in the older soil microbiology literature suggested that few, if any bacteria, existed in the C horizon (Alexander 1961). As mentioned in Section 8.1.1 the C soil horizon marks the beginning of the unsaturated subsurface zone, which supports far fewer bacteria than the B soil horizon. However, the numbers of bacteria usually do not continue to diminish with depth. Instead, microbial abundance typically increases substantially at the water table and just above it in the capillary fringe zone. We speculate that these interface zones between the unsaturated and saturated zones may be the site of relatively dynamic mixing of oxygen and recently recharged nutrients in shallow unconfined aquifers. Both horizontal saturated flow and vertical seasonal fluctuations of the water table may contribute to this effect.

As one bores deeper through the water table into the saturated zone, the abundance and potential activities of microorganisms generally remain high relative to the unsaturated zone. In a given locale, both the moisture regime and the type of geological strata below the water table may be highly varied, e.g., sediments high in clay of low transmissivity, beds of crystalline or porous rock with varying degrees of fracturing, and zones of highly transmissive sand and gravel may be present. A generalized relationship between sediment stratigraphy and bacterial abundance is shown in Figure 8.2. Highly transmissive saturated zones (high-yield aquifers containing water that was recharged relatively recently) typically show microbial abundances and metabolic activity potentials that are two to four orders of magnitude greater than those of hydrologically nontransmissive zones, whose waters and nutrients may be relatively old and depleted. Thus, depth *per se* has not been found to govern the abundance and activity of bacteria in the saturated zone; rather, the hydrological, physical, and geochemical properties of each stratum appear to govern the population density and degree of metabolic activity of its own community.

It should be noted here that endospore-forming bacteria have not been reported in significant numbers in any subsurface microbiological studies yet described, though they have been found in low numbers in some samples (Balkwill 1989). The apparent scarcity of these organisms is surprising given the known capacity of the endospore for long-term survival (see Sections 8.3.3.2 and 8.3.3.4) (Henis 1987).

The reader should realize that the generalizations shown in Figure 8.2 are based primarily on our own observations and those of our collaborators who worked on the sediment profile samples from the deep microbiology boreholes at the SRS in the south-eastern Atlantic coastal plain of the United States (Balkwill 1989; Francis *et al.* 1989; Fredrickson *et al.* 1989; Hicks & Fredrickson 1989; Jones *et al.* 1989; Madsen & Bollag 1989; Phelps *et al.* 1989b; Sinclair & Ghiorse 1989). We also considered results from our studies on vertical profile sediments from a series of buried valley aquifers (Sinclair *et al.* 1990), a shallow river flood plain aquifer in the midwestern United States (Beloin *et al.* 1988) and a shallow sandy aquifer in the north-eastern United States (Madsen *et al.* 1991). Given the limited number of sampling sites, we recognize that the generalized scenario of Figure 8.2 may change as our understanding is sharpened by data from new sites. However, we expect most of the significant features of this model will be confirmed for different hydrogeological conditions as they are encountered in the future. It is interesting to note that variations similar to those in Figure 8.2 also may be observed in shallow aquifers even when the sediments are of homogeneous sandy texture (Smith *et al.* 1991a). In this case, variations in the abundance and activity of bacteria seem to have been caused by chemical gradients in a polluted aquifer in Cape Cod, Massachusetts.

8.3.3.2 Filamentous spore-forming microorganisms (actinomycetes and fungi)

Both actinomycetes (Gram-positive, filamentous bacteria) and fungi (predominantly filamentous eukaryotes) are spore-forming microorganisms that are well adapted to environments subjected to wide variations in water availability and low-nutrient conditions (Alexander 1977; Henis 1987; Brock & Madigan 1988). Actinomycetes have not been detected in significant abundance in any subsurface microbiological studies reported to date. Approximately 4% of 1200 bacteria isolated from Atlantic coastal plain sediments were morphologically similar to actinomycetes; these were restricted largely to transmissive, sandy strata (Balkwill 1989). In examining a shallow sandy aquifer system in the north-eastern United States, Madsen *et al.* (1991) found that actinomycetes were restricted to the surface soil and unsaturated zone. Numbers of actinomycetes declined from 10^6 colony forming units (CFU) g^{-1} in the surface soil to below the detection limit ($10\,CFU\,g^{-1}$) in zones deeper than 3 m. Federle *et al.* (1990) detected phospholipid fatty acids characteristic of actinomycetes in the unsaturated zone of a sandy aquifer system to a depth of 13 m but not below. Other investigations examining a range of microbiological types in subsurface sediments from both humid (Sinclair *et al.* 1990) and arid (Colwell 1989) climates have specifically searched for actinomycetes and failed to find them. These results indicated that spore formation *per se* may not be a major fitness trait for bacteria inhabiting groundwater habitats. This conclusion is supported by the general lack of endospore-forming bacteria in subsurface sediments as well (see Section 8.3.3.4).

Much like actinomycetes, the abundance of fungi in subsurface habitats appears to be low. Low numbers of fungi ($<50\,CFU\,g^{-1}$) were detected in many subsurface samples from depths down to 216 m in the Atlantic coastal plain system (Sinclair & Ghiorse 1989). In shallow sandy aquifer sediments in the north-eastern United States, Madsen *et al.* (1991) noted that fungi declined from

10^5 CFU g^{-1} in the surface soil to below detection (<10 CFU g^{-1}) at depths greater than 2 m. In another sandy aquifer, Federle *et al.* (1990) identified characteristic fungal phospholipid fatty acids whose abundance varied in both unsaturated and saturated zones of the subsurface profiles to a depth of 19 m. In the deeper buried valley aquifers of north-eastern Kansas studied by Sinclair *et al.* (1990), fungi were not detected by plate counts in any samples. This finding was attributed to the generally elevated pH of the ground water in the area, which may have discouraged colonization by fungi.

It should be noted that fungal and actinomycete metabolism would be included in the radiorespirometric assays listed in Table 8.2 as well as in the metabolic potential assessments shown in Figure 8.2. Most metabolic potential studies do not attempt to identify the contribution of any particular group within the community. Fungi and actinomycetes could be very active in aerobic heterotrophic metabolism within the larger subsurface microbial community. Thus, the contributions of these organisms to potential aerobic metabolic activity may be significant, even when they are initially present at low numbers. The trend indicated by published reports shows that the abundance of actinomycetes and fungi in groundwater habitats is generally low, but their presence in low numbers may be important when conditions change to favor their aerobic heterotrophic metabolism.

8.3.3.3 Protozoa

Protozoa are unicellular eukaryotic microorganisms with diverse morphological and physiological characteristics (Stanier *et al.* 1986; Fenchel 1987; Foissner 1987; Sleigh 1989; Finlay 1990). Many protozoa are known to survive nutrient starvation and other environmental stresses by producing cysts. Photosynthetic and parasitic protozoa are unlikely to have any significance except as transients in the subsurface habitat; however, the importance of the remaining types of free-living phagotrophic protozoa should not be underestimated. Ingestion of bacterial prey is the dominant mode of nutrition for free-living protozoa in aquatic and terrestrial habitats (Fenchel 1987) and it is in this predatory capacity that protozoa have been shown to accelerate the cycling of carbon and other nutrients (Stout 1980; Fenchel 1987). Because protozoan biomass usually reflects the rate at which they are able to graze on their bacterial prey (Stout 1980; Fenchel 1987; Foissner 1987), the population density of protozoa may serve as an indirect measure of *in situ* bacterial growth rate. To the degree that bacterial growth rate can be tied to specific types of metabolism, i.e., denitrification, methane oxidation, and biodegradation of organic pollutants, protozoan biomass becomes an indirect means of determining *in situ* microbial processes (Madsen *et al.* 1991). In this regard, the presence of protozoa in the subsurface may be very important in documenting the expression of metabolic potential in groundwater habitats (see Section 8.3.6).

Early studies of the abundance and distribution of protozoa in deep soil horizons (Waksman 1916; Sandon 1927) indicated that, like bacterial numbers, protozoan numbers decline drastically with depth in horizons beneath the soil surface. These observations were confirmed by the recent studies of Sinclair & Ghiorse (1987) and Beloin *et al.* (1988) which show that in a shallow aquifer system in Oklahoma, the abundance of cyst-forming protozoa declined to below the detection limit (0.2 protozoa g^{-1}) in the unsaturated zone. However, protozoa reappeared in samples from the capillary fringe zone. Cores from 7–8 m deep in the aquifer contained 50–100 protozoan cysts g^{-1} of sediment. Protozoa also were detected in other samples from various subsurface depths. Numbers were lowest in clay confining layers and highest in the coarse-textured sands and gravels in which bacterial abundance also was elevated. Sinclair & Ghiorse (1987) noted that nearly all of the protozoa in their samples were in the cyst form rather than in the vegetative form. The dominant populations of subsurface protozoa were flagellates and amoebae. Ciliates were never found in sediment samples from depths greater than 0.5 m. These investigations enumerated protozoa with a most probable number (MPN) technique that relies upon predation of bacterial prey. Additional evidence for protozoa in the subsurface has been obtained by analyzing phospholipid fatty acids extracted directly from sediments. Federle *et al.* (1986; 1990) found fatty acids characteristic of protozoa in subsurface saturated zones in Alabama and Wisconsin.

These observations have been confirmed in studies of other shallow and deep subsurface sites (Sinclair & Ghiorse 1989; Sinclair *et al.* 1990; Madsen *et al.* 1991). In general, three common themes have emerged in the protozoology of subsurface sediments:

1 a modest population density of 1–100 cyst-forming protozoa g^{-1} may be expected to occur in shallow pristine aquifer sediments and in some deep aquifer sediments containing bacterial population densities of 10^6–10^7 g^{-1};

2 the abundance of protozoa is related to sediment texture – high numbers of protozoa are restricted to sandy transmissive zones containing abundant bacteria;

3 cyst-forming flagellates and amoebae appear to be the principal types of protozoa in subsurface ecosystems.

8.3.3.4 Morphology and life cycle considerations

One property of a microbial cell that will have enormous influence on its distribution in nature is morphology (size and shape). Most microorganisms display some degree of morphological variability during their life cycles. This variability is strongly related to environmental conditions, especially nutrient concentrations. Thus, starving (non-growing) bacteria, fungi, and protozoa tend to be much smaller than their actively growing counterparts. Furthermore, dormant or resting stages (spores, cysts, sclerotia, akinetes, virions, etc., which provide a means of dispersal and survival under adverse environmental conditions) are included in the life cycle of many microorganisms (Henis 1987).

Considering all of the possible microorganisms that could colonize groundwater habitats, it is appropriate to discuss the distribution patterns of the major microbiological groups by noting some obvious absences. The results of all subsurface microbiological studies to date show that bacteria with simple life cycles are by far the most widely distributed organisms in groundwater habitats. Filamentous spore-forming bacteria (actinomycetes), endospore-forming bacteria, and cyst-forming myxobacteria, all of which are common in surface soil, are rare in the subsurface. Also rare are the somewhat larger spore-forming filamentous fungi that are common in soil. Protozoa, which require the presence of smaller prey species for long-term survival, tend to be limited to subsurface sediments that contain an adequate number of bacterial prey as well as sufficient porosity to allow grazing activity. As mentioned in Section 8.3.1, algae that form various kinds of resting stages may be present in some subsurface sediments, but those that depend entirely on phototrophic nutrition will not persist. Thus, groundwater habitats generally do not contain algae unless the geological stratum is hydrologically connected to surface water. Budding and appendaged bacteria, arguably some of the most widely distributed and durable bacteria in other oligotrophic but open aquatic environments, are not widely reported in groundwater habitats. Likewise, spirochetes and other spiral-shaped bacteria, commonly found in organic-rich marine or freshwater sediments with steep oxygen gradients, have not been reported in the subsurface.

It must be recognized that the lack of data documenting the above mentioned microbiological groups in the subsurface may reflect the use of methods which fail to detect these microorganisms. Indeed, Hirsch & Rades-Rohkohl (1983) did find a very wide diversity of budding and other bacteria in well water from a sandy aquifer in northern Germany when they employed the proper enrichment procedures. But an alternative explanation for the absence of particular microorganisms from groundwater habitats is simply that its slow-changing, low-carbon characteristics fail to match the dynamic environmental and nutritional conditions required by many microorganisms dwelling in other aquatic sediments and surface soil. For this reason, perhaps it should be expected that groundwater habitats support a distinctive microflora of their own, reflecting the physical and chemical properties of the subsurface milieu at each given location.

8.3.4 Physiological status of subsurface microorganisms

Regardless of the microbiological habitat examined, counts of viable cells on agar media are usually one to three orders of magnitude lower than microscopic (total) counts performed on the same sample (Ferguson *et al.* 1984; Roszak & Colwell 1987; Lee & Fuhrman 1990). This discrepancy is cause for an important conceptual and operational distinction between culturable and nonculturable microorganisms. The distinction reflects both the physiological requirements and metabolic status of microorganisms in nature. Each culture medium and set of incubation conditions is selective to some degree and therefore able to meet the physiological requirements of only part of the culturable component of a given microbial community. For a given culture medium, the organisms that fail to grow either have different physiological needs than those that do grow or are nonculturable. Nonculturability may have two distinct physiological bases: cells can be dormant, i.e., they may require special resuscitative events prior to regaining an ability to grow, or they may be nonviable, i.e., intact and, therefore, detectable microscopically; but unable to reproduce regardless of the growth conditions applied by the microbiologist (Hoppe 1978; Fuhrman & Azam 1982; Henis 1987; Roszak & Colwell 1987).

It is important to note here that the above definition of nonviability (and by default, the definition of viability) is based on microbiologists' inability to provide proper conditions for growth in laboratory media. In nature, the proper conditions may be provided by changes in any number of unknown environmental factors that can support growth. Also this definition of nonviability does not take into account the ability of bacterial cells to maintain some of their enzymatic activities and physiological functions but lose the ability to reproduce. Thus, the concept of nonviability and death among bacterial cells in any natural setting is somewhat arbitrary (see Henis 1987 for further discussion of the arbitrary nature of the terms: survival, dormancy, and death as applied to bacteria).

Understanding the dichotomies of culturable versus nonculturable and dormant versus nonviable microorganisms is important in assessing and predicting the responsiveness of subsurface microbial communities to perturbations, such as pollution events, to groundwater habitats. Groundwater nutrient concentrations are usually very low and flow velocities can be very slow, with hydraulic conductivities sometimes of the order of centimeters to meters per year (Freeze & Cherry 1979). Thus, it is reasonable to assume correspondingly slow rates of metabolism and growth (or dormancy, under extremely low-nutrient conditions) as the usual physiological status of subsurface microorganisms. However, accurate information on *in situ* metabolic rates and other direct measures of the physiological status of microorganisms in groundwater habitats are extremely difficult to obtain because of disturbance artifacts and uncertainties in interpreting field data (see Section 8.3.6).

Criteria for evaluating the physiological status of microorganisms in their natural settings (Karl 1986; White 1986) have been applied to subsurface samples. These include a wide variety of laboratory-oriented procedures: cell size (Ghiorse & Balkwill 1983; Hirsch & Rades-Rohkohl 1983), cell structures indicative of active growth or dormancy (Ghiorse & Balkwill 1983), accumulation of storage polymers such as poly-β-hydroxyalkanoates (Ghiorse & Balkwill 1983; Balkwill *et al.* 1988; Bengtsson 1989), adenosine triphosphatase (ATP) levels (Wilson *et al.* 1986; Balkwill *et al.* 1988; Beloin *et al.* 1988; Jensen 1989), various stains for viability (Marxsen 1988; King & Parker 1988), extraction of polar lipid fatty acids (White *et al.* 1983; Federle *et al.* 1990), incorporation of ^{3}H-thymidine into cell constituents (Thorn & Ventullo 1988; Phelps *et al.* 1989b), and use of time courses of substrate mineralization or other activity measures to infer both substrate turnover times (Ventullo & Larson 1985; Armstrong *et al.* 1991), and a state of metabolic readiness (Francis *et al.* 1989; Kuhn *et al.* 1989; Hutchins *et al.* 1991). (For additional references see Table 8.2.)

Taken in total, there are no clear trends in these studies to indicate the overall physiological status of subsurface microbial communities. Albeit for pristine aquifers, most of the evidence suggests that metabolic activity is very low. A large portion of the studies cited in Table 8.2 were designed to reveal only the presence or potential activity of microorganisms in subsurface samples. The results suggest that subsurface bacteria possess the ability to adapt quickly from nutrient-limited to nutrient-sufficient conditions. This nutritional adaptability may be a consistent trait of many of the culturable microorganisms in aquifers (Wilson *et al.* 1983; Thorn & Ventullo 1988; Balkwill *et al.* 1988; Bengtsson 1989). Only occasionally have adaptive time lags indicative of metabolic dormancy been observed (Swindoll *et al.* 1988a; Armstrong *et al.* 1991; Madsen *et al.* 1991). To the contrary, many authors have indicated that aquifer organisms appeared to be poised for rapid utilization of added substrates

(Federle & Pastwa 1988; Madsen & Bollag 1989; Phelps *et al.* 1989b). In addition, several field studies have obtained sufficient evidence to prove that subsurface microorganisms can be metabolically active *in situ* (see Section 8.3.6).

In all probability, as is true in other habitats, the physiological status of microorganisms in the terrestrial subsurface is highly variable over time and space reflecting site-specific environmental conditions. As indicated above (see Section 8.3.3 and Figure 8.2) microbial abundance and metabolic activity in subsurface zones is expected to vary depending on environmental properties, such as the water regime, pore-water chemistry, mineral texture and type, and the availability of nutrients (see Section 8.3.7). The physiological status of the subsurface microbial community will follow the dictates of the physical and chemical properties of the subsurface sediments.

8.3.5 Attached versus unattached microorganisms

Can the attached microbial community of an aquifer be represented by the unattached community in the ground water pumped from a well? This question was raised by microbiologists during early inquiries on the nature of the subsurface microflora. It has generated debate among microbiologists concerned with other habitats as well (Costerton *et al.* 1986). The answer to this question has major implications for subsurface microbiological sampling methodologies. Drilling to aseptically obtain aquifer core samples is expensive, laborious, and destructive of the habitat. Each core is unique. Furthermore, both spatial variability and replication are problematic and the typically low population density of microorganisms generates a need for large quantities of sample material that may be in short supply. Deep aquifer sampling introduces severe restraints from drilling fluid contamination and its control (Beeman & Suflita 1989; Phelps *et al.* 1989a). By contrast, although installing a permanent well may be expensive, laborious, and destructive of the groundwater habitat, once installed and properly developed and tested, a well can supply large representative samples of ground water for repeated future samplings. Conventional well-water samples have been used to infer abundance and potential activity of bacteria in an aquifer formation in several of the investigations listed in Table 8.2 (Marxsen 1988; King & Parker 1988; Jensen 1989; Aamand *et al.* 1989; Hirsch & Rades-Rohkohl 1990; Armstrong *et al.* 1991); however, problems of microbiological contamination (see Section 8.3.1) and physical/chemical disturbance of the aquifer must be considered when interpreting results. In some shallow homogeneous hydrogeological settings, multilevel wells have been used for vertical profile sampling of ground water (Smith *et al.* 1991a). Thus, for purely logistical reasons, using well-water samples to probe the function of groundwater microbial communities is highly desirable. Moreover, knowing the extent to which the unattached ground water and attached sediment communities are alike or different would have major implications for managing biodegradation processes in aquifers. The relative mobility of organic contaminants and biodegrading microorganisms needs to be known before predictions of the fate of a pollutant can be made.[5]

To rigorously compare communities of attached and unattached subsurface microorganisms, one must scrutinize the structure and metabolic function of each community. Conventional viable cell enumeration methods followed by isolation and identification of representative microorganisms constitute one approach. However, such conventional methods are inherently imprecise, due to uncertainties associated with selective viable cell enumeration procedures. Even if a comprehensive array of population density, diversity, and activity measurements were employed, apparent differences between attached and unattached communities may reflect the effects of solid surfaces on the behavior of microorganisms (Costerton *et al.* 1986; van Loosdrecht *et al.* 1990) rather than genetically distinctive population differences between the communities. In theory, the most incisive comparison would utilize the conventional microbiological techniques in tandem with molecular techniques, such as DNA fingerprinting (Jeffreys *et al.* 1985; Kirby 1990), 16S rRNA sequencing (Pace *et al.* 1986; Amann *et al.* 1990; Ward *et al.* 1990), or DNA hybridization (Lee & Fuhrman 1990) of nucleic acids extracted directly from sediment and water (Ogram *et al.* 1987; Holben *et al.* 1988; Steffan *et al.* 1988; Weller & Ward 1989; Somerville *et al.* 1989; Hilger & Myrold 1991; Tsai & Olson 1991). However, application of the molecular techniques, especially nucleic acid extraction, may present new problems (see Chapter 2). For example, they may be biased by unequal lysis efficiency and thus be selective for some microorganisms while ignoring others. The extraction efficiency for each type of subsurface sample would also need to be known. These and other technical problems have slowed the rigorous application of molecular techniques in groundwater studies, although progress is being made in developing the methods (D.L. Balkwill & S. Nierzwicki-Bauer personal communication, 1992).

Most studies of attached versus unattached groundwater micro-

[5] A variety of dispersive, advective, sorption, and solute reactions governing the transport of organic compounds in ground water have been described (Brusseau *et al.* 1989; Rao *et al.* 1989; Khondaker *et al.* 1990); however, much less information is available about transport of microorganisms in the subsurface (Scholl *et al.* 1990; Gannon *et al.* 1991b; Harvey & Garabedian 1991).

flora have been done using conventional comparison techniques. Harvey *et al.* (1984) found that unattached groundwater bacteria represented only a small proportion of the total number of cells in sandy sediments in a highly transmissive, Cape Cod, Massachusetts, aquifer, unless levels of metabolizable organic carbon and other nutrients were high. This observation was supported by Smith & Duff (1988), working at the same site, and by Federle *et al.* (1990) at another highly transmissive aquifer, who found that levels of denitrification and biodegradative activity, respectively, were lower in well water than in adjacent aquifer sediments. Similarly, Hazen *et al.* (1991) reported that microbial populations in well waters at the SRS were lower and less physiologically diverse than corresponding sediment-borne populations. Insight into the shift of microorganisms from attached to unattached status was obtained by Bengtsson (1989), who found that increasing the nutrient level in a flow-through system also increased the proportion of unattached cells. Major distinctions between sediment-bound and unattached bacteria were documented by Kölbel-Boelke *et al.* (1988b) who performed a battery of physiological and morphological tests on bacterial isolates from water and sediment from the same borehole.

The available information suggests that an unattached well-water community may not be a valid surrogate for the indigenous aquifer microflora which largely consists of attached bacteria. Unattached and attached groundwater communities appear to be different, though they probably represent overlapping components of a dynamic community. The view of Hirsch & Rades-Rohkohl (1988) provides an appropriate summary. These authors pointed out that the majority of microorganisms in pristine subsurface environments are likely to live attached to particles and that a large proportion of the naturally attached microflora is likely to have motile, unattached stages. They concluded that there is likely to be a constant exchange of organisms between the attached and unattached communities. We would add that the rate and extent of exchange is likely to be regulated by the availability of electron donors and acceptors, inorganic and organic nutrients, and other environmental factors affecting growth, attachment, and life cycles of the subsurface bacteria.

8.3.6 *In situ* microbial activity

Despite the critical nature of microbiological processes for ecosystem maintenance in general (Odum 1971; Atlas & Bartha 1987), and groundwater chemistry in particular (Ghiorse & Wilson 1988), methodological limitations impose severe constraints upon attempts on determine the extent to which microbial communities are active *in situ* (Bull 1980; Karl 1986; Madsen 1991). The same methodological problems confront efforts in all habitats addressed by environmental microbiologists.[6] These problems become particularly acute for groundwater habitats due to their inaccessibility (Madsen *et al.* 1991). Environmental microbiologists must contend with their own version of the Heisenberg Uncertainty Principle (Gibbs 1976; Madsen 1991): the closer a given process is examined, the more likely artifacts are to be imposed on measurements of that process. Put more concretely, determination of microbial activity in disturbed, displaced environmental samples incubated in the laboratory is likely to be quantitatively, even qualitatively, different from the same determination if it could be done *in situ* (Venrick *et al.* 1977; Karl 1986). At issue are biotransformation rates, mass balances, and distinction of biotic from abiotic processes. For example, if an organic contaminant disappears from a field site, how does one know why or how? Possible mechanisms of contaminant disappearance include dilution, dispersion, volatilization, photolysis, sorption, and biodegradation. Contrast this ambiguity with the tidiness of a laboratory biodegradation experiment in which a test compound is converted to carbon dioxide in aquifer sediments confined in a sealed flask. The interpretability gained by the latter approach is counter-balanced by a lack of relevance to *in situ* processes. Results of laboratory flask assays provide physiological information, a measure of "biodegradation potential," but they are not a sound basis for extrapolation back to the field setting.

What then is known about true *in situ* microbial activity in the groundwater habitat? Precious little – far less than the references tabulated in Table 8.2 would seem to indicate. The intent of Table 8.2 is to document the presence, abundance, and diversity of subsurface microorganisms. Documenting *in situ* microbial activity in the subsurface is an entirely different matter in which at least two rigorous criteria must be met:

1 measurements at a field site, usually on groundwater samples, must indicate that a change has occurred (over time or space);

2 microbiological causes of that change must be uniquely distinguished from other potential causes.

The references in Table 8.3 are the result of our search for reports meeting the two criteria mentioned above. Carbon dioxide production, denitrification, methanogenesis, methane oxidation, and biodegradation are represented. In documenting *in situ* subsurface microbial activity, the investigations listed in Table 8.3 followed a similar strategy:

[6] Methodological artifacts are probably minimized in lake-water studies when assays for microbial activity are performed using bottles incubated *in situ* for only brief time periods. Unlike groundwater sediment, lake water has no rigid mineral matrix whose disruption influences resident microorganisms. But even when conditions are optimal for lake-water studies "bottle effects" can be troublesome (Bower *et al.* 1987).

Table 8.3 *In situ* microbial activity in the subsurface habitat

Type of activity	Site	Evidence	References
Carbon dioxide production	Anaerobic deep coastal plain aquifer (southern Maryland)	Known flow path, change in groundwater chemistry, isotopic fractionation, microscopic and viable counts of bacteria	Chapelle *et al.* (1987)
	Coastal aquifer (south-eastern South Carolina)	Known flow path, change in groundwater chemistry, isotopic fractionation, microscopic and viable counts of bacteria	Chapelle *et al.* (1988)
	Anaerobic deep coastal plain aquifer (south-eastern South Carolina)	Known flow path, change in groundwater chemistry, geochemical modeling of inorganic reactions, laboratory incubation of subsurface sediments amended with ^{14}C-acetate and -glucose	Chapelle & Lovley (1990)
	Ogallala aquifer (Texas)	Vertical profiles of gases, isotopic fractionation	Wood & Petraitis (1984)
Denitrification	Jurassic limestone and chalk aquifers (UK)	Known flow path, change in groundwater chemistry, microbial counts of denitrifiers	Foster *et al.* (1985)
	Alluvial aquifer	Vertical and seasonal profiles in water chemistry, ^{15}N isotopic fractionation, N_2O production in field cores using acetylene-block method	Fustec *et al.* (1991)
	Shallow coastal well waters (Cape Cod, Massachusetts)	Known flow path, detection of N_2O in water-chemistry profiles, microscopy, ^{3}H-glucose uptake by bacteria	Smith *et al.* (1991a)
	Shallow sand aquifer (Ontario)	Known flow path, change in groundwater chemistry relative to bromide tracer, microbial counts of denitrifiers	Trudell *et al.* (1986)
	Limestone ground water (eastern UK)	Known flow path, change in groundwater chemistry, isotopic fractionation	Wilson *et al.* (1990)
Methanogenesis	Shallow anoxic aquifer (Oklahoma)	Field installed gas collection devices sampled over time, laboratory incubation aquifer material, MPN enumeration of methanogens and sulfate reducers	Beeman & Suflita (1990)
	Neocene-Quarternary deep waters associated with industrial oil and gas deposits (Caspian Sea)	Isotopic fractionation in waters from varying depths, enumeration of methanogens, laboratory incubation of water samples converting ^{14}C-bicarbonate and -acetate to $^{14}CH_4$	Belyaev & Ivanov (1983)

continued on p. 196

Table 8.3 *Continued*

Type of activity	Site	Evidence	References
	Eocene aquifer (east-central Texas)	Known flow path, change in groundwater chemistry, isotopic fractionation	Grossman *et al.* (1989)
Methane oxidation	Shallow coastal aquifer (Cape Cod, Massachusetts)	Known flow path, loss of methane relative to conservative tracer injected into water at field site, conversion of injected $^{13}CH_4$ to ^{13}C-inorganic carbon	Smith *et al.* (1991b)
Biodegradation of polycyclicaromatic hydrocarbons	Shallow unconfined aquifer (north-eastern USA)	Metabolic adaptation to polycyclicaromatic hydrocarbons, enhanced protozoan biomass indicating *in situ* microbial growth in a contaminant plume	Madsen *et al.* (1991)
Biodegradation of chlorinated ethenes	Shallow confined aquifer (central California)	Known flow path, mass balances, cometabolic process stimulated by pulses of O_2 and CH_4, replicated in time	Semprini *et al.* (1990)

1 key chemical criteria for the microbiological process of interest were identified, e.g., aerobic respiration consumes oxygen and produces carbon dioxide;
2 a field system was characterized to allow measurement and interpretation of chemical data;
3 biotic processes were distinguished from abiotic ones;
4 in many cases confirmatory microbiological evidence was obtained, e.g., isolation and/or enumeration of microorganisms which catalyze the process of interest.

The work by Chapelle & Lovley (1990) (see Table 8.3) is particularly insightful because it confronted (but did not completely solve) the problem of attributing changes in water chemistry to microbial activity along flow paths tens of kilometers long. Furthermore, they found that estimates of carbon turnover based on laboratory incubation of subsurface sediments were much higher than the *in situ* process as determined using geochemical analyses and computer modeling. Smith *et al.* (1991b) (see Table 8.3) adeptly demonstrated *in situ* methane oxidation in a shallow saturated aquifer by injecting methane, ^{13}C-methane and conservative tracers and monitoring these analytes using multilevel sampling devices in down-gradient target wells. Not only did the concentration of methane decrease relative to the conservative tracer, but a pulse of ^{13}C-enriched dissolved inorganic carbon proved that ^{13}C-methane had been converted to

$^{13}CO_2$ (Smith *et al.* 1991b). By examining data from a comprehensive set of biomass, community structure, and biodegradation rate assays in subsurface samples along a gradient of contaminant concentrations, Madsen *et al.* (1991) (see Table 8.3) linked enhanced numbers of protozoa to *in situ* growth and enhanced activity of contaminant-adapted bacteria. Semprini *et al.* (1990) (see Table 8.3) were able to directly measure *in situ* biodegradation at a highly instrumented field site because the confined aquifer allowed mass balances to be constructed and because oxidation of chlorinated ethenes is a cometabolic process that was uniquely distinguishable from abiotic reactions. These investigations show that the methodological limitations that have impaired studies of *in situ* microbial activity in the subsurface can be surmounted by creative applications of conventional techniques. We anticipate accelerated progress on this important problem in future, especially towards measuring rates of *in situ* biotransformation processes.

8.3.7 Nutrient limitations

Heterotrophic bacteria appear to dominate the microflora of most subsurface ecosystems (see Section 8.3.3). Heterotrophs use organic compounds as sources of carbon and energy, thus organic carbon should be expected to be the most limiting nutrient for growth and activity of subsurface bacteria. Limitation of heterotrophic activity by inorganic nutrients is also possible. Limitation by inorganic nutrients has been found most commonly in surface environments where the organic carbon is high compared to the biomass of bacteria (Fenchel & Harrison 1976; Fogg 1988). For example, vascular plant detritus (from decaying leaves and wood) has carbon/nitrogen ratios of the order of 100:1 to 1000:1, while ratios in bacterial biomass are about 5:1 (Fenchel & Blackburn 1979). Thus in carbon-rich environments, inorganic nitrogen should be assimilated during heterotrophic decomposition. Enrichment experiments have shown that decomposition of carbon-rich substrates is stimulated by adding inorganic nitrogen and phosphorus (Valiela 1984; Webster & Benfield 1986; Fogg 1988). Likewise, it has long been known that addition of inorganic nutrients will stimulate mineralization rates of petroleum hydrocarbons under aerobic conditions (Bartha & Atlas 1977; Westlake *et al.* 1978). As a consequence of such studies, nutrient enrichment has become a basic strategy for bioremediation of toxic organic compounds in surface soil (Morgan & Watkinson 1989). The same strategy has been applied to shallow aquifers (Lee *et al.* 1988).

Despite the optimism over nutrient enrichment as a bioremediation strategy, little is known about the nutrient limitations that may control microbial growth in uncontaminated shallow aquifer systems (see discussion in Armstrong *et al.* 1991). Enrichment experiments have been conducted using shallow aquifer sediments, mostly with the purpose of identifying nutrient effects on the biodegradation of organic contaminants. Swindoll *et al.* (1988a) amended subsurface sediments with inorganic nutrients and observed changes in biodegradation and adaptation rates, but no single nutrient appeared to be most limiting. Thomas *et al.* (1989) amended subsurface sediments with different nitrogen and carbon sources and measured rates of polycyclic aromatic hydrocarbon mineralization; but no increases in mineralization rates were noted for any of the amendments. In contrast, Mercado *et al.* (1988) reported a possible stimulation of *in situ* denitrification by injecting a combination of sucrose and phosphate directly into an aquifer. In the latter investigation, it is unclear which of the two nutrients (phosphorus or carbon) may have been stimulatory. Armstrong *et al.* (1991) amended well water from a hazardous waste site with inorganic sources of nitrogen, phosphorus, potassium, and sulfur. Nitrogen, potassium, and phosphorus enhanced toluene mineralization in water from a contaminated well, with potassium and phosphorus being most stimulatory. Bengtsson (1989) also studied the response to nutrient enrichment of bacteria from a shallow aquifer. Phosphorus was indicated as a limiting nutrient. These reports notwithstanding, the current literature on this topic is too scant and the problems too complex (see discussion below) to allow generalizations to be made at this time.

In deeper aquifers, the supply of readily utilizable carbon and energy sources may be extremely small. In theory, only the most refractory organic compounds will survive the long and complex pathways through biologically active subsoil and subsurface sediments (Ghiorse & Wilson, 1988). Thus, as indicated above for shallow aquifers, organic carbon probably is the most limiting nutrient in deep aquifers. However, inorganic phosphorus, nitrogen, and other mineral nutrients, e.g., potassium, sulfur, and iron, must also be considered. Results from the analysis of pore water from deep boreholes at the SRS (Levine & Ghiorse 1991) indicated that phosphorus concentrations were typically very low while concentrations of dissolved organic carbon, nitrogen, and other elements appeared to be sufficient to support the standing crop of bacteria. However, the SRS deep aquifer pore-water data may have been influenced by drilling fluid intrusion. Nevertheless, the experimental evidence of Phelps & White (1991) also suggests that phosphorus may be a limiting nutrient in the SRS aquifers. These investigators added inorganic nitrogen, sulfur, and phosphorus sources to sediments and compared the resulting rates of ^{14}C-acetate incorporation into lipids (an estimation of microbial growth rates) with the rates measured for untreated sediments. Incorporation was consistently stimulated by adding phosphorus as phosphate. By contrast, nitrogen and sulfur amendments only occasionally stimulated ^{14}C-acetate incorporation in the sediments. These results support the tentative conclusion of Levine & Ghiorse (1991) that, despite the likely abundance of phosphate minerals in aquifer sediments (Matthess 1982; Ghiorse & Wilson 1988), phosphorus could be the most limiting nutrient in the SRS aquifers. However, as indicated above, no firm conclusions can yet be drawn from the existing data.

Further complexities of the nutrient limitation problem must be recognized. For example, the nutrient supply is more likely to be influenced by proximity to the surface and rapid recharge rates in shallow unconfined aquifers, than in deeper confined aquifers. Also, seasonal nutrient fluxes might be expected in shallow unconfined aquifers, but not in deeper ones. Moreover, nutrient availability in both shallow and deep aquifers will be influenced by local solid phase geochemical conditions, mineralogy, organic and inorganic coatings on mineral grains, surface pH, and redox conditions. Thus, predictions based on well-water or single-time-point pore-water chemistry alone probably cannot provide the best information on the nutrients available to the microflora in a given aquifer. This emphasizes the need for assay methods to determine nutrient limitations directly in aquifer sediments from a specific site at a specific depth. In our laboratory, a flow-through microcosm system is being developed for this purpose by S.N. Levine and D.T. Rudnick. The system includes an aseptic subcoring device to obtain relatively undisturbed subsurface samples and the use of aseptic flow conditions to allow for comparison of amended and aseptic control microcosms (Rudnick *et al.* 1992). The long-term aim of this project is to develop a system to determine the most limiting nutrient(s) for heterotrophic bacteria in an aquifer.

8.3.8 Nutrient cycling

In keeping with the focus of this article on ecosystem processes, it is appropriate to speculate on large-scale nutrient cycling in subsurface ecosystems. Bormann & Likens (1967; 1979) have presented a comprehensive definition of nutrient cycling in forested ecosystems which links the physical transport of nutrient elements to their temporal exchange with biomass and other pools within the system. Figure 8.3 is based on figures presented by Bormann & Likens (1967; 1979) but modified to emphasize the importance of hydrological flow and microorganisms in the groundwater habitat. The key role played by microorganisms in nutrient cycling is the enzymatic conversion of complex, high molecular weight carbon compounds, e.g., poly-

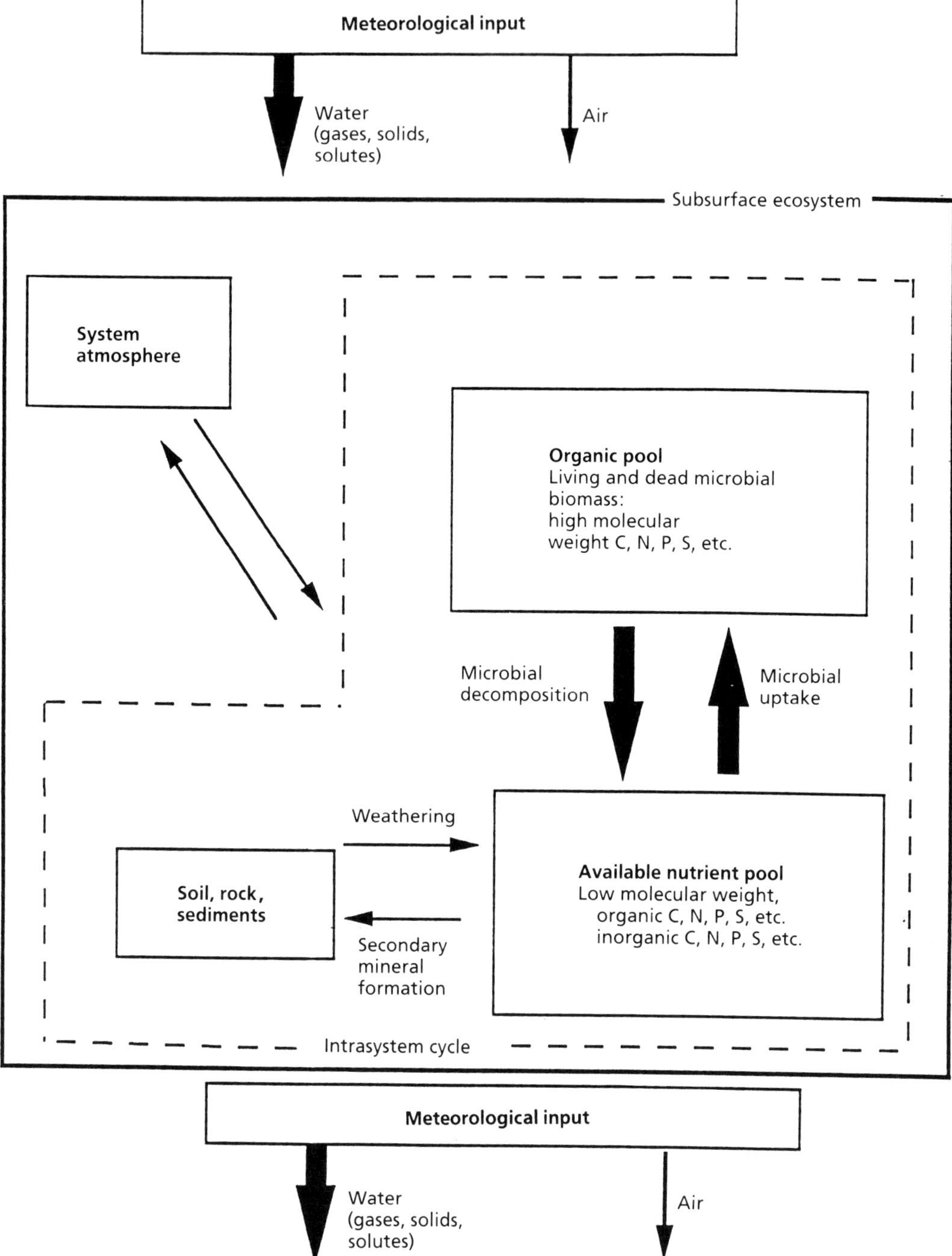

Figure 8.3 Nutrient cycling in subsurface ecosystems. Water flow through the subsurface is the dominant mechanism by which materials are transported within and exchanged between subsurface and adjacent ecosystems. Microorganisms are the principal inhabitants of the groundwater habitat. Delivery of organic substances to subsurface sediments, their microbiological colonization, and removal of by-products of microbial metabolism are all dependent upon hydrological flow. Adapted from Bormann & Likens (1967; 1979).

saccharides, lipids, protein, lignin, or humic material, either to simple low molecular weight components, e.g., sugars, fatty acids or amino acids or phenolic compounds, or to inorganic compounds, e.g., CO_2, NH_3, SO_4^{2-}, PO_4^{3-}, in the available nutrient pool. This catalysis is driven largely by the physiological need of heterotrophic microorganisms for carbon and energy. The other nutrient elements (nitrogen, phosphorus, sulfur, etc.) accompany carbonaceous materials during their transformation.

Primary production is unlikely to occur in groundwater ecosystems (see Section 8.2.2); therefore, autochthonous carbon compounds for fueling ecosystem function are restricted to those which comprise the microorganisms themselves or are contained in the subsurface rocks and sediments. Much of the autochthonous carbon may be unavailable for metabolism. Thus, allochthonous DOC, the majority of which may consist of the recalcitrant fraction of organic substances derived from surface plant material (Ghiorse & Wilson 1988), delivered through hydrological recharge and groundwater flow, may govern subsurface microbial metabolism (see Section 8.2.2). By definition, nutrient cycling describes the passage of an element through a series of steps that return it to its original state (see Figure 8.3). Evidence arguing for completion of the simplest (two-step) case of a nutrient cycle must document both the decomposition and uptake shown in Figure 8.3.

In marine and freshwater habitats, as well as terrestrial habitats, the most obvious evidence for microbiological cycling of nutrient elements is a balanced steady state between pools of detritus and living biomass. For instance, aquatic macrophytes grow each year and the previous year's primary production decays. Meanwhile, a portion of the limiting nutrients, such as phosphorus, remain within the system, becoming available for each new generation of macrophytes through a complex series of reactions (Wetzel 1983). The linkages between microbiological decomposition and higher plant uptake and growth are obvious and attributable to distinctive residents of the aquatic community. Similar observations cannot be made in the subsurface habitat because it is opaque, inaccessible, spatially heterogeneous, and its biota is entirely microscopic. Instead, more detailed and demanding types of investigations, involving identification of flow paths, changes in water chemistry, and unique processes attributable solely to microorganisms, must be carried out.

Based on chemical analyses of groundwater samples, Baedecker & Back (1979) assembled a plausible case for multiple transformations of carbon and nitrogen in sediments beneath a landfill. These results not withstanding, studies of nutrient cycles in groundwater habitats are very rare. This should come as no surprise, given the dearth of information available on *in situ* microbial activity discussed above (see Section 8.3.6). In the simplest case, each microbial activity listed in Table 8.3 represents one-half of the nutrient cycling process (uptake or decomposition). The task of measuring two complementary processes becomes extremely challenging, because the steps may occur simultaneously or may be separated by unknown scales of time and distance. Advancements in determining cyclic metabolic processes in subsurface ecosystems will depend upon improvements in techniques for defining and measuring both hydrological flow paths and microbiologically catalyzed fluxes between nutrient pools. One approach that has promise would apply the mass-balance concept of Bormann & Likens (1967; 1979) to the subsurface. It is theoretically possible to estimate the fraction of total annual photosynthesis that escapes the soil and is processed in the groundwater habitat by measuring pore-water chemistry at different distances along defined hydrological flow paths.

8.4 Summary and perspectives

Groundwater microbiology is subject to the vagaries of both environmental microbiology and hydrogeology. The traditional methodological limitations of environmental microbiology (that impose artifacts or otherwise hinder interpretation of experiments probing *in situ* processes) are exacerbated by the inaccessibility and heterogeneity of the subsurface. Furthermore, merging microbiology with hydrogeology is a formidable challenge because the usual temporal and spatial units used by hydrogeologists (months and years; meters and kilometers) do not match the units traditionally used by microbiologists (minutes and hours; micrometers and millimeters). Thus, hydrological data describing water flow in the subsurface are often not germane to small-scale controls of microbial activity that have proven influential in studies of microorganisms in other habitats and in pure culture.

Groundwater microbiology has progressed markedly since the 1970s. No longer is it a surprise to find microorganisms in groundwater sediments, even sediments hundreds of meters deep are assumed to contain viable microbial communities. Within the limits of temperature, water availability, and other environmental factors that confine living cells in the biosphere, groundwater habitats are expected to contain primarily bacteria, fungi, and protozoa. The abundance and diversity of subsurface microorganisms can be expected to fluctuate according to the geochemical and hydrogeological properties of the sediment or the ground water or both. Groundwater habitats have been found to support microbiological populations representing a broad array of metabolic capabilities, some of which can be expected to be expressed *in situ*.

In preparing this chapter we have attempted to identify areas in need of future research. Some of these have been pointed out in the appropriate sections above. It is fitting to conclude the chapter by listing several unexplored areas not included in previous sections.

1 We know virtually nothing about microbiological succession and development of climax communities in the groundwater habitat. Colonization has been examined by Hirsch & Rades-Rohkohl (1990), and metabolic adaptation has also been documented (Wilson *et al.* 1985; Federle & Pastwa 1988; Thomas *et al.* 1989; Madsen *et al.* 1991). These studies form a basis for practical goals like the future development of bioremediation protocols for polluted groundwater habitats. However, solutions for applied pollution problems are apt to be limited by insufficient basic knowledge of community development processes. More research is needed on the mechanisms which govern succession and climax of subsurface communities so that a complete understanding of groundwater ecosystems can be approached.

2 What is the true nature of the unsaturated zone habitat? This habitat is poorly understood, both physiologically and historically, in terms of the origin and activity of resident microorganisms. Perhaps this portion of the subsurface would be better understood if viewed not as an aquatic habitat, but rather as an atmospheric (aerobiological) milieu that may be inhabited by microorganisms capable of living on vapor-phase water and nutrients. We know precious little about the transformations and uptake of vapor-phase substances by microorganisms living in unsaturated zones.

3 Distinct relationships between geochemical and microbiological characteristics of the subsurface habitat need to be established in order to predict and extrapolate one from the other. If unequivocal rules can be proven which tie geochemical and geological conditions to microbiological processes, then the large amount of existing hydrogeological information might be used to infer the microbiological status of areas not yet explored by microbiologists. This chapter is testimony to the challenges of obtaining information about subsurface microorganisms. An additional critical step in establishing biogeochemical relationships is improving methods for extracting and analyzing pore waters from subsurface sediments. Geochemists face substantial problems in attempting to understand the chemistry of groundwater microhabitats. For example, accurate measurement of pore-water oxygen has thus far been impossible because of technical problems associated with sampling and analysis of extruded pore water. Development of new methods is desperately needed in this important area of biogeochemistry.

4 Many questions pertaining to *in situ* genetic exchange, response to selective pressures, and development of novel microorganisms in

subsurface environments are yet to be addressed. These questions directly affect our understanding of biotic resiliency and mechanisms of community development. Such issues may be very important for designing future approaches to managing the subsurface habitat.

5 Development of new methodologies for probing subsurface microbial activity in a field setting will be crucial for advancing subsurface microbiology toward the goal of understanding *in situ* microbial processes. Innovative technological approaches are needed that can combine mass-balance measurements with physiological activity and nutrient-limitation measurements, and molecular probe technology. Creative application of such methods combined with hydrogeological and geochemical studies may someday allow for accurate quantitative determination of microbiological processes in field settings where they really count.

Acknowledgments

Current support for deep subsurface microbiology research in our laboratory is provided through Grant No. DE-FG02-90ER1023 from the US Department of Energy, Office for Health and Environmental Research, Subsurface Science Program. Helpful suggestions on possible nutrient limitations in subsurface ecosystems were made by Suzanne N. Levine and David T. Rudnick. We thank Patti Lisk for secretarial assistance.

References

Aamand J., Jørgensen C., Arvin E. & Jensen B.K. (1989) Microbial adaptation to degradation of hydrocarbons in polluted and unpolluted groundwater. *J. Contam. Hydrol.* **4**, 299–312.

Adrian N.R. & Suflita J.M. (1990) Reductive dehalogenation of a nitrogen heterocyclic herbicide in anoxic aquifer slurries. *Appl. Environ. Microbiol.* **56**, 292–294.

Aelion C.M. & Bradley P.M. (1991) Aerobic biodegradation potential of subsurface microorganisms from a jet fuel-contaminated aquifer. *Appl. Environ. Microbiol.* **57**, 57–63.

Alexander M. (1961) *Introduction to Soil Microbiology*, 1st edn. John Wiley & Sons, New York.

Alexander M. (1977) *Introduction to Soil Microbiology*, 2nd edn. John Wiley & Sons, New York.

Alexander M. (1981) Biodegradation of chemicals of environmental concern. *Science* **211**, 132–138.

Alvarez-Cohen L. & McCarty P.L. (1991) Effects of toxicity, aeration, and reductant supply on trichloroethylene transformation by a mixed methanotrophic culture. *Appl. Environ. Microbiol.* **57**, 228–235.

Amann R.I., Binder B.J., Olson R.J., Chisholm S.W., Devereux R. & Stahl D.A. (1990) Combination of 16S rRNA-targeted oligonucleotide probes with flow cytometry for analyzing mixed microbial populations. *Appl. Environ. Microbiol.* **56**, 1919–1925.

Armstrong A.Q., Hodson R.E., Hwang H.-M. & Lewis D.L. (1991) Environmental factors affecting toluene degradation in groundwater at a hazardous waste site. *Environ. Toxicol. Chem.* **10**, 147–158.

Atlas R.M. & Bartha R. (1987) *Microbial Ecology*, 2nd edn. Benjamin Cummings, Menlo Park.

Baedecker M.J. & Back W. (1979) Hydrogeological processes and chemical reactions at a landfill. *Ground Water* **17**, 429–437.

Balkwill D.L. (1989) Numbers, diversity, and morphological characteristics of aerobic, chemoheterotrophic bacteria in deep subsurface sediments from a site in South Carolina. *Geomicrobiol. J.* 7, 33–52.

Balkwill D.L. (1990) Deep-aquifer microorganisms. In Labeda D.P. (ed.) *Isolation of Biotechnological Organisms From Nature*, pp. 183–211. McGraw-Hill, New York.

Balkwill D.L., Leach F.R., Wilson J.T., McNabb J.F. & White D.C. (1988) Equivalence of microbial biomass measures based on membrane lipid and cell wall components, adenosine triphosphate, and direct counts in subsurface aquifer sediments. *Microb. Ecol.* **16**, 73–84.

Balkwill D.L., Fredrickson J.K. & Thomas J.M. (1989) Vertical and horizontal variations in the physiological diversity of the aerobic chemoheterotrophic bacterial microflora in deep southeast coastal plain subsurface sediments. *Appl. Environ. Microbiol.* **55**, 1058–1065.

Barker C. (1979) *Organic Geochemistry in Petroleum Exploration*. American Association Petroleum Geologists. Department of Educational Activities, Tulsa.

Bartha R. & Atlas R.M. (1977) The microbiology of aquatic oil spills. *Adv. Appl. Microbiol.* **22**, 225–226.

Beeman R.E. & Suflita J.M. (1989) Evaluation of deep subsurface sampling procedures using serendipitous microbial contaminants as tracer organisms. *Geomicrobiol. J.* 7, 223–233.

Beeman R.E. & Suflita J.M. (1990) Environmental factors influencing methanogenesis in a shallow anoxic aquifer: a field and laboratory study. *J. Ind. Microbiol.* 5, 45–58.

Beloin R.M., Sinclair J.L. & Ghiorse W.C. (1988) Distribution and activity of microorganisms in subsurface sediments of a pristine study site in Oklahoma. *Microb. Ecol.* **16**, 85–97.

Belyaev S.S. & Ivanov M.V. (1983) Bacterial methanogenesis in underground waters. In Hallberg R. (ed.) *Environmental Biogeochemistry*, No. 35, pp. 273–280. Ecological Bulletins Publishing House/FRN, Stockholm.

Bengtsson G. (1989) Growth and metabolic flexibility in groundwater bacteria. *Microb. Ecol.* **18**, 235–248.

Bengtsson G. & Annadotter H. (1989) Nitrate reduction in a groundwater microcosm determined by ^{15}N gas chromatography–mass spectrometry. *Appl. Environ. Microbiol.* **55**, 2861–2870.

Bitton G. & Gerba C.P. (eds) (1984) *Groundwater Pollution Microbiology*. John Wiley & Sons, New York.

Bone T.L. & Balkwill D.L. (1988) Morphological and cultural comparison of microorganisms in surface soil and subsurface sediments at a pristine study site in Oklahoma. *Microb. Ecol.* **16**, 49–64.

Bormann F.H. & Likens G.E. (1967) Nutrient cycling. *Science* **155**, 424–429.

Bormann F.H. & Likens G.E. (1979) *Pattern and Process in a Forested Ecosystem*. Springer Verlag, New York.

Bower P.M., Kelly C.A., Fee E.J., Shearer J.A., DeClercq D.R. & Schindler D.W. (1987) Simultaneous measurement of primary production by whole lake and bottle radiocarbon additions. *Limnol. Oceanogr.* **32**, 299–313.

Brady N.C. (1984) *The Nature and Properties of Soils*, 9th edn. Macmillan Inc., New York.

Breen A., Stahl D.A., Flesher B. & Sayler G. (1989) Characterization of *Pseudomonas geomorphus*: a novel groundwater bacterium. *Microb. Ecol.* **18**, 221–233.

Brock T.D. & Madigan M.T. (1988) *Biology of Microorganisms*, 5th edn. Prentice-Hall, Englewood Cliffs.

Brockman F.J., Denovan B.A., Hicks R.J. & Fredrickson J.K. (1989) Isolation and characterization of quinoline-degrading bacteria from subsurface sediments. *Appl. Environ. Microbiol.* **55**, 1029–1032.

Brockman F.J., Kieft T.L., Fredrickson J.K., Bjornstad B.N., Li S.W., Spangenburg W. & Long P.E. (1992) Microbiology of vadose zone paleosols in South-Central Washington State. *Microb. Ecol.* **22**, 279–301.

Brusseau M.L., Jessup R.E. & Rao P.S.C. (1989) Modeling the transport of solutes influenced by multiprocess nonequilibrium. *Wat. Resources Res.* **25**, 1971–1988.

Bull A.T. (1980) Biodegradation: some attitudes and strategies of microorganisms and microbiologists. In Ellwood D.C., Medger J.N., Latham M.J., Lynch J.M. & Slater J.H. (eds) *Contemporary Microbial Ecology*, pp. 107–136. Academic Press, London.

Chapelle F.H. & Lovley D.R. (1990) Rates of microbial metabolism in deep coastal plain aquifers. *Appl. Environ. Microbiol.* **56**, 1865–1874.

Chapelle F.H., Zelibor J.L. Jr, Grimes D.J. & Knobel L.L. (1987) Bacteria in deep coastal plain sediments of Maryland: A possible source of CO_2 to groundwater. *Wat. Resources Res.* **23**, 1625–1632.

Chapelle F.H., Morris T., McMahon P.B. & Zelibor J.L. Jr (1988) Bacterial metabolism and the $\delta^{13}C$ composition of ground water, Floridian aquifer system, South Carolina. *Geology* **16**, 117–121.

Chiang C.Y., Salanitro J.P., Chai E.Y., Colthart J.D. & Klein C.L. (1989) Aerobic biodegradation of benzene, toluene, and xylene in a sandy aquifer – data analysis and computer modeling. *Ground Water* **27**, 823–834.

Colwell F.S. (1989) Microbiological comparison of surface soil and unsaturated subsurface soil from a semiarid high desert. *Appl. Environ. Microbiol.* **55**, 2420–2423.

Costerton J.W., Nickel J.C. & Ladd T.I. (1986) Suitable methods for the comparative study of free-living and surface-associated bacterial populations. In Poindexter J.S. & Leadbetter E.R. (eds) *Bacteria in Nature*, vol. 2, pp. 49–84. Plenum, New York.

Dagley S. (1977) Microbial degradation of organic compounds in the biosphere. In Scott A.F. (ed.) *Survey of Progress in Chemistry*, vol. 8, pp. 121–170. Academic Press, New York.

Davis S.N. & DeWiest R.J.M. (1966) *Hydrogeology*. John Wiley & Sons, New York.

Di-Ruggiero J. & Gounot A.M. (1990) Microbial manganese reduction mediated by bacterial strains isolated from aquifer sediments. *Microb. Ecol.* **20**, 53–63.

Ehrlich P.R., Ehrlich A.H. & Holdren J.P. (1977) *Ecoscience: Population, Resources, Environment*. Freeman and Co., San Francisco.

Federle T.W. & Pastwa G.M. (1988) Biodegradation of surfactants in saturated subsurface sediments: a field study. *Ground Water* **26**, 761–770.

Federle T.W., Dobbins D.C., Thornton-Manning J.R. & Jones D.D. (1986) Microbial biomass, activity, and community structure in subsurface soils. *Ground Water* **24**, 365–374.

Federle T.W., Ventullo R.M. & White D.C. (1990) Spatial distribution of microbial biomass, activity, community structure, and the biodegradation of linear alkylbenzene sulfonate (LAS) and linear alcohol ethoxylate (LAE) in the subsurface. *Microb. Ecol.* **20**, 297–313.

Fenchel T. (1987) *Ecology of Protozoa*. Science Tech. Publishers, Madison.

Fenchel T. & Harrison P. (1976) The significance of bacterial grazing and mineral cycling for the decomposition of particulate detritus. In Anderson J.M. & Macfadyen A. (eds) *The Role of Terrestrial and Aquatic Organisms in Decomposition Processes*. pp. 285–299. Blackwell Scientific Publications, Oxford.

Fenchel T. & Blackburn T.H. (1979) *Bacteria and Mineral Cycling*. Academic Press, London.

Ferguson R.L., Buckley E.N. & Palumbo A.V. (1984) Response of marine bacterioplankton to differential filtration and confinement. *Appl. Environ. Microbiol.* **47**, 49–55.

Finlay B.J. (1990) Physiological ecology of free-living protozoa. *Adv. Microb. Ecol.* **11**, 1–31.

Fliermans C.B, Phelps T.J., Ringelberg D., Mikell A.T. & White D.C. (1988)

Mineralization of trichloroethylene by heterotrophic enrichment cultures. *Appl. Environ. Microbiol.* **54**, 1709–1714.

Fogg K. (1988) The effects of added nitrogen on the rate of decomposition of organic matter. *Biol. Rev.* **663**, 433–462.

Foissner W. (1987) Soil protozoa: fundamental problems, ecological significance, adaptations in ciliates and testaceans, bioindicators, and guide to the literature. *Prog. Protistol.* **2**, 69–212.

Foster S.S.D., Kelly D.P. & James R. (1985) The evidence for zones of biodenitrification in British aquifers. In Caldwell D.E., Brierly J.A. & Brierly C.A. (eds) *Planetary Ecology*, pp. 356–369. VanNostrand Reinhold, New York.

Francis A.J., Slater J.M. & Dodge C.J. (1989) Denitrification in deep subsurface sediments. *Geomicrobiol. J.* **7**, 103–116.

Fredrickson J.K., Hicks R.J., Li S.W. & Brockman F.J. (1988) Plasmid incidence in bacteria from deep subsurface sediments. *Appl. Environ. Microbiol.* **54**, 2916–2923.

Fredrickson J.K., Garland T.R., Hicks R.J., Thomas J.M., Li S.W. & McFadden K.M. (1989) Lithotrophic and heterotrophic bacteria in deep subsurface sediments and their relation to sediment properties. *Geomicrobiol. J.* **7**, 53–66.

Fredrickson J.K., Brockman F.J., Workman D.J., Li S.W. & Stevens T.O. (1991a) Isolation and characterization of a subsurface bacterium capable of growth on toluene, naphthalene, and other aromatic compounds. *Appl. Environ. Microbiol.* **57**, 796–803.

Fredrickson J.K., Balkwill D.L., Zachara J.M., Li S.-M.W., Brockman F.J. & Simmons M.A. (1991b) Physiological diversity and distributions of heterotrophic bacteria in deep cretaceous sediments of the Atlantic coastal plain. *Appl. Environ. Microbiol.* **57**, 402–411.

Freeze R.A. & Cherry J.A. (1979) *Groundwater*. Prentice-Hall, Englewood Cliffs.

Fuhrman J.A. & Azam F. (1982) Thymidine incorporation as a measure of heterotrophic bacterioplankton production in marine surface waters: evaluation and field results. *Mar. Biol.* **66**, 109–120.

Fustec E., Mariott A., Grillo X. & Sajus J. (1991) Nitrate removal by denitrification in alluvial ground water: role of a former channel. *J. Hydrol.* **123**, 337–354.

Gannon J.T., Manilal V.B. & Alexander M. (1991a) Relationship between cell surface properties and transport of bacteria through soil. *Appl. Environ. Microbiol.* **57**, 190–193.

Gannon J.T., Tan Y., Baveye P. & Alexander M. (1991b) Effect of sodium chloride on transport of bacteria in a saturated aquifer material. *Appl. Environ. Microbiol.* **57**, 2497–2501.

Gerba C.P. (1983) Virus survival and transport in groundwater. *Dev. Ind. Microbiol.* **24**, 247–251.

Ghiorse W.C. & Balkwill D.L. (1983) Enumeration and morphological characterization of bacteria indigenous to subsurface environments. *Dev. Ind. Microbiol.* **24**, 213–224.

Ghiorse W.C. & Wilson J.T. (1988) Microbial ecology of the terrestrial subsurface. *Adv. Appl. Microbiol.* **33**, 107–172.

Gibbs C.F. (1976) Methods and interpretation in measurement of oil biodegradation rate. In Sharpley J.M. & Kaplan A.M. (eds) *Proceedings of the Third International Biodegradation Symposium*, pp. 127–140. Applied Science Publishers, London.

Gilbert J., Dole-Olivier M.-J., Marmonier P. & Vervier P. (1990) Surface water–groundwater ecotones. In Naiman R.J. & Décamps H. (eds) *The Ecology and Management of Aquatic–Terrestrial Ecotones*, pp. 199–225. UNESCO, Paris and Parthenon Publishing Group, Carnforth.

Gross M.G. (1987) *Oceanography*, 4th edn. Prentice-Hall, Englewood Cliffs.

Grossman E.L., Coffman B.K., Fritz S.J. & Wada H. (1989) Bacterial production of methane and its influence on ground-water chemistry in east-central Texas aquifers. *Geology* **17**, 495–499.

Harvey R.W. & Garabedian S.P. (1991) Use of colloid filtration theory in modeling movement of bacteria through a contaminated sandy aquifer. *Environ. Sci. Technol.* **25**, 178–185.

Harvey R.W., Smith R.L. & George L. (1984) Effect of organic contamination upon microbial distributions and heterotrophic uptake in a Cape Cod, Mass., aquifer. *Appl. Environ. Microbiol.* **48**, 1197–1202.

Hazen T.C. (1992) Test plan for *in situ* bioremediation demonstration of the Savannah River integrated demonstration project DOE/OTD TTP No.: SR 0566-01 (U). Report No. WSRC-RD-91-23. Westinghouse Savannah River Co., Aiken.

Hazen T.C., Jiménez L., López de Victoria G. & Fliermans C.B. (1991) Comparison of bacteria from deep subsurface sediment and adjacent groundwater. *Microb. Ecol.* **22**, 293–304.

Henis Y. (ed.) (1987) *Survival and Dormancy of Microorganisms*. John Wiley & Sons, New York.

Henry S.M. & Grbic-Galic D. (1990) Effect of mineral media on trichloroethylene oxidation by aquifer methanotrophs. *Microb. Ecol.* **20**. 151–169.

Henry S.M. & Grbic-Galic D. (1991) Influence of endogenous and exogenous electron donors and trichloroethylene oxidation toxicity on trichloroethylene oxidation by methanotrophic cultures from a groundwater aquifer. *Appl. Environ. Microbiol.* **57**, 236–244.

Hickman G.T., Novak J.T., Morris M.S. & Rebhun M. (1989) Effects of site variations on subsurface biodegradation potential. *J. Wat. Pollut. Control. Fed.* **61**, 1564–1575.

Hicks R.J. & Fredrickson J.K. (1989) Aerobic metabolic potential of microbial populations indigenous to deep subsurface environments. *Geomicrobiol. J.* **7**, 67–77.

Hilger A.B. & Myrold D.D. (1991) Method for extraction of *Frankia* DNA from soil. *Agric. Ecosyst. Environ.* **34**, 107–113.

Hirsch P. & Rades-Rohkohl E. (1983) Microbial diversity in a groundwater aquifer in northern Germany. *Dev. Ind. Microbiol.* **24**, 183–200.

Hirsch P. & Rades-Rohkohl E. (1988) Some special problems in the determination of viable counts of groundwater microorganisms. *Microb. Ecol.* **16**, 99–113.

Hirsch P. & Rades-Rohkohl E. (1990) Microbial colonization of aquifer sediment exposed in a groundwater well in northern Germany. *Appl. Environ. Microbiol.* **56**, 2963–2966.

Hirsch P. *et al.* (1992) Microbiology. In Matthess G., Frimmel F., Hirsch P., Schulz H.C. & Usdowski H.-E. (eds) *Progress in Hydrogeochemistry*, pp. 308–412. Springer Verlag, New York.

Holben W.E., Jansson J.K., Chelm B.K. & Tiedje J.M. (1988) DNA probe method for the detection of specific microorganisms in the soil bacterial community. *Appl. Environ. Microbiol.* **54**, 703–711.

Hoppe H.-G. (1978) Relationships between active bacteria and heterotrophic potential in the sea. *Neth. J. Sea Res.* **12**, 78–98.

Hutchins S.R. (1991) Biodegradation of monoaromatic hydrocarbons by aquifer microorganisms using oxygen, nitrate, or nitrous oxide as the terminal electron acceptor. *Appl. Environ. Microbiol.* **57**, 2403–2407.

Hutchins S.R., Sewell G.W., Kovacs D.A. & Smith G.A. (1991) Biodegradation of aromatic hydrocarbons by aquifer microorganisms under denitrifying conditions. *Environ. Sci. Technol.* **25**, 68–76.

Hynes H.B.N. (1983) Groundwater and stream ecology. *Hydrobiologia* **100**, 93–99.

Jannasch H.W. & Mottl M.J. (1985) Geomicrobiology of deep-sea hydrothermal vents. *Science* **229**, 717–725.

Jeffreys A.J., Wilson V. & Thein S.L. (1985) Individual-specific "fingerprints" of human DNA. *Nature* **316**, 76–79.

Jensen B.K. (1989) ATP-related specific heterotrophic activity in petroleum contaminated and uncontaminated groundwater. *Can. J. Microbiol.* **35**, 814–818.

Jiménez L. (1990) Molecular analysis of deep-subsurface bacteria. *Appl. Environ. Microbiol.* **56**, 2108–2113.

Jones R.E., Beeman R.E. & Suflita J.M. (1989) Anaerobic metabolic processes in deep terrestrial subsurface. *Geomicrobiol. J.* 7, 117–130.

Karl D.M. (1986) Determination of *in situ* microbial biomass, viability, metabolism, and growth. In Poindexter J.S. & Leadbetter E.R. (eds) *Bacteria in Nature*, vol. 2, pp. 85–176. Plenum, New York.

Kelly D.P. (1981) Introduction to the chemolithotrophic bacteria. In Starr M.P. Stolp H., Trüper H.G., Balows A. & Schlegal H.G. (eds) *The Prokaryotes*, vol. 1, pp. 997–1004. Springer Verlag, Berlin.

Keswick B.H. (1984) Sources of groundwater pollution. In Bitton G. & Gerba C.P. (eds) *Groundwater Pollution Microbiology*, pp. 39–64. John Wiley & Sons, New York.

Khondaker A.N., Al-Layla R.I. & Husain T. (1990) Groundwater contamination studies – the state-of-the-art. *Crit. Rev. Environ. Cont.* **20**, 231–256.

King L.K. & Parker B.C. (1988) A simple, rapid method for enumerating total viable and metabolically active bacteria in groundwater. *Appl. Environ. Microbiol.* **54**, 1630–1631.

Kirby L.T. (1990) *DNA Fingerprinting – An Introduction*. Stockton Press, New York.

Kjeldsen P., Kjølholt J., Schultz B., Christensen T.H. & Tjell J.C. (1990) Sorption and degradation of chlorophenols, nitrophenols and organophosphorus pesticides in the subsoil under landfills – laboratory studies. *J. Contam. Hydrol.* **6**, 165–184.

Klecka G.M., Davis J.W., Gray D.R. & Madsen S.S. (1990) Natural bioremediation of organic contaminants in ground water: Cliffs-Dow superfund site. *Ground Water* **28**, 534–543.

Kölbel-Boelke J., Tienken B. & Nehrkorn A. (1988a) Microbial communities in the saturated groundwater environment I. Methods of isolation and characterization of heterotrophic bacteria. *Microb. Ecol.* **16**, 17–29.

Kölbel-Boelke J., Anders E.-M. & Nehrkorn A. (1988b) Microbial communities in the saturated groundwater environment II. Diversity of bacterial communities in a Pleistocene sand aquifer and their *in vitro* activities. *Microb. Ecol.* **16**, 31–48.

Konopka A. & Turco R. (1991) Biodegradation of organic compounds in vadose zone and aquifer sediments. *Appl. Environ. Microbiol.* **57**, 2260–2268.

Kuenen J.G. & Bos P. (1989) Habitats and ecological niches of chemolitho(auto)trophic bacteria. In Schlegel H.G. & Bowien B. (eds) *Autotrophic Bacteria*, pp. 53–80. Springer Verlag, New York.

Kuhn E.P., Suflita J.M., Rivera M.D. & Young L.Y. (1989) Influence of alternate electron acceptors on the metabolic fate of hydroxybenzoate isomers in anoxic aquifer slurries. *Appl. Environ. Microbiol.* **55**, 590–598.

Kuhn E.P., Townsend G.T. & Suflita J.M. (1990) Effect of sulfate and organic carbon supplements on reductive dehalogenation of chloroanilines in anaerobic aquifer slurries. *Appl. Environ. Microbiol.* **56**, 2630–2637.

Kuznetsov S.I. (1970) *The Microflora of Lakes*. University of Texas Press, Austin.

Kuznetsov S.I., Ivanov M.W. & Lyankova N.N. (1963) In Broneer P.T. (trans.); Oppenheimer C.H. (ed.) *Introduction to Geological Microbiology*. McGraw-Hill, New York.

Leahy J.G. & Colwell R.R. (1990) Microbial degradation of hydrocarbons in the environment. *Microbiol. Rev.* **54**, 305–315.

Lee S. & Fuhrman J.A. (1990) DNA hybridization to compare species compositions of natural bacterioplankton assemblages. *Appl. Environ. Microbiol.* **56**, 739–746.

Lee M.D., Thomas J.M., Borden R.C., Bedient P.B., Ward C.H. & Wilson J.T. (1988) Biorestoration of aquifers contaminated with organic compounds. *CRC Crit. Rev. Environ. Cont.* **18**, 29–89.

Levine S.N. & Ghiorse W.C. (1991) Environmental factors affecting distribution and abundance of bacteria, fungi and protozoa in subsurface sediments of the Upper Atlantic Coastal Plain, USA. In Fliermans C.B. & Hazen T.C. (eds) *Proceedings of the First International Symposium on Microbiology of the Deep Subsurface*. January 15–19, 1990, Orlando, pp. 5 31 to 5-45. Westinghouse Savannah River Co. Information Services, Aiken.

Lindqvist R. & Bengtsson G. (1991) Dispersal dynamics of groundwater bacteria. *Microb. Ecol.* **21**, 49–72.

Longley G. (1981) The Edwards aquifer: Earth's most diverse groundwater ecosystem? *Int. J. Speleol.* **11**, 123–128.

Lovley D.R., Chapelle F.H. & Phillips E.J. P. (1990) Fe(III)-reducing bacteria in deeply buried sediments of the Atlantic coastal plain. *Geology* **18**, 954–957.

Madsen E.L. (1991) Determining *in situ* biodegradation: facts and challenges. *Environ. Sci. Technol.* **25**, 1662–1673.

Madsen E.L. & Bollag J.M. (1989) Aerobic and anaerobic microbial activity in deep subsurface sediments from the Savannah River Plant. *Geomicrobiol. J.* **7**, 93–101.

Madsen E.L., Sinclair J.L. & Ghiorse W.C. (1991) *In situ* biodegradation: microbiological patterns in a contaminated aquifer. *Science* **252**, 830–833.

Major D.W., Mayfield C.I. & Barker J.F. (1988) Biotransformation of benzene by denitrification in aquifer sand. *Ground Water* **26**, 8–14.

Marxsen J. (1988) Investigations into the number of respiring bacteria in groundwater from sandy and gravelly deposits. *Microb. Ecol.* **16**, 65–72.

Matthess G. (Harvey J.C., trans.) (1982) *Properties of Groundwater*. John Wiley & Sons, New York.

McCoy E.L. & Hagedorn C. (1979) Quantitatively tracing bacterial transport in saturated soil systems. *Wat. Air Soil Pollut.* **11**, 467–479.

McMahon P.B. & Chapelle F.H. (1991) Microbial production of organic acids in aquitard sediments and its role in aquifer geochemistry. *Nature* **349**, 233–235.

McMahon P.B., Williams D.F. & Morris J.T. (1990) Production and carbon isotopic composition of bacterial CO_2 in deep coastal plain sediments of South Carolina. *Ground Water* **28**, 693–702.

McNabb J.F. & Dunlap W.J. (1975) Subsurface biological activity in relation to ground-water pollution. *Ground Water* **13**, 33–44.

McNabb J.F. & Mallard G.E. (1984) Microbiological sampling in the assessment of groundwater pollution. In Bitton G. & Gerba C.P. (eds) *Groundwater Pollution Microbiology*, pp. 235–260. John Wiley & Sons, New York.

Mercado A., Libhaber M. & Soares M.I.M. (1988) *In situ* biological groundwater denitrification: concepts and preliminary field tests. *Wat. Sci. Tech.* **20**, 197–209.

Miller R.E., Randtke S.J., Hathaway L.R. & Denne J.E. (1990) Organic carbon and THM formation potential in Kansas groundwaters. *J. Am. Well Water Assoc.* **March**, 49–62.

Morgan P. & Watkinson R.J. (1989) Hydrocarbon degradation in soils and methods for soil biotreatment. *CRC Crit. Rev. Biotechnol.* **8**, 305–333.

Nybakken J.W. (1988) *Marine Biology*, 2nd edn. Harper & Row, New York.

Obenhuber D.C. & Lowrance R. (1991) Reduction of nitrate in aquifer microcosms by carbon additions. *J. Environ. Qual.* **20**, 255–258.

Odum E.P. (1971) *Fundamentals of Ecology*, 3rd edn. W.B. Saunders, Philadelphia.

Ogram A., Sayler G.S. & Barkay T. (1987) The extraction and purification of microbial DNA from sediments. *J. Microbiol. Methods* **7**, 57–66.

Pace N.R., Stahl D.A., Lane D.J. & Olsen G.J. (1986) The analysis of natural microbial populations by ribosomal RNA sequences. *Adv. Microbial Ecol.* **9**, 1–55.

Pederson K. & Ekendahl S. (1990) Distribution and activity of bacteria in deep granitic groundwaters of southeastern Sweden. *Microb. Ecol.* **20**, 37–52.

Phelps T.J. & White D.C. (1991) Activities of microorganisms in deep unconsolidated eastern Coastal Plain sediments. In Fliermans C.B. & Hazen T.C. (eds) *Proceedings of the First International Symposium on Microbiology of the Deep Subsurface*. January 15–19, 1990, Orlando, pp. 4-125 to 4-135. Westinghouse Savannah River Co. Information Services, Aiken.

Phelps T.J., Fliermans C.B., Garland T.R., Pfiffner S.M. & White D.C. (1989a) Methods for recovery of deep terrestrial subsurface sediments for microbiological studies. *J. Microbiol. Methods* **9**, 267–279.

Phelps T.J., Raione E.G., White D.C. & Fliermans C.B. (1989b) Microbial activities in deep subsurface environments. *Geomicrobiol. J.* 7, 79–91.

Powelson D.K., Simpson J.R. & Gerba C.P. (1990) Virus transport and survival in saturated and unsaturated flow through soil columns. *J. Environ. Qual.* **19**, 396–401.

Ramanand K. & Suflita J.M. (1991) Anaerobic degradation of *m*-cresol in anoxic aquifer slurries: carboxylation reactions in a sulfate-reducing bacterial enrichment. *Appl. Environ. Microbiol.* 57, 1689–1695.

Rao P.S.C., Lee L.S., Nkedi-Kizza P. & Yalkowsky S.H. (1989) Sorption and transport of organic pollutants at waste disposal sites. In Gerstl Z., Chen Y., Mingelgrin U. & Yaron B. (eds) *Toxic Organic Chemicals in Porous Media*, pp. 177–192. Springer Verlag, Berlin.

Raven P.H. & Johnson G.B. (1989) *Biology*, 2nd edn. Times Mirror/Mosby College Publishing, Boston.

Reynolds P.J., Sharma P., Jenneman G.E. & McInerney M.J. (1989) Mechanisms of microbial movement in subsurface materials. *Appl. Environ. Microbiol.* 55, 2280–2286.

Ridgway H.F., Safarik J., Phipps D., Carl P. & Clark D. (1990) Identification and catabolic activity of well-derived gasoline-degrading bacteria from a contaminated aquifer. *Appl. Environ. Microbiol.* **56**, 3565–3575.

Roszak D.B. & Colwell R.R. (1987) Survival strategies of bacteria in the natural environment. *Microbiol. Rev.* **51**, 365–379.

Rudnick D.T., Levine S.N. & Ghiorse W.C. (1992) A flow through microcosm for the testing of nutrient limitations in groundwater sediments. In Simons J. & Stanford J. (eds) *Proceedings of the First International Conference on Ground-Water Ecology.* American Water Resources Association, Bethesda. (In press.)

Sandon H. (1927) *The Composition and Distribution of the Protozoan Fauna of the Soil*, pp. 39–43 and 222–223. Oliver & Boyd, Edinburgh.

Sargent K.A. & Fliermans C.B. (1989) Geology and hydrology of deep subsurface microbiology sampling sites at the Savannah River plant, South Carolina. *Geomicrobiol. J.* 7, 3–13.

Scholl M.A., Mills A.L., Herman J.S. & Hornberger G.M. (1990) The influence of mineralogy and solution chemistry on the attachment of bacteria to representative aquifer materials. *J. Contam. Hydrol.* **6**, 321–326.

Semprini L., Roberts P.V., Hopkins G.D. & McCarty P.L. (1990) A field evaluation of *in-situ* biodegradation of chlorinated ethenes. Part 2, results of biostimulation and biotransformation experiments. *Ground Water* **28**, 715–727.

Sewell G.W. & Gibson S.A. (1991) Stimulation of the reductive dechlorination of tetrachloroethene in anaerobic aquifer microcosms by the addition of toluene. *Environ. Sci. Technol.* **25**, 982–984.

Sinclair J.L. (1991) Eukaryotic microorganisms in subsurface environments. In Hazen T.C. & Fliermans C.B. (eds) *Proceedings of the First International Symposium on the Deep Subsurface.* January 15–19, 1990, Orlando, pp. 3-35 to 3-45. Westinghouse Savannah River Co. Information Services, Aiken.

Sinclair J.L. & Ghiorse W.C. (1987) Distribution of protozoa in subsurface sediments of a pristine groundwater study site in Oklahoma. *Appl. Environ. Microbiol.* **53**, 1157–1163.

Sinclair J.L. & Ghiorse W.C. (1989) Distribution of aerobic bacteria, protozoa, algae, and fungi in deep subsurface sediments. *Geomicrobiol. J.* 7, 15–31.

Sinclair J.L., Randtke S.J., Denne J.E., Hathaway L.R. & Ghiorse W.C. (1990) Survey of microbial populations in buried-valley aquifer sediments from northeastern Kansas. *Ground Water* **28**, 369–377.

Slater J.M. & Capone D.G. (1987) Denitrification in aquifer soil and nearshore marine sediments influenced by groundwater nitrate. *Appl. Environ. Microbiol.* 53, 1292–1297.

Sleigh M.A. (1989) *Protozoa and Other Protists*. Edward Arnold, London.

Smith R.L. & Duff J.H. (1988) Denitrification in a sand and gravel aquifer. *Appl. Environ. Microbiol.* 54, 1071–1078.

Smith R.L., Harvey R.W. & LeBlanc D.R. (1991a) Importance of closely spaced vertical sampling in delineating chemical and microbiological gradients in groundwater studies. *J. Contam. Hydrol.* 7, 285–300.

Smith R.L., Howes B.L. & Garabedian S.P. (1991b) *In situ* measurement of methane oxidation in groundwater by using natural-gradient tracer tests. *Appl. Environ. Microbiol.* 57, 1997–2004.

Somerville C.C., Knight I.T., Straube W.L. & Colwell R.R. (1989) Simple, rapid method for direct isolation of nucleic acids from aquatic environments. *Appl. Environ. Microbiol.* 55, 548–554.

Stanier R.Y., Ingraham J.L., Wheelis M.L. & Painter P.R. (1986) *The Microbial World*, 5th edn. Prentice-Hall, Englewood Cliffs.

Steffan R.J., Goksøyr J., Bej A.K. & Atlas R.M. (1988) Recovery of DNA from soils and sediments. *Appl. Environ. Microbiol.* 54, 2908–2915.

Stout J.D. (1980) The role of protozoa in nutrient cycling and energy flow. *Adv. Microb. Ecol.* 4, 1–49.

Suflita J.M., Liang L. & Saxena A. (1989) The anaerobic biodegradation of *o*-, *m*-, and *p*-cresol by sulfate-reducing bacterial enrichment cultures obtained from a shallow anoxic aquifer. *J. Ind. Microbiol.* 4, 255–266.

Swindoll C.M., Aelion C.M. & Pfaender F.K. (1988a). Influence of inorganic and organic nutrients on aerobic biodegradation and on the adaptation response of subsurface microbial communities. *Appl. Environ. Microbiol.* 54, 212–217.

Swindoll C.M., Aelion C.M., Dobbins D.C., Jiang O., Long S.C. & Pfaender F.K. (1988b.) Aerobic biodegradation of natural and xenobiotic organic compounds by subsurface microbial communities. *Environ. Toxicol. Chem.* 7, 291–299.

Thomas J.M., Lee M.D., Scott M.J. & Ward C.H. (1989) Microbial ecology of the subsurface at an abandoned creosote waste site. *J. Ind. Microbiol.* 4, 109–120.

Thorn P.M. & Ventullo R.M. (1988) Measurement of bacterial growth rates in subsurface sediments using the incorporation of tritiated thymidine into DNA. *Microb. Ecol.* 16, 3–16.

Thorstenson D.C., Fisher D.W. & Croft M.G. (1979) The geochemistry of the Fox Hills–Basal Hell Creek aquifer in southwestern North Dakota and northwestern South Dakota. *Wat. Resources Res.* 15, 1479–1498.

Trudell M.R., Gillham R.W. & Cherry J.A. (1986) An *in situ* study of the occurrence and rate of denitrification in a shallow unconfined sand aquifer. *J. Hydrol.* 83, 251–268.

Tsai Y.-L. & Olson B.H. (1991) Rapid method for direct extraction of DNA from soil and sediments. *Appl. Environ. Microbiol.* 57, 1070–1074.

Valiela I. (1984) *Marine Ecological Processes*. Springer Verlag, New York.

van Beelen P. & Fleuren-Kemila A.K. (1989) Enumeration of anaerobic and oligotrophic bacteria in subsoils and sediments. *J. Contam. Hydrol.* 4, 75–284.

van Beelen P., Fleuren-Kemilä A.K., Huys M.P.A., van Montfort A.C.P. & van Vlaardingen P.L.A. (1991) The toxic effects of pollutants on the mineralization of acetate in subsoil microcosms. *Environ. Toxicol. Chem.* 10, 775–789.

van der Leeden F., Troise F.L. & Todd D.K. (1990) *The Water Encyclopedia*, 2nd edn. Lewis Publishers, Washington DC.

van Krevelen D.W. (1984) Organic geochemistry – old and new. *Org. Geochem.* 6, 1–10.

van Loosdrecht M.C.M., Lyklema J., Norde W. & Zehnder A.J.B. (1990) Influence of interfaces on microbial activity. *Microbiol. Rev.* 54, 75–87.

Venrick E.L., Beers J.R. & Heinbokel J.F. (1977) Possible consequences of containing microplankton for physiological rate measurements. *J. Exp. Mar. Biol. Ecol.* **26**, 55–76.

Ventullo R.M. & Larson R.J. (1985) Metabolic diversity and activity of heterotrophic bacteria in ground water. *Environ. Toxicol. Chem.* **4**, 759–771.

Waksman S.A. (1916) Studies on soil protozoa. *Soil Sci.* **1**, 135–152.

Ward C.H., Giger G. & McCarty P.L. (eds) (1985) *Ground Water Quality.* John Wiley & Sons, New York.

Ward D.M., Weller R. & Bateson M.M. (1990) 16S rRNA sequences reveal numerous uncultured microorganisms in a natural community. *Nature* **345**, 63–65.

Ward T.E. (1986) Aerobic and anaerobic biodegradation of nitrilotriacetate in subsurface soils. *Ecotoxicol. Environ. Safety* **11**, 112–125.

Webster J.R. & Benfield E.F. (1986) Vascular plant breakdown in freshwater ecosystems. *Ann. Rev. Ecol. Syst.* **17**, 567–594.

Weller R. & Ward D.M. (1989) Selective recovery of 16S rRNA sequences from natural microbial communities in the form of cDNA. *Appl. Environ. Microbiol.* **55**, 1818–1822.

Westlake D.W.S., Jobson A.M. & Cook F.D. (1978) *In situ* degradation of oil in a soil of the boreal region of the Northwest Territories. *Can. J. Microbiol.* **24**, 254–260.

Wetzel R.G. (1983) *Limnology*, 2nd edn. W.B. Saunders, New York.

White D.C. (1986) Quantitative, physicochemical characterization of bacterial habitats. In Poindexter J.S. & Leadbetter E.R. (eds) *Bacteria in Nature*, vol. 2, pp. 177–203. Plenum, New York.

White D.C., Smith G.A., Gehron M.J., Parker J.H., Findlay R.H., Martz R.F. & Fredrickson H.L. (1983) The groundwater aquifer microbiota: biomass, community structure, and nutritional status. *Dev. Ind. Microbiol.* **24**, 201–211.

Wilson G.B., Andrews J.N. & Bath A.H. (1990) Dissolved gas evidence for denitrification in the Lincolnshire limestone groundwaters, eastern England. *J. Hydrol.* **113**, 51–60.

Wilson J.T., McNabb J.R., Wilson B.H. & Noonan M.J. (1983) Biotransformation of selected organic pollutants in ground water. *Dev. Ind. Microbiol.* **24**, 225–233.

Wilson J.T., McNabb J.F., Cochran J.W., Wang T.H., Tomson M.B. & Bedient P.B. (1985) Influence of microbial adaptation on the fate of organic pollutants in ground water. *Environ. Toxicol. Chem.* **4**, 721–726.

Wilson J.T., Miller G.D., Ghiorse W.C. & Leach F.R. (1986) Relationship between the ATP content of subsurface material and the rate of biodegradation of alkylbenzenes and chlorobenzene. *J. Contam. Hydrol.* **1**, 163–170.

Wood W.W. & Petraitis M.J. (1984) Origin and distribution of carbon dioxide in the unsaturated zone of the southern high plains of Texas. *Wat. Resources Res.* **20**, 1193–1208.

Zobell C. (1946) *Marine Microbiology.* Chronicle Botanica, Waltham.

Zobell C. (1952) Part played by bacteria in petroleum formation. *J. Sed. Petrol.* **22**, 42–49.

Peter Westermann

Wetland and swamp microbiology

9.1 Introduction

Wetlands are important components of the biosphere and play a significant role in the global cycling and geochemical balance of carbon, nitrogen, and sulfur.

The diversity of swamp and wetland types caused by regional, geological, topographical, and climatic differences is huge. The terminology is similarly complex and often confusing. Several types of classification have been proposed based on hydrological, ecological, morphological, and chemical characteristics (Zoltai *et al.* 1975; Cowardin *et al.* 1979; Carter 1986). Generally, wetlands are considered poorly drained areas at least periodically saturated or covered with water. Wetlands covered by forest are normally called swamps. Water-saturated soil or sediments with shrubby or herbaceous vegetation, with little or no standing water are called marsh or mire. Marshes generally consist of a mixture of inorganic and organic matter of sedimentary origin, the decomposition activity is high, and pH alkaline to slightly acidic. Mires or peatlands, including both bogs and fens, are peat producing ecosystems. Bogs are very acidic peats dominated by *Sphagnum* species and other mosses and are often ombrotrophic, implying a water supply dominated by atmospheric precipitation resulting in low salinity and buffering capacity. Fens consist of alkaline, neutral, or mildly acidic peats. These minerotrophic ecosystems are similar to swamps and marshes, but are generally richer in mineral salts due to water supply from the surroundings as runoff and/or ground water.

Wetlands and swamps are formed by a variety of ways:

1 where rivers, lakes, or the sea floods adjacent lower lying land, either naturally, or as a consequence of human activity, e.g., paddy soils;

2 where the drainage of a soil is poor due to an impermeable rock bed, permafrost, or a loam layer, and where annual rainfall exceeds evaporation;

3 through accumulation of organic matter in shallow parts of lakes,

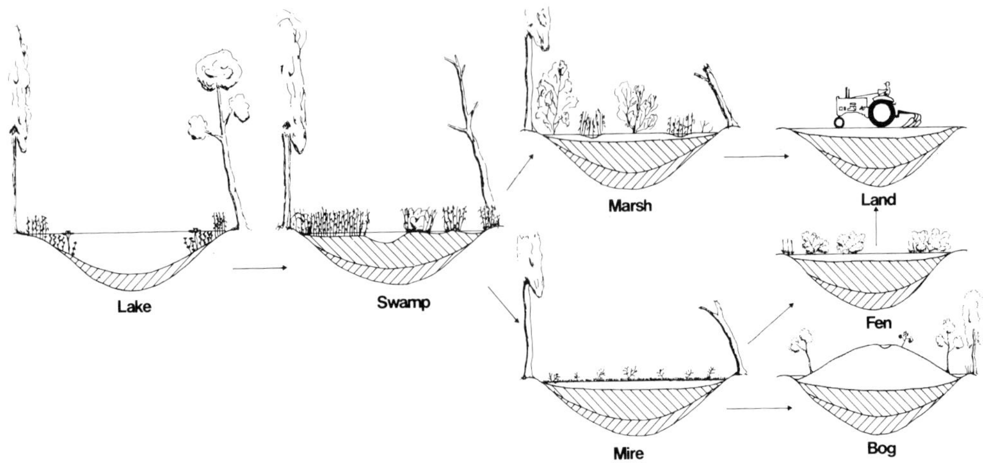

Figure 9.1 Development of a lake into various types of wetlands.

or shallow watered lakes, leading to a succession as shown in Figure 9.1.

Whether a swamp develops into a marsh or a mire depends upon climatic conditions. In moderately dry regions, most of the organic matter is aerobically degraded upon sedimentation, resulting in marsh formation, where structural remains are not immediately recognizable. In regions where the precipitation and the humidity is high and the temperature is low, primary production often exceeds decomposition leading to mire formation. In these temperate to arctic zones of the world, such conditions often encourage the colonization and eventual dominance of *Sphagnum* species. Due to the cation exchange capacity of these mosses, salt concentration and pH of the interstitial water rapidly drops (Clymo 1967) decreasing the rate of decomposition.

Many wetlands can be considered as extremely shallow lakes with little or no water column above the sediment. However, major differences should be taken into consideration before extended comparisons are made; lakes of a certain depth are often thermally and chemically stratified during the productive season resulting in anaerobiosis in the hypolimnion, and hence dominance of anaerobic metabolism of organic matter in the whole sediment column. In waterlogged wetlands, the water column and the top of the sediment is aerobic resulting in 70–80% mineralization of organic matter by aerobic microorganisms (Kjøller & Struwe 1980). In many lake ecosystems, algae constitute the major input of organic matter to the sediment, while most inputs to wetlands are of macrophytic origin.

Macrophytes contain an average of 20–30% lignin compared to the absence of lignous compounds in algae. Lignin is degraded very slowly under anaerobic conditions. As a result, the overall rates of anaerobic carbon metabolism are often much slower in wetlands than in sediments, despite the fact that volatile solids might constitute 80% of the dry weight of wetland sediments.

9.2 Aerobic mineralization processes

The aerobic zone of water-logged soil is generally limited to the water column plus the uppermost millimeters to centimeters of the sediment (King 1990a; King *et al.* 1990). Furthermore the proximate part of the rhizosphere of living roots from plants inhabiting the surface of the wetland is aerobic as a consequence of oxygen excretion by the roots (Holzapfel-Pschorn *et al.* 1985). Besides its importance as the major site of litter degradation, the aerobic environment plays an important role in the cycling of nitrogen and sulfur, which will be discussed below.

When plant litter is immersed in water, water-soluble organic substances are lost by leaching, even without the activity of microorganisms. The leachate, consisting mainly of amino acids, sugars, nonvolatile and volatile aliphatic acids, is released from within 7 days for alder and oak leaves to more than 12 weeks for intact oak twigs (Nykvist 1963; Willoughby & Archer 1973). These leached compounds are rapidly metabolized both by the resident phylloplane and by the senescent leaf microflora, and also by the aquatic microflora able to utilize the substrate.

Further degradation rates of the plant material depend on the ratios of major polymers (lignin, cellulose, and hemicellulose) in the deposits, the structure of the lignocellulose, and the physicochemical character of the wetland (Moran *et al.* 1989). Eighty-five per cent of *Spartina* plant material in litter cylinders was degraded after 2 months in a natural salt marsh creek, while only 35% of *Carex* plant material was degraded in the acid Okefenokee swamp after 1 year (Moran *et al.* 1989). This difference was considered to be due to pH differences rather than to compositional differences of the plants (Benner *et al.* 1985). Hemicellulose is generally degraded at an equal or faster rate than cellulose, in most wetlands (Suberkropp *et al.* 1976; Hedges *et al.* 1985). The degradation rates of the two polysaccharide polymers, however, have been found to exceed lignin degradation rates by two- to five-fold (Crawford *et al.* 1977; Benner *et al.* 1984a; 1985). In the case of rapid and heavy deposition, anaerobic conditions may occur early during decomposition, slowing down the depolymerization rate.

9.2.1 Population structure

Fungi have been characterized as the most important group of organisms responsible for decomposition of plant material in forest and agricultural soils (Anderson & Domsch 1975) and in aerobic freshwater sediments (Kaushik & Hynes 1971). In acid peatlands, Latter *et al.* (1967) showed that microfungi play an important role in aerobic mineralization processes, as only fungi could be isolated on a cellulose medium.

Fungi recovered from alder swamps, peatlands, and freshwater sediments include species of *Alternaria*, *Cylindrocarpon*, *Cladosporium*, *Penicillium*, *Fusarium*, and *Trichoderma* and Hyphomycetes such as *Alatospora*, *Tetracladium*, *Helicodendron*, and *Helicoon* species (Kaushik & Hynes 1971; Given & Dickinson 1975; Kjøller & Struwe 1980). In a study of substrate utilization by 1900 microfungi isolated throughout the year from an alder swamp, Kjøller & Struwe (1980) showed an increase in pectinolytic and cellulolytic fungi after litter fall. Chitin and gelatine were attacked more slowly, resulting in later population maxima of fungi with these degradation capabilities. Eleven strains of microfungi isolated from an acid peatland were capable of utilizing an array of aromatic compounds, hydrocarbons, and sugars (Williams & Crawford 1983a). Five isolates could degrade ^{14}C-labeled spruce lignocellulose, but at a slower average rate than 25 bacterial isolates from the same habitat.

Despite our perception of a less important role, bacteria occurred in relatively high numbers (10^9–$10^{10}\,g^{-1}$ dry litter) in litter from an alder swamp (Kjøller *et al.* 1985). Amylolytic and proteolytic bacteria were found in the highest numbers following litter fall, while the highest number of chitinolytic bacteria were found some months later paralleling the functional population shifts of microfungi. No pectinolytic or cellulolytic bacteria were found, confirming the important role of microfungi in degradation of these compounds.

Peatlands have previously been characterized as environments low in microbial activity due to the low pH and the refractory nature of available organic substrates (Latter *et al.* 1967). Processes such as methanogenesis were previously considered impossible in acid peats (Willoughby 1974). Despite the low pH (3.7), levels of microbial biomass and turnover rates of simple organic molecules in acid peatlands were found equal to, or higher than, those of other wetland ecosystems (Murray & Hodson 1984; 1985). Acridine orange direct counts of bacteria from surface interstitial water of a Minnesota peatland revealed 10^7–$10^8\,ml^{-1}$ with a high substrate utilization diversity (Williams & Crawford 1983a), which is in the same range as the total counts found in the neutral alder swamp described above. Aerobic bacterial genera isolated from peatlands include

Achromobacter, *Arthrobacter*, *Bacillus*, *Cytophaga*, *Chromobacterium*, *Micrococcus*, *Pseudomonas*, *Actinomyces*, and *Streptomyces* (Given & Dickinson 1975; Wheatley *et al.* 1976).

9.2.2 Aerobic mineral cycling processes

As a consequence of close associations with anaerobic conditions, aerobic microorganisms are important in mineral and carbon cycling in wetlands, both for the local ecosystem, and for the global cycles.

Ammonium is produced in the anoxic layers by nitrate-ammonification and from degradation of nitrogen-containing organic compounds. When ammonia diffuses to the aerobic surface, it is oxidized by autotrophic nitrifying bacteria to nitrite and finally to nitrate as in other aerobic systems. This process has been well studied in rice paddies, where nitrification is undesirable because of significant loss of nitrate by leaching or by denitrification (Patrick & Reddy 1978). Focht & Verstrate (1977) compared denitrification and nitrification rates and found a higher capacity for denitrification than for nitrification in surface soils, indicating that the rate of nitrification was limiting *in situ* denitrification rates. This also corresponds to the low numbers of ammonium oxidizers ($10^2\,g^{-1}$ soil) recovered from rice paddy top soil (Watanabe & Furusaka 1980), compared to the number of denitrifying bacteria (10^4–$7 \times 10^6\,g^{-1}$ soil) (Gamble *et al.* 1977; Sethunathan *et al.* 1983).

From the point of view of plant nutrition, nitrogen and phosphorus are the most important and also the most limiting components in freshwater wetlands. Williams & Crawford (1983b) found that addition of a phosphorus–nitrogen mixture to acid peats stimulated ^{14}C-labeled sphagnum and benzoate mineralization significantly, indicating a mineral limitation of the microbial activity. Sulfur is rarely a limiting nutrient and is sometimes present in quantities that lead to a domination of sulfate respiration in the anaerobic layers (Westermann & Ahring 1988). Depending upon the thickness of the overlying oxic layer and water column, reduced sulfur species formed by dissimilatory reduction of sulfate and metabolism of sulfur-containing amino acids, may be oxidized chemically or microbiologically to sulfate or elemental sulfur. The biological oxidation is either carried out by chemoautotrophic bacteria, such as *Thiobacillus* and *Beggiatoa* species (Sethunathan *et al.* 1983), or anaerobically by photoautotrophic bacteria utilizing the reduced sulfur species as electron donors (Burke *et al.* 1974). As hydrogen sulfide is toxic to plant roots, these bacterial-oxidation processes occurring within the oxidized rhizosphere are important for optimal growth of wetland plants (Joshi & Hollis 1977). Studies of sulfur cycling in peatbogs (Wieder & Lang 1988; Urban *et al.* 1989) revealed that virtually all reduced sulfur species reaching the oxic surface layers are oxidized, resulting in negligible atmospheric releases.

Methane oxidation has been studied mostly in lake waters and sediments (Lidstrom & Somers 1984; Frenzel *et al.* 1990), while reports from wetlands are mainly concerned with paddy soils and peatlands. Methane produced in the reduced layers of wetlands diffuses into the oxidized surface layers and may here undergo oxidation to carbon dioxide with molecular oxygen as electron acceptor (DeBont *et al.* 1978). Yavitt *et al.* (1990) found a methane oxidation rate in peatlands of approximately 80% of the methane production rate, resulting in a net methane release to the atmosphere. King *et al.* (1990) showed that methane oxidation consumed up to 91% of the methane produced in Florida Everglades peat sediment. This amount corresponded to 14–20% of the total sediment oxygen uptake, while almost no oxidation occurred in marl sediments from the same wetland. Most of the probable number enumerations of methane oxidizing and methanogenic bacteria from acid peatland revealed lower numbers of methane oxidizing bacteria (10^3–$10^4\,ml^{-1}$) than methanogenic bacteria (10^3–$10^6\,ml^{-1}$) (Williams & Crawford 1983a). In the aerobic surface layers of rice paddies, DeBont *et al.* (1978) recovered 2–18×10^6 methane oxidizers g^{-1} dry soil. King (1990b) and King *et al.* (1990) showed that the rate of methane oxidation in the surface of a wetland algal mat was dependent upon available light. Phototrophic oxygen production increased with increasing light intensity leading to higher oxygen concentrations and, hence, higher methane oxidation rates in the mat. Although not verified, the rate of other microbial oxidation processes in the aerobic–anaerobic interface might be similarly dependent upon light intensity and penetration.

Microbial oxidation of dissolved ferrous iron to insoluble ferric compounds occurs mainly in acidic environments, such as the edge of bogs, as ferrous iron is rapidly oxidized by molecular oxygen at conditions of neutral pH. Ferrous iron can, however, form soluble complexes with organic molecules if present. Bacterial oxidation of the organic ligands will lead to oxidation and hence precipitation of ferric iron. The importance of iron oxidation serving as an energy source under neutral pH conditions is still disputed, although chemoautotrophic growth of *Gallionella ferruginea* in a mineral medium with ferrous sulfide as electron donor, has been demonstrated (Kucera & Wolfe 1957). Many of the commercially utilized iron-ore beds of the world are ancient bog-iron deposits.

9.3 Anaerobic metabolism

Depending upon the available amount of easily degradable carbon, oxygen consumption exceeds the oxygen diffusion rate at a depth from a few millimeters to several centimeters below the surface of the wetland sediment, resulting in a shift in the microbial metabolism from aerobic to anaerobic. Under aerobic conditions, the degrading community consists of bacteria, fungi, and protozoa together with larger meio- and macrofauna. A common feature of all these aerobic organisms is their ability to mineralize organic molecules completely to carbon dioxide. Under anaerobic conditions, bacteria are mainly responsible for biological mineralization, while protozoa and fungi are considered less important. Furthermore, anaerobic mineralization of organic molecules demands the interaction of a complex microbial food web, where the product of one microbial group serves as substrate for the subsequent microbial group, and where the consumption of a product regulates its type and formation rate. Finally, organic carbon may be mineralized either completely to carbon dioxide by oxidative processes or to methane and carbon dioxide by combined oxidation–reduction processes, depending upon the availability of inorganic electron acceptors in the ecosystem.

Figure 9.2 shows the main features of anaerobic metabolism, and the microbial groups responsible for the degradation of organic matter if various electron acceptors are present (see Zehnder 1988 for a detailed description). If no external electron acceptors are present, at least four entirely different groups of organisms are active in the conversion of complex organic molecules to methane, carbon dioxide, and water. The fermentative bacteria hydrolyze the polymers to mono- and oligomers which are then absorbed and fermented to an array of products such as alcohols, short-chained fatty acids, hydrogen, and carbon dioxide. Volatile fatty acids and alcohols are then oxidized by hydrogen-producing syntrophic bacteria, leading to the formation of hydrogen, acetate, and carbon dioxide. The syntrophic bacteria can only metabolize the fermentative products if the partial pressure of hydrogen is kept low by carbon dioxide-reducing methanogenic bacteria, as this proton reducing oxidation process is energetically unfavorable under standard conditions. Acetate is also produced in large amounts directly by the fermentative bacteria when the hydrogen partial pressure is low, and is finally split into methane and carbon dioxide by aceticlastic methanogens. Acetogenic bacteria might also produce acetate from hydrogen and carbon dioxide, or even higher fatty acids from acetate and other C_1 and C_2 compounds (Dolfing 1988).

9.3.1 Electron acceptors for anaerobic carbon metabolism

The succession of terminal anaerobic metabolism of wetlands is similar to that in freshwater and marine sediments (Sørensen *et al.* 1979; Jones & Simon 1981; Winfrey *et al.* 1981), in that it can be predicted thermodynamically. The metabolic activity yielding the

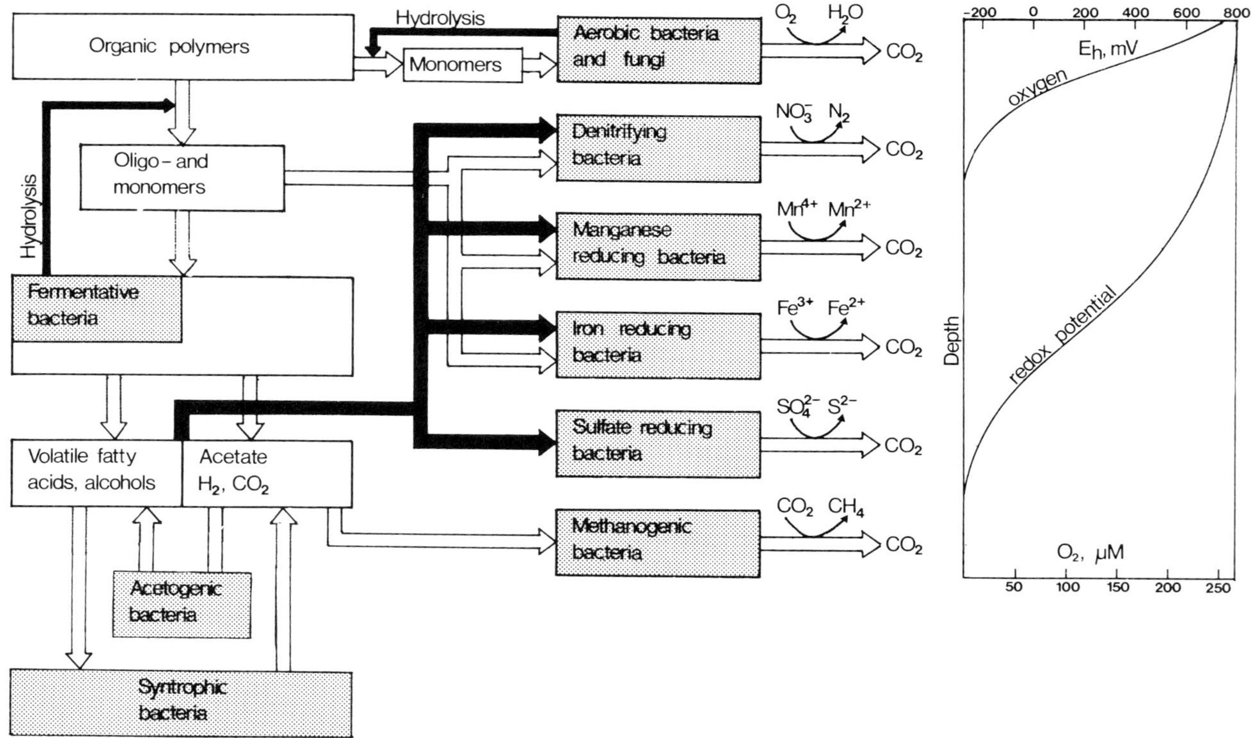

Figure 9.2 Main pathways in the degradation of organic matter in wetlands, and the corresponding oxygen concentrations and redox potentials. The redox potentials represent standard conditions at pH 7 for the various reactions, and will often be lower in nature. Redox potentials from Zehnder & Stumm (1988).

highest amount of energy from a common electron donor will dominate at a particular depth if the electron acceptor is present. These criteria result in a sequential reduction of O_2, NO_3^-, Mn^{4+}, Fe^{3+}, SO_4^{2-}, and HCO_3^- as the redox potential and hence the oxidizing capacity of the half reaction decreases (Zehnder & Stumm 1988). The significance of these respiratory processes in a particular ecosystem is dependent upon the concentration of the specific electron acceptor and the amount of available organic matter. In some wetlands, such as salt marshes, all processes except methanogenesis occur at significant rates, although spatially separated in different strata or microniches; in most freshwater systems, the concentration of inorganic electron acceptors in relation to available carbon is low, resulting mainly in methanogenic respiration.

Although bacteria capable of aerobic denitrification have been isolated (Robertson & Kuenen 1984), respiration with nitrate as electron acceptor is generally initiated when the oxygen concentration is below $10\,\mu mol\,l^{-1}$. The bacteria responsible for this process are Gram-negative anaerobes and facultative anaerobes such as *Pseudomonas* species. Nitrogen and nitrous oxide are the major products of denitrification at the positive redox potentials found in the surface layers, while the relative significance of nitrate reduction to ammonia increases at negative redox potentials (Sørensen 1978; Jones & Simon 1981). Manganese reduction from Mn^{3+} and Mn^{4+} oxides linked to oxidation of organic substrates occurs under aerobic and anaerobic conditions (Trimble & Ehrlich 1968; Lovley & Phillips 1988; Lovley 1991). This type of respiration is carried out by a wide range of aerobic and facultatively anaerobic bacterial groups, but also microfungi have been shown capable of dissimilative manganese reduction (Ghiorse 1988).

Below the zone of denitrification and manganese reduction, iron reduction might take place, utilizing ferric iron as an electron acceptor. Similar to dissimilatory manganese and nitrate reduction, many different microorganisms from aerobic, facultatively anaerobic and obligately anaerobic bacterial groups are able to perform this type of metabolism. Both soil, freshwater, and marine sediment studies have suggested that denitrifying bacteria are mainly responsible for observed iron reduction (Komatsu *et al.* 1978; Sørensen 1982; Jones *et al.* 1983). Lovley & Goodwin (1988), however, showed that the inhibition of iron reduction upon nitrate addition observed by these authors, might be a consequence of competition for common electron donors between two groups of organisms, rather than preferential nitrate reduction by iron reducers.

Sulfate reduction occurs in strata below or microniches within the more oxidized zones at redox potentials below $-120\,mV$ (Connell &

Patrick 1968; Widdel 1988). The bacteria responsible for sulfate reduction belong to several genera of obligate anaerobes present in virtually all anaerobic environments. Most sulfate reducers can grow with sulfite or thiosulfate as electron acceptors, and several species are able to ferment some organic substrates, either alone or syntrophically with methanogens as hydrogen scavengers in the absence of sulfate (Bryant *et al.* 1977). Some strains of sulfate reducers are able to utilize nitrate as electron acceptors forming ammonia but never molecular nitrogen (McCready *et al.* 1983).

When other inorganic electron acceptors are depleted, methanogenesis is the terminal process responsible for electron displacement into methane, as long as available precursors are present. In marine sediments and saline wetlands such as salt marshes and salterns, methylated amines occur in significant amounts as a degradation product of choline and glycine betaine, but also from the reduction of trimethylamine oxide, a metabolite and excretory product of many marine organisms. Oremland *et al.* (1982) showed that methylated amines were noncompetitive methanogenic substrates in sulfate-rich sediments and probably responsible for methane production observed in the upper layers of salt-marsh sediments. Methanogenesis is performed by a specialized group of obligately anaerobic archaebacteria with a very narrow substrate utilization range.

As a consequence of the shallow-watered nature of wetlands, light easily penetrates the water column and the uppermost millimeter of sediment where anaerobic microniches might be present. Both non-sulfur purple bacteria ($10^6\,g^{-1}$ soil) and green sulfur bacteria (10^2–$10^3\,g^{-1}$ soil) have been recovered from rice paddies (Watanabe & Furusaka 1980), but their role in carbon and sulfide metabolism is so far unknown. The nitrogen-fixing capability of many anaerobic phototrophic bacteria may be unimportant in rice paddies, as they were found to constitute only 2% of the nitrogen-fixing biomass of the surface soil (Watanabe & Furusaka 1980).

Several abiotic reduction processes occur in anaerobic environments – often indirectly mediated by microbial activity. Both iron and manganese might be reduced by biogenic sulfide and reduced organic compounds, and part of the microbiological research on iron and manganese reduction has been concerned with verification of the active role of microorganisms in these processes.

9.3.2 Hydrolysis and fermentation

Degradation of plant material and hydrolytic products, such as glucose, has been studied in both rice paddies and flooded soils. These investigations have focused on production of short-chained volatile fatty acids as compounds toxic to rice roots (Rao & Mikkelsen 1977), and on general metabolism and gas formation from various polymers (Yamane & Sato 1963; 1964; 1967; Bell 1969). Cellulose degradation has been verified in several environments and both bacteria and fungi capable of hydrolytic cleavage of cellulose have been isolated (Colberg 1988). Most organic polymers (protein, pectin, cellulose, and starch) were found to be degradable and resulted in increases in both carbon dioxide and methane production, while lignin had no significant effect on gas production in a 1-month experiment (Yamane & Sato 1963). The fermentation processes responsible for glucose, starch, and gelatine degradation were found to be less sensitive to temperature reduction than the subsequent processes in the degradation chain, resulting in an increased accumulation of volatile fatty acids with decreasing temperature. In contrast, the temperature sensitivity of hydrolysis and fermentation of more complex cellulose polymers corresponded with the sensitivity of subsequent metabolism, resulting in no accumulation of intermediary fermentation products (Yamane & Sato 1967).

Lignin was previously considered inert under anaerobic conditions (Hackett *et al.* 1977), while lignin-related and lignin-derived aromatic substrates were found degradable (Colberg & Young 1982; Zeikus *et al.* 1982). Long-term incubation (1–18 months) of wooden blocks in fresh, brackish, and sea-water anaerobic muds, however, showed that lignified cells were degraded by rod-shaped bacteria (Holt & Jones 1983). Benner *et al.* (1984b), finally confirmed anaerobic lignin degradation by showing that added ^{14}C-labeled polymeric lignin was slowly but significantly degraded in several wetland types (salt marsh, acidic freshwater marsh, and mangrove swamp).

The presence of fermentative *Clostridium* species has been verified in both wet soils and peatlands (Skinner 1968; Given & Dickinson 1975). Goodwin & Zeikus (1987) isolated three species of hydrolytic anaerobes able to grow at pH 5 from a peat bog. The bacteria were able to ferment glucose to either lactate (a *Lactobacillus* sp.) or acetate, ethanol, hydrogen, and carbon dioxide (a *Clostridium* sp. and a strain of *Sarcina ventriculi*). Other genera of hydrolytic bacteria are presumably also present in wetlands, as found in freshwater sediments where a diverse population of Gram-negative hydrolyzing bacteria are active (Jones 1971).

In an investigation of the vertical distribution of heterotrophic bacteria in an alder swamp, Kjøller *et al.* (1985) found a decreasing number with depth. In the surface layers, the numbers ranged from 5×10^6 to $4 \times 10^7 g^{-1}$ dry litter, while the number was 10^4 to $2 \times 10^4 g^{-1}$ at a depth of 25 cm. This is in the same range as observed in Minnesota peatland pore water at a depth of 20 cm

(3.5×10^1 bacteria ml^{-1} able to grow on a mixture of acetate, glucose, and glutamate) (Williams & Crawford 1983a). Pectinolytic ($4.3 \times 10^2 ml^{-1}$) and glucose-utilizing ($4.3 \times 10^4 ml^{-1}$) bacteria were recovered in an acid peat bog at pH 5 (Goodwin & Zeikus 1987). Increasing pH from 5 to 7 increased the recovery from four to 10 times, indicating a pH limitation under *in situ* conditions.

9.3.3 Metabolism of fermentation products

To my knowledge, no syntrophic bacteria have been isolated from wetlands that are able to further oxidize fermentation products to acetate, under simultaneous reduction of protons to molecular hydrogen. Metabolism of volatile fatty acids has, however, been verified in both rice paddies and temperate swamps (Yamane & Sato 1967; Takai 1970; Westermann 1990). Bacteria capable of metabolizing virtually all known fermentation products have been isolated and characterized from other natural and synthetic anaerobic environments (Dolfing 1988). Many of these organisms are probably also present in wetlands.

Formation of acetate from hydrogen and carbon dioxide (acetogenesis) has been considered relatively unimportant in most environments. Phelps & Zeikus (1984) showed that a mildly acidic lake sediment contained more hydrogen-consuming acetogenic bacteria than hydrogenotrophic methanogens, indicating an important role of acetogenesis in this environment. This was confirmed by isotopic measurements of methane precursors. Only 4% of the methane produced was derived from hydrogen, in contrast to the normally observed 20–30%. Instead, acetate was the precursor of almost all methane produced. As many wetlands, such as peat bogs and mires, are acidic, a similar pattern might be observed in such environments.

9.3.4 Denitrification

As a part of the nitrogen cycle, denitrification is a well-studied process in rice paddies and natural swamp ecosystems. Ammonia present in ammonia fertilizers or applied manure is easily oxidized by aerobic nitrifying bacteria to nitrite and nitrate, which then serve as electron acceptors for denitrifying bacteria. Up to 30–40% of soil nitrogen was estimated lost by denitrification in flooded soils (Broadbent & Tusneem 1971).

The most common denitrifying bacteria in natural ecosystems are aerobes belonging to the genera *Pseudomonas* and *Alcaligenes*. Dissimilatory reduction of nitrate to ammonia is mainly carried out by enterobacteria, *Clostridium* and *Bacillus* species which are typically obligate and facultative anaerobes with fermentative metabolism (Tiedje 1988). In wetlands, only a few characterizations of dissimilatory nitrate-reducing bacteria have been performed. Gamble *et al.* (1977) examined the denitrifier population in five rice soils and found *Pseudomonas* species as the most dominant bacterium. Similarly *Pseudomonas* species were dominant in an alder swamp (Struwe & Kjøller 1989). The population size of denitrifying bacteria in paddy soils was enumerated by Gamble *et al.* (1977) and reviewed by Sethunathan *et al.* (1983), as 2×10^3–7×10^6 bacteria g^{-1} of soil. This is comparable to a number of 8×10^4–3×10^5 denitrifying bacteria ml^{-1} of soil found in a temperate alder swamp (Westermann & Ahring 1986), and in the same range as found in freshwater sediments and arable soil (Tiedje 1988). As previously described, nitrate is only present in the superficial layers of wetlands and the denitrification rate, therefore, falls rapidly with depth. The denitrification rate in temperate swamps ranges between 0.5 and $5\,mg\,m^{-2}\,d^{-1}$ (Westermann & Ahring 1986; Struwe & Kjøller 1989) which is somewhat lower than observed values from freshwater lake sediments at the same latitude ($64\,mg\,m^{-2}\,d^{-1}$) (Andersen 1977). Table 9.1 compares activity measurements from various types of wetlands calculated on a volumetric basis.

9.3.5 *Manganese and iron reduction*

Manganese and iron reduction has mainly been studied in rice paddies. As previously mentioned, much research effort has been concentrated on proving that the reduction processes actually are microbially mediated (Watanabe & Furusaka 1980; Burdige & Nealson 1985; Lovley & Phillips 1988). There is very little quantitative data available on microbial iron/manganese reduction or the bacteria responsible for the processes in wetlands. However, studies in marine and brackish environments (Lovley & Phillips 1986; Lovley 1991) indicate that the role of iron and manganese reduction in carbon mineralization might be significant in wetlands with high mineral inputs.

9.3.6 *Sulfate reduction*

Sulfate reduction has been studied in most wetland types. In rice paddies the formation of toxic hydrogen sulfide might inhibit the growth of the rice plants. Exudates from roots have been shown to stimulate sulfate reduction in the rhizosphere while the oxidized layer in the proximate vicinity of the roots stimulates sulfide oxidation, and hence leads to some protection for the roots (Sethunathan *et al.* 1983). In most rice paddies and temporarily waterlogged soils, hydrogen sulfide rarely accumulates in toxic concentrations due to precipitation, mainly as iron sulfide (Bloomfield 1969; Widdel 1988). Peat wetlands play an important role in the local control of sulfur species and acidity of drain water, and therefore acidity of weakly buffered streams and lakes. The majority of the sulfur in peatlands originates from wet and dry atmospheric deposition. Under waterlogged anaerobic conditions, sulfate is retained in the bog by dis-

Table 9.1 Methanogenesis, sulfate reduction, and denitrification in wetlands

	Activity (nmol d^{-1} g^{-1} dry weight)		
Wetland type	**Methanogenesis**	**Sulfate reduction**	**Denitrification**
Low carbon, neutral	6 (15°C)[1]	23 (15°C)[1]	0.16 (15°C)[1]
Peaty wetland, neutral	160 (10°C)[2]	130 (10°C)[2]	2 (10°C)[2]
Acid peatlands	1600 (27°C)[3] 110–720 (8°C)[4]	670–1670 (8°C)[4]	96 (24°C)[5]
Rice paddies	250–950 (17–30°C)[6]	No data available	850–4800 (>20°C)[7]
Oligotrophic lake sediments	658 (10°C)[8]	707 (10°C)[8]	12–180 (20°C)[9]
Eutrophic lake sediments	7920 (12–16°C)[10]	1224 (12–16°C)[10]	0–28[11]
Marine and salt-marsh sediments	<1 (23°C)[12]	744–24 000 (25–32°C)[13,14]	0–240 (3–18°C)[15]

Activity rates are calculated from the following references.
[1] Westermann & Ahring (1986).
[2] Westermann & Ahring (1988).
[3] King *et al.* (1981), assuming a dw of 15%.
[4] Wieder *et al.* (1990).
[5] Gordon *et al.* (1986), assuming a dry weight (dw) of 15%. The rate is potential as 25 μmol l^{-1} nitrate was added. Peat only slightly acidic (pH 6.2–6.7).
[6] Conrad *et al.* (1987).
[7] Garcia (1974).
[8] Lovley & Klug (1983), assuming a dw of 10%.
[9] Christensen & Sørensen (1986), assuming a dw of 10%.
[10] Lovley *et al.* (1982), assuming a dw of 10%.
[11] Chan & Knowles (1979), assuming that denitrification occurs in the uppermost centimeter of the sediment.
[12] Jones & Paynter (1980).
[13] Skyring *et al.* (1979), assuming a dw of 25%.
[14] Howarth & Teal (1979), assuming a dw of 25%.
[15] Sørensen *et al.* (1978), assuming a dw of 25%.

similatory sulfate reduction followed by incorporation into the organic matrix (Brown 1986). During sulfate reduction, protons are consumed leading to an increase in alkalinity of the pore water (Widdel 1988). In dry periods, chemical and microbiological reoxidation of precipitated sulfides and organic sulfur to H_2SO_4 can result in the death of fish when the acid is flushed out following rainfall (Gosling & Baker 1980).

The population structure of sulfate-reducing bacteria in freshwater wetlands is largely unknown. Both *Desulfovibrio* and *Desulfotomaculum* strains have been isolated from rice paddies (Watanabe & Furusaka 1980) but presumably a diverse population is present, as in freshwater sediments. Sulfate-reducing bacteria are able to metabolize an array of compounds such as hydrogen, volatile fatty acids, alcohols, amino acids, sugars, and aromatic compounds with sulfate as electron acceptor. Many sulfate reducers can grow fermentatively or in syntrophic relationships with hydrogen-utilizing methanogens in the absence of sulfate (Bryant *et al.* 1977; Widdel 1988). Therefore, enumerated populations of sulfate reducers should be considered as potentially, rather than necessarily, performing sulfate reduction *in situ*. The population size in rice paddies ranges from 6×10^1 to $7 \times 10^5\,g^{-1}$ soil, decreasing with depth (Sethunathan *et al.* 1983) which is similar to numbers and distribution in a temperate alder swamp (10^2–$10^5\,ml^{-1}$ mud) (Westermann & Ahring 1986) and in freshwater sediments (Cappenberg 1974).

Based upon atmospheric deposition rates, sulfate reduction rates in *Sphagnum* bogs have been considered insignificant in carbon cycling and energy flow (Urban *et al.* 1989). Spratt *et al.* (1987), however, showed a sulfate reduction rate in a New Jersey swamp similar to rates observed in eutrophic lake sediments (1–173 $\mu mol\,l^{-1}\,d^{-1}$ in the swamp, 13–140 $\mu mol\,l^{-1}\,d^{-1}$ in the lake sediments) (King & Klug 1982). Similarly high rates of sulfate reduction have been observed in two peatlands (Wieder *et al.* 1990) and in an alder swamp (Westermann & Ahring 1986) (104–146 $\mu mol\,l^{-1}\,d^{-1}$ in the peatlands, 2–108 $\mu mol\,l^{-1}\,d^{-1}$ in the alder swamp). The sulfate concentration in these environments is typically in the range of 100 to 500 $\mu mol\,l^{-1}$. The high sulfate reduction rate observed at the low *in situ* concentration can be explained by a rapid turnover of the reduced sulfur pool. This is a consequence of the close interaction between oxic and anoxic layers in wetlands (Wieder *et al.* 1990).

9.3.7 *Methanogenesis*

Methane production is the terminal anaerobic mineralization process in environments where the concentration of inorganic electron acceptors, other than carbon dioxide is low. Freshwater wetlands have been considered one of the most important sources of atmospheric methane (Cicerone & Oremland 1988), which will be discussed in the next section. Methane production has been studied in rice paddies, peatlands, and natural swamps. The production rates found in these environments range from far below those found in oligotrophic lake sediments to rates close to eutrophic lake sediments (see Table 9.1). Several authors have observed that optimum pH of methanogenesis in acid peats is above *in situ* pH values. Goodwin & Zeikus (1987) found that increasing pH from *in situ* (4.7) to optimum (5.2) in a peat bog sediment increased methanogenesis by a factor of 1.7. A similar increase and pH optimum was observed by Williams & Crawford (1984) and King *et al.* (1981) for acid peatlands. Interestingly, methanogenesis in acid peats seems to have a pH optimum well below optimum values of known methanogenic species (Boone & Whitman 1988). Svensson (1984) found that methanogens from acid peats were able to adapt to low temperatures, resulting in a decreased temperature optimum for growth upon prolonged storage at low temperature. A similar pH-adaptation mechanism might occur after prolonged incubation at low pH.

The knowledge of population structure of methanogens in wetlands is sparse and restricted to neutral wetlands. Mayer & Conrad (1990) found 10 times as many hydrogenophilic as aceticlastic methanogens in paddy soils. In my laboratory, a *Methanothrix* sp. was highly enriched, and *Methanobacterium formicicum* and a *Methanosarcina* sp. were isolated, from an alder swamp (P. Westermann unpublished), indicating a methanogenic diversity which probably is similar to freshwater lake sediments.

Several studies of methanogens in neutral wetlands have shown population sizes in the same order of magnitude; 10^5–10^6 methanogens ml^{-1} sediment were found in an alder swamp (Westermann & Ahring 1986), while 10^2–10^6 were found g^{-1} of rice paddy soil (Garcia *et al.* 1974). Similar numbers were found by Mayer & Conrad (1990) in flooded paddy soils (10^5–10^6 methanogens g^{-1} soil). In acid-bog sediments, Goodwin & Zeikus (1987) recovered only 10^2–10^3 methanogens ml^{-1} of sediment while Williams & Crawford (1983a) recovered a methanogenic population from acid peatlands similar to neutral wetlands (10^4–$10^6\,ml^{-1}$ interstitial water).

As previously mentioned, the amount of electrons proceeding through sulfate reduction or methanogenesis in the wetland sediment is dependent upon the concentration of sulfate. In a study of sulfate reduction and methane production in an alder swamp, Westermann & Ahring (in press) showed that the majority of acetate, hydrogen, and propionate was mineralized by sulfate-reducing bacteria while 28% of the acetate and 37% of the H_2 was metabolized by methanogens (Figure 9.3). Similar to most other freshwater environments, the majority of the methane produced was derived from acetate. Fourteen per cent of the methane was derived from unknown precursors.

9.3.8 Atmospheric interactions

Concentrations of methane in the atmosphere have been rising at the rate of $1\%\,yr^{-1}$ at least since 1950 (Rasmussen & Khalil 1981; Ramanathan *et al.* 1985). As methane absorbs 25 times as much

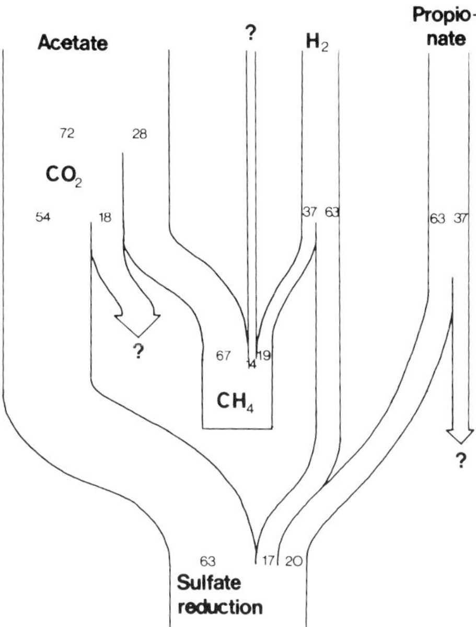

Figure 9.3 The role of sulfate reduction and methanogenesis in conversion of important fermentation products in an alder swamp sediment. All values are percentages.

infrared radiation as carbon dioxide, molecule for molecule, the relative importance of methane as a greenhouse gas increases with the rapid atmospheric increase compared to the much slower carbon dioxide increase. Flooded soils and freshwater wetlands are considered some of the most important sources of atmospheric methane (Baker-Blocker *et al.* 1977). However, they have also been shown capable of consuming atmospheric methane by microbial methane oxidation in dryer periods when the water table decreases (Harriss *et al.* 1982; Moore & Knowles 1990; Whalen & Reeburgh 1990; Yavitt *et al.* 1990). The fluxes of methane vary with temperature, and the highest activity has not surprisingly been demonstrated in tropical and subtropical marshes and swamps (Table 9.2). Despite a productive season estimated to only 100 days, methane emissions from tundra and boreal regions were calculated to account for up to 23% of global wetland emissions, due to the large geographical extent of these ecosystems (Sebacher *et al.* 1986). The major sources of biogenic methane to the atmosphere are considered to be natural wetlands and rice paddies followed by ruminants (Watson *et al.* 1991). Holzapfel-Pschorn *et al.* (1986) showed that the aerenchyma and intercellular gas space system of rice plants that supplies the roots with oxygen, acted as methane transport shunts from the anoxic soil to the atmosphere. Consequently, methane-oxidizing

Table 9.2 Emission of methane from various wetlands

Locality	Temperature range (°C)	Flux ($mg\,m^{-2}\,d^{-1}$)	Period
Sudd (Sudan)[1]	23.3–32.2	1090	Annual
Everglades (Florida)[1]	20.0–27.8	627	Annual
Uganda swamps[1]	21.1–22.2	463	Annual
Michigan swamp[1]	13–22	356	July–Sept
Italian rice paddy[3]	19–27	279	May–Aug
Pripett Marshes (USSR)[1]	−6.1–18.9	109	Annual
Temperate swamp (Virginia)[2]	4–23	41–44	Annual
Boreal and tundra mires[4,5]	−17.5–16	4–289	Annual

[1] Baker-Blocker *et al.* (1977), data estimated from a temperature model.
[2] Wilson *et al.* (1989).
[3] Holzapfel-Pschorn *et al.* (1986).
[4] Sebacher *et al.* (1986).
[5] Svensson (1976).

bacteria in the surface layers of the paddies are bypassed, resulting in a 1.8 times higher methane emission from rice covered paddies compared with a similar but unplanted field.

Temperature has been shown to have a marked effect on the formation of methane in all types of wetlands (Baker-Blocker *et al.* 1977; Conrad *et al.* 1987; Westermann & Ahring 1987). An increase in wetland temperatures upon global warming, therefore, might lead to an increase in methane production rates, and also methane oxidation rates in wetlands and soil, in an unknown numerical relationship. A better understanding of the regulation of both biogenic and abiogenic methane production and consumption is therefore needed to obtain more precise models of the impact of climatic changes.

Other reduced gases such as ammonia, nitrous oxide and hydrogen sulfide are also emitted from freshwater wetlands and might contribute to acid rain (Gorham *et al.* 1984), but are probably of minor importance compared with the anthropogenic release of nitrous oxide and marine releases of hydrogen sulfide.

9.4 Conclusion

Freshwater wetlands cover vast areas and are highly diverse. They are an important source of atmospheric methane and, in the case of rice paddies, provide a major food supply for more than 40% of the human population. Yet, compared to salt marshes and lake sediments, the basic microbial metabolism and dynamics in freshwater wetlands are poorly understood. The research reviewed in this chapter indicates that more differences exist compared to lake sediments than outlined in the introduction. Sulfate reduction was shown to be a very important terminal process in peatlands and neutral swamps, resembling brackish-water sediments rather than lake sediments. Other types of anaerobic respiration, such as manganese and iron reduction, are still poorly investigated in most wetland types, and further knowledge of the microbial activity will probably lead to even more complex wetland type divisions as seen for lake ecosystems.

The wetland consists of very narrow interfaces between most elements of the biosphere, from trees and shrubs, macro- and meiofauna, to deeply reduced anoxic strata, often within a distance of millimeters. The complex nature of these very heterogeneous environments is a challenge, but also a cause of methodological frustrations to the researcher. The importance of wetlands in the global cycling of elements and the availability of more refined methods such as microelectrodes, mass spectrometers, and molecular tools, e.g., gene probes and immunotechniques, will hopefully stimulate further microbiological research in wetland ecosystems.

References

Andersen J.M. (1977) Rates of denitrification of undisturbed sediment from six lakes as a function of nitrate concentration, oxygen and temperature. *Arch. Hydrobiol.* **80**, 147–159.

Anderson J.P.E. & Domsch K.H. (1975) Measurement of bacterial and fungal contributions to respiration of selected agricultural and forest soils. *Can. J. Microbiol.* **21**, 314–322.

Baker-Blocker A., Donahue T.M. & Mancy K.H. (1977) Methane flux from wetland areas. *Tellus.* **29**, 245–250.

Bell R.G. (1969) Studies on the decomposition of organic matter in flooded soil. *Soil Biol. Biochem.* **1**, 105–116.

Benner R., Maccubbin A.E. & Hodson R.E. (1984a) Preparation, characterization, and microbial degradation of specifically radiolabelled (^{14}C)-lignocellulose from marine and freshwater macrophytes. *Appl. Environ. Microbiol.* **47**, 381–398.

Benner R., Maccubbin A.E. & Hodson R.E. (1984b) Anaerobic biodegradation of the lignin and polysaccharide components of lignocellulose and synthetic lignin by sediment microflora. *Appl. Environ. Microbiol.* **47**, 998–1004.

Benner R., Moran M.A. & Hodson R.E. (1985) Effects of pH and plant source on lignocellulose biodegradation rates in two wetland ecosystems, the Okefenokee Swamp and Georgia salt marsh. *Limnol. Oceanogr.* **30**, 489–499.

Bloomfield C. (1969) Sulphate reduction in waterlogged soils. *J. Soil Sci.* **20**, 207–221.

Boone D.R. & Whitman W.B. (1988) Proposal of minimal standards for describing new taxa of methanogenic bacteria. *Int. J. Syst. Bacteriol.* **38**, 212–219.

Broadbent F.E. & Tusneem M.E. (1971) Losses of nitrogen from some flooded soils in tracer experiments. *Soil Sci. Soc. Am. Proc.* **35**, 922–928.

Brown K.A. (1986) Formation of organic sulfur in anaerobic peat. *Soil Biol. Biochem.* **18**, 131–140.

Bryant M.P., Campbell L.L., Reddy C.A. & Crabil M.R. (1977) Growth of *Desulfovibrio* in lactate or ethanol media low in sulfate in association with H_2-utilizing methanogenic bacteria. *Appl. Environ. Microbiol.* **33**, 1162–1169.

Burdige D.J. & Nealson K.H. (1985) Microbial manganese reduction by enrichment cultures from coastal marine sediments. *Appl. Environ. Microbiol.* **50**, 491–497.

Burke M.E., Gorham E. & Pratt D.C. (1974) Distribution of purple photosynthetic bacteria in wetland and woodland habitats of central and northern Minnesota. *J. Bacteriol.* **117**, 826–833.

Cappenberg Th.E. (1974) Interrelations between sulfate-reducing and methane producing bacteria in bottom deposits of a freshwater lake. I. Field observations. *Antonie van Leeuwenhoek*. **40**, 285–295.

Carter V. (1986) An overview of the hydrologic concerns related to wetlands in the United States. *Can. J. Bot.* **64**, 364–374.

Chan Y.-K. & Knowles R. (1979) Measurement of denitrification in two freshwater sediments by an *in situ* acetylene inhibition method. *Appl. Environ. Microbiol.* **37**, 1066–1072.

Christensen P.B. & Sørensen J. (1986) The variation of denitrification activity in plant-covered littoral sediment from Lake Hampen, Denmark. *Appl. Environ. Microbiol.* **51**, 1174–1179.

Cicerone R.J. & Oremland R.S. (1988) Biogeochemical aspects of atmospheric methane. *Glob. Biogeochem. Cycles* **2**, 299–327.

Clymo R.S. (1967) Control of cation concentrations, and in particular of pH, in *Sphagnum* dominated communities. In Golterman H.L. & Clymo R.S. (eds) *Chemical Environment in the Aquatic Habitat*, pp. 273–284. Noord-Hollandsche Uitgevers Maatschappig, Amsterdam.

Colberg P.C. (1988) Anaerobic microbial degradation of cellulose, lignin, oligolignols, and monoaromatic lignin derivatives. In Zehnder A.J.B. (ed.) *Biology of Anaerobic Microorganisms*, pp. 333–372. John Wiley & Sons, New York.

Colberg P.C. & Young L.Y. (1982) Biodegradation of lignin-derived molecules under anaerobic conditions. *Can. J. Microbiol.* **28**, 886–889.

Connell W.E. & Patrick W.H. Jr (1968) Sulfate reduction in soil: effects of redox potential and pH. *Science* **159**, 86.

Conrad R., Schütz H. & Babbel M. (1987) Temperature limitation of hydrogen turnover and methanogenesis in anoxic paddy soil. *FEMS Microbiol. Ecol.* **45**, 281–289.

Cowardin L.M., Carter V., Golet F.C. & LaRoc E.T. (1979) *Classification of Wetlands and Deepwater Habitats of the United States*. US Fish. Wildl. Serv. Off. Biol. Serv. FWS/OBS-79/31.

Crawford D.L., Crawford R.L. & Pometto A.L. (1977) Preparation of specifically radiolabelled ^{14}C-(lignin)- and ^{14}C-(cellulose)-lignocelluloses and their decomposition by the microflora of soil. *Appl. Environ. Microbiol.* **33**, 1247–1251.

DeBont J.A.M., Lee K.K. & Bouldin D.F. (1978) Bacterial oxidation of methane in rice paddy. *Ecol. Bull.* **26**, 91–99.

Dolfing J. (1988) Acetogenesis. In Zehnder A.J.B (ed.) *Biology of Anaerobic Microorganisms*, pp. 417–468. John Wiley & Sons, New York.

Focht D.D. & Verstrate W. (1977) Biochemical ecology of nitrification and denitrification. In Alexander M. (ed.) *Advances in Microbial Ecology*, vol. 1, pp. 135–214. Plenum, New York.

Frenzel P., Thebrath B. & Conrad R. (1990) Oxidation of methane in the oxic surface layer of a deep lake sediment (Lake Constance). *FEMS Microbiol. Ecol.* **73**, 149–158.

Gamble T.N., Betlach M.R. & Tiedje J.M. (1977) Numerically dominant denitrifying bacteria from world soils. *Appl. Environ. Microbiol.* **33**, 926–939.

Garcia J.L. (1974) Reduction de l'oxyde nitreux dans les sols de rizieres du Sénegal: mesure de l'activité denitrificante. *Soil Biol. Biochem.* **6**, 79–84.

Garcia J.L., Rimbault M., Jacq V., Rinaudo G. & Roger P.A. (1974) Activités microbiennes dans les sols des rizieres du Sénegal: relations avec les caracteristiques physico-chimiques et influence de la rhizosphere. *Rev. Ecol. Biol. Sol.* **11**, 169–185.

Ghiorse W.C. (1988) Microbial reduction of manganese and iron. In Zehnder A.J.B. (ed.) *Biology of Anaerobic Microorganisms*, pp. 305–332. John Wiley & Sons, New York.

Given P.H. & Dickinson C.H. (1975) Biochemistry, microbiology of peats. In Paul E.A. & McLaren A.D. (eds) *Soil Biochemistry*, vol. 3, pp. 123–211. Marcel Dekker, New York.

Goodwin S. & Zeikus J.G. (1987) Ecophysiological adaptations of anaerobic bacteria to low pH: analysis of anaerobic digestion in acidic bog sediments. *Appl. Environ. Microbiol.* **53**, 57–64.

Gordon A.S., Cooper W.J. & Scheidt D.J. (1986) Denitrification in marl and peat sediment in the Florida Everglades. *Appl. Environ. Microbiol.* **52**, 987–991.

Gorham E., Bayley S.E. & Schindler D.W. (1984) Ecological effects of acid deposition upon wetlands: a neglected field in "acid rain" research. *Can. J. Fish. Aquat. Sci.* **41**, 1256–1268.

Gosling L.M. & Baker S.J. (1980) Acidity fluctuations at a broadland site in Norfolk. *J. Appl. Ecol.* **17**, 479–490.

Hackett W.F., Connors W.J., Kirk T.K. & Zeikus J.G. (1977) Microbial decomposition of synthetic ^{14}C-labelled lignins in nature: lignin biodegradation in a variety of natural materials. *Appl. Environ. Microbiol.* **33**, 43–51.

Harriss R.C., Sebacher D.F. & Day F.P.Jr (1982) Methane flux in the Great Dismal Swamp. *Nature* **297**, 673–674.

Hedges J.I., Cowie G.L. & Ertel J.R. (1985) Degradation of carbohydrates and lignins in buried woods. *Geochim. Cosmochim. Acta* **49**, 701–711.

Holt D.M. & Jones E.B.G. (1983) Bacterial degradation of lignified wood cells in anaerobic aquatic habitats. *Appl. Environ. Microbiol.* **46**, 722–727.

Holzapfel-Pschorn A., Conrad R. & Seiler W. (1985) Production, oxidation, and emission of methane in rice paddies. *FEMS Microbiol. Ecol.* **31**, 343–351.

Holzapfel-Pschorn A., Conrad R. & Seiler W. (1986) Effects of vegetation on the emission of methane from submerged paddy soil. *Plant and Soil* **92**, 223–233.

Howarth R.W. & Teal J.M. (1979) Sulfate reduction in a New England salt marsh. *Limnol. Oceanogr.* **24**, 999–1013.

Jones J.G. (1971) Studies on freshwater bacteria: factors which influence the population and its activity. *J. Ecol.* **59**, 593–613.

Jones J.G. & Simon B.M. (1981) Differences in microbial decomposition processes in profundal and littoral lake sediments with particular reference to the nitrogen cycle. *J. Gen. Microbiol.* **123**, 297–312.

Jones J.G., Gardener S. & Simon B.M. (1983) Bacterial reduction of ferric iron in a stratified eutrophic lake. *J. Gen. Microbiol.* **129**, 131–139.

Jones W.J. & Paynter M.J.B. (1980) Populations of methane-producing bacteria and *in vitro* methanogenesis in salt marsh and estuarine sediments. *Appl. Environ. Microbiol.* **39**, 864–871.

Joshi M.M. & Hollis J.P. (1977) Interaction of *Beggiatoa* and rice plant: detoxification of hydrogen sulfide in the rice rhizosphere. *Science* **195**, 179–180.

Kaushik N.K. & Hynes, H.B.N. (1971) The fate of dead leaves that fall into streams. *Arch. Hydrobiol.* **68**, 465–515.

King G.M. (1990a) Dynamics and controls of methane oxidation in a Danish wetland sediment. *FEMS Microbiol. Ecol.* **74**, 309–324.

King G.M. (1990b) Regulation by light of methane emissions from a wetland. *Nature* **345**, 513–514.

King G.M. & Klug M.J. (1982) Comparative aspects of sulfur mineralization in sediments of a eutrophic lake basin. *Appl. Environ. Microbiol.* **43**, 1406–1412.

King G.M., Berman T. & Wiebe W.J. (1981) Methane formation in the acidic peats of Okefenokee Swamp, Georgia. *Am. Midl. Nat.* **105**, 386–389.

King G.M., Roslev P. & Skovgaard H. (1990) Distribution and rate of methane oxidation in sediments of the Florida Everglades. *Appl. Environ. Microbiol.* **56**, 2902–2911.

Kjøller A. & Struwe S. (1980) Microfungi of decomposing red alder leaves and their substrate utilization. *Soil Biol. Biochem.* **12**, 425–431.

Kjøller A., Struwe S. & Vestberg K. (1985) Bacterial dynamics during decomposition of alder litter. *Soil Biol. Biochem.* **17**, 463–468.

Komatsu Y., Takagi M. & Yamaguchi M. (1978) Participation of iron in denitrification in waterlogged soil. *Soil Biol. Biochem.* **10**, 21–26.

Kucera S. & Wolfe R.S. (1957) A selective enrichment method for *Gallionella ferruginea*. *J. Bacteriol.* **74**, 344–349.

Latter P.M., Craig J.B. & Heal O.W. (1967) Comparative studies on the microbiology of four moorland soils in the northern pennines. *J. Ecol.* **55**, 445–464.

Lidstrom M.E. & Somers L. (1984) Seasonal study of methane oxidation in Lake Washington. *Appl. Environ. Microbiol.* **47**, 1255–1260.

Lovley D.R. (1991) Dissimilatory Fe(III) and Mn(IV) reduction. *Microbiol. Rev.* **55**, 259–287.

Lovley D.R. & Klug M.J. (1983) Sulfate reducers can outcompete methanogens at freshwater sulfate concentrations. *Appl. Environ. Microbiol.* **45**, 187–192.

Lovley D.R. & Phillips E.J.P. (1986) Organic matter mineralization with reduction of ferric iron in anaerobic sediments. *Appl. Environ. Microbiol.* **51**, 683–689.

Lovley D.R. & Goodwin S. (1988) Hydrogen concentrations as an indicator of the predominant terminal electron accepting reactions in aquatic sediments. *Geochim. Cosmochim. Acta* **52**, 2993–3003.

Lovley D.R. & Phillips E.J. (1988) Novel mode of microbial energy metabolism: organic carbon oxidation coupled to dissimilatory reduction of iron and manganese. *Appl. Environ. Microbiol.* **54**, 1472–1480.

Lovley D.R., Dwyer D.F. & Klug M.J. (1982) Kinetic analysis of competition between sulfate reducers and methanogens for hydrogen in sediments. *Appl. Environ. Microbiol.* **43**, 1373–1379.

Mayer H.P. & Conrad R. (1990) Factors influencing the population of methanogenic bacteria and the initiation of methane production upon flooding of paddy soil. *FEMS Microbiol. Ecol.* **73**, 103–112.

McCready R.G.L., Gould W.D. & Cook F.D. (1983) Respiratory nitrate reduction by *Desulfovibrio* sp. *Arch. Microbiol.* **135**, 182–185.

Moore T.R. & Knowles R. (1990) Methane emissions from fen, bog and swamp peatlands in Quebec. *Biogeochemistry* **11**, 45–61.

Moran M.A., Benner R. & Hudson R.E. (1989) Kinetics of microbial degradation of vascular plant material in two wetland ecosystems. *Oecologia* **79**, 158–167.

Murray R.E. & Hodson R.E. (1984) Microbial biomass and utilization of dissolved organic matter in the Okefenokee Swamp ecosystem. *Appl. Environ. Microbiol.* **47**, 685–692.

Murray R.E. & Hodson R.E. (1985) Annual cycle of bacterial secondary production in five aquatic habitats of the Okefenokee Swamp ecosystem. *Appl. Environ. Microbiol.* **49**, 650–655.

Nykvist N. (1963) Leaching and decomposition of watersoluble organic substances from different types of leaf and needle litter. *Stud. Forest Suecica.* **3**, 1–31.

Oremland R.S., Marsh L.M. & Polcin S. (1982) Methane production and simultaneous sulfate reduction in anoxic salt marsh sediments. *Nature* **296**, 143–145.

Patrick W.H. Jr & Reddy K.R. (1978) Fate of fertilizer nitrogen in a flooded rice soil. *Soil Sci. Soc. Am. Proc.* **40**, 678–684.

Phelps T.J. & Zeikus J.G. (1984) Influence of pH on terminal carbon metabolism in anoxic sediments from a mildly acidic lake. *Appl. Environ. Microbiol.* **48**, 1088–1095.

Ramanathan V., Cicerone R.J., Singh H.B. & Kiehl J.T. (1985) Trace gas trends and their potential role in climate change. *J. Geophys. Res.* **90**, 5547–5566.

Rao D.N. & Mikkelsen, D.S. (1977) Effect of rice straw additions on production of organic acids in a flooded soil. *Plant and Soil* **47**, 303–311.

Rasmussen R.A. & Khalil M.A.K. (1981) Atmospheric methane: trends and seasonal cycles. *J. Geophys Res.* **86**, 9826–9832.

Robertson L.A. & Kuenen J.G. (1984) Aerobic denitrification: a controversy revived. *Arch. Microbiol.* **139**, 351–354.

Sebacher D.I., Harriss R.C., Bartlet K.B., Sebacher S.M. & Grice S.S. (1986) Atmospheric methane sources: Alaskan tundra bogs, an alpine fen, and subarctic boreal marsh. *Tellus* **38B**, 1–10.

Sethunathan N., Rao V.R., Adhya T.K. & Raghu K. (1983) Microbiology of rice soils. *CRC Crit. Rev. Microbiol.* **10**, 125–173.

Skinner F.A. (1968) The anaerobic bacteria of soil. In Cray T.R.G. & Parkinson D. (eds) *The Ecology of Soil Bacteria*, pp. 573–592. Liverpool University Press, Liverpool.

Skyring G.W., Oshrain R.L. & Wiebe W.J. (1979) Sulfate reduction rates in Georgia marshland soils. *Geomicrobiol. J.* **1**, 389–400.

Sørensen J. (1978) Capacity for denitrification and reduction of nitrate to ammonia in a coastal marine sediment. *Appl. Environ. Microbiol.* **35**, 301–305.

Sørensen J. (1982) Reduction of ferric iron in anaerobic marine sediment, and interaction with reduction of nitrate and sulfate. *Appl. Environ. Microbiol.* **43**, 319–324.

Sørensen J., Jørgensen B.B. & Revsbech N.P. (1979) A comparison of oxygen, nitrate, and sulfate respiration in coastal marine sediments. *Microb. Ecol.* **5**, 105–115.

Spratt H.G., Morgan M.D. & Good R.E. (1987) Sulfate reduction in peat from a New Jersey pinelands cedar swamp. *Appl. Environ. Microbiol.* **53**, 1406–1411.

Struwe S. & Kjøller A. (1989) Field determination of denitrification in water-logged forest soils. *FEMS Microbiol. Ecol.* **62**, 71–78.

Suberkropp K., Godschalk G.L. & Klug M.J. (1976) Changes in the chemical composition of leaves during processing in a woodland stream. *Ecology* **57**, 720–727.

Svensson B.H. (1976) Methane production in tundra peat. In Schlegel H.G., Gottschalk G. & Pfennig N. (eds) *Microbial Production and Utilization of Gases (H_2, CH_4, CO_2)*, pp. 135–139. E. Goltze KG, Göttingen.

Svensson B.H. (1984) Different temperature optimum for methane formation when enrichments from acid peat are supplemented with acetate or hydrogen. *Appl. Environ. Microbiol.* **48**, 389–394.

Takai Y. (1970) The mechanism of methane fermentation in flooded paddy soil. *Soil Sci. Plant Nutr.* **16**, 238–244.

Tiedje J.M. (1988) Ecology of denitrification and dissimilatory nitrate reduction to ammonium. In Zehnder A.J.B. (ed.) *Biology of Anaerobic Microorganisms*, pp. 179–244. John Wiley & Sons, New York.

Trimble R.B. & Ehrlich H.L. (1968) Bacteriology of manganese nodules. III. Reduction of MnO_2 by two strains of nodule bacteria. *Appl. Microbiol.* **16**, 695–702.

Urban N.R., Eisenreich S.J. & Grigal D.F. (1989) Sulfur cycling in a forested *Sphagnum* bog in northern Minnesota. *Biogeochemistry* 7, 81–109.

Watanabe I. & Furusaka C. (1980) Microbial ecology of flooded rice soils. In Alexander M. (ed) *Advances in Microbial Ecology*, vol. 4, pp. 125–168. Plenum, New York.

Watson R.T., Rodhe H., Oeschger H. & Siegenthaler U. (1991) Greenhouse gases and aerosols. In Houghton J.T., Jenkins G.J. & Ephraums J.J. (eds) *Climate Change. The IPCC Scientific Assessment*, pp. 1–40. Cambridge University Press, Cambridge.

Westermann P. (1990) The effect of temperature on butyrate degradation. In Belaich J.-P. (ed.) *Anaerobes Involved in Interspecies Hydrogen Transfer*, pp. 489–491. Plenum, New York.

Westermann P. & Ahring B.K. (1986) Terminal anaerobic carbon mineralization in an alder swamp. In Jensen V., Kjøller A. & Sørensen L.H. (eds) *Microbial Communities in Soil*. Elsevier, Amsterdam.

Westermann P. & Ahring B.K. (1987) Dynamics of methane production, sulfate reduction, and denitrification in a permanently waterlogged alder swamp. *Appl. Environ. Microbiol.* **53**, 2554–2559.

Westermann P. & Ahring B.K. (1988) Denitrification, methane production, and sulfate reduction in a permanently waterlogged forest swamp. *Arch. Hydrobiol. Beih.* **31**, 383–390.

Westermann P. & Ahring B.K. (in press) The role of sulfate reducing and methane producing bacteria in the metabolism of hydrogen and short-chained fatty acids in a freshwater swamp. In Adams D.D., Seitzinger S.P. & Crill P.M. (eds) *Cycling of Reduced Gases in the Hydrosphere*. E. Schweizerbart'sche Verlagsbuchhandlung, Stuttgart.

Whalen S.C. & Reeburgh W.S. (1990) Consumption of atmospheric methane by tundra soils. *Nature* **346**, 160–162.

Wheatley R.E., Greaves M.P. & Inkson R.H.E. (1976) The aerobic bacterial flora of a raised bog. *Soil Biol. Biochem.* **8**, 453–460.

Widdel F. (1988) Microbiology and ecology of sulfate- and sulfur-reducing bacteria. In Zehnder A.J.B. (ed.) *Biology of Anaerobic Microorganisms*, pp. 469–585. John Wiley & Sons, New York.

Wieder R.K. & Lang G.E. (1988) Cycling of inorganic and organic sulfur in peat from Big Run Bog, West Virginia. *Biogeochemistry* **5**, 221–242.

Wieder R.K., Yavitt J.B. & Lang G.E. (1990) Methane production and sulfate reduction in two Appalachian peatlands. *Biogeochemistry* **10**, 81–104.

Williams R.T. & Crawford R.L. (1983a) Microbial diversity of Minnesota peatlands. *Microb. Ecol.* **9**, 201–214.

Williams R.T. & Crawford R.L. (1983b) Effects of various physiochemical factors on microbial activity in peatlands: aerobic biodegradative processes. *Can. J. Microbiol.* **29**, 1430–1437.

Williams R.T. & Crawford R.L. (1984) Methane production in Minnesota peatlands. *Appl. Environ. Microbiol.* **47**, 1266–1271.

Willoughby L.G. (1974) Decomposition of litter in fresh water. In Dickinson C.H. & Pugh G.J.F. (eds) *Biology of Plant Litter Decomposition*, vol. 2, pp. 659–682. Academic Press, London.

Willoughby L.G. & Archer J.F. (1973) The fungal spora of a freshwater stream and its colonization pattern on wood. *Freshwat. Biol.* **3**, 219–239.

Wilson J.O., Crill P.M., Bartlett K.B., Sebacher D.I., Harriss R.C. & Sass R.L. (1989) Seasonal variation of methane emissions from a temperate swamp. *Biogeochemistry* **8**, 55–71.

Winfrey M.R., Marty D.G., Bianchi A.J.M. & Ward D.M. (1981) Vertical distribution of sulfate reduction, methane production, and bacteria in marine sediments. *Geomicrobiol. J.* **2**, 341–362.

Yamane I. & Sato K. (1963) Decomposition of plant constituents and gas formation in flooded soil. *Soil Sci. Plant Nutr.* **9**, 28–31.

Yamane I. & Sato K. (1964) Decomposition of glucose and gas formation in flooded soil. *Soil Sci. Plant Nutr.* **10**, 35–41.

Yamane I. & Sato K. (1967) Effect of temperature on the decomposition of organic substances in flooded soil. *Soil Sci. Plant Nutr.* **13**, 94–100.

Yavitt J.B., Downey D.M., Lancaster E. & Lang G.E. (1990) Methane consumption in decomposing *Sphagnum*-derived peat. *Soil Biol. Biochem.* **22**, 441–447.

Zehnder A.J.B. (ed.) (1988) *Biology of Anaerobic Microorganisms*. John Wiley & Sons, New York.

Zehnder A.J.B. & Stumm W. (1988) Geochemistry and biogeochemistry of anaerobic habitats. In Zehnder A.J.B (ed.) *Biology of Anaerobic Microorganisms*, pp. 1–39. John Wiley & Sons, New York.

Zeikus J.G., Wellstein A.L. & Kirk T.K. (1982) Molecular basis for the biodegradative recalcitrance of lignin in anaerobic environments. *FEMS Microbiol. Lett.* **15**, 193–197.

Zoltai S.C., Pollett F.C., Jeglum J.K. & Adams D.G. (1975) Developing a wetland classification for Canada. In Bernier B. & Winget C.H. (eds) *Forest Soils and Forest Land Management. Proceedings of the Fourth Forest Soils Conference*, pp. 497–511. Les Presses de l'Université Laval, Québec.

Robert W. Howarth

Microbial processes in salt-marsh sediments

10.1 Introduction

Salt marshes are ecosystems dominated by vascular plants (grasses) with their roots in soils which are regularly or permanently waterlogged with sea water. They are found along the coast throughout the temperate zone of the world and are particularly dominant features along the east and Gulf coasts of North America, where marshes frequently cover vast expanses behind barrier islands and in bays. Marshes tend to be quite flat, and their mean elevation is near or slightly below the mean high tide level. Thus, they are regularly flooded by tidal waters. Along much of the east coast of North America, tides flood completely over much of the surface of marsh soils twice each day, although lower amplitude tides result in much less regular and less frequent flooding in some areas. The diversity of grass species in salt marshes is quite low; the marshes of the east coast of North America are dominated by a single grass, *Spartina alterniflora*, at lower elevations. Other species such as *Spartina patens* and *Distichlis spicata* are found at slightly higher elevations where tidal flooding is less regular. In most marshes, *S. alterniflora* exhibits two distinct forms: a tall form near creeks where soil drainage is highest at low tide and a short form further away from creeks.

Microbial processes in salt marshes have received extensive study, and both rates and controls of processes in marshes are among the best known for any ecosystem. Nonetheless, much is still unknown, in part because of the great diversity of processes found. Almost every microbial process found in any ecosystem can be found in salt marshes. Also, as in terrestrial ecosystems, many microbial processes in marshes are tightly linked to the activity of the dominant vascular plants. The interactions between marsh grasses and microbial processes have proven difficult to study, particularly for processes in or near the grass rhizosphere (Laanbroek 1990).

A complete presentation of microbiological studies in salt marshes is beyond the scope of this chapter, which will concentrate on reviewing some of the major processes in marsh sediments. Rates for many microbial processes in salt-marsh sediments are often extremely high, in part because of the large inputs of organic matter. This chapter reviews the pathways whereby bacteria degrade this material or produce new material through chemoautotrophic production.

10.2 Sources of organic matter in marshes

Salt marshes are among the most productive ecosystems in the world, with rates of net primary production by the dominant grasses ranging from 460 to $3200\,g\,C\,m^{-2}\,yr^{-1}$ (Schubauer & Hopkinson 1984; Giblin & Wieder 1992). Algae on the surface of marsh sediments may produce another $80-190\,g\,C\,m^{-2}\,yr^{-1}$ in net primary production (Pomeroy *et al.* 1981), and in some marshes, production by chemoautotrophic sulphur-oxidizing bacteria may fix as much as another $275-500\,g\,C\,m^{-2}\,yr^{-1}$ (see Section 10.6) (Howarth 1984). Only a small fraction of this production is permanently accreted in the sediments of marshes (Howarth & Hobbie 1982), and the net export of organic matter from marshes to coastal waters is also a small percentage of the rate of primary production (Nixon 1980). Most of the organic matter produced in marshes is decomposed *in situ*, fueling intense rates of microbial processes.

At least half, and often much more, of all of the net primary production of marsh grasses occurs as the below-ground production of roots and rhizomes (Valiela *et al.* 1976; de la Cruz & Hackney 1977; Pomeroy *et al.* 1981; Good *et al.* 1982; Schubauer & Hopkinson 1984); such production typically ranges from 220 to $1680\,g\,C\,m^{-2}\,yr^{-1}$ (Good *et al.* 1982). These estimates may underestimate the true extent of below-ground production by grasses since they do not include sloughing of root material and excretion of dissolved organic matter (DOM) from roots and rhizomes. In terrestrial ecosystems, such losses can be as great as 50% of the rate of production by roots (McDougall & Rovira 1970; Head 1973). The excretion of DOM from marsh grasses may be much higher than is typically the case for terrestrial plants (Howarth & Hobbie 1982), since unlike terrestrial plants, the below-ground tissues of marsh grasses are often anoxic (Mendelssohn *et al.* 1981). When anoxic, plant metabolism proceeds through anaerobic fermentations, producing alcohols and low molecular weight fatty acids as products of this metabolism. These products can easily diffuse into the surrounding sediments (Mendelssohn *et al.* 1981). In addition to the below-ground production of marsh grasses, most of the production by chemoautotrophic bacteria in marshes probably occurs below-ground in the sediments (Howarth 1984). Thus, marshes have very large inputs of newly produced, relatively labile organic matter directly into their sediments.

10.3 Oxygen and aerobic respiration

When saturated with water, marsh sediments are anoxic below the top few millimeters (King 1988; Giblin & Wieder 1992). This is a result both of the high rate of oxygen consumption by marsh sediments and of the very slow rate of oxygen diffusion through waterlogged pores, which is some 10 000 times slower than diffusion in gas-filled pores (Ponnamperuma 1972). However, at low tide air can enter some of the pore spaces of marsh sediments, due both to lateral drainage of pore waters to creeks (Howarth *et al.* 1983; Anderson & Howell 1984; Agosta 1985) and to loss of sediment pore water during evapotranspiration by the marsh grasses (Dacey & Howes 1984; Morris & Whiting 1985). This greatly speeds the rate of oxygen diffusion in the sediment, both from the sediment surface and from the grass rhizosphere. Salt-marsh grasses, in common with many other wetland species, have hollow internal channels which allow oxygen diffusion in a gas phase from the air to the rhizosphere (Teal & Kanwisher 1961; Armstrong 1975), and oxygen diffusion from the plant roots into the sediment is probably faster when the sediment contains gas-filled pores (Howarth & Hobbie 1982). In fact, no oxygen may actually reach the sediment when it is completely waterlogged (Bedford *et al.* 1991). Within 5 m of a creek, gas-filled pore spaces can be found as deep as 10 cm in the sediment within 12 hr following a high tide (Howes *et al.* 1981; Howarth *et al.* 1983). The oxidation–reduction potential in these partially drained sediments is much higher than in waterlogged sediments, suggesting oxygen penetration (Howes *et al.* 1981). Actual data on oxygen concentrations in salt-marsh sediments have been reported by King (1988), who found oxygen penetrating only to depths of 2–5 mm. Oxygen occurred at greater depths before tidal flooding than after tidal flooding. In order to use oxygen microelectrodes in the marsh sediment, King (1988) was forced to add water above cores; this may have caused water to displace gas in some of the pores in the sediment cores, leading to an underestimation of oxygen penetration during low-tide periods.

Oxygen penetration into marsh sediments probably varies greatly from marsh to marsh, and even within a marsh, due to differences in frequency of tidal flooding, porosity of the sediment, and distance to creeks. In most salt marshes along the east coast of North America, tidal waters completely cover the surface of the sediments twice each day, replenishing waters which were lost through drainage and evapotranspiration at low tide. But in some areas with relatively small tidal amplitudes, tidal waters flood over the sediment only once a month or so during the unusually high spring tides or during storm tides. This is the case for salt marshes in Delaware, leading to quite dry sediments between flooding events (Lord & Church 1983). In marshes which flood completely twice each day, significant lowering of the water table within sediments only occurs within 5–10 m of creeks; at greater distances, sediments tend to remain waterlogged to within a centimeter or two of the surface.

Clearly when sediments are completely waterlogged and anoxic, decomposition occurs only through anaerobic processes. However, aerobic respiration using oxygen as the electron acceptor can occur when oxygen penetrates into partially drained sediments. Aerobic respiration can also always occur at the sediment surface and perhaps at the surface of grass roots in the rhizosphere, if oxygen actually diffuses out of the roots (Bedford *et al.* 1991). Unfortunately, it is not possible to directly measure rates of aerobic respiration in marsh sediments since oxygen is consumed both in aerobic respiration and in oxidation of reduced compounds such as sulfides. In principle, aerobic respiration can be estimated by difference if enough other processes are known. In the only study in a salt marsh specifically designed to do this, Howes *et al.* (1984) simultaneously measured sulfate reduction, carbon dioxide production, and oxygen uptake in cores sampled from a Massachusetts salt marsh in October. By assuming that all respiration used either oxygen or sulfate as the electron acceptor (see Section 10.4), they concluded that aerobic respiration produced 44% of the carbon dioxide. Comparing this respiration rate with their oxygen consumption data, Howes *et al.* (1984) estimated that roughly half of the oxygen was consumed in aerobic respiration, with the remainder used to oxidize reduced sulfur compounds. This is similar to the finding of Jørgensen (1977; 1982a) for shallow, subtidal marine sediments free of vascular plants. However, the study by Howes *et al.* (1984) purposefully excluded oxygen fluxes through the internal spaces of marsh grasses into the sediment, thereby underestimating total oxygen consumption and overestimating the percentage of oxygen consumption which represents aerobic respiration.

Oxygen uptake has been measured by a variety of techniques, using *in situ* chambers or cores incubated in the laboratory or in the field, and having either an air phase or water just above the sediment (Howarth & Hobbie 1982). There is some tendency to find higher rates when an air phase rather than water is used, perhaps because this favors more gas-filled pores within the surface layers of the sediment. Rates of oxygen uptake by the sediments in a variety of marshes from Nova Scotia to Louisiana tend to fall in the range of 5–65 mol O_2 m^{-2} yr^{-1}, with an average value of 17 mol O_2 m^{-2} yr^{-1} for measurements made with water over the sediments and an average value of 41 mol O_2 m^{-2} yr^{-1} for measurements made with air over the sediments (Howarth & Hobbie 1982; Howes *et al.*

1984). If we assume that half of this consumption represents aerobic respiration, then we estimate aerobic respiration as 2.5–33 mol O_2 $m^{-2} yr^{-1}$, or enough to oxidize 30–400 g C $m^{-2} yr^{-1}$. As noted above, aerobic respiration probably is in fact less than half the rate of oxygen uptake.

10.4 Rates of anaerobic processes

In the absence of oxygen, bacterial respiration can proceed using other electron acceptors. The energy available to bacteria carrying out respiration depends upon the particular electron acceptor used and generally decreases according to the following list of acceptors: oxygen, nitrate, manganese, iron, sulfate, and carbon dioxide (Ponnamperuma 1972, 1984; Berner 1980; Laanbroek 1990; see also Chapter 9). Generally, competition among bacteria for substrates will tend to lead to a domination by respiration that uses the more energetically favorable electron acceptors if they are available. Thus, respiration using nitrate reduction should become a dominant process only in the absence of oxygen, but should proceed preferentially to respiration using other electron acceptors. Iron and manganese should be the electron acceptors used preferentially after nitrate is depleted, and sulfate reduction should dominate only after these metals become unavailable as electron acceptors. Following this line of reasoning, the reduction of carbon dioxide to methane (methanogenesis) should dominate only in the absence of sulfate. However, this idealized separation of the use of electron acceptors does not always hold in nature, perhaps because competition among bacteria for substrates is not that severe and other factors are partially controlling bacterial activity. Sulfate reduction and methanogenesis often co-occur in marine sediments (Ivanov *et al.* 1989), which perhaps should not be surprising since the energy yields for the two processes are not that different. Perhaps more surprisingly, sulfate reduction has been found to occur in the presence of oxygen in microbial mats (Canfield & Des Marais 1991).

The use of various electron acceptors in anaerobic respiration in salt marsh sediments is discussed below in the order of energy yield.

10.4.1 Denitrification

Dissimilatory nitrate reduction (denitrification) occurs in salt marshes, but the rates tend to be limited by a relatively slow rate of supply of nitrate. Denitrification is a major sink for nitrogen in marshes, which is important because both salt marshes and the coastal marine ecosystems with which they interact are frequently nitrogen limited (Whitney *et al.* 1981; Chalmers 1982; Boynton *et al.* 1982; Howarth 1988). However, the consumption of organic carbon during denitrification appears to be quite small, of the order of

3–10 $g\,C\,m^{-2}\,yr^{-1}$ (Haines *et al.* 1977; Kaplan *et al.* 1979; DeLaune & Patrick 1980).

However, a methods problem may have led to an underestimation of denitrification rates in all studies in marshes. Since rates are limited by the supply of nitrate, any addition of nitrate will increase rates dramatically (Lindau & DeLaune 1991). It seems likely that at least some nitrification (the oxidation of ammonium to nitrate by bacteria) occurs in the rhizosphere of marsh grasses since oxygen and high concentrations of ammonium may coexist there. Any nitrate which is thus formed can rapidly diffuse into the adjacent anoxic sediments and be denitrified. In fact, the rhizosphere of marsh grasses seems like an ideal location for coupled rates of nitrification and denitrification (Kemp *et al.* 1982; Jenkins & Kemp 1984; Reddy *et al.* 1989; Lindau and DeLaune 1991). When denitrification is measured in sediment cores where the rhizosphere has been isolated from the above-ground portions of grasses, oxygen diffusion to the rhizosphere is halted, and nitrification ceases. To the extent denitrification proceeds via a close coupling with nitrification, denitrification would be underestimated (Lindau & DeLaune 1991).

10.4.2 Iron and manganese reduction

In some pure-culture studies and some sediments, iron-reducing bacteria have been found to outcompete sulfate-reducing bacteria for substrates (Lovley 1987), as predicted from thermodynamic considerations (Berner 1980). However, iron (III) can also be reduced by chemical reaction with hydrogen sulfide, and in salt-marsh sediments it appears that this chemical process dominates (Jacobson 1990). Jacobson (1990) added iron (III) to salt-marsh sediments and found no effect on the rate of sulfate reduction, indicating little if any competition between the processes of sulfate reduction and iron reduction. Adding molybdate to inhibit sulfate reduction also slowed the rate of iron reduction, apparently because less sulfide was available for the chemical reduction of iron. If iron reduction in the marsh was predominantly bacterial, the accumulation of fatty acids resulting from inhibition of sulfate reduction would have been expected to increase the rate of iron reduction (Jacobson 1990). Apparently sulfides from sulfate reduction are abundant enough in the marsh sediments to chemically react with iron (III) before it can become available to any iron-reducing bacteria (Jacobson 1990).

I am aware of no research on the use of oxidized manganese as an electron acceptor for bacterial respiration in salt-marsh sediments. However, as for iron, it seems likely that the manganese would be chemically reduced before becoming available to bacteria. In freshwater wetlands, the reduction of manganese is also thought to be largely a chemical process (Laanbroek 1990).

10.4.3 Sulfate reduction

Sulfate reduction rates in marsh sediments have been measured by a variety of techniques (Giblin & Wieder 1992), but the most commonly used and least ambiguous method is to measure the reduction of ^{35}S-labeled sulfate injected into cores. Most early studies using this technique only measured acid-volatile products of sulfate reduction, such as soluble sulfides (H_2S, HS^-) and iron monosulfides (FeS). However, ^{35}S reduced during sulfate reduction can very rapidly become incorporated into pools, such as pyrite (FeS_2) and elemental sulfur (S^0), which are not acid volatile; sulfate reduction rates can be underestimated if these pools are not measured (Howarth 1979; Howarth & Teal 1979). Failure to include these nonacid-volatile pools when measuring sulfate reduction in sulfide-rich, subtidal sediments usually results in only small errors (Howarth & Jørgensen 1984), but in salt marshes, pyrite and elemental sulfur can be the major short-term products of sulfate reduction. Failure to measure these can underestimate rates of sulfate reduction by two- to 10-fold or more (Howarth 1979; Howarth & Teal 1979; Howarth & Merkel 1984; King 1988), although in some cases the error is as small as 20% (Hines *et al.* 1989).

The original work showing the importance of pyrite as a short-term product of sulfate reduction assayed for ^{35}S in pyrite and elemental sulfur by first acidifying sediment and distilling off acid-volatile products, then carefully washing away unreduced ^{35}S-sulfate with repeated washes of water, and finally digesting the remaining sediment (Howarth 1979; Howarth & Teal 1979). This work generated some controversy since the incorporation of ^{35}S into pyrite could be overestimated if some unreduced ^{35}S-sulfate remained adsorbed to the sediment following the water washes (King 1983; Howes *et al.* 1984). Subsequently, a consensus has developed that the preferred way to assay for ^{35}S-labeled pyrite is by reducing sediments with chromium (II), a method not subject to any interference by unreduced ^{35}S-sulfate (Howarth & Jørgensen 1984; Howarth & Merkel 1984; King *et al.* 1985; Canfield *et al.* 1986; King 1988; Fossing & Jørgensen 1989; Hines *et al.* 1989). The use of the chromium reduction method in salt-marsh sediments yields results which are quite comparable to the original washing method (Howarth & Merkel 1984), supporting the data of Howarth & Teal (1979) and of Howarth & Giblin (1983). Until quite recently, there was some uncertainty as to whether pyrite actually formed rapidly in marsh sediments (Howarth 1979) or whether ^{35}S was being rapidly incorporated into pyrite by isotopic exchange (Howarth & Jørgensen 1984; King 1988). However, recently Fossing & Jørgensen (1990) have demonstrated that isotopic exchange does not occur between ^{35}S-labeled sulfide and pyrite, indicating that pyrite does indeed form in marshes on the time scale of a day or less, as originally proposed (Howarth 1979).

Rates of sulfate reduction in salt-marsh sediments tend to range from 6 to 75 $mol\,S\,m^{-2}\,yr^{-1}$ (Giblin & Wieder 1992). Since two moles of organic carbon are respired for every mole of sulfate reduced (Jørgensen 1977; Howarth & Teal 1979), sulfate reduction consumes 144–1800 $g\,C\,m^{-2}\,yr^{-1}$ in marshes. This is by far the major sink for organic carbon in marsh sediments (Howarth 1984; King 1988; Hines *et al.* 1989; Giblin and Wieder 1992). However, these rates of sulfate reduction may be overestimated to some extent since the rhizosphere of the marsh grasses is disrupted during sampling. By separating roots and rhizomes from the above-ground plant tissues, oxygen fluxes are stopped, perhaps switching some bacterial metabolism in the rhizosphere from aerobic to anaerobic. Also, fermentative metabolism of the plant roots themselves may increase, increasing the supply of organic substrates to the sulfate-reducing bacteria (Howarth & Hobbie 1982). However, a careful comparison of sulfate reduction rates in small cores (the usual technique) with rates measured by injecting ^{35}S label *in situ* into marsh sediments with an undisturbed rhizosphere yielded comparable results (Hines *et al.* in press).

It should be noted that sulfate-reducing bacteria themselves are able to use only a relatively limited number of organic substrates, principally low molecular weight alcohols and fatty acids. Other bacteria ferment the more complex organic compounds found in sediments, producing the substrates used by sulfate-reducing bacteria. Thus, it is an assemblage or community of bacteria, including both the fermenters and sulfate reducers, which overall decompose the organic matter. As might be expected, ammonium regeneration during these decomposition processes is associated with fermentation rather than sulfate reduction itself (Jacobson *et al.* 1987). The details of the functioning of this community in salt marshes have received relatively little study. In subtidal marine sediments, acetate is usually the major substrate for sulfate-reducing bacteria (Sorensen *et al.* 1981; Christensen & Blackburn 1982; Christensen 1984; Shaw *et al.* 1984), but in salt-marsh sediments, acetate appears to support only some 10% of the rate of sulfate reduction (J.E. Hobbie, A. Mann, & R.W. Howarth unpublished data). The major substrates remain unknown.

10.4.4 Methanogenesis

Actual rates of methane formation (methanogenesis) have not been measured in marsh sediments but are probably quite low. Fluxes of methane from the surface of sediments to the atmosphere have been measured in several marshes, and vary between 0.4 and

$40\,g\,C\,m^{-2}\,yr^{-1}$ (King & Wiebe 1978; Howarth & Hobbie 1982; Howes *et al.* 1985; Bartlett *et al.* 1987). Interestingly, both the highest and lowest fluxes are reported from the same marsh in Georgia, with the highest rates in the most waterlogged soils away from creek banks (King & Wiebe 1978). These atmospheric fluxes provide minimum rates of methane formation because some methane is probably oxidized aerobically at the sediment surface or anaerobically within the sediment. Both the aerobic and anaerobic oxidation of methane was found to be very small in a Georgia salt marsh (King 1978; King & Wiebe 1978). However, I am aware of no other studies on methane oxidation in salt marshes, and the generality of this result is uncertain; in many subtidal marine sediments, appreciable anaerobic oxidation of methane occurs (Ivanov *et al.* 1989). Nonetheless, available evidence suggests that atmospheric fluxes from marshes may not be much less than the rate of methane formation, and that methanogenesis is therefore only a minor part of the metabolism of salt-marsh sediments.

Methane fluxes from salt marshes are inversely correlated with the concentration of sulfate in the sediments (King & Wiebe 1980; Bartlett *et al.* 1987). At one time it was thought that this was just a result of sulfate-reducing bacteria outcompeting methanogenic bacteria for substrates. While the results of King (1978) suggest that this may indeed be the case in salt-marsh sediments, in many subtidal marine sediments high rates of methanogenesis and sulfate reduction can coexist; the relatively low fluxes out of sulfate-rich marine sediments are often a result of anaerobic methane oxidation rather than inhibition of methane formation (see review by Ivanov *et al.* 1989). Although the mechanism of this anaerobic methane oxidation remains unknown, it appears to be related to the sulfate concentration (Ivanov *et al.* 1989). Thus, the fluxes of methane out of low-sulfate freshwater sediments are much higher than the fluxes from salt marshes (see Chapter 9).

10.4.5 Summary of respiration rates in two marshes

Table 10.1 presents a summary of respiration rates using different electron acceptors in two particularly well-studied salt marshes, Sapelo Island (Georgia) and the Great Sippewissett Marsh (Massachusetts). For both marshes, estimates are for relatively waterlogged areas well away from creeks and dominated by stands of the short growth form of the salt-marsh grass *S. alterniflora*. For both marshes, it is assumed that 50% of oxygen consumption by the sediment represents aerobic respiration (Howes *et al.* 1984). Oxygen consumption data for Sippewissett are from Howes *et al.* (1984) and for Sapelo Island from Teal & Kanwisher (1961) and both exclude oxygen transport down the arenchymal spaces. Thus, aerobic re-

Table 10.1 Respiration rates in two salt marshes ($g\,C\,m^{-2}\,yr^{-1}$). See text for reference and assumptions

	Sapelo Island (Ga)	**Sippewissett (Ma)**
Aerobic respiration	390	390
Denitrification	10	3
Mn and Fe reduction	Negligible	Negligible
Sulfate reduction	850	1800
Methanogenesis	40	1–8

spiration may be underestimated, if any oxygen actually reaches the sediment through the rhizosphere (but see Bedford *et al.* 1991).

Denitrification appears to be a very minor sink for carbon in both marshes, accounting for an estimated respiration of $3\,g\,C\,m^{-2}\,yr^{-1}$ in Sippewissett (Kaplan *et al.* 1979) and $10\,g\,C\,m^{-2}\,yr^{-1}$ at Sapelo Island (Haines *et al.* 1977). Respiration using oxidized iron and manganese as electron acceptors has not been measured in either marsh, but as discussed above, such respiration is probably small since most metal reduction in salt marshes probably occurs from chemical reduction with sulfide (Jacobson 1990). Methanogenesis is also a minor sink of organic matter in both marshes, with methane fluxes from sediment to the atmosphere of approximately $40\,g\,C\,m^{-2}\,yr^{-1}$ at Sapelo Island (King & Wiebe 1978) and $1–8\,g\,C\,m^{-2}\,yr^{-1}$ at Sippewissett (Howarth & Teal 1979; Howes *et al.* 1985). Actual rates of methane formation must be somewhat higher than this, but probably not by much (King 1978; King & Wiebe 1978).

Sulfate reduction and associated fermentations are the dominant sink for organic matter in both marshes (see Table 10.1). Estimates for the Sippewissett Marsh are from Howarth & Teal (1979) and for the marsh at Sapelo Island from Howarth & Giblin (1983), Howarth & Merkel (1984), and R.W. Howarth (unpublished). From Table 10.1, we see that sulfate reduction accounts for 67% of all of the respiration in the marsh at Sapelo Island and 80% of all the respiration in the Great Sippewissett Marsh. This is a high percentage compared to many nearshore subtidal marine sediments, where sulfate reduction typically accounts for 25% of sediment respiration in waters of depths between 20 and 200 m and 50% of sediment respiration in more shallow waters (Jørgensen 1982a). However, it is consistent with the finding that sulfate reduction is an increasingly greater proportion of all respiration as the total respiration rate increases (Howarth 1984). Total rates of respiration in the sediments of these two marshes are some 15- to 25-fold greater than in the 20- to 200-m deep waters and some two- to three-fold greater than in the less than 20-m waters studied by Jørgensen (1982a). The rate of oxygen respiration in sediments is determined principally by the rate of oxygen diffusion, so as total respiration increases, an increasing proportion must be anoxic.

10.5 Controls on sulfate reduction and associated fermentations

Given the importance of sulfate reduction and associated fermentations in the metabolism of salt marshes, it is of interest to examine the controls on these processes. These controls have been extensively studied not only in salt marshes but in the sediments of freshwater wetlands, lakes, estuaries, and coastal waters. Temperature is one obvious control, with rates of sulfate reduction within a marsh increasing with seasonal increases in temperature. However, temperature alone does not explain differences in rates of sulfate reduction, either within or between marshes. For instance, sulfate reduction rates are probably higher in a Massachusetts marsh than in Georgia (Table 10.1), and sulfate reduction rates within marshes often show peak rates which are higher than can be explained by temperature (Howarth & Teal 1979; King 1988; Hines *et al.* 1989). Below I present information on two hypothesized controls on sulfate reduction/oxidation–reduction potentials and organic substrates – and conclude that the rate of supply of substrates is the major control on sulfate reduction in waterlogged salt-marsh sediments.

10.5.1 Oxidation–reduction potential

Oxidation–reduction potentials are frequently mentioned as a factor controlling sulfate reduction rates in wetlands and other sediments, with sulfate reduction occurring only at relatively low redox potential (Eh). The literature and textbooks suggest several cutoff values (Eh corrected to pH = 7) above which sulfate reduction would not be expected: +100 mV (Stevenson 1986); −120 to −180 mV (Ponnamperuma 1972; Gambrell & Patrick 1978; Patrick & Reddy 1978; Laanbroek 1990); and −214 mV (Ponnamperuma 1984; Schlesinger 1991). These cutoffs are estimated either from empirical studies of changes in Eh and metabolism which accompany waterlogging of sediments and soils (Gambrell & Patrick 1978; Patrick & Reddy 1978), or from thermodynamic calculations of the stability of sulfate and sulfide (Ponnamperuma 1984). However, measured oxidation–reduction potentials in the top 10–15 cm of salt-marsh sediments are generally in the range of +300 to −200 mV (Teal & Kanwisher 1961; Howes *et al.* 1981) when corrected to a pH value of 7 (Bohn 1971), and high rates of sulfate reduction occur even in the sediments with the higher Eh values (Howarth & Teal 1979; Howarth & Giblin 1983; Howes *et al.* 1984). Significant rates of sulfate reduction have also been found in the surface layers of some nearshore marine sediments having measured Eh values as high as +400 mV (Howarth & Jørgensen 1984).

It is important to distinguish between a theoretical oxidation–reduction potential and the potential as it is measured in marsh sediments. Measured Eh potentials in sediments often "have no precise thermodynamic significance" (Ponnamperuma 1972). In salt-marsh sediments, Eh electrodes respond principally to the electro-activity of sulfide ions, and high values often just reflect the low concentration of sulfide resulting from rapid precipitation of pyrite (Howarth *et al.* 1983). To some extent, the finding of sulfate reduction at high measured values of Eh could just reflect a discrepancy between the measured Eh and the "actual" potential,[1] as suggested by Ponnamperuma (1972) to explain a lack of correspondence between measured Eh and microbial processes in flooded rice fields.

It is also important to note that while thermodynamic considerations influence competition among bacteria using different electron acceptors, thermodynamics do not set absolute limits on bacterial processes unless the energy yield of a reaction is simply too low to obtain sufficient energy. Sulfate is thermodynamically unstable relative to sulfide at oxidation–reduction potentials above −120 to −180 mV (Ponnamperuma 1972; Patrick & Reddy 1978). However, the reduction of sulfate to sulfide, when coupled with the oxidation of organic matter to carbon dioxide, will still produce a net yield of energy for sulfate-reducing bacteria, and so sulfate reduction can occur. In fact, the energy yield of respiration through sulfate reduction is actually greater under more oxidizing conditions since the concentration of sulfide, one of the products of sulfate reduction, is lower. Of course, aerobic bacteria will gain appreciably more energy from respiring the same organic matter and so are likely to outcompete sulfate-reducing bacteria. Also, the enzyme systems of most sulfate-reducing bacteria are easily poisoned by oxygen, although active sulfate-reducing bacteria are found in the presence of oxygen in organic-rich microbial mats (Canfield & Des Marais 1991). In any event, it is clear that the oxidation–reduction potentials found in salt-marsh sediments do not prevent sulfate reduction.

10.5.2 Substrates for sulfate reduction

For marine sediments, organic carbon availability generally controls the rate of sulfate reduction when the sulfate concentration is greater than 5 mmol l^{-1} (Westrich & Berner 1984). At lower sulfate concentrations, sulfate can limit the process. In salt-marsh sediments,

[1] Salt-marsh sediments are unlikely to have a single, theoretical oxidation–reduction potential; such a potential would exist only at chemical equilibrium. Marsh sediments are far from equilibrium, as indicated by the high rates of chemical reactions and microbial processes. Any particular chemical couple (sulfate/sulfide; oxidized iron/reduced iron, etc.) will have its own oxidation–reduction potential; these may differ significantly if reactions are poorly coupled or kinetically constrained.

sulfate concentrations are seldom below 5 mmol l^{-1}, and so we would expect organic carbon availability rather than sulfate to regulate rates of sulfate reduction.

Sulfate reduction rates often vary by two- to three-fold within a given marsh, with rates being higher in the more waterlogged sediments away from creek banks where the short form of *S. alterniflora* dominates (Howarth & Giblin 1983; King 1988). Sulfate reduction is lower in the creekside sediments dominated by the tall form of *S. alterniflora* (Hines *et al.* 1989). Since the creekside sediments tend to be better drained and have more gas-filled pores, one might expect that sulfate reduction rates are lower there merely because aerobic respiration rates are higher. However, total rates of respiration as measured by fluxes of carbon dioxide are also lower in creekside sediments (Morris & Whiting 1986). The higher rates of sulfate reduction in the sediments underlying the short form of grass may at least in part be the result of more organic matter inputs to these sediments, perhaps because more roots and rhizomes are concentrated in the upper 10–15 cm of sediment in the short-grass stands (King 1988).

The metabolism of the grasses may also be a factor leading to higher sulfate reduction rates in more waterlogged sediments. The marsh grasses may excrete alcohols and fatty acids if anoxia in their roots and rhizomes force them to metabolize through fermentative processes (Howarth & Hobbie 1982). In an experiment to test this hypothesis large cores (30 cm high × 15 cm in diameter) were taken from a short-grass stand and kept in a greenhouse for 1 month. Water was allowed to drain from some cores and others were kept waterlogged. Of those kept waterlogged, some were completely stagnant, whereas in others, the pore waters were partially replaced each day by allowing some water to flow through the core from the surface. At the end of the experiment, sulfate reduction rates were highest in the completely stagnant, waterlogged cores, were intermediate in the waterlogged cores which had some daily water replacement (and were therefore less reducing), and were lowest in the partially drained cores (R.W. Howarth, R. Marino, S. Beard, & B.J. Peterson unpublished). The higher sulfate reduction rates in the more stagnant of the waterlogged cores seem likely to be the result of greater or more frequent anoxia within the plant tissues in these cores. This may have increased the fermentations by the grasses, with concomitant leakage of fermentation products, or it may have killed some root and rhizome material, producing a highly labile substrate for bacterial fermenters.

Sulfate reduction rates in sediments can change rapidly, apparently in response to changes in grass activity. Thus, Hines *et al.* (1989)

found that sulfate reduction rates in a New Hampshire marsh increased five-fold over a few weeks as marsh grasses began to grow more rapidly. Rates decreased equally dramatically as the grasses put their energy into flowering, indicating that the changes in sulfate reduction rates were due to release of dissolved organic carbon from the rhizosphere during periods of active growth (Hines *et al.* 1989). An infestation of grasses by flies in one year stopped grass growth in late June, resulting in a major decrease in sulfate reduction rates (Hines *et al.* 1989).

10.6 Chemoautotrophic production

During sulfate reduction, energy flow is decoupled from carbon cycling (Fenchel & Jørgensen 1977; Howarth & Teal 1980; Howarth 1984). Approximately 75% of the energy content of the organic matter which is decomposed is transferred and stored in hydrogen sulfide, or 740 kJ mol^{-1} of sulfide (Howarth 1984). When the sulfide and other reduced sulfur compounds are later oxidized, this energy is released. If the oxidation is merely a chemical one, the energy of oxidation is dissipated into the environment. However, many bacteria can catalyze the oxidation of reduced sulfur, either to detoxify the sulfide or to gain energy. A variety of chemoautotrophic bacteria, including many strains of *Thiobacillus* and *Beggiatoa*, can capture the energy of sulfur oxidations using oxygen or nitrate as the oxidant, and use the energy to fix carbon dioxide (Jørgensen 1982b; Kelly 1982; Nelson & Jannasch 1983; Howarth 1984). Although these bacteria produce new organic matter from carbon dioxide fixation, their production is a form of secondary production since the energy originated with the organic matter decomposed during sulfate reduction.

The rate of energy flow through reduced sulfur compounds is substantial in salt marshes. Assuming that rates of sulfate reduction range from 6 to 75 mol S $m^{-2} yr^{-1}$ (Giblin & Wieder 1992), rates of energy transfer from organic matter to sulfide range from 4440 to 55 500 kJ $m^{-2} yr^{-1}$ (for comparison, rates of net primary production by marsh grasses vary from 19 000 to 130 000 kJ $m^{-2} yr^{-1}$). The immediate fate of most of this sulfide and its embodied energy is precipitation in minerals, such as iron monosulfides (FeS) and pyrite (FeS_2) (Howarth 1979; 1984). However, the permanent rate of accretion of reduced sulfur in salt-marsh sediments is small relative to the rate of sulfate reduction and is generally much less than 1 mol S $m^{-2} yr^{-1}$ (Howarth 1984; King 1988; Giblin & Wieder 1992). That is, most of the reduced sulfur formed during sulfate reduction is reoxidized with a concomitant release of energy. Some reduced sulfur can be advected out of sediments to creeks in soluble forms such as thiosulfate and polythionates, but this transfer is much smaller than the rate of reoxidation of sulfur compounds within the sediments (Howarth *et al.* 1983; Howarth 1984). Approximately 95% of the reduced sulfur formed during sulfate reduction is reoxidized to sulfate within the sediments or at the sediment surface (Howarth *et al.* 1983; Howarth 1984).

The mechanisms whereby pyrite and other reduced sulfur compounds are oxidized within marsh sediments remain poorly known (Giblin & Wieder 1992). These oxidations may be associated with air entry into the sediment as a result of pore-water drainage and evapotranspiration. Oxidation also seems likely in the grass rhizosphere, perhaps due either to oxygen transport down the arenchymal tissues (Teal & Kanwisher 1961) or to enzymatic oxidations involving an organic oxidant (Howarth 1984), as has been suggested for rice (Vamos & Koves 1972; Armstrong 1975). A recent model of carbon, sulfur, and oxygen in marsh sediments points towards such organic oxidants as being more important than gaseous oxygen (Gardner 1990). Oxidized forms of iron and manganese may also serve as oxidants for the oxidation of reduced sulfur compounds (Howarth 1984; Jacobson 1990; Luther & Church 1992).

It is also not known if the oxidation of reduced sulfur compounds in marsh sediments is a purely chemical process or is microbially catalyzed. However, the very rapid rates of pyrite oxidation sometimes observed suggest a microbially mediated reaction (Luther & Church 1988; Giblin & Wieder 1992). Also, pyrite oxidation in marsh sediments apparently produces organic thiols as an intermediate oxidation product, and it seems unlikely these would be products of a purely chemical oxidation (Luther *et al.* 1986; Luther & Church 1992).

The actual extent of production by chemoautotrophic bacteria in marsh sediments has never been measured. Such a measurement would be technically very difficult since it would require that the functioning of the grass rhizosphere and sediment drainage were not disrupted. However, the possible rate of chemoautotrophic production can be estimated from the rate of sulfur oxidation and from laboratory studies on the efficiency of chemoautotrophy. Studies with a variety of strains of chemoautotrophic bacteria show that the energetic efficiency of production of new biomass as a percentage of energy available from the oxidation of sulfur varies from 21 to 37% when using oxygen or nitrate as oxidants (Kelly 1982; Howarth 1984). Assuming that the oxidation of reduced sulfur in marsh sediments is catalyzed by chemoautotrophic bacteria with these sorts of efficiencies, we can estimate that chemoautotrophic production in various salt marshes might range from 890 $kJ\,m^{-2}\,yr^{-1}$ (21% of the energy released in sulfur oxidation in a marsh with the lowest measured rates of sulfate reduction) to 20 000 kJ $m^{-2} yr^{-1}$ (37% of the energy released in sulfur oxidation in a marsh with the highest measured rates of sulfate reduction). Such production rates are equivalent to carbon fixation rates of 20–480 g C $m^{-2} yr^{-1}$ (Howarth 1984), or some 5 to 15% of the rate of net primary production.

Such estimates for chemoautotrophic production in marshes may seem high, particularly since they are not verified by direct measurements. In addition, the efficiency of chemoautotrophic production may well be lower than 21–37% if organic compounds or oxidized metals are the major oxidants rather than oxygen or nitrate. Furthermore, it is of course possible that the oxidation of reduced sulfur within marsh sediments is a totally chemical process which supports no chemoautotrophic production, or that the oxidation is a biologically catalyzed reaction which nonetheless yields no energy to chemoautotrophic bacteria; the sole benefit to the bacteria or other organisms may be to detoxify sulfides.

However, recent evidence from the Black Sea shows that chemoautotrophic production can indeed be a significant source of organic matter in natural ecosystems (Lein & Ivanov 1992). The Black Sea is a stratified water body with anoxic bottom waters overlaid by oxic surface waters. Sulfate reduction dominates decomposition in the bottom waters, and chemoautotrophic bacteria are abundant at the oxic–anoxic interface. Measurements of chemoautotrophic production suggest rates of some 22 g C $m^{-2} yr^{-1}$, the equivalent of 30% of all carbon inputs to the anoxic bottom waters or 10% of all organic carbon inputs to the entire Black Sea (Lein & Ivanov 1992).

Sulfide can also be oxidized anoxically in the presence of light by various photosynthetic bacteria which use sulfide as an electron donor. This process also has never been measured in salt marshes, although some of the estimates of production by algae on the sediment surface may in fact include production by photosynthetic bacteria. Such bacteria include the purple and green sulfur bacteria as well as some cyanobacteria which can use either sulfides or water as electron donors (Cohen *et al.* 1989). Photosynthetic bacteria are quite apparent on the surface sediments of many salt marshes.

10.7 Summary

Salt-marsh sediments have some of the highest rates of heterotrophic bacterial activity of any ecosystem. This activity is supported by large inputs of organic matter to the sediments. The major inputs probably are the below-ground production of grass roots and rhizomes, excretion of DOM from roots and rhizomes (particularly during times of rapid growth or when the below-ground tissues of the grasses are anoxic), and chemoautotrophic production by sulfur-oxidizing bacteria. These last two inputs remain unquantified, and their potential importance is only indirectly inferred.

When waterlogged, marsh sediments are anoxic below the top few millimeters, and most organic matter is decomposed by sulfate-reducing bacteria and associated fermentative bacteria. Sulfate reduction may make up 67–80% of the total rate of respiration in waterlogged marsh sediments. Even when sediments are partially drained, sulfate reduction remains a dominant process. Aerobic respiration is the second most important respiratory process in marsh sediments, but this is confined to the immediate surface sediments and perhaps to the grass rhizosphere. Denitrification, iron reduction, manganese reduction, and methanogenesis all appear to be minor processes in the marsh carbon cycle, although more research on iron and manganese reduction is required before this can be said with any certainty.

Rates of sulfate reduction are tightly tied to the rate of supply of organic matter substrate. Sulfate reduction occurs in marsh sediments even when the measured oxidation potential is quite high (as high as +300 mV, corrected to pH 7).

Because the metabolism of the grass rhizosphere is easily altered by sampling and incubation techniques, microbial processes in or near the rhizosphere have proven very difficult to quantify. Most of the remaining uncertainty in our understanding of microbial processes in salt-marsh sediments concerns the functioning of the rhizosphere.

References

Agosta K.A. (1985) The effect of tidally induced changes in the creek bank water table on porewater chemistry. *Estuarine, Coastal and Shelf Science* **21**, 389–400.

Anderson F.E. & Howell B.A. (1984) Dewatering of an unvegetated muddy tidal flat during exposure – desiccation or drainage? *Estuaries* 7, 225–232.

Armstrong W. (1975) Waterlogged soils. In Etherington J.R. (ed.) *Environment and Plant Ecology*, pp. 181–219. John Wiley & Sons, London.

Bartlett K.B., Bartlett D.S., Harris R.C. & Sebacher D.I. (1987) Methane emissions along a salt marsh salinity gradient. *Biogeochemistry* **4**, 183–202.

Bedford B.L., Bouldin D.R. & Beliveau B.D. (1991) Net oxygen and carbon dioxide balances in solutions bathing roots of wetland plants. *J. Ecol.* 79, 943–959.

Berner R.A. (1980) *Early Diagenesis*. Princeton University Press, Princeton.

Bohn H.L. (1971) Redox potentials. *Soil Science* **112**, 39–45.

Boynton W.R., Kemp W.M. & Keefe C.W. (1982) A comparative analysis of nutrients and other factors influencing estuarine phytoplankton production. In Kennedy V.S. (ed.) *Estuarine Comparisons*, pp. 69–90. Academic Press, New York.

Canfield D.W. & Des Marais D.J. (1991) Aerobic sulfate reduction in microbial mats. *Science* **251**, 1471–1473.

Canfield D.E., Raiswell R., Westrich J.T., Reaves C.M. & Berner R.A. (1986) The use of chromium reduction in the analysis of reduced inorganic sulfur in sediments and shales. *Chem. Geol.* **54**, 149–155.

Chalmers A.G. (1982) Soil dynamics and the productivity of *Spartina alterniflora*. In Kennedy V.S. (ed.) *Estuarine Comparisons*, pp. 231–242. Academic Press, New York.

Christensen D. (1984) Determination of substrates oxidized by sulfate reducing bacteria in intact cores of marine sediments. *Limnol. Oceanogr.* **29**, 189–192.

Christensen D. & Blackburn T.H. (1982) Turnover of ^{14}C labelled acetate in marine sediments. *Mar. Biol.* **71**, 113–119.

Cohen Y., Gorlenko V.M. & Bonch-Osmolovskaya E.A. (1989) Interactions of sulphur and carbon cycles in microbial mats. In Brimblecombe P. & Lein A. Yu. (eds) *Evolution of the Global Sulphur Cycle*, pp. 191–238. John Wiley & Sons, Chichester.

Dacey J.W. & Howes B.L. (1984) Water uptake by roots controls water table movement and sediment oxidation in short *Spartina* marsh. *Science* **224**, 487–489.

de la Cruz A.A. & Hackney C.T. (1977) Energy value, elemental composition, and productivity of belowground biomass of a *Juncus* tidal marsh. *Ecology* **58**, 165–170.

DeLaune R.D. & Patrick W.H. (1980) Nitrogen and phosphorus cycling in a Gulf Coast salt marsh. In Kennedy V.S. (ed.) *Estuarine Perspectives*, pp. 143–151. Academic Press, New York.

Fenchel T. & Jørgensen B.B. (1977) Detritus food chains of aquatic ecosystems: the role of bacteria. *Adv. Microb. Ecol.* **1**, 1–57.

Fossing H. & Jørgensen B.B. (1989) Measurement of bacterial sulfate reduction in sediments: evaluation of a single-step chromium reduction method. *Biogeochemistry* **8**, 205–222.

Fossing H. & Jørgensen B.B. (1990) Isotope exchange reactions with radiolabelled sulfur compounds in anoxic seawater. *Biogeochemistry* **9**, 223–245.

Gambrell R.P. & Patrick W.J. (1978) Chemical and microbiological properties of anaerobic soils and sediments. In Hook D.D. & Crawford R.M.M. (eds) *Plant Life in Anaerobic Environments*, pp. 375–423. Ann Arbor Science Publishers, Ann Arbor.

Gardner L.R. (1990) Simulation of the diagenesis of carbon, sulfur, and dissolved oxygen in salt marsh sediments. *Ecol. Mon.* **60**, 91–111.

Giblin A.E. & Wieder R.K. (1992) Sulphur cycling in marine and freshwater wetlands. In Howarth R.W., Stewart J.W.B. & Ivanov M.V. (eds) *Sulphur Cycling on the Continents: Wetlands, Terrestrial Ecosystems, and Associated Water Bodies*, pp. 85–117. John Wiley & Sons, Chichester.

Good R.E., Good N.F. & Frasco B.R. (1982) A review of primary production and decomposition dynamics of the belowground marsh component. In Kennedy V.S. (ed.) *Estuarine Comparisons*, pp. 139–157. Academic Press, New York.

Haines E., Chalmers A., Hanson B. & Sherr B. (1977) Nitrogen pools and fluxes in a Georgia salt marsh. In Wiley M.L. (ed.) *Estuarine Processes*, vol. 2, pp. 241–254. Academic Press, New York.

Head G.C. (1973) Shedding of roots. In Kozlowski T.T. (ed.) *The Shedding of Roots*, pp. 237–293. Academic Press, New York.

Hines M.E., Knollmeyer S.L. & Tugel J.B. (1989) Sulfate reduction and other sedimentary biogeochemistry in a northern New England salt marsh. *Limnol. Oceanogr.* **34**, 578–590.

Hines M.E., Banta, G.T., Giblin, A.E., Hobbie J.E. & Tugel J.B. (In press) Acetate concentrations and oxidation in salt marsh sediments. *Limnol. Oceanogr.*

Howarth R.W. (1979) Pyrite: its rapid formation in a salt marsh and its importance in ecosystem metabolism. *Science* **203**, 49–51.

Howarth R.W. (1984) The ecological significance of sulfur in the energy dynamics of salt marsh and coastal marine sediments. *Biogeochemistry* **1**, 5–27.

Howarth R.W. (1988) Nutrient limitation of net primary production in marine ecosystems. *Ann. Rev. Ecol. & Syst.* **19**, 89–110.

Howarth R.W. & Teal J.M. (1979) Sulfate reduction in a New England salt marsh. *Limnol. Oceanogr.* **24**, 999–1013.

Howarth R.W. & Teal J.M. (1980) Energy flow in a salt marsh ecosystem: the role of reduced inorganic sulfur compounds. *American Naturalist* **116**, 862–872.

Howarth R.W. & Hobbie J.E. (1982) The regulation of decomposition and heterotrophic microbial activity in salt marsh soils: a review. In Kennedy V.S. (ed.) *Estuarine Comparisons*, pp. 183–207. Academic Press, New York.

Howarth R.W. & Giblin A.E. (1983) Sulfate reduction in the salt marshes at Sapelo Island, Georgia. *Limnol. Oceanogr.* **28**, 70–82.

Howarth R.W., Giblin A.E., Gale J., Peterson B.J. & Luther G.W. (1983) Reduced sulfur compounds in the pore waters of a New England salt marsh. In Hallberg R.O. (ed.) *Environmental Biogeochemistry, Ecological Bulletin (Stockholm)* **35**, 135–152.

Howarth R.W. & Jørgensen B.B. (1984) Formation of ^{35}S-labelled elemental sulfur and pyrite in coastal marine sediments (Limfjorden and Kysing Fjord, Denmark) during short-term $^{35}SO_4^{2-}$ reduction measurements. *Geochim. Cosmochim. Acta* **48**, 1807–1818.

Howarth R.W. & Merkel S. (1984) Pyrite formation and the measurement of sulfate reduction in salt marsh sediments. *Limnol. Oceanogr.* **29**, 598–608.

Howes B.L., Howarth R.W., Teal J.M. & Valiela I. (1981) Oxidation–reduction potentials in a salt marsh: spatial patterns and interactions with primary production. *Limnol. Oceanogr.* **26**, 350–360.

Howes B.L., Dacey J.W.H. & King G.M. (1984) Carbon flow through oxygen and sulfate reduction pathways in salt marsh sediments. *Limnol. Oceanogr.* **29**, 1037–1051.

Howes B.L., Dacey J.W.H. & Teal J.M. (1985) Annual carbon mineralization and belowground production of *Spartina alterniflora* in a New England salt marsh. *Ecology* **66**, 595–605.

Ivanov M.V., Lein A. Yu., Reeburgh M.S. & Skyring G.W. (1989) Interaction of sulphur and carbon cycles in marine sediments. In Brimblecombe P. & Lein A. Yu. (eds) *Evolution of the Global Biogeochemical Sulphur Cycle*, pp. 125–179. John Wiley & Sons, Chichester.

Jacobson M.E. (1990) *Constraints on Sulfate Reduction in Nearshore Marine Sediments*. PhD thesis, State University of New York.

Jacobson M.E., Mackin J.E. & Capone D.G. (1987) Ammonium production in sediments inhibited with molybdate: implication for the sources of ammonium in anoxic marine sediments. *Appl. Environ. Microbiol.* **53**, 2435–2439.

Jenkins M.C. & Kemp W.M. (1984) The coupling of nitrification and denitrification in two estuarine sediments. *Limnol. Oceanogr.* **29**, 609–619.

Jørgensen B.B. (1977) The sulfur cycle of a coastal marine sediment (Limfjorden, Denmark). *Limnol. Oceanogr.* **22**, 814–832.

Jørgensen B.B. (1982a) Mineralization of organic matter in the sea bed: the role of sulphate reduction. *Nature* **296**, 643–645.

Jørgensen B.B. (1982b) Ecology of the bacteria of the sulphur cycle with special reference to anoxic–oxic interface environments. *Phil. Trans. R. Soc. Lond. B.* **298**, 543–561.

Kaplan W., Valiela I. & Teal J.M. (1979) Denitrification in a salt marsh ecosystem. *Limnol. Oceanogr.* **24**, 726–734.

Kelly D.P. (1982) Biochemistry of chemolithotrophic oxidation of inorganic sulphur. *Phil. Trans. R. Soc. Lond. B.* **298**, 499–528.

Kemp W.M., Wetzel R.L., Boynton W.R., D'Elia C.F. & Stevenson J.C. (1982) Nitrogen cycling and estuarine interfaces: some current concepts and research directions. In Kennedy V.S. (ed.) *Estuarine Comparisons*, pp. 209–230. Academic Press, New York.

King G.M. (1978) *The Nature of Methanogenesis in Soils of a Georgia Salt Marsh*. PhD thesis, University of Georgia.

King G.M. (1983) Sulfate reduction in Georgia salt marshes: an evaluation of pyrite formation by use of ^{35}S and ^{55}Fe tracers. *Limnol. Oceanogr.* **28**, 987–995.

King G.M. (1988) Patterns of sulfate reduction and the sulfur cycle in a South Carolina salt marsh. *Limnol. Oceanogr.* **33**, 376–390.

King G.M. & Wiebe W.J. (1978) Methane release from soils of a Georgia salt marsh. *Geochim. Cosmochim. Acta* **42**, 343–348.

King G.M. & Wiebe W.J. (1980) Regulation of sulfate concentrations and methanogenesis in salt marsh soils. *Est. Coast. Mar. Sci.* **10**, 215–223.

King G.M., Howes B.L. & Dacey W.H. (1985) Short-term endproducts of sulfate reduction in a salt marsh: the significance of acid volatile sulfide, elemental sulfur, and pyrite. *Geochim. Cosmochim. Acta* **49**, 1561–1566.

Laanbroek H.J. (1990) Bacterial cycling of minerals that affect plant growth in waterlogged soils: a review. *Aquatic Botany* **38**, 109–125.

Lein A. Yu. & Ivanov M.V. (1992) Interactions of the oxygen, sulphur, and carbon cycles in continental and marginal seas. In Howarth R.W., Stewart J.W.B. & Ivanov M.V. (eds) *Sulphur Cycling on the Continents: Wetlands, Terrestrial Ecosystems, and Associated Water Bodies*, pp. 189–213. John Wiley & Sons, Chichester.

Lindau C.W. & DeLaune R.D. (1991) Dinitrogen and nitrous oxide emission and entrapment in *Spartina alterniflora* saltmarsh soils following addition of N-15 labelled ammonium and nitrate. *Estuarine, Coastal and Shelf Science* **32**, 161–172.

Lord C.J. & Church T.M. (1983) The geochemistry of salt marshes: sedimentary ion diffusion, sulfate reduction, and pyritization. *Geochim. Cosmochim. Acta* **47**, 1381–1391.

Lovley D.L. (1987) Organic mineralization with reduction of ferric iron: a review. *Geomicrobiol. J.* **5**, 375–399.

Luther G.W. & Church T.M. (1988) Seasonal cycling of sulfur and iron in porewaters of a Delaware salt marsh. *Mar. Chem.* **23**, 295–309.

Luther G.W. & Church T.M. (1992) An overview of the environmental chemistry of sulfur in wetlands. In Howarth R.W., Stewart J.W.B. & Ivanov M.V. (eds) *Sulphur Cycling on the Continents: Wetlands, Terrestrial Ecosystems, and Associated Water Bodies*, pp. 125–144. John Wiley & Sons, Chichester.

Luther G.W., Church T.M., Scudlark J.R. & Cosman M. (1986) Inorganic and organic sulfur cycling in salt-marsh pore waters. *Science* **232**, 746–749.

McDougall B.M. & Rovira A.D. (1970) Movement of ^{14}C-photosynthate into roots of wheat seedlings and exudation of ^{14}C from intact roots. *New Phytol.* **69**, 37–46.

Mendelssohn I.A., Mckee K.L. & Patrick W.H. (1981) Oxygen deficiency in *Spartina alterniflora* roots: metabolic adaptation to anoxia. *Science* **214**, 439–441.

Morris J.T. & Whiting G.J. (1985) Gas advection in sediments of a South Carolina salt marsh. *Mar. Ecol. Prog. Ser.* **27**, 187–194.

Morris J.T. & Whiting G.J. (1986) Emission of gaseous carbon dioxide from salt-marsh sediments and its relation to other carbon losses. *Estuaries* **9**, 9–19.

Nelson D.C. & Jannasch H.W. (1983) Chemoautotrophic growth of a marine *Beggiatoa* in sulfide-gradient cultures. *Arch. Microbiol.* **136**, 262–269.

Nixon S.W. (1980) Between coastal marshes and coastal waters – a review of twenty years of speculation and research on the role of salt marshes in estuarine productivity and water chemistry. In Hamilton P. & Macdonald K.B. (eds) *Estuarine and Wetland Processes*, pp. 437–525. Plenum, New York.

Patrick W.H. & Reddy C.N. (1978) Chemical changes in rice soils. In *Soils and Rice*, pp. 361–379. International Rice Research Institute, Los Banos, Philippines.

Pomeroy L.R., Darley W.M., Dunn E.L., Gallagher J.L., Haines E.B. & Whitney D.M. (1981) Primary production. In Pomeroy L.R. & Wiegert R.G. (eds) *Ecology of a Salt Marsh*, pp. 40–67. Springer Verlag, New York.

Ponnamperuma F.N. (1972) The chemistry of submerged soils. *Advances in Agronomy* **24**, 29–96.

Ponnamperuma F.N. (1984) Effects of flooding on soils. In Kozlowski T.T. (ed.) *Flooding and Plant Growth*, pp. 9–45. Academic Press, Orlando.

Reddy K.R., Patrick W.H. & Lindau C.W. (1989) Nitrification–denitrification at the plant root–sediment interface in wetlands. *Limnol. Oceanogr.* **34**, 1004–1013.

Schlesinger W.H. (1991) *Biogeochemistry: An Analysis of Global Change.* Academic Press, San Diego.

Schubauer J.P. & Hopkinson C.S. (1984), Above- and belowground emergent macrophyte production and turnover in a coastal marsh ecosystem, Georgia. *Limnol. Oceanogr.* **29**, 1052–1065.

Shaw D.G., Alperin M.J., Reeburgh W.S. & McIntosh D.J. (1984) Biogeochemistry of acetate in anoxic sediments of Skan Bay, Alaska. *Geochim. Cosmochim. Acta* **48**, 1819–1825.

Sørensen J., Christensen D. & Jørgensen B.B. (1981) Volatile fatty acids and hydrogen as substrates for sulfate reducing bacteria in anaerobic marine sediments. *Appl. Environ. Microbiol.* **42**, 5–11.

Stevenson F.J. (1986) *Cycles of Soil: Carbon, Nitrogen, Phosphorus, Sulfur, Micronutrients.* John Wiley & Sons, New York.

Teal J.M. & Kanwisher J. (1961) Gas exchange in a Georgia salt marsh. *Limnol. Oceanogr.* **6**, 388–399.

Valiela I., Teal J.M. & Persson N.Y. (1976) Production and dynamics of experimentally enriched salt marsh vegetation: belowground biomass. *Limnol. Oceanogr.* **21**, 245–252.

Vamos R. & Koves E. (1972) Role of light in the prevention of the poisoning action of hydrogen sulphide in the rice plant. *J. Appl. Ecol.* **9**, 519–525.

Westrich J.T. & Berner R.A. (1984) The role of sedimentary organic matter in bacterial sulfate reduction: the G model tested. *Limnol. Oceanogr.* **29**, 236–249.

Whitney D.M., Chalmers A.G., Haines E.B., Hanson R.B., Pomeroy L.R. & Sherr B. (1981) The cycles of nitrogen and phosphorus. In Pomeroy L.R. & Wiegert R.G. (eds) *Ecology of a Salt Marsh*, pp. 163–181. Springer Verlag, New York.

Hugh W. Ducklow & Fuh-Kwo Shiah

Bacterial production in estuaries

11.1 Introduction

Estuaries are ecosystems which are characterized by intensive exchanges of water, energy, and materials across the margin between marine and freshwater environments. The nutrient-laden runoff from large watersheds is concentrated in, transported through, and processed by freshwater, brackish, and marine populations within estuaries, resulting in very high levels of primary and secondary production (Day *et al.* 1989). For example, a total drainage area of 179 000 km^2 supplies Chesapeake Bay (area 12 500 km^2) with $7.7 \times 10^{10} m^3$ of fresh water each year, delivering 10–40 g N $m^{-2} yr^{-1}$ to fuel primary production (Boynton *et al.* 1982). Allochthonous production imported from the land or surrounding marshes, and autochthonous production derived from new and recycled nutrients (Malone *et al.* 1988) together provide large inputs of organic matter for bacterial populations. Many estuaries are seasonally or permanently anoxic because the oxygen demand exerted by the aerobic decay of organic detritus exceeds the supply of oxygen (Day *et al.* 1989). In fact, production of organic matter exceeds consumption by microbes and higher organisms, leading to deposition and burial of organic matter and formation of anoxic sediments. Thus it would seem that there is always a surplus of organic matter for bacterial growth in estuarine waters but growth limitation by the biochemical nature (quality and physical form) of the organic matter must be as important as the amount present and its rate of supply.

There are several other chapters in this book which address aspects of bacterial production dynamics in aquatic ecosystems, so we will provide just a brief general synopsis here. Bacterioplankton in natural waters are small ($<0.5\,\mu m$), free living, and depend primarily on fluxes of dissolved organic carbon (DOC) compounds (Azam & Hodson 1977) to synthesize biomass. The conversion of organic matter into bacterial biomass is bacterial production. The development of sensitive, easily performed techniques for measuring bacterial abundance (Hobbie *et al.* 1977; Watson *et al.* 1977), production (Fuhrman & Azam 1980) and substrate concentrations (Mopper & Lindroth 1982; Fuhrman & Bell 1985) has stimulated a large amount of research on the trophodynamics of bacterioplankton in marine and freshwater systems over the past 15 years. This has led to the development and wide acceptance of the paradigm of the microbial loop. The term was coined by Azam *et al.* (1983), but its roots lie further back, particularly in syntheses by Pomeroy (1974) and Williams (1981). Strictly speaking the loop refers to the capture by bacteria of small molecular weight dissolved organic matter (DOM) released from plankton, which would otherwise be "lost" from the food web. Following incorporation into bacterial biomass, this material can be passed back through the food web via ingestion by bacteriovores, but the ultimate fate of organic matter processed through the microbial loop remains unclear (Ducklow *et al.* 1986; Sherr *et al.* 1987; Ducklow 1991). Because of generally low bacterial conversion efficiencies (Bjornsen 1986) and increased losses at each trophic level (Pomeroy & Wiebe 1988), the loop is broken. None-theless, bacteria are an important means for re-entry of carbon into microbial food webs.

The study of bacterial growth dynamics in estuaries is a surprisingly young field. Early evidence for the importance of bacteria in the estuarine budget came from Hobbie *et al.* (1968). Attention tended to focus on the mean, and seemingly invariant levels of bacterial abundance in various aquatic systems, relative to variations over 4–5 orders of magnitude for activity measurements (Hobbie &

Table 11.1 Estimates of bacterial abundance in estuarine waters, emphasizing early reports. Refer to Table 11.2 for other articles citing bacterial abundance and biomass in estuaries

Location	Abundance (10^9 cells l^{-1})	Reference
Sapelo Island, USA	5	Wiebe & Pomeroy (1972)
Humber, UK	0.5–5.6	Goulder (1976)
Kiel Bight, Ger.	0.5–5.6	Zimmerman (1978)
Essex, USA	2.0–8.3	Wright (1978)
Newport, USA	2.0–18.4	Palumbo & Ferguson (1978)
Palo Alto marshes, USA	7.7–35.4	Harvey & Young (1980)
Coastal Georgia, USA	3.7–21.4	Newell & Christian (1981)
Woods Hole ponds, USA	1.8–14	Kirchman (1983)
York River mouth, USA	1–8	Ducklow (1982)
Baltic Sea proper	2.6–6.4	Gast (1985)
Delaware Bay, USA	0.1–8	Coffin & Sharp (1987)
Hudson Estuary, USA	2.5–8	Hudak *et al.* (1988)
Baltic Sea proper	1.1–4.4	Heinanen (1991)
Chesapeake Bay, USA	0.5–25	This chapter
Open sea	0.2–2	Ducklow (in press)

Wright 1979). But it was also apparent quite early that bacterial abundance in estuaries and salt marshes was very high (Table 11.1). Away from the influence of estuarine runoff (Hanson *et al.* 1988; Malone & Ducklow 1990) bacterial abundance in the ocean seldom exceeds 2×10^9 cells per liter, whereas it commonly does so in estuaries. For example, in over 1100 samples taken from the upper 100 m of various oceanic locations, only 5% were $>2 \times 10^9$ cells per liter, whereas in 2000 samples from mid-Chesapeake Bay, 91% surpassed that level (Ducklow 1992). The fundamental reason for the accumulation of bacterial populations in estuaries was described by Hobbie & Cole (1984). They showed in a series of mesocosms supplemented with increasing loadings of nitrogen and phosphorus, that bacterial abundance and production increased as a direct consequence of increased primary production driven by nutrient enrichment. Thus the final cause of high bacterial biomass levels in estuaries seems clear, but the proximate causes, or mechanisms translating nutrient inputs through primary production into bacterial biomass are not. Why does eutrophication of estuaries and other aquatic ecosystems culminate in microbial biomass and anoxia, rather than fish and shellfish? Answering this important question requires a better understanding of the physical, ecological, and biochemical processes governing bacterial growth in estuarine systems.

In this chapter we review research on the biomass and production of bacterioplankton in estuaries, and present a synthesis of data we have collected over the past 8 years in Chesapeake Bay. The emphasis will be on the physical and hydrodynamic factors influencing bacterial variability in estuaries.

11.2 Bacterial production in estuaries

A recent compilation of data emphasizes the high bacterial productivity of estuaries among global marine ecosystems (Table 11.2). Estuarine bacterial production rates derived from recent studies are rivalled only by a few estimates from salt marshes and the coastal ocean, and many observations in the coastal zone are from inshore locations strongly influenced by terrestrial runoff (e.g., Hanson *et al.* 1988). Table 11.2 also suggests that bacterial production processes have been studied almost as frequently in estuaries as in all other systems combined. Yet we would argue that even in estuaries, only a small amount of research has been published in the accessible literature, and much remains to be done, especially regarding seasonal coverage of whole estuarine systems. Many of the studies addressed only limited areas, and most sampling has been carried out in spring and summer. In the following discussion we summarize some of the notable studies which illustrate different aspects of the estuarine cycle of bacterial production.

Table 11.2 Mean bacterial production in surface waters of marine habitats. From Ducklow & Carlson (in press)

Habitat	Bacterial production ($mg\,C\,m^{-3}\,d^{-1}$)	Number of estimates
Estuaries	42–134	45
Salt marshes	97–116	6
Coastal plumes	50–73	7
Fjords	2–40	3
Coastal ocean	25–119	25
Coastal upwelling zones	19–46	9
Continental shelves	4–18	7
Coral reefs	15–52	6
Open ocean	12–15	28
Southern ocean	2–4	17

In Table 11.3 we present a summary of determinations of bacterial production rates in estuaries, categorized by the geographical locations in which they were measured. In some cases we have further subdivided the estimates by season or method used. Sampling has favored small estuaries or marsh systems and the inshore habitats of larger estuaries. The Baltic Sea, the world's largest body of brackish water and a permanently stratified fjordic estuary of low to moderate salinity (0–15‰), has been the most intensively studied. There are local and national research programs in most of the surrounding nations (Wulff 1990), though the majority of observations come from coastal areas, subfjords, and mesocosms.

11.2.1 Bacterial and primary production

Larsson & Hagstrom (1979; 1982) and Hagstrom *et al.* (1979) provided some of the first observations of the seasonal cycle of bacterial production. Although sampling was carried out at just two stations in the coastal zone on the northern edge of the Baltic proper, these studies remain some of the most comprehensive examinations of bacterial and primary production dynamics in the estuarine literature. They showed that both bacterial production and biomass were limited for the most part to the period of active photosynthesis, extending from mid-April to mid-October (Larsson & Hagstrom 1982). At their more offshore station, which was less influenced by nutrient enrichment from a sewage treatment plant, bacterial biomass exhibited two peaks, following the spring and fall phytoplankton blooms, respectively. At the more eutrophic inshore station, biomass and production peaked in early July, about a month before the maximum annual temperature, and were less tightly tied to the primary production cycle. By comparing total bacterial production estimates (from the frequency of dividing cells, Hagstrom *et al.*

Table 11.3 Estimates of bacterial production in estuarine waters, 1974–1991. Values from the literature were averaged and grouped by season and location. All units were converted to mg C m^{-3} d^{-1} using conversion factors given in reports or using standard factors of 2×10^{-14} g C per cell and 2×10^{18} cells per mole if factor values were missing. Average production values were derived from minimum–maximum values in the Table. No integrations over days or seasons were made, except for the upper mesohaline Chesapeake Bay average for 1984–1989, derived from data presented elsewhere in this chapter

Location	Method	Bacterial production mg C m^{-3} d^{-1}	Reference
Baltic Sea	*Avg.*	*16–42*	
Coastal habitats	*Avg.*	*15–60*	
Kiel Bight, Ger.	SWCULT	10–520	Meyer-Reil (1977)
Kiel Fjord, Ger.	^{14}C REL	1–70	Wolter (1982)
Asko, Sweden	^{14}C REL	4.7	Larsson & Hagstrom (1979)
Himmerfjorden	FDC	4.6	Larsson & Hagstrom (1982)
Randers Fjord	^{14}C REL	19–24	Jensen (1983)
Limfjord, Oct	FDC	8.5–14.5	Riemann *et al.* (1984)
Limfjord, Oct	DARK CO_2	11	Riemann *et al.* (1984)
Limfjord, Oct	TDR	3.8–10.6	Riemann *et al.* (1984)
Limfjord, Sept	FDC	2.6–7.0	Riemann *et al.* (1984)
Limfjord, Sept	DARK CO_2	17	Riemann *et al.* (1984)
Limfjord, Sept	TDR	6.1–23.9	Riemann *et al.* (1984)
Mesocosm SW Finland	TDR	21–38	Kuuppo-Leinikki (1990)
Roskilde Fjord	TDR	88	Sand-Jensen *et al.* (1990)
Finnish bay	TDR	9.1	Wikner *et al.* (1990)
Open Baltic	*Avg.*	*17*	
Bothnian Sea	TDR	13.6	Wikner *et al.* (1990)
Bothnian Sea	TDR	1.8–6.1	Heinanen (1991)
Northern Baltic	TDR	33.1	Lignell (1990)
Northern Baltic	DARK CO_2	105	Lignell (1990)
Baltic proper	TDR	2.8–4.5	Heinanen (1991)
Baltic proper	FDC	3	Larsson & Hagstrom (1982)
Gulf of Finland, spring	TDR	4.5	Kuosa & Kivi (1989)
Gulf of Finland, summer	TDR	2.93	Kuosa & Kivi (1989)
Gulf of Finland, fall	TDR	0.52	Kuosa & Kivi (1989)
Gulf of Finland, winter	TDR	0.51	Kuosa & Kivi (1989)
Chesapeake Bay, USA	*Avg.*	*75*	
York River	TDR	7–75	Ducklow (1982)
Upper mesohaline, fall	TDR	2.6–34	Malone *et al.* (1986)
Chesapeake Bay	LEU	33.12	Chin-Leo & Kirchman (1988)
Chesapeake Bay	TDR	40.5	Chin-Leo & Kirchman (1988)
Lower mesohaline, spring	TDR	52	Jonas & Tuttle (1990)
Lower mesohaline, summer	TDR	132	Jonas & Tuttle (1990)
Upper mesohaline 1984–1989	TDR	75 (–384)	This chapter
Sapelo Island, Ga.	*Avg.*	*119*	
Sapelo Island	TDR	223.2	Pedrós-Alió & Newell (1989)
Sapelo open estuary	TDR	37	Sherr *et al.* (1989)
Sapelo tidal creek	TDR	131	Sherr *et al.* (1989)
Sapelo estuary	TDR	85	Griffith *et al.* (1990)
Other temperate	*Avg.*	*53–156*	
Buzzards Bay, USA	TDR	48–96	Kirchman *et al.* (1982)

continued on p. 266

Table 11.3 *Continued*

Location	Method mg C m^{-3} d^{-1}	Bacterial production	Reference
Essex estuary, USA	SWCULT	100–150	Wright & Coffin (1984)
Oosterschelde, Neth.	TDR	0.5–7	Laanbroek *et al.* (1985)
Oosterschelde, Neth.	TDR	15–90	Laanbroek & Verplanke (1986)
Narragansett, USA	TDR	336	Hobbie & Cole (1984)
Ems-Dollard	TDR	24.6	Admiraal *et al.* (1985)
Delaware Bay, USA	CYCLHX	9.53–44.9	Coffin & Sharp (1987)
Delaware Bay, USA	TDR, LEU	4.8–38	Kirchman & Hoch (1988)
St Lawrence (upper)	TDR, SWCULT	4–34	Painchaud *et al.* (1987)
Hudson River, USA	TDR	20–175	Hudak *et al.* (1988)
San Francisco Bay, USA	TDR	15–76	Hollibaugh (1988)
Loire, France	TDR	57–803	Relaxans *et al.* (1988)
Subtropical & tropical	*Avg.*	*362–2797*	
Port Hacking, Aust.	^{14}C REL	2.4	Wiebe & Smith (1977)
Bietri Bay, Ivory Coast	TDR	264–2184	Torreton *et al.* (1989)
Tampa Bay, USA	TDR	70–309	Paul *et al.* (1985)
Tampa Bay, USA	TDR	19–264	Paul *et al.* (1988)
Dona Paula, India	^{14}C REL	7–38	Gomes & Do (1991)
Global estuaries	*Avg.*	*42–134*	

^{14}C REL, incorporation of ^{14}C-DOC released by phytoplankton; CYCLHEX, changes in cell numbers after treatment with antibiotic; DARK CO_2, anapleurotic CO_2 uptake; FDC, frequency of dividing cells; LEU, leucine incorporation; SWCULT, changes in cell numbers during incubation; TDR, thymidine incorporation.

1979) with incorporation of exudates released from phytoplankton (determined by ^{14}C release with differential filtration), they estimated that exudation from photosynthesizing phytoplankton contributed about 50% of the total bacterial requirements over the annual cycle.

Other workers have also addressed the linkage between phytoplankton and bacteria by measuring the release of carbon from phytoplankton as DOC and its incorporation by bacteria using differential filtration techniques. Jorgensen (1982) reported that phytoplankton in a Danish subestuary released 8–42% of the carbon fixed during photosynthesis, but that the extracellular release supplied only about 3–10% of the ambient amino acid pools preferred by bacteria. Wolter (1982) studied the annual cycle of exudate release and its reincorporation in the Kiel Fjord, and demonstrated large variations in both exudate release and incorporation rates. In common with Larsson & Hagstrom (1982), her observations suggested a greater bacterial dependence on direct exudate release in spring and early summer compared to later in the growth season. This conclusion was supported by Kuosa & Kivi (1989) who also estimated (though without direct exudation measurements) that exudation appeared to supply only about half the annual bacterial carbon requirement, with the largest discrepancies in summer. Lignell (1990) found that phytoplankton exudation in Finnish waters further north was about half the rates estimated earlier by Larsson & Hagstrom (1982), and supplied only 8–25% of the bacterial carbon demand. Malone *et al.* (1991) estimated that in the mesohaline reaches (salinity 8–18‰) of Chesapeake Bay, exudation met about 25% of the carbon demand of the bacteria in spring and 40% in summer.

These ^{14}C-release experiments indicate that about 50% or less of the bacterial production is supported by flows of dissolved substrates more or less directly from phytoplankton to bacteria. Such conclusions should not in themselves be taken to mean that primary production is insufficient to meet the bacterial demand. Short-term determinations of the transfer of labeled carbon do not address the flux of DOC from nonphytoplankton sources like decaying detritus, and diffusion out of fecal pellets (Jumars *et al.* 1989), nor do they include the potentially important supply of substrates from sloppy feeding on large phytoplankton by zooplankton (Lampert 1978; Riemann *et al.* 1986). Also, it is not clear to what extent the incubation techniques employed in the classical ^{14}C-release/size-fractionation experiments account for releases of DOM by protozoan grazers (Andersson *et al.* 1985), which dominate herbivory in most environments. All trophic exchanges and other physiological processes result in release of DOM to the surrounding water, and there is great potential for *in situ* production of DOM compounds by other mechanisms than direct release (Williams 1984; Jumars *et al.* 1989).

However, given some of the low estimates of the amount of bacterial production supported by phytoplankton, it is conceivable that primary production is insufficient to meet the bacterial demand in estuaries. Smith *et al.* (1989) have hypothesized that some estuaries have annual net system heterotrophy as a result of inputs and bacterial utilization of terrestrially derived dissolved organic nitrogen. We can examine annual estimates for bacterial and primary production (Table 11.4) to judge if this is the case. When total primary and bacterial production estimates are compared from studies in which the seasonal cycles were adequately covered to provide good annual estimates, the ratio of bacteria to primary production averaged 17%, which is about half of the cross-system average reported by Cole *et al.* (1988). Viewed in this way, bacterial production in estuaries seems low relative to primary production even though it is high in absolute terms. The reason for this might be related to the efficiency with which bacterioplankton convert organic matter into biomass. The conversion, or growth efficiency was once believed to be high (Payne & Wiebe 1978), and indeed, most of the reports already discussed used a canonical value of 50% to derive utilization (total uptake) rates from production data. With a conversion

Table 11.4 Ratio of annual bacterial production to annual primary production in the surface (euphotic) layers of estuaries. This compilation is limited to studies in which the seasonal cycle has been sampled with sufficient coverage to make an annual estimate with some certainty. The averages are taken directly from cited publications as reported

Location	Year	Ratio	Reference
Coastal Baltic proper	1978	18–24%	Larsson & Hagstrom (1982)
Delaware Bay	1985	23%	Coffin & Sharp (1987)
Northern Baltic	1986	15%	Kuosa & Kivi (1989)
Northern Baltic	1985	14%	Lignell (1990)
Mid-Chesapeake Bay[1]	1987–1989	9%	Malone *et al.* (1991)
Mid-Chesapeake Bay[2]	1984–1989	22%	This chapter[2]

[1] Based on monthly determinations of exudate incorporation and accompanying bacterial production estimates.
[2] Based on comprehensive seasonal measurements of bacterial and primary production. For primary production see Malone & Conley (submitted).

efficiency of about 50%, estuarine bacterioplankton would be utilizing an average of 34% of the primary production on an annual basis. However more recent determinations of bacterial conversion efficiencies suggest that mean values based on naturally occurring dissolved substrates are only about 20% (Bjornsen 1986; Billen & Fontigny 1987) and 10–20% for particulate resources (Newell & Linley 1984; Pomeroy *et al.* 1984). The conditions under which these experiments were conducted might exclude utilization of high-quality easily utilized DOM. If such material supports a large share of the bacterial production, overall efficiencies might be closer to 50%. But if these revised estimates for conversion efficiency are realistic, estuarine bacterioplankton in the euphotic zone alone might consume 45–120% of the annual primary production. Since bacterioplankton production in the bottom layer also consumes primary production, though at a lower rate, the data do imply that *in situ* primary production may not be sufficient to meet the bacterial demand on an annual basis, and must certainly be less than adequate over shorter periods at times during the year. High bacterial production can be reconciled with demands by other consumers through recycling within the food web (Fasham 1985; Scavia 1988; Strayer 1988).

11.2.2 Effects of the estuarine circulation

In the open sea, where there are no external sources of organic matter for pelagic bacterioplankton, the bacterial production must be sustained by *in situ* processes of DOC supply arising from trophodynamic exchanges within the food web (e.g., Fasham *et al.* 1990). However, estuaries receive inputs of dissolved and particulate organic matter from rivers and marshes, and bacterial metabolism could

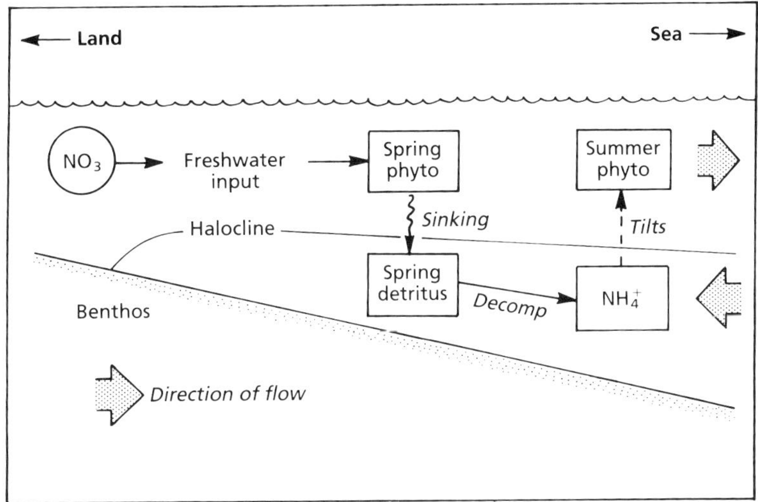

Figure 11.1 Conceptual cartoon of the effects of the two-layered estuarine circulation on primary production. New nutrients (nitrate) supplied in the inflow of river water are assimilated by phytoplankton downstream. Phytoplankton sink into the bottom layer and are carried back upstream, decay and return regenerated nutrients, e.g., ammonium, to the surface layer via a range of physical mixing processes. Summer production supported by recycled nutrients is coupled to spring production on the seasonal timescale of about 2–4 months.

potentially draw on these resources in addition to those produced by phytoplankton-based food webs *in situ*. To evaluate the potential for allochthonous support of bacterioplankton production in estuaries, we consider aspects of the estuarine circulation.

The principal types of estuarine circulation are discussed in Dyer (1973) and Day *et al.* (1989). Estuarine circulation is driven by a complex interplay of forcing mechanisms, all of which have been shown to be important in regulating biological processes. Some examples will be given below. In the most familiar examples of partially mixed coastal plain estuaries and fjords, the mean circulation pattern is characterized by a seaward outflow of fresh water in the surface layer and a landward inflow of sea water in the bottom layer, driven by pressure and density gradients (Figure 11.1). Tidal motions generate eddies which move sea water from the bottom layer into the surface, where it is entrained into the outflow, and increases the salinity. The seaward flow in the surface layer must be enhanced by this entrainment in order that the river input of fresh water can be discharged, and the compensating inflow in the bottom is also enhanced, leading to sometimes intense recirculation of water within the estuary (Dyer 1973).

A few studies of microbes have implicitly or explicitly addressed circulation patterns in estuaries. One question concerns the fate of bacteria supplied in the freshwater input. Do these riverine and terrestrial soil bacteria survive and multiply as they move into the saltier reaches of estuaries? Many investigators have reported gradients of decreasing bacterial abundance with increasing salinity (e.g., Palumbo & Ferguson 1978; others reviewed in Painchaud *et al.* 1987). Painchaud *et al.* (1987) analyzed bacterial dynamics in the strongly river flow-dominated St Lawrence estuary with two-dimensional box models and salinity balances, concluding that the freshwater bacteria died rapidly, and that upstream flow of estuarine-tolerant bacteria in the bottom layer balanced losses of freshwater populations. They also suggested that midestuary maxima in bacterial abundance could be explained by rapid growth of halotolerant cells in the low salinity reaches of the St Lawrence, and predation by microflagellates further downstream. This is the only study of its kind, and should serve as a model for other investigations. Valdés & Albright (1981) also concluded that it was unlikely for inflowing freshwater bacterial populations to survive within the estuary.

Tidal motions move plankton populations back and forth and create complex circulation patterns which require careful studies to sample and understand properly. Wright & Coffin (1983) studied bacterial distributions in the Essex estuary, Massachusetts, which was small enough to permit comprehensive spatial coverage over the tidal cycle and throughout the year. They showed that in spring a peak in abundance develops in the midestuary, with lower numbers up- and downstream. This was attributed to the tidal regime, which allowed bacterial populations to be retained within the estuary over several tidal cycles, and exposed them to contact with surrounding salt-marsh waters, which could supply growth substrates and/or more bacteria.

Most investigators sample along the central axes of estuaries and describe longitudinal distribution patterns. Tidal motions also interact with wind- and gravitation-driven circulations to generate vertical and horizontal distribution patterns. Most estuaries are usually vertically stratified owing to the two-layered salinity distribution, but tidal stirring can, in some cases, provide sufficient energy to overturn and destratify the water column (Haas 1977). Ducklow (1982) showed that bacterial standing stocks and production were stimulated during a single spring tidal destratification event in the York River, a subestuary of Chesapeake Bay. In contrast, Ray *et al.* (1989) showed in subsequent studies that autotrophic cyanobacteria decreased during destratification events in the York River as a consequence of mixing effects on light limitation and grazing. Horizontal patterns were studied along a crossestuary transect in Chesapeake Bay (T2 in Figure 11.5) in summer 1984 (see below)

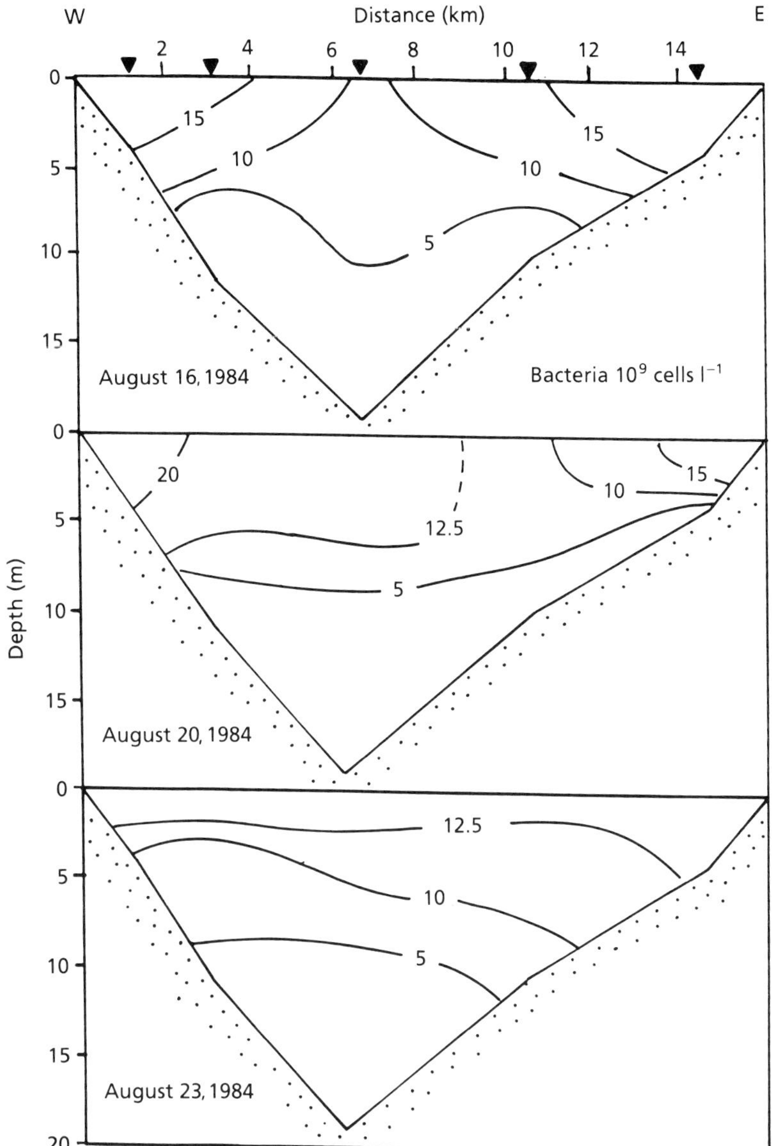

Figure 11.2 Time series of transverse bacterial distributions across transect T2 (see Figure 11.5) in mid-Chesapeake Bay. Triangles show stations. Bacterial abundance is highest on the Bay flanks as a result of pycnocline tilting. The pycnocline is horizontal on August 16, 1984 following a tilt to the west, tilted up sharply to the east on August 20, 1984, and relaxed back to the west in equilibrium with the Coriolis force by August 23, 1984. From Malone *et al.* (1986).

(Malone *et al.* 1986). Interactions between wind forcing and the gravitational circulation produce an internal transverse oscillation or tilting of the pycnocline (vertical density gradient) which also serves to redistribute water properties across the estuary. Malone *et al.* (1986) showed that tilting produced by changes in wind speed and

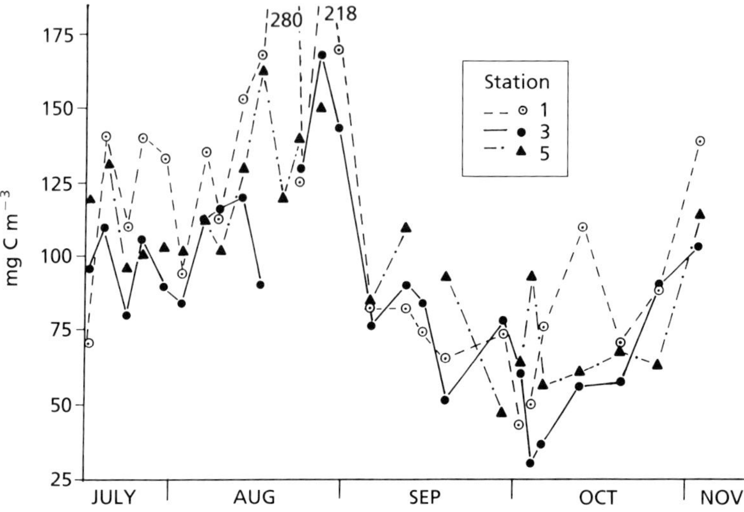

Figure 11.3 Time series of bacterial biomass at stations 1, 3, and 5 (west, center, east) on T2 in Chesapeake Bay, July–November 1984. Sampling was carried out about every 3–5 days at five stations. In response to tilting events and advective accumulations of phytodetritus, bacterial concentrations are maintained at high levels on the flanks, and bloom on a timescale of about 10 days.

direction appeared to trigger phytoplankton blooms on the flanks of the mid-Bay. These blooms or advective accumulations of phytoplankton biomass were succeeded by large densities of bacterioplankton exceeding 20×10^9 cells per liter (Figure 11.2) and also resulted in generally higher bacterial densities on the edges compared to the center of the Bay (Figure 11.3).

Hydrodynamic effects clearly influence plankton distributions and processes in a variety of ways, but the most important effect of the two-layered flow is that it serves to concentrate and retain particles within the central regions of the estuary. The cartoon in Figure 11.1 shows the fate of inorganic nutrients supplied in the freshwater input, which are taken up by phytoplankton flowing downstream in the surface layer. Phytoplankton, especially the larger-celled and chain-forming diatoms produced in spring, sink into the bottom layer and are transported back up the estuary. As the water warms the phytodetritus is decomposed and subsequent ammonium regeneration supports recycled production during the annual peak of primary production in summer (Malone *et al.* 1988). Thus we see that nutrient supply and primary production are decoupled in space and time, and that production can be enhanced by elements of the estuarine circulation. The same is true for bacterial production as shown in Figure 11.4, a conceptual model drawn from some of the papers discussed above, and our own studies. Some fraction of the bacterial production in the surface layer is supported by the flows

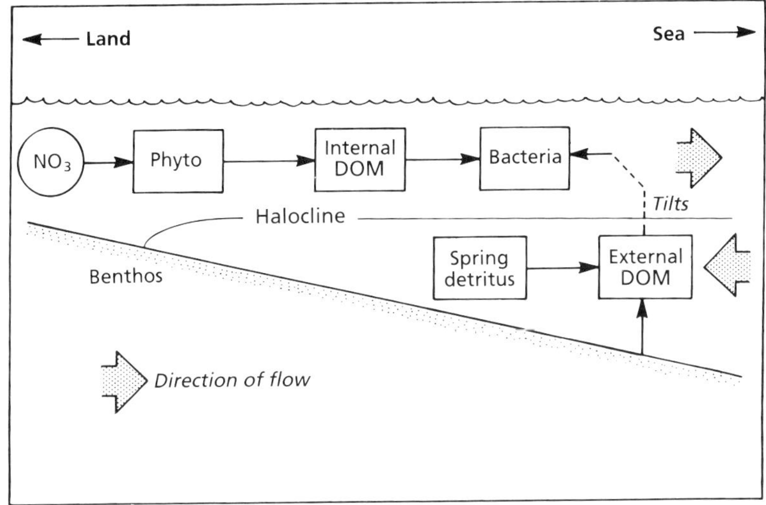

Figure 11.4 Conceptual cartoon of the effects of the two-layered estuarine circulation on bacterial production. Bacterial production in the surface layer is sustained by a mixture of exogenous organic matter supplied in the freshwater flow and produced *in situ* by primary production and trophic exchanges. Phytodetritus arising from the sinking of the spring bloom decays over 2–4 months and supplies ammonium and DOM to the surface layer, supporting additional bacterial production. Thus bacterial production in summer is sustained in part by primary production which occurred months before.

of DOM within the food web, but an additional fraction might be fueled by DOM supplied from the bottom layer by decay of phytodetritus produced weeks to months previously. The two-layered estuarine circulation, by concentrating and conserving organic matter in midestuary, provides a mechanism by which bacterial and primary production are coupled over timescales much longer than the scale of hours–days usually addressed in incubation experiments.

11.3 Bacterioplankton dynamics in Chesapeake Bay

Over the past 8 years we have been investigating bacterioplankton production dynamics in the mesohaline regime (about 8–18‰ salinity range) of the mid-Chesapeake Bay. Here we provide an overview of the annual cycle of bacterial biomass and production in the context of the circulation and hydrography of the mid-Bay region. Chesapeake Bay, the largest estuary in the USA (Figure 11.5), is a partially mixed coastal plain estuary in the valley of the preglacial Susquehanna River, created during the marine transgression over the last 18 000 years. About half the freshwater flow is supplied by the River Susquehanna, with the remainder supplied mostly by the Potomac and James rivers, and nonpoint source runoff entering in the mid- and lower-estuary (Taft *et al.* 1980). Malone *et al.* (1986; 1988) have described the cycle of primary production in the Bay, and suggested

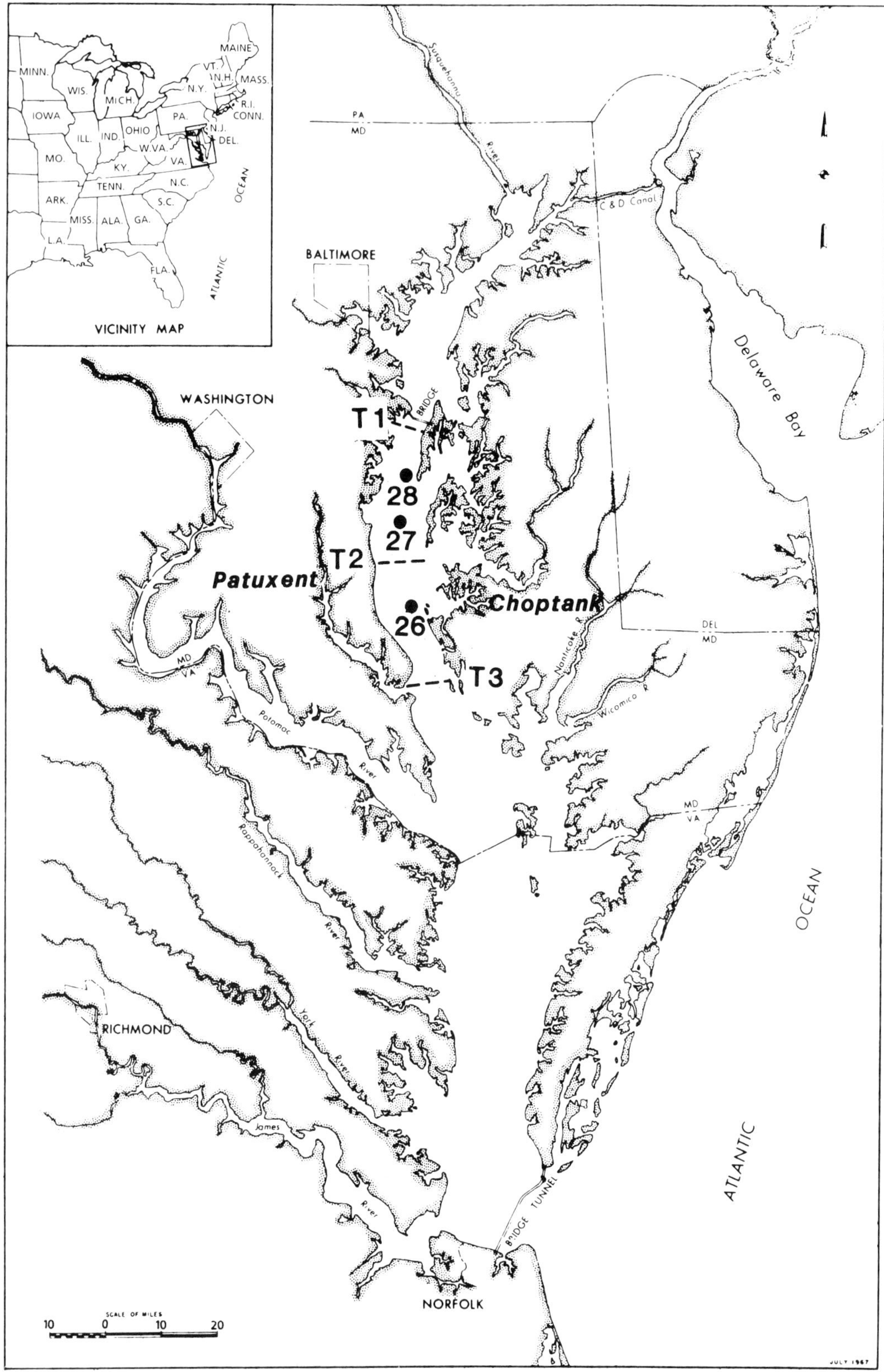

Figure 11.5 Map of Chesapeake Bay. The mesohaline mid-Bay region is located between the Bay Bridge south of Baltimore (T1) and the mouth of the Potomac River south of T3. T1–3 and stations 26–28 have been sampled repeatedly since 1984.

how it is driven by the spring delivery of nutrients and summer nutrient recycling as outlined above.

Chesapeake Bay exhibits annually recurring summer anoxia in the mesohaline region (Officer *et al.* 1984; Seliger *et al.* 1985), which is driven by a combination of water column and benthic decomposition of spring phytodetritus (Malone *et al.* 1988). An important scientific and policy question concerns the physical, ecological, and biochemical mechanisms which link springtime inputs of land-derived nutrients, primary producers, and heterotrophic processes which consume oxygen each summer. Other questions which we will suggest answers to are listed below.

1 What is the magnitude of bacterial production in the mid-Bay?

2 What is the role of physical dispersion (dilution) in controlling net accumulations of bacterial biomass?

3 What other factors govern the annual bacterioplankton cycle?

Since 1984, we have measured bacterial abundance and production, along with chlorophyll, oxygen, nutrients, temperature, and salinity along cross-Bay transects of stations between the Bay Bridge and the Patuxent River (see Figure 11.5). Our sampling design has emphasized the periods of the onset of anoxia in May, peak anoxia in July–August and the reoxygenation of the bottom layer following the seasonal destratification in September–October. However, over the years we have observed most of the annual cycle, with the exception of December–January when portions of the study area may be ice-covered. Samples were collected at the surface and bottom at all stations, and at 1–5 intermediate depths at the stations in midchannel, reaching a maximum of about 40 m near the bridge and 20 m at our main transect (see Figure 11.2). The methods are described in detail elsewhere (see Malone *et al.* 1986; Malone & Ducklow 1990). Briefly, we enumerated bacteria using acridine orange-epifluorescence microscopy (Hobbie *et al.* 1977) and estimated bacterial production from the incorporation of ^{3}H-thymidine into cold trichloroacetic acid (TCA) insoluble cell fractions (Fuhrman & Azam 1980). For the carbon-based estimates, we converted cell counts to biomass using a factor of 2×10^{-14} g C per cell, and thymidine incorporation to production using 2×10^{18} cells produced per mole incorporated. The latter figure is slightly greater than the "theoretical" value of the thymidine conversion factor, but may correct for nonsaturation of incorporation caused by using 5 nmol l^{-1} concentrations in rich estuarine waters (Ducklow & Kirchman 1983). The data reported here include over 2000 individual cell counts and 1200 paired determinations of abundance and thymidine incorporation.

The annual cycle of bacteria in mesohaline Chesapeake Bay appears

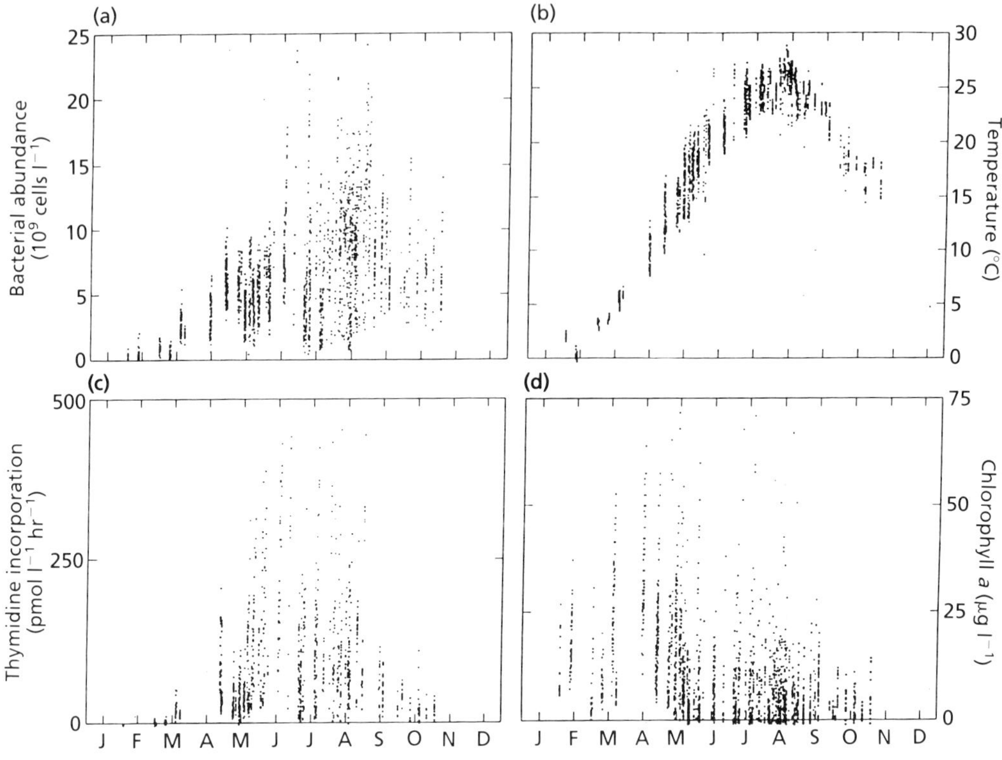

Figure 11.6 Data collected in 1984–1989 in mid-Chesapeake Bay at all stations and depths: (a) bacterial abundance; (b) temperature; (c) thymidine incorporation; and (d) chlorophyll *a*.

to be closely tied to the annual cycle of temperature which varies little from year to year. Figure 11.6 shows all the individual data collected in the mid-Bay in 1984–1989. These figures suggest several aspects which a comprehensive theory of estuarine bacterial dynamics (cf. Painchaud *et al.* 1987) should attempt to explain. The most striking aspect is the large seasonal signal. Bacterial abundance varies over 20-fold from about $0.5-1 \times 10^9$ cells per liter in late winter to over 2×10^{10} cells per liter in July–August (Figure 11.6a). Thymidine incorporation rates vary more, over nearly three orders of magnitude, from $0.5-500\,pmol\,l^{-1}\,hr^{-1}$ (Figure 11.6c). One question, first noted by Hobbie & Wright (1979), concerns the relatively narrow range of abundances, given the large range of production rates. Clearly there are other factors besides growth which regulate the standing stocks of bacteria, and we will discuss the balance between production and removal below. Another striking aspect of these data is the large variations in abundance and production observed within months. This variability, which can be as large as the entire seasonal range, is an expression of the vertical stratification

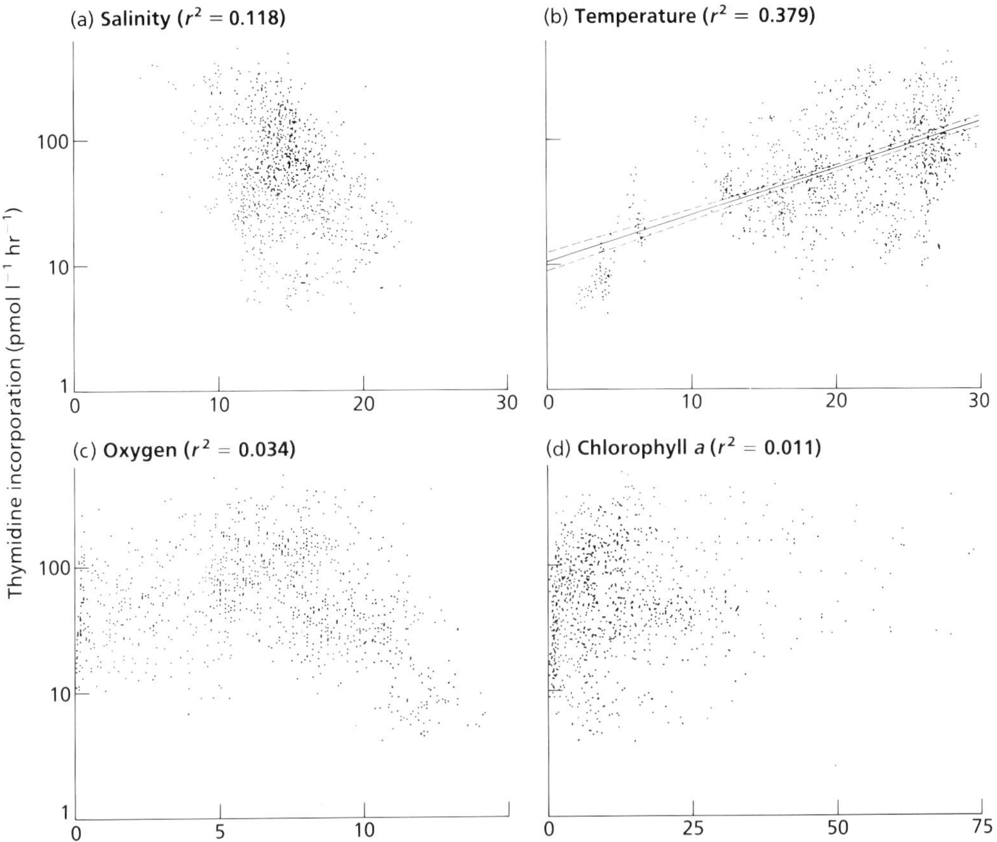

Figure 11.7 Regressions of thymidine incorporation on: (a) salinity (‰); (b) temperature (°C); (c) ppm dissolved oxygen; and (d) μg l^{-1} chlorophyll *a*. All the regressions save chlorophyll are significant at $P < 0.01$, owing to the large sample sizes.

of the water column in summer, when bacterial levels are high in the surface and very low in the bottom layer (see Figure 11.2). Some of the variability evident in Figures 11.6a and c is spatial and some is interannual. Malone *et al.* (1986) explained the spatial (lateral) variability in terms of pycnocline tilting. Interannual variations in primary production are related to variations in freshwater and nitrate inputs (Malone *et al.* 1988). Comparison of Figures 11.6a and c and d indicates that there is no direct relationship between bacteria and chlorophyll (see also Figure 11.7d), but the phasing of the peaks suggests that bacteria respond to phytoplankton with a lag of up to about 3–4 months, i.e., on a seasonal timescale, as suggested in Figures 11.1 and 11.4. Interannual variations in bacterial stocks and production may be related to changes in this coupling driven by variations in streamflow.

Relationships between bacterial production and other properties are shown in Figure 11.7. There is a weak but significant negative correlation with salinity (see Figure 11.7a), which reflects primarily the occurrence of high salinities during the colder months and in the bottom layer. This relationship, observed strictly within the mid-Bay, is not indicative of the supply of bacteria in the freshwater input, or their dieoff at higher salinity. Temperature is clearly the major single factor influencing bacterial production (see Figure 11.7b), explaining about 40% of the variability in the whole data set (cf. Wright & Coffin 1983). The nonlinear relationship with oxygen (see Figure 11.7c) is a consequence of somewhat lower activity in the hypoxic bottom water in summer, and the high solubilities of oxygen in cold waters. Most of the observations of oxygen >11 ppm are from temperatures <5°C. Thus both the oxygen and salinity plots reflect the seasonal correlation between bacterial growth and temperature, but the three properties together explain about 50% of the variability in thymidine incorporation rates. There is no relationship between bacterial production and chlorophyll (see Figure 11.7d).

As suggested above, circulation and hydrography may strongly influence bacterial distribution and dynamics in estuaries. The mean annual cycles for 1984–1989 are shown in Figure 11.8. The plots of mean monthly abundance and incorporation in the surface (0–5 m) and bottom (>5 m) layers show that the water column is vertically homogeneous with respect to these properties until after seasonal stratification (represented here by the mean top–bottom difference in salinity, Figure 11.8c) sets up in response to the arrival of the spring freshet (Figure 11.8d) in April–May. The increases in bacterial abundance and especially thymidine incorporation slow abruptly in the bottom layer after April. Since bacterial levels continue to increase for 3–4 months in the surface, there is increasing divergence between the two layers. Figure 11.9, gridded from all the data shown in Figures 11.6 and 11.7, shows that bacterial abundance, incorporation, and temperature are uniform in the water column until April–May, then increasingly stratified in summer. The strong correspondence between temperature and bacterial activity is again evident. Integration of the surface curve in Figure 11.8b yields a mean annual production of $137\,g\,C\,m^{-2}\,yr^{-1}$.

Hydrodynamic control of bacterial abundance in the mid-Bay can be evaluated by comparing specific growth rates with mean dilution rates in the surface layer. The dilution rate is an estimate of the rate of losses of conservative substances due to nontidal seaward flows of fresh water. It is highest in late winter and early spring when the annual spring freshet (see Figure 11.8d) delivers melt water from the watershed into the Bay, and lowest in mid–late summer and fall when precipitation and streamflows are minimal. The highest

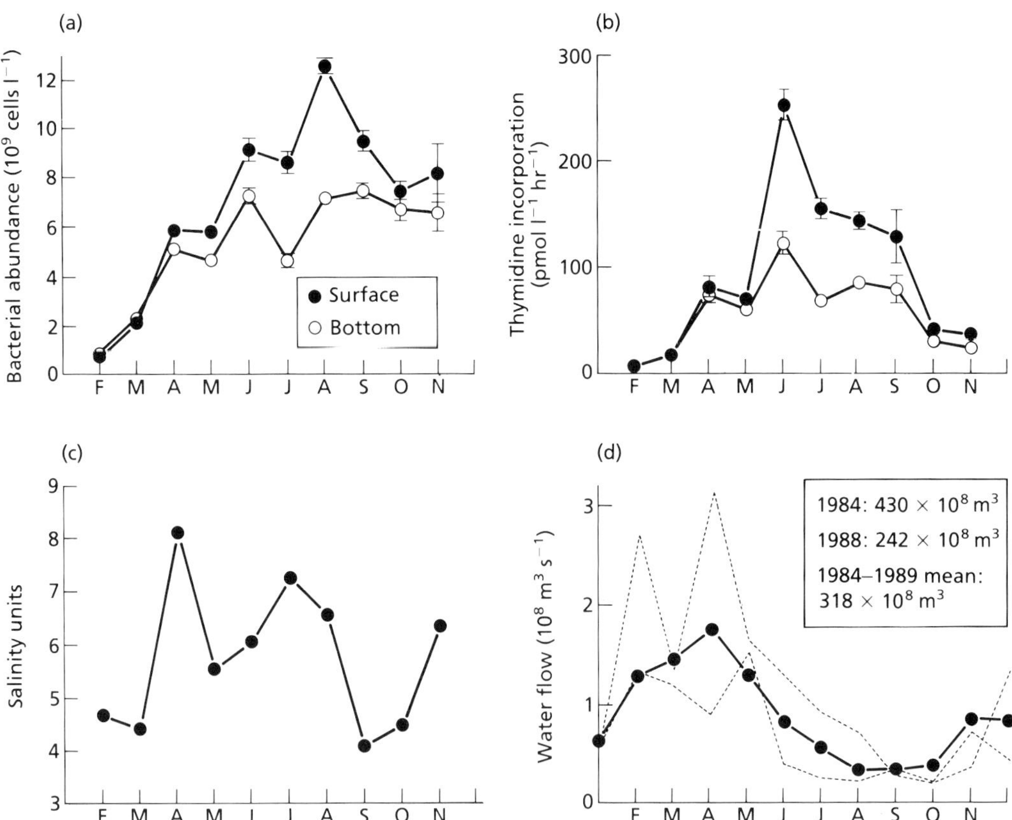

Figure 11.8 Monthly mean properties for mid-Chesapeake Bay, 1984–1989. (a) and (b) bacterial abundance and thymidine incorporation in the surface (0–5 m) and bottom (>5 m) layers. Bars are standard errors, which in some cases are smaller than symbols. (c) Mean difference between mean bottom and surface salinity. (d) Monthly mean freshwater flow in the Susquehanna River. Dotted lines line the 1984 peak and 1988 drought year flows. The 1984, 1988, and 1984–1989 mean annual discharge totals are shown in m^3.

monthly values of the dilution rate for 1984–1989 (a period encompassing annual freshwater flows among the century's highest and lowest), were $0.21\,d^{-1}$ in March–April 1987, and $0.26\,d^{-1}$ in February 1988 and March 1989. Specific growth rates of bacteria are always in excess of dilution rates (Table 11.5), suggesting that bacteria in the mid-Bay are always supplied with sufficient substrate concentrations to enable them to increase population size. Net observed rates of increase, however, are much lower. In summer, bacterial abundance increases only slowly, if at all, in spite of high growth rates. This is presumably because dilution, with predation by bacteriovores nearly balances growth. The large discrepancy between growth and dilution rates seems to indicate that dilution is not a very important mechanism limiting bacterial accumulation in the mid-Bay

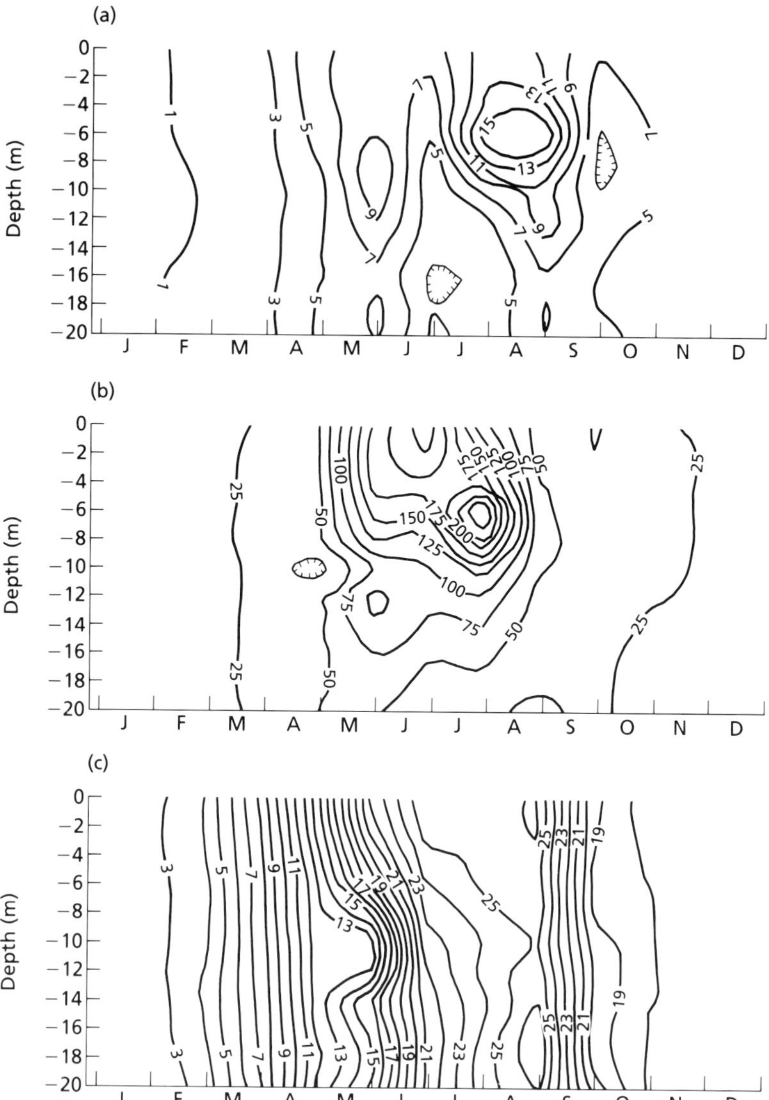

Figure 11.9 Contour plots gridded from the 1984–1989 data in center channel stations on T1–3 and 26–28. (a) Bacterial abundance in 10^9 cells per liter. (b) Thymidine incorporation in $\text{pmol l}^{-1}\text{hr}^{-1}$. (c) Temperature (°C). Note the correspondence between the $50\,\text{pmol l}^{-1}\text{hr}^{-1}$ and 13°C isopleths in May, which survive the averaging process.

region. This may be so in summer, but in spring when dilution rates are high, and bacteria are increasing faster than the combined forces of predation and dilution can remove them, circulation must exert an influence nearly equal to the predators. Circulation may provide stronger constraints on bacterial populations in winter. We have not observed the final stages of decline of abundance to the February–

Table 11.5 Mean specific rates (per day) of dilution[1] due to freshwater flow and entrainment, growth, and net changes in abundance[2] for the mid-Bay surface layer, 1984–1989

Month/year	Dilution rate	Specific growth rate	Net increase
Feb–Apr	0.18	0.43	0.04
May–June	0.09	0.76	0.01
July–Aug	0.05	0.72	0.01
Sept–Nov	0.04	0.37	−0.01

[1] Calculated from $D = Q_s/V$, where $Q_s = Q_f\,(S_b/[S_b - S_s])$; V, volume of mid-Bay surface layer; Q_f, freshwater input (see Figure 11.8d); S_b, S_s, bottom and surface salinity. From Malone *et al.* (1988).
[2] Calculated from the mean abundances shown in Figure 11.8a, using $u = (\ln B_i - \ln B_{i-1})/30$ d, where B_{i-1} and B_i are successive monthly abundances.

March level. Pomeroy & Deibel (1986) suggested that temperatures below 1°C strongly inhibit bacterial growth. During the annual temperature minimum in Chesapeake Bay, bacterial growth may nearly cease, allowing bacterial populations to be washed out of the estuary, even at low dilution rates. Closer examination of interannual differences in flow, dilution, bacterial growth, and removal, may clarify the average picture sketched here.

11.4 Conclusion and prospects

In general, our analyses of the effects of circulation, stratification, and pycnocline tilting in Chesapeake Bay support the observations of Wright & Coffin (1983), Painchaud *et al.* (1987), and others who argued that estuarine bacterial dynamics are the result of a complex interplay of physical and biological factors. Circulation supplies exogenous organic matter to estuarine bacterial populations, concentrates and conserves locally produced organic matter within the estuary and provides mechanisms for longer-term couplings of bacterial production to internal sources of organic matter. As a result of these factors, we hypothesize that bacteria in Chesapeake Bay and possibly other estuaries are seldom, if ever, limited by substrate supply, even though local, contemporaneous primary production may not always be adequate to supply the daily bacterial requirement for maintenance and growth. Two lines of evidence support this tentative hypothesis.

The strongest evidence for lack of substrate limitation is the strong correlation between thymidine incorporation rates (production rates) and temperature (see Figures 11.7–9). In any season, increases in temperature result in increases in incorporation, and decreases in temperature are accompanied by decreases in incorporation. We have observed the same effect during 1990–1991 throughout the Bay, at *in situ* temperatures ranging from about 2–30°C, and in short term incubations over the temperature range 5–30°C (F. Shiah unpublished). This effect is only possible if no other factor is limiting bacterial growth. In any season, temperature is the proximal factor limiting bacterial growth, and we have little evidence for any form of substrate limitation (e.g., no correlation with chlorophyll in Figure 11.7). However, the annual range of production rates may be set by the rates of organic matter supply, and supply rates might explain differences between systems. The other main argument against substrate limitation is the persistently high specific growth rates maintained by bacteria even at relatively low temperatures (see Table 11.5). Rates of 0.3–0.4 d^{-1} imply mean generation times of 1.6–1.8 days, which compare very favorably with the high end of the range of growth rates measured in the open sea (Ducklow in press). It is difficult to envision how growth rates this high at low temperatures could be substrate limited.

This hypothesis has important implications for management of water quality in Chesapeake Bay and other estuaries. As controls on nutrient discharge to Bay waters are imposed, primary production may decrease (cf. Boynton *et al.* 1982) and the supply of organic matter available to bacteria will be lessened. In addition, the concentrations of readily assimilable DOM may decrease to rate-limiting levels for bacterial growth. Thus rates of bacterial respiration (oxygen consumption) as well as the absolute amount of oxygen-demanding matter will be lessened. Currently, bacterial growth and substrate limitation in the Bay are yield limited, in the sense proposed by Caperon *et al.* (1971). That is, they can respond to increased inputs of nutrients only by increasing the biomass. According to the model proposed by Billen *et al.* (1990), yield-limited bacterial populations are controlled by bottom–up factors: increases in resources result in increases in stocks, as evidenced by correlations between production and abundance. We seek conditions in estuaries where populations are top–down controlled (Billen *et al.* 1990), by decreasing their substrate supply through nutrient management strategies.

Acknowledgments

Research reported in this paper was supported by the University of Maryland Sea Grant College, the US Environmental Protection Agency, the NOAA Coastal Ocean Program Office, and the NSF Land-Margin Ecosystem Research Program. We are grateful to Helen Quinby, Kathryn Gloersen, and a large number of colleagues and students at Horn Point Environmental Laboratory for assistance in collecting data and to Tom Malone for discussions and loans of data. We thank Craig Carlson for help in compiling Table 11.3, and John Hobbie and David Kirchman for comments on an earlier draft.

References

Admiraal W., Beukema J. & Es F.B.V. (1985) Seasonal fluctuation in the biomass and metabolic activity of bacterioplankton and phytoplankton in a well-mixed estuary: the Ems-Dollard (Wadden Sea). *J. Plank. Res.* 7, 877–890.

Andersson A., Lee C., Azam F. & Hagström Å. (1985) Release of amino acids and inorganic nutrients by heterotrophic marine microflagellates. *Mar. Ecol. Prog. Ser.* **23**, 99–106.

Azam F. & Hodson R.E. (1977) Size distribution and activity of marine microheterotrophs. *Limnol. Oceanogr.* **22**, 492–501.

Azam, F., Fenchel T., Field J.G., Gray J.S., Meyer-Reil L.A. & Thingstad F. (1983) The ecological role of water-column microbes in the sea. *Mar. Ecol. Prog. Ser.* **10**, 257–263.

Billen G. & Fontigny A. (1987) Dynamics of a *Phaeocystis*-dominated spring bloom in Belgian coastal waters. II. Bacterioplankton dynamics. *Mar. Ecol. Prog. Ser.* **37**, 249–257.

Billen G., Servais P. & Becquevort S. (1990) Dynamics of bacterioplankton in oligotrophic and eutrophic aquatic environments: bottom–up or top–down control? *Hydrobiologia* **207**, 37–42.

Bjornsen P.K. (1986) Bacterioplankton growth yield in continuous seawater cultures. *Mar. Ecol. Prog. Ser.* **30**, 191–196.

Boynton W.R., Kemp M. & Keefe C.R. (1982) A comparative analysis of nutrients and other factors influencing estuarine primary production. In Kennedy V.S. (ed.) *Estuarine Perspectives*, pp. 69–90. Academic Press, New York.

Caperon J., Cattell S.A. & Krasnick G. (1971) Phytoplankton kinetics in a subtropical estuary: eutrophication. *Limnol. Oceanogr.* **16**, 599–607.

Chin-Leo G. & Kirchman D.L. (1988) Estimating bacterial production in natural waters from the simultaneous incorporation of thymidine and leucine. *Appl. Environ. Microbiol.* **54**, 1934–1939.

Coffin R.B. & Sharp J.H. (1987) Microbial trophodynamics in the Delaware Estuary. *Mar. Ecol. Prog. Ser.* **41**, 253–257.

Cole J.J., Pace M.L. & Findlay S. (1988) Bacterial production in fresh and saltwater ecosystems: a cross-system overview. *Mar. Ecol. Prog. Ser.* **43**, 1–10.

Day J.W.Jr, Hall C.A.S., Kemp W.M. & Yanez-Arancibia A. (1989) *Estuarine Ecology*. John Wiley & Sons, New York.

Ducklow H.W. (1982) Chesapeake Bay nutrient and plankton dynamics. I. Bacterial biomass and production during spring tidal destratification in the York River, Virginia, estuary. *Limnol. Oceanogr.* 27, 651–659.

Ducklow H.W. (1991) The passage of carbon through microbial foodwebs: results from flow network models. *Mar. Microbial Foodwebs* 5, 1–16.

Ducklow H.W. (1992) Factors regulating bottom–up control of bacterial biomass in open ocean plankton communities. *Archiv für Hydrobiologie* (in press).

Ducklow H.W. & Kirchman D. (1983) Bacterial dynamics and distribution during a spring diatom bloom in the Hudson River plume. *J. Plank. Res.* 5, 333–355.

Ducklow H.W. & Carlson C.A. (1992) Oceanic bacterial production. *Adv. Microb. Ecol.* **12** (in press).

Ducklow H.W., Purdie D.A., Williams P.J. Le B. & Davies J.M. (1986) Bacterioplankton: a sink for carbon in a coastal plankton community. *Science* **232**, 865–867.

Dyer K.R. (1973) *Estuaries: A Physical Introduction*. John Wiley & Sons, New York.

Fasham M.J.R. (1985) Flow analysis of materials in the marine euphotic zone. In Ulanowicz R.E. & Platt T. (eds) *Ecosystem Theory for Biological Oceanography. Can Bull. Fish. Aquat. Sci.* **213**, 139–162.

Fasham M.J.R., Ducklow H.W. & McKelvie S.M. (1990) A nitrogen based model of plankton dynamics in the oceanic mixed layer. *J. Mar. Res.* **48**, 1–49.

Fuhrman J.A. & Azam F. (1980) Bacterioplankton secondary production estimates for coastal waters of British Columbia, Antarctica, and California. *Appl. Environ. Microbiol.* **39**, 1085–1095.

Fuhrman J.A. & Bell T.M. (1985) Biological considerations in the measurement of dissolved free amino acids in seawater and implications for chemical and microbiological studies. *Mar. Ecol. Prog. Ser.* **25**, 13–21.

Gast V. (1985) Bacteria as a food source for microzooplankton in the Schlei Fjord and Baltic Sea with special reference to ciliates. *Mar. Ecol. Prog. Ser.* **22**, 107–120.

Gomes H. & Do R. (1991) Heterotrophic utilization of extracellular products of phytoplankton in a tropical estuary. *J. Plank. Res.* **13**, 487–498.

Goulder R. (1976) Relationships between suspended solids and standing crops and activities of bacteria in an estuary during a spring-neap tidal cycle. *Oecologia* **24**, 83–90.

Griffith P.C., Douglas D.J. & Wainright S.C. (1990) Metabolic activity of size-fractionated microbial plankton in estuarine, nearshore, and continental shelf waters of Georgia. *Mar. Ecol. Prog. Ser.* **59**, 263–270.

Haas L.W. (1977) The effect of the spring-neap tidal cycle on the vertical salinity structure of the James, York and Rappahannock Rivers, Virginia, USA. *Est. Coast. Mar. Sci.* **5**, 483–496.

Hagstrom A., Larrson U., Horstedt P. & Normark S. (1979) Frequency of dividing cells, a new approach to the determination of bacterial growth rates in aquatic environments. *Appl. Environ. Microbiol.* **37**, 805–812.

Hanson R.B., Pomeroy L.R., Blanton J.O., Biddanda B.A., Wainwright S., Bishop S.S., Yoder J.A. & Atkinson L.P. (1988) Climatological and hydrographic influences on nearshore food webs off the southern United States: bacterioplankton dynamics. *Cont. Shelf Res.* **8**, 1321–1344.

Harvey R.W. & Young L.Y. (1980) Enrichment and association of bacteria and particulates in salt marsh surface water. *Appl. Environ. Microbiol.* **39**, 894–899.

Heinanen A.P. (1991) Bacterial numbers, biomass and productivity in the Baltic Sea: a cruise study. *Mar. Ecol. Prog. Ser.* **70**, 283–290.

Hobbie J.E. & Wright R.T. (1979) An assessment of quantitative measurements of aquatic microbes. *Arch. Hydrobiol. Beih. Ergebn. Limnol.* **13**, 85–95.

Hobbie J.E. & Cole J.J. (1984) Response of a detrital food web to eutrophication. *Bull. Mar. Res.* **35**, 357–363.

Hobbie J.E., Crawford C.C. & Webb K.L. (1968) Amino acid flux in an estuary. *Science* **159**, 1463–1464.

Hobbie J.E., Daley R.J. & Jasper S. (1977) Use of Nuclepore filters for counting bacteria by fluorescence microscopy. *Appl. Environ. Microbiol.* **33**, 1225–1228.

Hollibaugh J.T. (1988) Limitation of the [3H]thymidine method for estimating bacterial productivity due to thymidine metabolism. *Mar. Ecol. Prog. Ser.* **43**, 19–30.

Hudak J.P., McDaniel J., Lee S. & Fuhrman J.A. (1988) Mineralization potentials of aromatic hydrocarbons by estuarine microorganisms: variations with season, location and bacterioplankton production. *Mar. Ecol. Prog. Ser.* **47**, 97–102.

Jensen L.M. (1983) Phytoplankton release of extracellular organic carbon, molecular weight composition, and bacterial assimilation. *Mar. Ecol. Prog. Ser.* **11**, 39–48.

Jonas R.B. & Tuttle J.H. (1990) Bacterioplankton and organic carbon dynamics in the lower mesohaline Chesapeake Bay. *Appl. Environ. Microbiol.* **56**, 747–757.

Jorgensen N.O.G. (1982) Heterotrophic assimilation and occurrence of dissolved free amino acids in a shallow estuary. *Mar. Ecol. Prog. Ser.* **8**, 145–159.

Jumars P.A., Penry D.L., Baross J.A., Perry M.J. & Frost B.W. (1989) Closing the microbial loop: dissolved organic carbon pathway to heterotrophic bacteria from incomplete ingestion, digestion and absorption in animals. *Deep-Sea Res.* **36**, 483–495.

Kirchman D. (1983) The production of bacteria attached to particles suspended in a freshwater pond. *Limnol. Oceanogr.* **28**, 858–872.

Kirchman D.L. & Hoch M.P. (1988) Bacterial production in the Delaware Bay estuary estimated from thymidine and leucine incorporation rates. *Mar. Ecol. Prog. Ser.* **45**, 169–178.

Kirchman D., Ducklow H. & Mitchell R. (1982) Estimates of bacterial growth from changes in uptake rates and biomass. *Appl. Environ. Microbiol.* **44**, 1296–1307.

Kuosa H. & Kivi K. (1989) Bacteria and heterotrophic flagellates in the pelagic carbon cycle in the northern Baltic Sea. *Mar. Ecol. Prog. Ser.* **53**, 93–100.

Kuuppo-Leinikki P. (1990) Protozoan grazing on planktonic bacteria and its impact on bacterial populations. *Mar. Ecol. Prog. Ser.* **63**, 227–238.

Laanbroek H.J. & Verplanke J.C. (1986) Tidal variations in bacterial biomass, productivity and oxygen uptake rates in a shallow channel in the Oosterschelde basin, The Netherlands. *Mar. Ecol. Prog. Ser.* **29**, 1–5.

Laanbroek H.J., Verplanke J.C., de Visscher P.R.M. & de Vuyt R. (1985) Distribution of phyto- and bacterioplankton growth and biomass parameters, dissolved inorganic nutrients and free amino acids during a spring bloom in the Oosterschelde basin, The Netherlands. *Mar. Ecol. Prog. Ser.* **25**, 1–11.

Lampert W. (1978) Release of dissolved organic carbon by feeding zooplankton. *Limnol. Oceanogr.* **25**, 982–990.

Larsson U. & Hagstrom A. (1979) Phytoplankton exudate release as an energy source for the growth of pelagic bacteria. *Mar. Biol.* **52**, 199–206.

Larsson U. & Hagstrom A. (1982) Fractionated phytoplankton production, exudate release and bacterial production in a Baltic eutrophication gradient. *Mar. Biol.* **67**, 57–70.

Lignell R. (1990) Excretion of organic carbon by phytoplankton: its relation to algal biomass, primary productivity and bacterial secondary productivity in the Baltic Sea. *Mar. Ecol. Prog. Ser.* **68**, 85–99.

Malone T.C. & Ducklow H.W. (1990) Microbial biomass in the coastal plume of Chesapeake Bay: phytoplankton–bacterioplankton relationships. *Limnol. Oceanogr.* **35**, 296–312.

Malone T.C. & Conley D.J. (1992) Trends in nutrient loading and eutrophication: a comparison of Chesapeake Bay and Hudson River ecosystems (submitted).

Malone T.C., Kemp W.M., Ducklow H.W., Boynton W.R., Tuttle J.H. & Jonas R.B. (1986) Lateral variation in the production and fate of phytoplankton in a partially stratified estuary. *Mar. Ecol. Prog. Ser.* **32**, 149–160.

Malone T.C., Crocker L.H., Pike S.E. & Wendler B.W. (1988) Influences of river flow on the dynamics of phytoplankton production in a partially stratified estuary. *Mar. Ecol. Prog. Ser.* **48**, 235, 249.

Malone T.C., Ducklow H.W., Peele E.R. & Pike S.R. (1991) Picoplankton carbon fluxes in Chesapeake Bay. *Mar. Ecol. Prog. Ser.* **78**, 11–22.

Meyer-Reil L.-A. (1977) Bacterial growth and biomass production. In Rheinheimer G. (ed.) *Microbial Ecology of a Brackish-Water Environment*, pp. 223–235. Springer Verlag, Berlin.

Mopper K. & Lindroth P. (1982) Diel and depth variations in dissolved free amino acids and ammonium in the Baltic Sea determined by shipboard HPLC analyses. *Limnol. Oceanogr.* **27**, 336–347.

Newell R.C. & Linley E.A.S. (1984) Significance of microheterotrophs in the decomposition of phytoplankton: estimates of carbon and nitrogen flow based on the biomass of plankton communities. *Mar. Ecol. Prog. Ser.* **16**, 105–119.

Newell S.Y. & Christian R.R. (1981) Frequency of dividing cells as an estimator of bacterial productivity. *Appl. Environ. Microbiol.* **42**, 23–31.

Officer C.B., Biggs R.B., Taft J.L., Cronin L.E., Tyler M.A. & Boynton W.R. (1984) Chesapeake Bay anoxia: origin, development and significance. *Science* **223**, 22–27.

Painchaud J., Lefaivre D. & Therriault J.-C. (1987) Box model analysis of bacterial fluxes in the St. Lawrence Estuary. *Mar. Ecol. Prog. Ser.* **41**, 241–252.

Palumbo A.V. & Ferguson R.L. (1978) Distribution of suspended bacteria in the Newport River estuary, North Carolina. *Est. Coast. Mar. Sci.* **7**, 521–529.

Paul J.H., Jeffrey W.H. & DeFlaun M. (1985) Particulate DNA in subtropical oceanic and estuarine planktonic environments. *Mar. Biol.* **90**, 95–101.

Paul J.H., DeFlaun M.F., Jeffrey W.H. & David A.W. (1988) Seasonal and diel variability in dissolved DNA and in microbial biomass and activity in a subtropical estuary. *Appl. Environ. Microbiol.* **54**, 718–727.

Payne W.J. & Wiebe W.J. (1978) Growth yield and efficiency in chemosynthetic microorganisms. *Ann. Rev. Microbiol.* **32**, 155–184.

Pedrós-Alió C. & Newell S.Y. (1989) Microautoradiographic study of thymidine uptake in brackish waters around Sapelo Island, Georgia, USA. *Mar. Ecol. Prog. Ser.* **55**, 83–94.

Pomeroy L.R. (1974) The ocean's food web: a changing paradigm. *BioScience* **24**, 499–504.

Pomeroy L.R. & Deibel D. (1986) Temperature regulation of bacterial activity during the spring bloom in Newfoundland coastal waters. *Science* **233**, 359–361.

Pomeroy L.R. & Wiebe W.J. (1988) Energetics of microbial food webs. *Hydrobiologia* **159**, 7–18.

Pomeroy L.R., Hanson R.B., McGillivary P.A., Sherr B.F., Kirchman D. & Deibel D. (1984) Microbiology and chemistry of fecal products of pelagic tunicates: rates and fates. *Bull. Mar. Sci.* **35**, 426–439.

Ray R.T., Haas L.W. & Sieracki M.E. (1989) Autotrophic picoplankton dynamics in a Chesapeake Bay sub-estuary. *Mar. Ecol. Prog. Ser.* **52**, 273–285.

Relaxans J.C., Meybeck M., Billen G., Brugeaille M., Etcheber H. & Somville M. (1988) Algal and microbial processes involved in particulate organic matter dynamics in the Loire estuary. *Estuarine Coastal and Shelf Science* **27**, 625–644.

Riemann B., Nielsen P., Jeppesen M., Marcussen B. & Fuhrman J.A. (1984) Diel changes in bacterial biomass and growth rates in coastal environments, determined by means of thymidine incorporation into DNA, frequency of dividing cells (FDC), and microautoradiography. *Mar. Ecol. Prog. Ser.* **17**, 227–235.

Riemann B., Jorgensen N.O., Lampert W. & Fuhrman J.A. (1986) Zooplankton-induced changes in dissolved free amino acids and in production rates of freshwater bacteria. *Microb. Ecol.* **12**, 247–258.

Sand-Jensen K., Jensen L.M., Marcher S. & Hansen M. (1990) Pelagic metabolism in eutrophic coastal waters during a late summer period. *Mar. Ecol. Prog. Ser.* **65**, 63–72.

Scavia D. (1988) On the role of bacteria in secondary production. *Limnol. Oceanogr.* **33**, 1220–1224.

Seliger H.H., Boggs J.A. & Biggley W.H. (1985) Catastrophic anoxia in the Chesapeake Bay in 1984. *Science* **228**, 70–73.

Sherr B.F., Sherr E.B. & Pedrós-Alió C. (1989) Simultaneous measurement of bacterioplankton production and protozoan bacterivory in estuarine water. *Mar. Ecol. Prog. Ser.* **54**, 209–219.

Sherr E.B., Sherr B.F. & Albright L.J. (1987) Bacteria: link or sink? *Science* **235**, 88.

Smith S.V., Hollibaugh J.T., Dollar S.J. & Vink S. (1989) Tomales Bay, California: a case for carbon-limited nitrogen cycling. *Limnol. Oceanogr.* **34**, 37–52.

Strayer D. (1988) On the limits to secondary production. *Limnol. Oceanogr.* **33**, 1217–1220.

Taft J.L., Taylor W.R., Hartwig E.D. & Loftus R. (1980) Seasonal oxygen depletion in Chesapeake Bay. *Estuaries* **3**, 242–247.

Torreton J.-P., Guiral D. & Arfi R. (1989) Bacterioplankton biomass and production during destratification in a monomictic eutrophic bay of a tropical lagoon. *Mar. Ecol. Prog. Ser.* **57**, 53–67.

Valdés M. & Albright L.J. (1981) Survival and heterotrophic activities of Fraser River and Strait of Georgia bacterioplankton within the Fraser River plume. *Mar. Biol.* **64**, 231–241.

Watson S.W., Novitsky T.J., Quinby H.L. & Valois F.W. (1977) Determination of bacterial number and biomass in the marine environment. *Appl. Environ. Microbiol.* **33**, 940–946.

Wiebe W.J. & Pomeroy L.R. (1972) Microorganisms and their association with aggregates and detritus in the sea: a microscopic study. *Mem. 1st. Ital. Idrobiol.* **29**(Suppl.), 325–352.

Wiebe W.J. & Smith D.F. (1977) Direct measurement of dissolved organic carbon release by phytoplankton and incorporation by microheterotrophs. *Mar. Biol.* **42**, 213–233.

Wikner J., Rassoulzadegan F. & Hagstrom A. (1990) Periodic bacteriovore activity balances bacterial growth in the marine environment. *Limnol. Oceanogr.* **35**, 313–325.

Williams P.J. le B. (1981) Incorporation of microheterotrophic processes into the classical paradigm of the planktonic food web. *Kieler Meeresforsch.* **5**, 1–28.

Williams P.J. le B. (1984) Bacterial production in the marine food chain: the emperor's new suit of clothes? In Fasham M. (ed.) *Flows of Energy and Materials in Marine Ecosystems: Theory and Practice*, pp. 271–299. Plenum, New York.

Wolter K. (1982) Bacterial incorporation of organic substances released by natural phytoplankton populations. *Mar. Ecol. Prog. Ser.* 7, 287–295.

Wright R.T. (1978) Measurement and significance of specific activity in the heterotrophic bacteria of natural waters. *Appl. Environ. Microbiol.* **36**, 297–305.

Wright R.T. & Coffin R.B. (1983) Planktonic bacteria in estuaries and coastal waters of northern Massachusetts: spatial and temporal distribution. *Mar. Ecol. Prog. Ser.* **11**, 205–216.

Wright R.T. & Coffin R.B. (1984) Factors affecting bacterioplankton density and productivity in salt marsh estuaries. In Klug M.J. & Reddy C.A. (eds) *Current Perspectives in Microbial Ecology*, pp. 485–494. American Society for Microbiology, Washington DC.

Wulff F. (ed.) (1990) *Large-Scale Environmental Effects and Ecological Processes in the Baltic Sea*. Swedish Environmental Protection Agency Report 3849, Stockholm.

Zimmerman R. (1978) Estimation of bacterial number and biomass by epifluorescence microscopy and scanning electron microscopy. In Rheinheimer G. (ed.) *Microbial Ecology of a Brackish-Water Environment*, pp. 103–120. Springer Verlag, Berlin.

12

Staffan Kjelleberg, Klas B.G. Flärdh, Thomas Nyström, & David J.W. Moriarty

Growth limitation and starvation of bacteria

12.1 Introduction

The oceans contain 10^{22} liters of water, and as there are 10^7 to 10^9 bacteria per liter, a significant portion of the world's bacteria (10^{30}) live in the water column of the oceans. A large proportion of the remaining bacteria live in the ocean sediments, which cover 70% of the world's surface. The ultimate source of energy for feeding most of them is sunlight, and as this is received by only the surface waters, the majority of bacteria in water and sediments would be expected to experience growth-limiting or starving conditions for varying periods of time. Most oceans are oligotrophic, i.e., the concentration and rate of supply of available organic matter is generally low, so even bacteria in surface waters may be short of food. The hypothesis that marine bacteria are limited by carbon and the physiological mechanisms by which they survive will be analyzed in this chapter.

From an ecological viewpoint, we will examine the evidence for food (or organic carbon) limitation on bacterial growth and the responses of bacteria to oligotrophy. We will discuss factors that may limit numbers of cells, biomass or cell size, growth rates, and species composition in different habitats.

From a microbiological viewpoint the intermittent or unbalanced growth that may characterize bacteria in natural ecosystems will be addressed. We will discuss the physiological and molecular pathways employed by bacteria in response to stationary phase and starvation. Important questions refer to whether marine bacteria display a morphogenetic life-cycle in response to changes in energy and nutrient availability, whether the physiology of the nongrowing cell differs from that of growing cells, and if marine bacteria utilize a true differentiation program. If so, the identity of genes and function of their products essential for the starvation-induced differentiation program as well as those involved in adapting the cell to efficient recovery from starvation should be sought. It is worth emphasizing that the biochemistry of bacterial growth may require no more than one-third of the 4000-protein coding capacity of the chromosome of most bacteria (Neidhardt *et al.* 1990).

Given the relatively long generation times that exist in any natural aquatic ecosystem and the enormous degree of heterogeneity on the microbes scale, unbalanced growth may be normal for members of the bacterial community. This includes periods of growth at various growth rates, interspersed periods of nongrowth, starvation, recovery, and regrowth. As will be stressed in this chapter, this life cycle allows for the ability of the bacterium to compete, persist, colonize other habitats, and proliferate. To perceive marine bacteria as continuously being in steady-state growth or permanently starved is equally simplistic. This chapter offers some insight into the rapidly emerging picture of how bacteria enter the nonreplicative state, undergo a differentiation program to allow for optimal adjustment to this state, and how the starved cell has elaborate ways to recover and regrow.

12.2 Ecological aspects of bacterial growth and survival

The structure of the habitat in which marine bacteria live has a controlling influence on their supply of food and their responses to it. There are five main types of habitat:

1 the water column where bacteria live as individual free cells;
2 particles and other structures in the water to which bacteria attach;
3 the air–water interface;
4 external and internal surfaces of animals and plants;
5 the sediments.

Each of these habitats has a complex internal structure. In sediments, there are gradients in the concentration of oxygen and organic and inorganic nutrients; these are especially complex in the rhizosphere of seagrasses and mangroves. There are considerable differences between coastal and open ocean environments. In the open ocean, the water column is structured; there may be a thermocline with, in some cases, greater production of phytoplankton above, and in other places greater productivity below the thermocline.

There are major differences between water masses in the photic zone, the upper mixed layer, and the deep ocean. In the Pacific Ocean, residence times for water masses are 1000 or more years for the deep water compared to 2–10 or so years for the Antarctic intermediate water (Marshall 1979). The low temperature in the deep ocean slows rates of metabolism, so organic matter would last longer than at the surface, but bacteria traveling with these water masses would need to have strategies for surviving long periods with little nutrient input. When one considers the volume of water in the deep ocean compared to the photic zone, it is clear that the concentration of organic matter that is available to bacteria must be low because it is all synthesized from carbon dioxide in the photic zone, either immediately above or in a distant, polar ocean. This is discussed in more detail in Section 12.2.2.

The two most important factors that control the growth of bacteria in the sea are the concentration of organic substrates that are readily available and temperature. In this chapter we will primarily discuss the effect of the former. The effect of substrate concentration is modified by the biochemical composition of the organic matter. The quality of organic matter as a food source varies depending upon the particular species of bacteria utilizing it. However, few studies have been carried out on the species composition of bacteria that are active in the decomposition of organic matter at any particular time or place in the sea (Painting *et al.* 1989). The main ways in which we can observe the response of bacteria to substrate concentration and quality in natural communities are by examining changes in bacterial numbers, size or biomass, growth rates and productivity, and species diversity or species composition. Of these, productivity, i.e., the rate of production of cell biomass, is a good indicator of trophic status. It is directly related to the flux of organic carbon through the heterotrophic bacteria. It is clear, however, that the next generation of techniques should also allow us to determine which bacterial species are growing at a given time and how the species composition of the bacterial community changes over the short term as controlling factors such as temperature and nutrient concentrations and composition change.

12.2.1 Abundance and biomass

The principal source of organic nutrients for bacteria in the mixed layer is the phytoplankton, so bacterial abundance should be directly correlated to the abundance of algae in the sea. Indeed, there are reports of positive correlations between chlorophyll concentration and bacterial abundance, but there are many exceptions (e.g., Bird & Kalff 1984; Ducklow 1984). Billen (1990) observed a delay in the response of bacterial abundance to phytoplankton growth, both during the spring bloom in temperate waters and during the summer in the Antarctic. He concluded that the bacterial abundance and production were not responses to release of dissolved organic matter (DOM) during growth of the phytoplankton, but rather to release of macromolecules after death and lysis of the phytoplankton. The proportion of primary production that is secreted as dissolved compounds is low compared to that incorporated into algal biomass (Sharp 1981). So it would be expected that bacterial abundance and production would be higher when a phytoplankton bloom was decaying than when the phytoplankton were actively growing. Again, the inferred, simple relationship is not always found; it is modified by other factors that cause algal death, e.g., heavy grazing on the algae by zooplankton.

Numbers by themselves are not a good indication of the trophic status of bacteria because, as discussed below, a response to starvation by bacteria may be rapid division and the formation of small cells (Morita 1982). Thus, numbers of bacteria may increase during starvation and, conversely, they may decrease when food is plentiful due to grazing by animals. The biomass of bacteria, i.e., the size of the bacterial cells, is an indicator of the trophic status of the bacteria because starved bacteria are smaller than well-fed bacteria. There is a consistent trend for the bacteria in an oligotrophic environment to be smaller than those in eutrophic environments, but the differences are not large (van Duyl & Kop 1988; Fuhrman *et al.* 1989; van Duyl *et al.* 1990).

Starved bacteria are small, but they increase in size before dividing when incubated with nutrients (see Section 12.3.1) (Morita 1982). Most marine bacteria are small and pass through filters with pore sizes of 0.8–1 μm; it is commonly observed that they then increase in size when incubated with organic substrates, which suggests that they are starved or living at suboptimal concentrations of organic nutrients (e.g., Cho & Azam 1988; Kirchman *et al.* 1991). These small bacteria represent a high proportion of the biomass in the oceans and they are major agents in biogeochemical cycling of elements (Cho & Azam 1988; 1990). It should be stated that other factors, such as selective grazing on large cells, may also contribute to the predominance of small bacteria in the sea.

In addition to the effect of grazing, numbers will decrease in response to viral infection, which can be severe at times (Heldal & Bratbak 1991). It would seem likely, however, that if one or a few species of bacteria were growing rapidly in response to nutrient enrichment and then were lysed by viruses, other species resistant to the viruses would then respond to the nutrients. Thus the total bacterial abundance is probably controlled by nutrient concentration, even though the species composition may change over a short time period.

Quantitative models have been developed to explain the control of bacterial biomass by organic substrate concentration and grazing in planktonic systems (Wright & Coffin 1984; Billen *et al.* 1988; Wright 1988). In a steady-state system, bacterial biomass is directly related to the concentration of organic matter, but no real systems are in a steady-state condition. The concentration of organic matter fluctuates considerably and the structure of the environment or the bacterial habitat on a microscale is also quite variable. For example, bacteria associated with particles, particularly particles derived from phytoplankton, will obviously experience greater, but presumably transient, concentrations of organic matter than those free in the water column and away from particles. When grazing animals are present, the release of organic matter from phytoplankton would be greater than in their absence. Thus interactions of grazers with phytoplankton and bacteria are important in determining whether bacterial production is closely linked to primary production in the water column.

We conclude that in pelagic ecosystems, the heterotrophic bacteria are mostly small and limited by organic nutrients, and their population density does not vary greatly. The important controlling factors are organic substrate concentration and grazing, although death and lysis due to viral infection may be important on some occasions.

Bacterial numbers in sediment vary considerably between environments and are controlled not only by the rates of organic matter input, but also by grazing. Evidence that bacterial population density in surface sediments is controlled by grazing is shown by low numbers and rapid specific growth rates (Figure 12.1) (Alongi 1989; Moriarty *et al.* 1990; Moriarty *et al.* 1991). Deeper in sediments, population density and production are controlled by the supply of organic matter (Alongi *et al.* 1989; Moriarty *et al.* 1991). In deep waters where the supply of organic matter is limited, it has been found that bacterial population density at the sediment surface was high, but productivity, and therefore specific growth rates, were very low, indicating not only that grazing was low, but that the bacteria were starved (Moriarty *et al.* 1991).

The abundance of bacteria below the zone of bioturbation in sediment decreases exponentially (Figure 12.2a). It is likely that once bacteria use up their endogenous sources of organic matter most

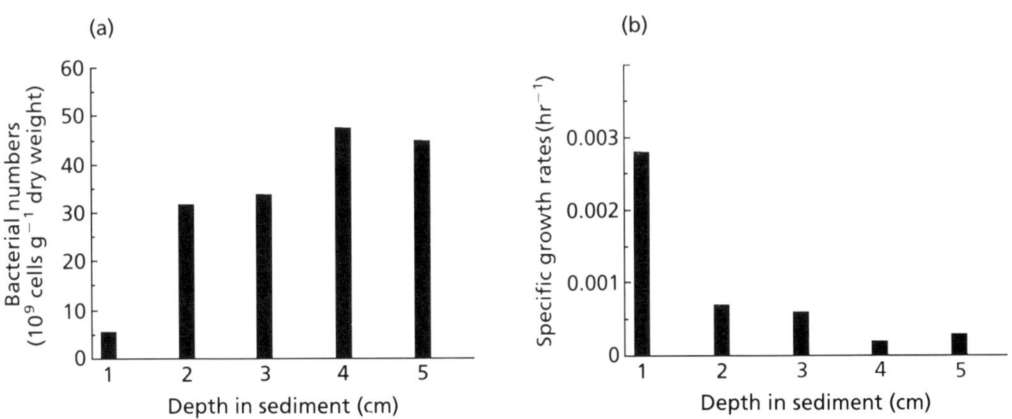

Figure 12.1 Variations in bacterial population density and specific growth rate with depth in sediment in tropical seagrass beds. From Moriarty *et al.* (1991).

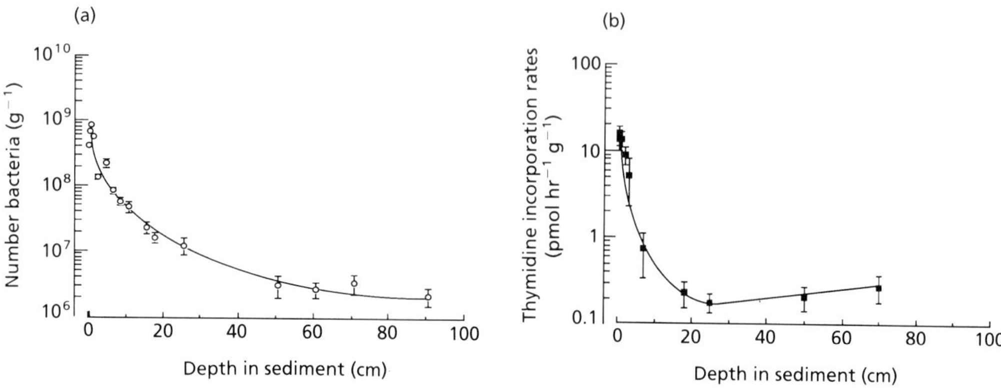

Figure 12.2 Bacterial population density and rates of DNA synthesis (measured as rate of tritiated thymidine incorporation) in sediment at 400 m water depth on the eastern Australian continental margin. From Moriarty *et al.* (1991).

eventually die. Some bacteria are found deep within the sediments and this raises interesting questions as to whether those bacteria simply survive or whether they are able to grow and divide very slowly, using organic matter that was buried with them, perhaps several thousand years ago, or dissolved organic compounds that diffuse slowly from the surface. Bacteria buried below the surface of marine sediments have functioning biosynthetic pathways and the capacity for growth, even if they do not actually grow as inferred from the rates of DNA synthesis (Figure 12.2b), and only begin to grow after a lag phase of about 12 hours following substrate addition (Moriarty *et al.* 1991). Deep within sediments where there are no animals to graze on bacteria, death and lysis is the final fate, and thus in these sediments supply of organic matter is directly correlated to both productivity and numbers of bacteria.

12.2.2 Specific growth rates and productivity

In Section 12.3, the strategies bacteria have evolved to adapt to starvation and rapid recovery when nutrients become available are discussed. The measuring of productivity and specific growth rates can be used to determine where and when bacteria are in a non-growth phase in natural environments. Bacterial productivity is the rate of increase in biomass and is directly related to the availability of nutrients. The specific growth rate is the number of generations occurring per unit time (the cell production rate divided by the number of cells present). Changes in specific growth rate occur in response to fluctuations in nutrient supply to the bacteria, but the specific growth rate itself is not necessarily directly related to the trophic status of the environment, i.e., whether it is eutrophic or oligotrophic. It is possible for the specific growth rate to be as fast in an oligotrophic as it is in a eutrophic environment, if the bacterial numbers are lower in the oligotrophic environment.

Bacterial productivity can be determined from the rate of increase in bacterial numbers and size of cells in samples of sea water from which grazing animals have been removed by filtration. The process of filtering, however, changes relationships between bacteria and their source of organic matter and disrupts the bacterial community when some are attached to particles. Productivity can now also be determined in almost undisturbed natural samples from the rate of tritiated thymidine incorporation into DNA (Moriarty 1986).

Some discussion of methods for measuring bacterial production in natural environments is necessary here because our interpretations of ecological processes are dependent on the presumed accuracy of the data generated. Of the several different methods that have been developed for determining bacterial production in the aquatic environment, the thymidine and leucine methods are currently the most favored. The accuracy is subject to technical problems and conversion factors, which have been discussed by Moriarty (1990) and Riemann & Bell (1990). Provided due attention is paid to technical matters, including measurement of radioactivity in DNA only and not all macromolecules, the thymidine method does, in principle, give the most accurate result. The leucine method is accurate when bacteria are rapidly growing and growth is balanced, but this does not always occur in natural systems, when rates of protein synthesis can be much greater than rates of DNA synthesis (Chin-Leo & Kirchman 1990; Riemann & Bell 1990; Moriarty *et al.* 1991). Protein synthesis occurs during starvation, whereas DNA synthesis shuts down after the completion of initiated rounds of replication (see Section 12.3).

The productivity of bacteria is a better indicator of trophic status than abundance because it is more directly related to the turnover of organic carbon than either specific growth rate or changes in numbers. Bacterial production in the water column is dependent on the concentration and availability of dissolved organic compounds and thus is related to the rate of primary production by phytoplankton or benthic plants (e.g., Wright 1984; Moriarty *et al.* 1990). In the oligotrophic Sargasso Sea, bacterial productivity ranged from 0.8 to $6\,\mathrm{mg\,C\,l^{-1}\,d^{-1}}$, whereas in summer in Chesapeake Bay, it ranged from about 8 to 70 and over tropical seagrass beds from 20 in winter to 200 in summer (Ducklow 1982; Fuhrman *et al.* 1989; Moriarty *et al.* 1990).

Bacteria in natural environments experience fluctuating nutrient concentrations on time scales varying from hours to seasons. Seasonal and daily variations in bacterial productivity, growth rates, and numbers have been found in many environments (Moriarty 1986). In the North Sea and Dutch Wadden Sea, there were noticeable variations in bacterial productivity and numbers throughout the year, with lower values in winter than at other seasons (van Duyl & Kop 1988; van Duyl *et al.* 1990). In temperate environments it is difficult to separate the effects of temperature from substrate concentration. In coastal environments, where there is also input of nutrients from sources other than the phytoplankton, the analysis of factors controlling bacterial growth is further complicated.

In the Celtic Sea, Joint & Pomroy (1987) found that a large proportion of bacterial production was correlated with temperature and not with primary production. However, Lancelot & Billen (1984) have found that bacterial production was correlated directly with phytoplankton production and in particular the release of DOM during the phytoplankton bloom in the North Sea. In water upwelling from the deep ocean, initial bacterial cell size was small, however, as phytoplankton blooms developed there was a rapid increase in productivity (Painting *et al.* 1989).

Dissolved organic compounds from phytoplankton turn over rapidly in the surface waters of the North Atlantic during the spring bloom (Kirchman *et al.* 1991). The growth efficiency or cell biomass yield was low (2–9%), which indicates that abundance would not increase markedly and suggests that the bacterial production may have been limited by other factors, perhaps phosphate concentration. Below the mixed layer in the ocean, which is usually 100–200 m depth, bacteria are dependent on organic matter falling through the water column as debris, e.g., fecal pellets, and in feces from migrating animals and on organic matter carried by water masses that have sunk from the surface at convergence zones. The extent to which bacteria in the ocean depths are supplied by organic matter from the surface varies considerably between seas. Cho & Azam (1988) estimated that 17% of primary production settled to depths of 1000 m from the photic zone in the north Pacific gyre compared to 40% in the Santa Monica basin. In some parts of the sea, feces from migrating zooplankton and fish may be the main source; in other parts, transport of organic matter from polar convergence zones may be more important. Sorokin *et al.* (1985) concluded that as the biomass and production of bacteria was higher in the Antarctic intermediate water and the deep water masses than in water masses above these, there must be a stock of organic matter available to bacteria that was derived from the polar seas. If the bacteria were dependent solely on a vertical flux of organic matter, then productivities would decrease continuously with depth; the substantially higher abundance and production of bacteria indicate that there must

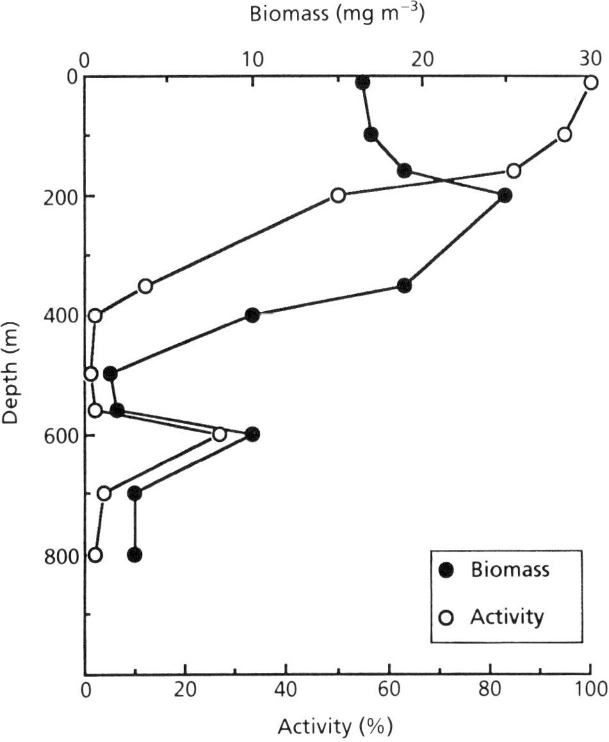

Figure 12.3 Bacterial biomass and relative production or metabolic activity (determined from the rate of dark CO_2 fixation) in water at the South Tropical Convergence in the Indian Ocean. From Sorokin *et al.* (1985).

be horizontal advection of organic matter from polar zones (Figure 12.3).

In general, it is clear that bacterial production is controlled to a large extent by primary production in the water column (Cole *et al.* 1988). The relationship varies over short time and space scales and so bacterial productivity is not necessarily directly proportional to primary productivity at a given time or place (Ducklow 1984). Furthermore, Chin-Leo & Kirchman (1990) point out that the time scales for bacterial growth vary, depending on whether the source of organic matter is exudate material from normal phytoplankton during photosynthesis, or pulses of organic matter from decaying or stressed phytoplankton, or release by excretion from zooplankton, or lysis of cells during feeding by zooplankton. Bacterial growth may thereby become "intermittent" or "unbalanced," i.e., the rate of DNA synthesis or cell division does not occur at the same rate as synthesis of other cellular constituents such as protein.

Bacterial production in sediment is regulated by the supply of organic matter (Moriarty 1989). Bacterial productivity in coral reef and seagrass sediments is closely correlated to the productivity of

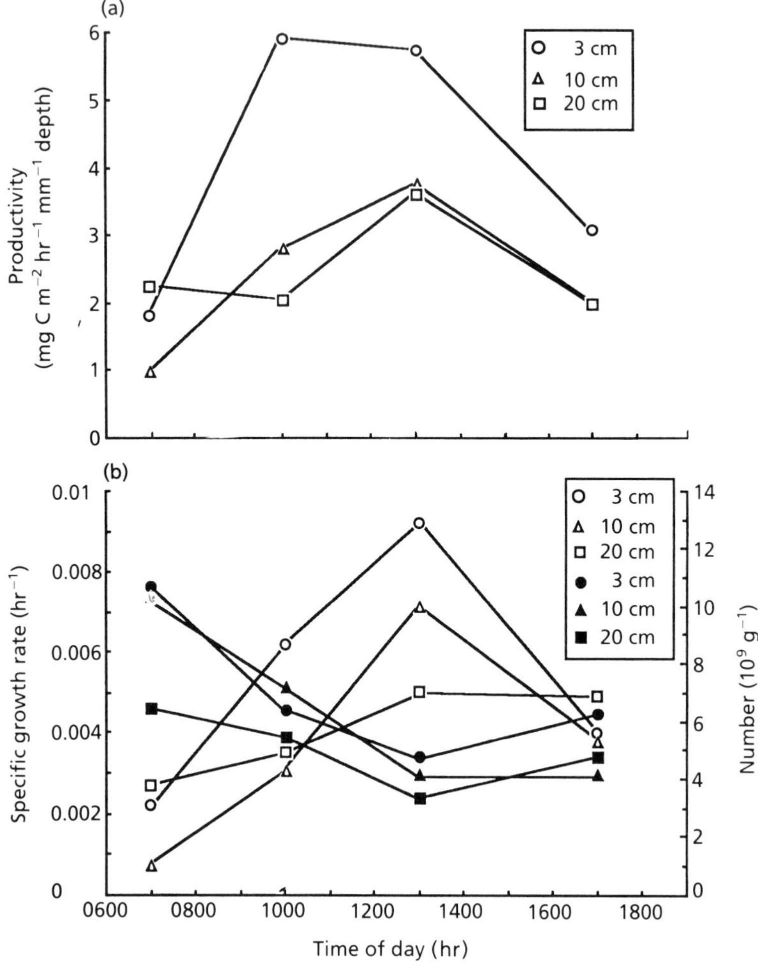

Figure 12.4 (a) Bacterial productivity; (b) specific growth rates (clear symbols) and population density (solid symbols) in sediment on tropical seagrass beds of northern Australia at different times of the day. From Pollard & Moriarty (1991).

seagrasses and benthic algae on daily and seasonal scales (Moriarty & Pollard 1982; Moriarty *et al.* 1985; Moriarty & Hansen 1990; Moriarty *et al.* 1990; Pollard & Moriarty 1991). On diurnal time scales, bacterial productivity and specific growth rates in seagrass rhizosphere sediment at 3 cm depth varied with photosynthetic rates of the primary producers, but below the rhizosphere at 20 cm depth, there was little variation (Figure 12.4). Bacterial productivity and primary productivity were correlated in coral reef sediments where a seasonal variation was also apparent (Moriarty *et al.* 1985).

These results indicate that bacteria in sediment are limited by organic carbon substrates. There are periods varying from hours to months in surface sediments when organic substrates are permanently

limiting, and the only bacteria that can survive are those that have developed strategies for maintenance and growth at very low concentrations of organic nutrients.

12.2.3 *Species diversity*

There is a great variety of bacterial species in the water column and sediments and these have been identified by various cultural techniques (Sieburth 1979) as well as more recently by 16S rRNA gene cloning and sequencing (Giovannoni *et al.* 1990; Schmidt *et al.* 1991). Early work with culture or viable count techniques has been discounted because only a few of the bacteria present grew on the media used. It may be suggested, however, that at any one time only one or a few species would have their appropriate nutrient conditions met and that viable count techniques do indeed indicate the active populations in some environments at certain times (e.g., Painting *et al.* 1989).

Bacterial communities respond to limitations in the availability of organic substrates by changes in species diversity. In water upwelling from the deep ocean, the species composition was diverse with vibrios being dominant, but as phytoplankton blooms developed, the diversity decreased and pseudomonads became dominant (Painting *et al.* 1989). When the bloom decayed the species diversity increased, and in addition to vibrios, flavobacteria, and enterobacters were common. This last change in species composition occurred when more refractory macromolecules were the main source of organic matter. Thus, changes in the type of organic matter and its concentration lead to changes in the species of bacteria, with different rates of growth. Obviously, we will need to know what species of bacteria are present in order to fully understand and predict interactions of phytoplankton and bacteria.

Lee & Fuhrman (1990) used DNA hybridization to compare species composition of bacteria in the sea. They found considerable differences in the composition of the bacterial community in coastal waters and in the Sargasso Sea, and between seasons in coastal waters. Although studies on water column samples have not been done, it is worth mentioning that the heterogeneity of the bacterial community in a marine coastal sediment has been estimated by determining the reassociation of single-strand DNA. The major part of DNA isolated from the bacterial fraction of sediment is very heterogeneous, corresponding to at least 4000 completely different bacterial genomes (Torsvik *et al.* submitted).

12.2.4 *Particles*

The term marine snow is used for aggregates of organic matter in the sea. These particles are found throughout the surface waters of the oceans and are enriched in organic matter and microorganisms.

There are other particles in the sea, e.g., the fragments of decaying phytoplankton and the fecal pellets of zooplankton. Bacteria attached to the surfaces of particles are generally larger than those free in the water column and are more metabolically active. This indicates that bacteria attached to particles are not as limited in food supply as those free in the water column (see Chapter 13) (Kirchman 1983; Simon *et al.* 1990).

Many nutrient-limited or starved bacteria in the water column could be species that normally grow on or within organic aggregates and other particles. It might be expected that small invertebrates, such as copepods, in which food has a short passage time would not have a distinct gut microflora, but in fact they do (Sochard *et al.* 1979). Perhaps those bacteria that only grow slowly or not at all in the water column are species that grow rapidly in the guts of animals after being ingested together with the food of the animal. When excreted in the fecal pellet these bacteria would have a continued advantage over those free in the water because they would have a concentrated food supply of phytoplankton fragments. These bacteria could then be viewed as transient members of the water populations rather than transient members of the gut microflora (Moriarty 1990).

12.3 Physiological and molecular aspects of starvation and stress resistance

As is evident from the preceding sections, it is possible to isolate small bacterial cells from marine environments that respond to addition of nutrients by increasing their size and undergoing cell division. These cells respond to starvation by forming ultramicrocells similar in size to free-living marine bacteria (Torrella & Morita 1981; Tabor *et al.* 1981; Kjelleberg *et al.* 1987). Thus, some marine bacteria exhibit a nutrient-regulated life-cycle with ultramicrocells developing as a result of exogenous nutrient limitation.

12.3.1 Morphogenetic life-cycles

A closer examination reveals that this life-cycle includes pronounced morphological variations during growth, starvation, and recovery, some of which are species specific. For example, while defined as nondifferentiating bacteria, several marine *Vibrio* and *Pseudomonas* strains have been shown to respond to nutrient limitation by morphological alterations, concomitant to the miniaturization of the cell. The periplasmic space is enlarged, apparently as a result of unequal shrinkage of the outer and cytoplasmic membranes (Novitsky & Morita 1976; Mårdén *et al.* 1985). Increased surface roughness, as revealed by a highly textured outer membrane, the formation of outer membrane vesicles, and increased surface hydrophobicity and adhesion have been reported during the initial phase of the miniaturization process. Furthermore, formation of appendages, including fimbriae-like structures, which appear to mediate adhesion and cell

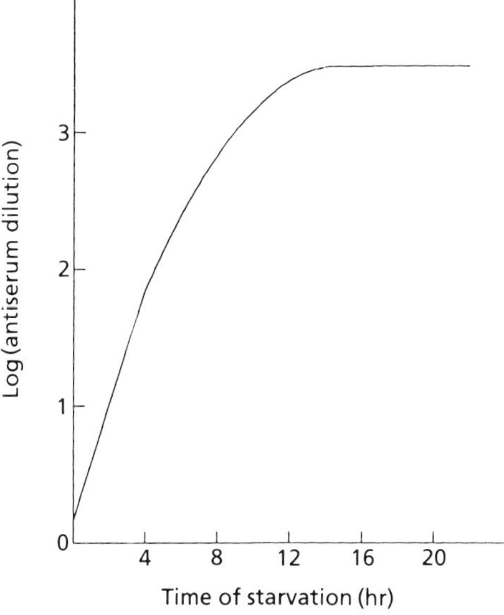

Figure 12.5 Detection of starvation-specific antigens in multiple-nutrient starved *Vibrio* sp. S14. Schematic presentation of the titration of whole cells with homologous rabbit antibodies raised against whole cells that were multiple-nutrient starved for 24 hr. The antiserum dilution that gave the optimal antigen/antibody ratio for agglutination at each starvation time is shown. Antibodies against growth antigens were removed prior to the titration experiment by absorption with growing cells. Redrawn from Albertson *et al.* (1987).

aggregation, have been observed (Dawson *et al.* 1981; Kjelleberg & Hermansson 1984; Mårdén *et al.* 1985; Hermansson *et al.* 1987; J. Östling unpublished observations). Stationary-phase and starvation-induced modulation of the exopolysaccharide configuration in *Pseudomonas* sp. S9 also strongly affects the adhesive properties of the cell (Wrangstadh *et al.* 1990). Starvation-induced changes in antigenicity, due to new proteins in the outer membrane, have been reported for a *Vibrio* strain (Figure 12.5) (Albertson *et al.* 1987).

The size difference between a rapidly growing cell and a non-growing cell is pronounced. For example, growing *Vibrio* cells may have a cell volume of 5.94 μm^3 while starved cells are as small as 0.05 μm^3 (Moyer & Morita 1989a). The cell volume of *Pseudomonas* sp. S9 may decrease from around 2 μm^3 to less than 0.1 μm^3 after being starved of energy and nutrients for 24 hours. The miniaturization is due to reductive cell division, i.e., completion of ongoing rounds of replication and subsequent cell division without biomass increase and a continuous size reduction due to decrease in endogenous material (Nyström & Kjelleberg 1989). Extensive degradation of cell constituents such as storage compounds and phospholipids may affect the morphology during the miniaturization process (Oliver & Stringer 1984; Hood *et al.* 1986; Malmcrona-Friberg *et al.* 1986).

It should be stated that miniaturization is an active process. Energy derived from respiration is required for size reduction of starved marine vibrios and the introduction of mutations leading to defects of the respiratory chain prevents further size reduction (Kjelleberg *et al.* 1983; Smigielski *et al.* 1990).

A "shut down" or "growth precursor" cell is a survival form suggested to be especially suited for organisms living in environments in which energy and nutrient depletion frequently prevents growth, and where long periods of time may elapse between periods of nutrient excess (Dow & Whittenbury 1980). While evidence for the "shut down" stage has primarily been demonstrated for aquatic prosthecate bacteria and species of cyanobacteria, Dow *et al.* (1983) hypothesized that such cell cycle stages, are also expressed by morphologically "conventional" forms of bacteria. Several observations support this possibility. Slow growing *Escherichia coli* grow polarly (Donachie & Begg 1970) as do bacteria entering the shut down state, and lysates from slow growing *E. coli* cultures contain two nucleoids, one in a relaxed state and one in a condensed format similar to the nucleoid of the shut down cell (Whittenbury & Dow 1977; C.S. Dow unpublished observations). Slow growing bacteria or those during the initial phase of starvation, undergo unequal division similar to the first morphological manifestations of the sporulation process in *Bacillus* (Dow *et al.* 1983; S. Kjelleberg unpublished observations).

12.3.2 Endogenous metabolism of nongrowing bacteria

Bacterial cells deprived of exogenous carbon and energy sources have to exploit endogenous resources to meet their maintenance energy requirements. The endogenous metabolism has been defined as the total metabolic reactions that occur when the bacterial cell is deprived of compounds or elements that may serve specifically as exogenous substrates (Dawes & Ribbons 1964). Studies on some nonmarine bacteria have indicated that starvation survival may be dependent upon the regulation of endogenous metabolism in accordance with the energy of maintenance requirements (Thomas & Batt 1969; Gray 1976; Chapman & Gray 1981). No universal patterns or predictable sequences of utilization of intracellular constituents have, however, been found. The exact role of storage polymers, such as polyhydroxybutyrate and glycogen, in the survival of the bacterial cell is unclear, although the presence of such polymers has in some cases been shown to enhance survival (reviewed by Dawes & Senior 1973; Dawes 1976; 1985; Preiss 1984). The utilization of intracellular substrates by marine bacteria exposed to starvation conditions is indicated by: disappearance of polyhydroxybutyrate; 88.7% loss of total carbohydrates and 99.8% loss of the total lipids during 7 days of multiple-nutrient starvation in *Vibrio cholera*; and significant decreases in lipids and fatty acids in other marine vibrios (Oliver & Stringer 1984; Hood *et al.* 1986; Malmcrona-Friberg *et al.* 1986). Results on changes in RNA and protein content vary. Extensive degradation as well as comparatively small decreases or even increases on a per cell basis have been reported (Jones & Rhodes-Roberts 1981; Amy *et al.* 1983a; Hood *et al.* 1986; Nissen 1987; Moyer & Morita 1989b). The degradation of endogenous material results in a continuous size reduction that, in addition to the reductive cell division contributes to the formation of ultramicrocells (see Section 12.3.1).

While bacteria are likely to undergo relatively short-term life-cycles of growth, starvation, and recovery, as pointed out by Morita (1982; 1985) the remarkable longevity of starved marine bacteria should also be considered. Marine bacteria have been shown to survive in the laboratory without a population decrease for periods of 5–7 years (Hood *et al.* 1986; Nissen 1987). *Vibrio* Ant-300 cells reduced their endogenous metabolism to 0.0017% hr^{-1} (net amount of $^{14}CO_2$ evolved expressed as a percentage of the total labeled cellular carbon) in 7 days of starvation and remained at that level for 28 days (Novitsky & Morita 1977). If extrapolated to residence times of 20–30 years commonly observed for natural sea waters, this low endogenous rate is still too high provided no further input of organic matter occurs (Morita 1985). It has been questioned whether the energy of maintenance has any bearing on the long-term survival of marine microorganisms and if starved marine bacteria can enter a true dormant state (Morita 1982; 1985). Jones & Rhodes-Roberts (1981) stated that it was not possible to relate any metabolic properties or maintenance requirements to the remarkable capacity for prolonged survival under starvation conditions observed for two marine bacteria. As discussed in Section 12.3.4 a different approach must be used to explain the efficient adaptation of bacteria to nutrient-depleted conditions.

12.3.3 Starvation-induced physiological alterations

The initial phase of adaptation to exogenous nutrient deprivation is not that of a gradual decrease in metabolic rates. When studied in detail, a multiphasic pattern of metabolic activities was revealed in the marine bacterium *Vibrio* sp. S14 when exposed to multiple-nutrient starvation. Based on biochemical and 16S RNA sequencing analysis this strain resides in the core of *Vibrio* genus (Albertson *et al.* 1990a; Zabkar S. *et al.* in preparation). The immediate response of rapidly growing cells to the nutrient deprivation was a shutdown of RNA, protein, and peptidoglycan synthesis, and a pronounced increase in the rate of proteolysis (Figure 12.6). The signal molecule guanosine 5′-diphosphate 3′-diphosphate (ppGpp)

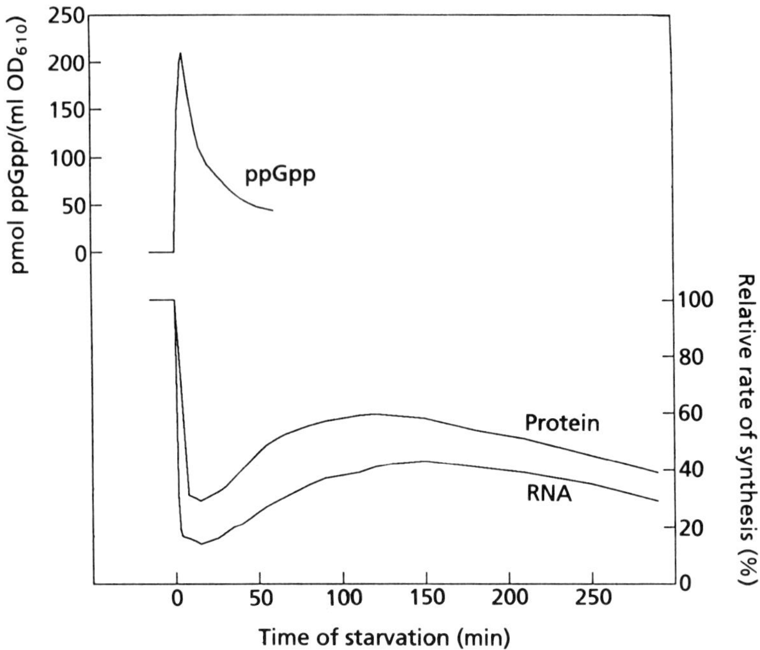

Figure 12.6 Occurrence of the stringent response at the onset of multiple-nutrient starvation in *Vibrio* sp. S14. Schematic presentation of the effect of multiple-nutrient starvation on the rates of total RNA and protein synthesis and the size of the intracellular pool of guanosine 5′-diphosphate 3′-diphosphate (ppGpp). Redrawn from Nyström *et al.* (1990b).

was rapidly accumulated at the onset of starvation, indicating that the stringent response was provoked (Figure 12.6). The stringent response network occurs in bacteria in response to amino acid or energy limitation (Neidhardt *et al.* 1990). After 30 min the stringent control seemed to be relieved as the ppGpp level decreased, and relative increases in the rates of macromolecular syntheses, as well as in the respiratory activity were observed. Subsequent to the first few hours of starvation, the rate of endogenous metabolism decreased gradually, eventually becoming undetectable (reviewed by Nyström *et al.* 1990a).

In an attempt to define a number of biological processes that contribute to the maintenance requirements during carbon-limited growth, Stouthamer *et al.* (1990) suggested that processes related to the stringent response, such as ppGpp formation, translational proofreading, and protein turnover, made significant contributions to the maintenance demand, and this was increasingly important at very slow growth. The initial phase of nongrowth may thus, in addition to the subsequent temporary increase in macromolecular synthesis, represent a period of relatively high energy costs, during which the degradation of polyhydroxybutyrate could serve as a carbon and energy source in *Vibrio* sp. S14 (Malmcrona-Friberg *et al.* 1986). The observed relatively high metabolic activities during the initial phase of the starvation response may be a reflection of an active reorganization of cell constituents, i.e., differentiation. Smigielski *et al.* (1990) recently demonstrated that the starvation response of *Vibrio fluvialis* is blocked when insufficient energy is available for cellular reorganization. When respiration was blocked, miniaturization was prevented and the viability during starvation was reduced. Further support for the idea of an active initial period of metabolic and structural reorganization was derived using inhibitors of the proton flow, the electron transport chain, and membrane-bound ATPase (Kjelleberg *et al.* 1983).

After the first period of starvation, cells enter a phase of long-term survival in which the rate of endogenous metabolism is low and there are small alterations in morphology and amounts of endogenous constituents. Energy-requiring processes are probably reduced to a minimum. One observation related to this is the increase in the mean mRNA half-life in energy- and nutrient-starved *Vibrio* S14, which was due both to an overall mRNA stabilization and the presence of extremely long-lived starvation-specific messages (Albertson *et al.* 1990c). The cellular composition of the reorganized cells is different from their growing counterparts. For example, drastic changes in fatty acid profiles, increased D-alanine content, and reduction of nucleoid volume have been reported for marine *Vibrio* strains, in addition to the alterations in protein set-up discussed in Section 12.3.4 (Guckert *et al.* 1986; Hood *et al.* 1986; Malmcrona-Friberg *et al.* 1986; Moyer & Morita 1989b). Volumes of intracellular water decreased from 1.02 to 0.52 ml mg^{-1} bacterial dry weight during 24 hour starvation of *V. fluvialis*, indicating that dehydration may be an important feature of starved bacterial cells (Smigielski *et al.* 1989). Strikingly, low water contents were also observed by Simon & Azam (1989) in bacterial cells harvested from sea water.

It is important to realize that starved bacteria, whether the period of nongrowth is short or long term, respond rapidly to nutrients. All open-ocean bacteria studied by Amy & Morita (1983a) were able to metabolize ^{14}C-glutamic acid when it was added at low concentrations to cultures that had been starved for up to 9 months. When nutrients are added to starved bacteria, growth and cell division resume, but only after a lag phase which increases with starvation time (Amy *et al.* 1983b; Albertson *et al.* 1990d). An ordered pattern of protein synthesis was observed during the lag or "maturation" phase, that possibly allows the cells to escape from the differentiated, starvation-survival state and regrow (Figure 12.7). Some starvation-specific proteins were repressed, some maturation-specific proteins

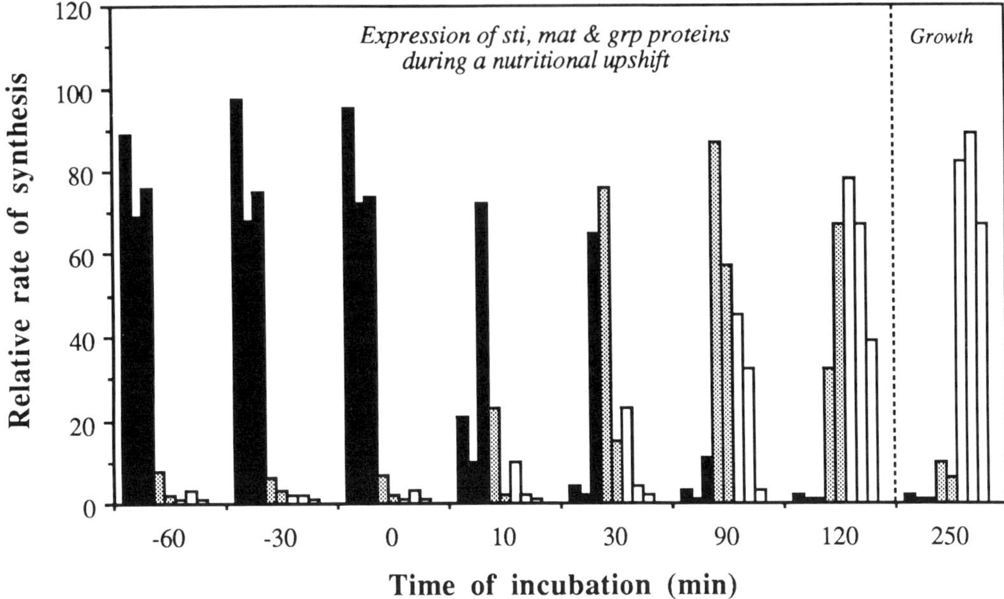

Figure 12.7 Relative rate of synthesis at 3 sti (starvation-inducible) proteins (■), 2 mat (maturation-specific) proteins (▨), and 3 grp ("growth-related") proteins (□) during starvation (−60, −30 min), maturation (0–120 min), and regrowth (250 min). Relative rates of synthesis of individual proteins were obtained by computer-assisted microdensitometry of two-dimensional autoradiograms. Cells were pulse-labeled for 10 min with ^{35}S-methionine at intervals during multiple-nutrient starvation and during recovery after addition of glucose minimal medium. From Nyström *et al.* (1990a).

were transiently expressed at different times during the lag phase, and some growth-related proteins, repressed during starvation, were resynthesized during or after the maturation period (Albertson *et al.* 1990d).

The ability to rapidly and efficiently utilize nutrient inputs may depend on the conservation of a functional protein synthesis machinery during starvation. Addition of glucose minimal medium after 24, 120, or 200 hr of energy and nutrient starvation of *Vibrio* S14 cells resulted within the first minute in a several-fold increase in the rate of protein synthesis (Albertson *et al.* 1990d). It was demonstrated that *de novo* RNA synthesis was not required for the initial rapid increase in protein synthesis, indicating that rRNA and ribosomes may exist in excess in the starved cells. In fact, 50% of the ribosomes present at the onset of carbon starvation are stable for at least 3 days in this organism, at which time the rate of protein synthesis is less than 1% of that recorded at the onset of starvation (Figure 12.8) (Flärdh *et al.* 1992). Slowly growing carbon-limited *E. coli* cells contained an excess of stable RNA that was not involved

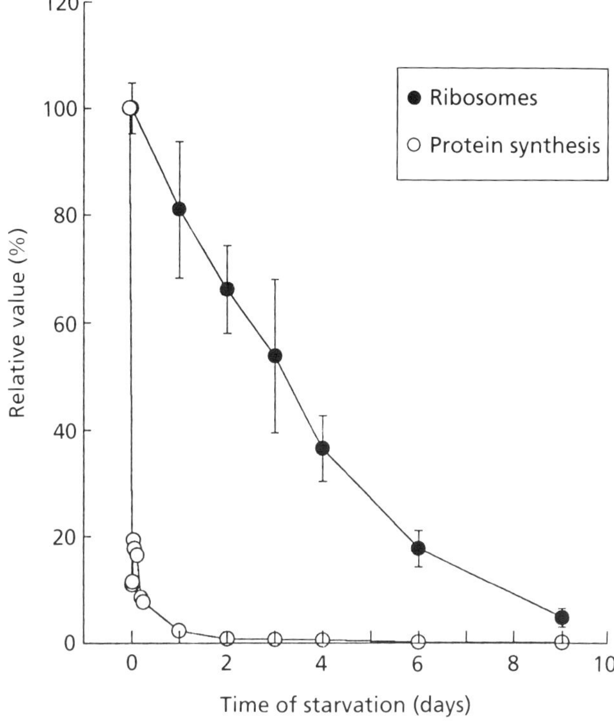

Figure 12.8 Relative amount of ribosomes and rate of protein synthesis during carbon starvation of *Vibrio* sp. S14. The amount of ribosomes was determined as the RNA in cell lysates that sedimented at 150 000 × **g**. The rate of protein synthesis was determined by pulse-labeling with ^{3}H-leucine. Redrawn from Flärdh *et al.* (1992).

in making proteins, but was suggested to increase the cells' ability to rapidly respond to sporadic inputs of nutrients (Koch 1971).

12.3.4 Do marine bacteria utilize a true differentiation program?

It has been established that unperturbed protein synthesis is a prerequisite for maximal viability of heterotrophic bacteria during starvation (Reeve *et al.* 1984a; Nyström *et al.* 1990b). Two-dimensional gel electrophoresis of pulse-labeled polypeptides revealed a sequential synthesis of proteins in response to either multiple or individual starvation conditions (Dow *et al.* 1983; Groat *et al.* 1986; Spector *et al.* 1986; Nyström *et al.* 1990b). Relatively few studies on alterations in protein composition or pattern of protein synthesis during starvation of marine bacteria have been reported (Amy & Morita 1983b; Jouper-Jaan *et al.* 1986; Nyström *et al.* 1988; 1990a,b; Smigielski *et al.* 1989). Taken together and in particular based on the rather extensive studies by Nyström *et al.* (1990a,b) on *Vibrio* S14, it is suggested that marine bacteria synthesize starvation-specific proteins that can be divided into several temporal classes with respect to their time of appearance during starvation. Notably, such sequential changes in the protein composition persisted during 120 hr in *Vibrio* S14 and new proteins were suggested to appear after 3 months of nongrowth of *Vibrio* sp. Ant 300 (Amy & Morita 1983b).

Considering the studies on *Vibrio* S14 as a potential model for the starvation-induced protein expression, several key features are apparent. The starvation-induced proteins could be divided into at least two categories: those transiently synthesized during starvation and those whose synthesis extended throughout the entire starvation period studied. Their synthesis may coincide with the generation of crossprotection against different stress conditions (Nyström *et al.* 1990a). The synthesis of some late starvation-induced proteins, in different late temporal classes, was dependent on protein synthesis during the initial stringent control phase (Nyström *et al.* 1990b). This is reminiscent of the sporulation process which, through sequentially ordered expression of development specific genes, leads to the generation of resistant spores (Losick & Kroos 1989). The proteins synthesized during the initial phase (4 hr), during which time 41 of a total of 66 starvation-induced proteins were synthesized, in particular, include those that may be essential for long-term survival.

Two-dimensional gel electrophoresis analysis allows for the identification of the proteins synthesized at the onset of different individual starvation and stress conditions. These results may be combined with data from physiological experiments on the phenotypic responses as well as the viability of individually starved or stressed cells, in order to elucidate different regulatory pathways of the starvation response. The individual starvation responses of *Vibrio* S14 have been studied in terms of survival and development of crossprotection against other stresses, and compared to the response when several nutrients are removed. It was revealed that carbon rather than nitrogen or phosphorus starvation is the determinant for the time-dependent development of miniaturization, extended survival and a resistant cell state (Nyström *et al.* 1992). Accordingly, the overlap between sets of proteins induced by various stress and individual starvation conditions was most pronounced for carbon starvation and individual stress inducers such as heat, nalidixic acid, or $CdCl_2$. Interestingly, it has been shown that marine bacteria in the subarctic Pacific are carbon and/or energy deprived rather than limited by nitrogen or phosphorus (Kirchman 1990).

Detailed molecular studies of carbon starvation genes of *Vibrio* S14 are aimed at identifying the subclass of essential survival genes. These studies encompass two main approaches. The genes are cloned via reverse genetics after identification of the corresponding protein on two-dimensional gel electrophoresis autoradiograms. The specific protein is extracted from the gel, N-terminal amino acid sequenced, and a synthetic oligonucleotide probe is made from the deduced DNA sequence. The second approach is based on the construction of a series of transposon insertion strains that are induced at different times of starvation (Östling *et al.* 1991). A suitable transposon carries a promotorless reporter gene whose product can be monitored. In this case, the expression of the reporter gene is directed by a starvation promoter. This allows for the cloning of starvation-specific genes by using transposon sequences as probes.

Although several events, strikingly similar to the sporulation of some true differentiating bacteria, mark the starvation-induced response in marine bacteria, there are obvious principal differences in the two pathways. Protein synthesis is detectable in marine vibrios even when they are starved for a long time (Nyström *et al.* 1990a; Flärdh *et al.* 1992). Furthermore, starved marine vibrios, like other heterotrophic bacteria, do not appear to be committed to form ultramicrocells. At any time during the differentiation process, nutrients suffice as activation agents. Starved cells respond to nutrient addition by an instantaneous increase in protein synthesis and cell size, and apparent repression of starvation-induced proteins (Albertson *et al.* 1990d). As stated above, these results support the notion of different physiological stages of growth and nongrowth in the life cycle of bacteria in natural environments.

An additional program displayed by several marine bacteria is the formation of a "viable but nonculturable" cell-state (Roszak & Colwell 1987). These ultramicrocells are metabolically active as measured by direct viable count techniques, but are in a nonreplicative mode and do not form colonies on solid media or grow in liquid media.

12.3.5 Possible identity and function of genes/proteins essential for the differentiation program

Evidently, a series of proteins and their functions may be suggested to be hallmarks of the differentiation program. Experimental evidence is generally lacking, however, and the following discussion on possible roles of starvation proteins reflects this paucity of data.

The stringent control appears to be essential for starvation survival of *E. coli* (Mach *et al.* 1989). The expression of starvation-induced genes that are stringent control-dependent was demonstrated in *Salmonella typhimurium* (Spector *et al.* 1988). The initial phase of the starvation response of *Vibrio* S14 includes a rapid accumulation of ppGpp. This organism was shown to develop starvation resistance to a greater extent when exposed to conditions that induced the stringent control, compared to conditions that did not (Figure 12.9) (Nyström *et al.* 1990b). Thus, it is possible that proteins that belong to the stringent control regulon are important for starvation survival.

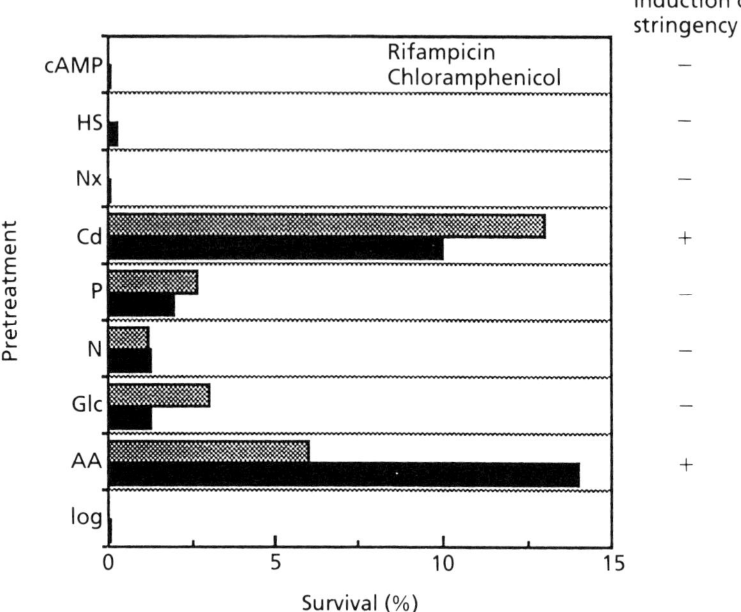

Figure 12.9 Correlation between induction of stringent control and acquisition of starvation resistance in *Vibrio* sp. S14. The effect of various pretreatments (1 hr) on the development of starvation resistance was studied by determination of the number of colony forming units during the subsequent multiple-nutrient starvation in the presence of chloramphenicol or rifampicin. Survival is expressed as viability after 48 hr of multiple-nutrient starvation relative to the viability at the onset of starvation. Induction of stringent control during the 1 hr pretreatment, as demonstrated by ppGpp accumulation and the kinetics of total RNA synthesis, is indicated. The pretreatments were: AA, amino acid starvation; Glc, glucose starvation; N, ammonium starvation; P, phosphate starvation; Cd, $CdCl_2$ exposure; Nx, nalidixic acid exposure; HS, heat shock (40°C); cAMP, addition of 5 $mmol\,l^{-1}$ cAMP; log, control culture without pretreatment. Redrawn from Nyström *et al.* (1990b).

Proteins produced during the first 4 hr of nutrient and energy depletion appear to be essential for cell integrity during periods of survival under starvation conditions (Nyström & Kjelleberg 1989). The peptidoglycan synthesized at the onset of starvation is resistant to cell wall-degrading enzymes compared to that produced during growth. *De novo* protein synthesis is necessary for this protection, making starved cells resistant to the activity of autolytic enzymes during nongrowth. An increased cellular resistance may also result from the increase in cell wall material during starvation. While peptidoglycan degradation is prevented by proteins synthesized during the first hour of starvation, cell wall synthesis continues at relatively high rates for many hours subsequent to the downshift. These alterations may enhance the predation resistance of small ultramicrocells (see discussions in Enger *et al.* 1990; Nyström *et al.* 1990a).

The initial phase of the starvation response includes a significant increase in the degree of proteolysis (Nyström *et al.* 1988). Proteases belonging to both stringent control-dependent and -independent regulons may be involved in this response (Nyström *et al.* 1990b). Peptidase-deficient mutants of *E. coli* and *S. typhimurium* lose viability during carbon starvation (Reeve *et al.* 1984b).

The induction of crossprotection against heat and the significant overlap between carbon starvation and heat shock proteins, point to an interesting role for members of this stress regulon (Nyström *et al.* 1990b; Nyström *et al.* 1992). Specifically, a 70 kDa protein, that is recognized by antibodies raised against the *E. coli* DnaK protein, is induced after the initial phase of starvation in several marine vibrios and is subsequently produced at increasingly greater rates of synthesis during a period of 48 hr of energy and nutrient starvation (L. Holmqvist, A. Jouper-Jaan, D. Weichart, D. Nelson, and S. Kjelleberg in preparation). Recently, several interesting properties of DnaK have been discovered. It has been shown to bind to unfolded peptides, which among other things may improve protein export, enhance degradation by proteases, prevent misfolding, and accelerate correct protein folding (Ellis & Hemmingson 1989; Phillips & Silhavy 1990; Skowyra *et al.* 1990). Obviously, all these processes would be essential for efficient adaptation of starving cells. For example, the DnaK protein may assist in degrading growth-related proteins or other proteins that are deleterious to any given phase of the starvation program. The export of exoenzymes (see Section 12.3.6) during starvation could be facilitated by DnaK. The DnaK protein affects the synthesis of carbon starvation and Pex ("postexponential") proteins in *E. coli* (Spence *et al.* 1990). The induction of these proteins has been inferred to be essential for starvation survival in *E. coli* (Matin 1990). Furthermore, a DnaK mutant was found to be highly susceptible to killing by carbon starvation (Spence *et al.* 1990). DnaK is suggested to belong to the class of macrophage-induced proteins that confer phagocytosis resistance in *S. typhimurium* (Weiss 1990). It will be interesting to find out whether protection of nongrowing marine bacteria against degrading enzymes inside predators, is offered by stress proteins such as DnaK. It is expected that several essential stationary phase and starvation genes and their products will be identified in the near future. For example, in *E. coli* a new sigma factor, RpoS, has been identified as essential for induction of a series of stationary phase regulated genes and survival during starvation (Lange & Hengge-Aronis 1991). The regulatory molecule of the heat shock regulon, Sigma 32, has also been suggested to be involved in enhanced starvation survival and even protection against other stress inducers during starvation (Jenkins *et al.* 1991).

12.3.6 Recovery-related functions of the differentiation program

The active state of starved marine bacteria, including nonreplicating cells, is evidenced by the wide range of recovery-related functions of starvation-induced proteins. Several sophisticated means of enhancing the ability for substrate scavenging are employed at different times after the induction of the starvation response.

In striking similarity to the sporulation response of some true differentiating bacteria, an increased expression and excretion of exoproteases has been implicated to coincide with the initial phase of starvation (Albertson *et al.* 1990a). Similarly, the induction of exoproteases during stationary rather than log phase is known for several marine strains. Clearly, exoenzyme synthesis occurs during prolonged periods of starvation of several marine bacterial strains and the induction of such enzymes, both carbohydrate-degrading enzymes and exoproteases, by addition of the polymer substrate, takes place even after extended starvation periods (Davis 1992). Bacteria may clearly play a role in utilization of polymeric substrates in the marine environment, even after long periods of nutrient deprivation. This would be of particular relevance in the vicinity of adsorbed macromolecules (see Chapter 13).

Starved bacteria take up small compounds, released by hydrolysis of polymers, very efficiently. The presence of high affinity uptake systems would allow bacteria to scavenge solutes at concentrations that do not promote growth, but may be used for maintenance requirements. The binding and transport of the amino acid arginine by *Vibrio* sp. Ant 300, found to be maintained during prolonged starvation, have been studied in detail (Geesey & Morita 1979; 1981; Faquin & Oliver 1984). The induction of proteins mediating such uptake has also been reported. Mårdén *et al.* (1987) noted that *Vibrio* S14 cells subjected to osmotic shock for release of the periplasmic content regained the capacity for binding-protein-dependent leucine uptake after prolonged starvation. The high affinity binding protein in the periplasmic space was identified and shown to be induced during the initial phase of starvation. Davis & Robb (1985) found that *de novo* protein synthesis was required before mannitol was transported into starved marine *Pseudomonas*. After being starved for 72 hr, glucose uptake in *Vibro* S14 switches from dependence on a low- to high-affinity binding protein (Table 12.1) (Albertson *et al.* 1990b). In two *Vibrio* strains the high affinity uptake of glutamic acid appeared after 3 and 5 weeks (Geesey & Morita 1979; Davis 1985).

Apparently, chemotactic responses during starvation may direct the cells towards microzones with a high concentration of dissolved organic carbon compounds. Arginine is clearly an attractant for *Vibrio* Ant 300 starved for a long time; the positive chemotactic

Table 12.1 Binding affinity for glucose uptake during 168 hr of multiple-nutrient starvation of *Vibrio* sp. S14. Apparent K_m values were determined graphically by direct linear plots and Lineweaver–Burk plots. From Albertson *et al.* (1990b)

Time of starvation (hr)	Apparent K_m (μmol l^{-1})
0	4.6
2	4.0
4	3.8
24	4.0
72	3.6
120	0.56
144	0.60
168	0.55

response is induced after 21 days multiple starvation (Torrella & Morita 1982). Modulations in the chemotactic response to glucose and several amino acids by *Vibrio* S14 were reported to take place during the initial 24 hr of starvation (Malmcrona-Friberg *et al.* 1990). Mannitol has been shown to be a chemotactic attractant for *Vibrio* and *Pseudomonas* strains studied by Davis (1985). A pronounced chemotactic response by *Vibrio* S14 cells starved for a short time was found to exist towards a gradient of breakdown products generated by a *Flavobacterium* sp. immobilized in alginate agar beads (Malmcrona-Friberg *et al.* 1991). In several of these reports it is suggested that the chemosensory system is linked to that of uptake and transport. Binding proteins may clearly serve also as chemotactic receptors. The starvation-induced alterations in cell surface characteristics that allow for an increase in the ability of starved cells to adhere to different substrata (see Section 12.3.1) obviously also improves the utilization of surface-localized material (Marshall 1989). Thus, recovery-related proteins may alter the capacity by starved cells for adhesion, chemotaxis, uptake, and transport, as well as synthesis and excretion of exoenzymes.

12.4 Conclusions

Heterotrophic marine bacteria are frequently depleted of carbon and energy in their natural habitat (Morita 1985; Wheeler & Kirchman 1986; Kirchman 1990; Lewis & Gattie 1990). At any given time, the observed bacterial production is likely to be mediated by a limited number of species, i.e., by a small fraction of the highly diverse bacterioplankton community (Å. Hagström personal communication, 1991). While most bacteria therefore include phases of non-growth in their life-cycles (Roszak & Colwell 1987), relatively little is known about the physiology and genetics of the starvation-survival response.

The evidence available to date infer the existence of differentiation programs during nongrowth as well as during upshift from the starvation-survival state to growth conditions. A morphogenetic life-cycle has been observed during these growth-limiting and growth-promoting conditions and specific genes are expressed in an orderly fashion during the different phases of the physiological down- and upshift states. Nondifferentiating bacteria respond immediately to nutritional downshifts and starvation by synthesizing new proteins that assist the cells to survive by making them more efficient in taking up and using organic compounds at very low concentrations. Furthermore, alterations to the cell structure make the cells more resistant to various forms of stress and possibly predation. The starved, nongrowing cells are metabolically active and are primed for rapid recovery when nutrients become available. They participate in carbon cycling in the ocean, i.e., they are not dormant and the nongrowing state is a regular state of the natural life-cycle of bacteria.

Information on the identity and function of starvation- and recovery-specific proteins as well as an understanding of the physiology of starved cells will aid in explaining the nature of the differentiation program, the development of a starvation- and stress-resistant cell, and the ecological adaptation of bacteria in their natural habitat. Based on improved molecular and physiological information we will be in a better position to suggest methodological approaches to study the physiological state(s) of mixed natural populations. This will hopefully enable us to enter the next phase of marine microbial ecology in which the diverse physiological responses to any given situation can be addressed, rather than the net response by the bacterial community.

Acknowledgments

The authors are grateful to the editor, Dr Tim Ford, for active and cheerful support during the writing of this chapter. Very useful criticisms and suggestions were provided by Drs Hugh Ducklow and Kevin Marshall, as well as an anonymous reviewer. The stimulating discussions with Hugh Ducklow on various aspects on bacteria and growth-limiting conditions proved particularly useful in preparing this chapter.

References

Albertson N.H., Jones G.W. & Kjelleberg S. (1987) The detection of starvation-specific antigens in two marine bacteria. *J. Gen. Microbiol.* **133**, 2225–2231.

Albertson N.H., Nyström T. & Kjelleberg S. (1990a) Synthesis and regulation of exoproteases during the starvation of two marine bacteria. *Appl. Environ. Microbiol.* **56**, 218–223.

Albertson N.H., Nyström T. & Kjelleberg S. (1990b) Starvation-induced modulations in binding protein-dependent glucose transport by the marine *Vibrio* sp. S14. *FEMS Microbiol. Lett.* **70**, 205–210.

Albertson N.H., Nyström T. & Kjelleberg S. (1990c) Functional mRNA half-lives in the marine *Vibrio* sp. S14 during starvation and recovery. *J. Gen. Microbiol.* **136**, 2195–2199.

Albertson N.H., Nyström T. & Kjelleberg S. (1990d) Macromolecular synthesis during recovery of the marine *Vibrio* sp. S14. *J. Gen. Microbiol.* **136**, 2201–2207.

Alongi D.M. (1989) The fate of bacterial biomass and production in marine benthic food chains. In Hattori T., Ishida Y., Maruyama Y., Morita R.Y. & Uchida A. (eds) *Recent Advances in Microbial Ecology*, pp. 355–359. Japan Scientific Societies Press, Tokyo.

Amy P.S. & Morita R.Y. (1983a) Starvation-survival patterns of sixteen freshly isolated open-ocean bacteria. *Appl. Environ. Microbiol.* **45**, 1109–1115.

Amy P.S. & Morita R.Y. (1983b) Protein patterns of growing and starved cells of a marine *Vibrio* sp. *Appl. Environ. Microbiol.* **45**, 1748–1752.

Amy P.S., Pauling C. & Morita R.Y. (1983a) Starvation-survival processes of a marine *Vibrio*. *Appl. Environ. Microbiol.* **45**, 1041–1048.

Amy P.S., Pauling C. & Morita R.Y. (1983b) Recovery from nutrient starvation by a marine *Vibrio* sp. *Appl. Environ. Microbiol.* **45**, 1685–1690.

Billen G. (1990) Delayed development of bacterioplankton with respect to phytoplankton: a clue for understanding their trophic relationships. *Arch. Hydrobiol. Beih. Ergebn. Limnol.* **34**, 191–201.

Billen G., Servais P. & Fontigny A. (1988) Growth and mortality in bacterial population dynamics of aquatic environments. *Arch. Hydrobiol. Beih. Ergebn. Limnol.* **31**, 173–183.

Bird D.F. & Kalff J. (1984) Empirical relationships between bacterial abundance and chlorophyll concentration in fresh and marine waters. *Can. J. Fish. Aquat. Sci.* **41**, 1015–1023.

Chapman S.J. & Gray T.R.G. (1981) Endogenous metabolism and macromolecular composition of *Arthrobacter globiformis*. *Soil Biol. Biochem.* **13**, 11–18.

Chin-Leo G. & Kirchman D.L. (1990) Unbalanced growth in natural assemblages of marine bacterioplankton. *Mar. Ecol. Prog. Ser.* **63**, 1–8.

Cho B.C. & Azam F. (1988) Major role of bacteria in biogeochemical fluxes in the ocean's interior. *Nature* **332**, 441–443.

Cho B.C. & Azam F. (1990) Biogeochemical significance of bacterial biomass in the ocean's euphotic zone. *Mar. Ecol. Prog. Ser.* **63**, 253–259.

Cole J.J., Findlay S. & Pace M.L. (1988) Bacterial production in fresh and saltwater ecosystems: a cross-system overview. *Mar. Ecol. Prog. Ser.* **43**, 1–10.

Davis C.L. (1985) *Physiological and Ecological Studies of Mannitol Utilizing Marine Bacteria*. PhD thesis, University of Cape Town.

Davis C.L. (1992) Production of laminarinase and alginase by marine bacteria after starvation. *FEMS Microbiol. Ecol.* **86**, 349–356.

Davis C.L. & Robb F.T. (1985) Maintenance of different mannitol uptake systems during starvation of oxidative and fermentative marine bacteria. *Appl. Environ. Microbiol.* **50**, 743–748.

Dawes E.A. (1976) Endogenous metabolism and the survival of starved prokaryotes. In Gray T.R.G. & Postgate J.R. (eds) *The Survival of Vegetative Microbes*, pp. 19–53. Cambridge University Press, Cambridge.

Dawes E.A. (1985) Starvation, survival and energy reserves. In Fletcher M. & Floodgate G.D. (eds) *Bacteria in their Natural Environments*, pp. 43–79. Academic Press, London.

Dawes E.A. & Ribbons D.W. (1964) Some aspects of the endogenous metabolism of bacteria. *Bacteriol Rev.* **28**, 126–149.

Dawes E.A. & Senior P.J. (1973) The role and regulation of energy reserve polymers in micro-organisms. *Adv. Microb. Physiol.* **10**, 135–266.

Dawson M.P., Humphrey B.A. & Marshall K.C. (1981) Adhesion: a tactic in the survival strategy of a marine *Vibrio* during starvation. *Curr. Microbiol.* **6**, 195–199.

Donachie W.D. & Begg K.J. (1970) Growth of the bacterial cell. *Nature* **277**, 1220–1224.

Dow C.S. & Whittenbury R. (1980) Procaryotic form and function. In Ellwood D.C., Hedger J.N., Latham M.J., Lynch J.M. & Slater J.H. (eds) *Contemporary Microbial Ecology*, pp. 391–417. Academic Press, London.

Dow C.S., Whittenbury R. & Carr N.G. (1983) The "shut down" or "growth precursor" cell – an adaptation for survival in a potentially hostile environment. In Slater J.H., Wittenbury R. & Wimpenny J.W.T. (eds) *Microbes in their Natural Environments. Society for General Microbiology Symposium*, **34**, 187–247. Cambridge University Press, Cambridge.

Ducklow H.W. (1982) Chesapeake Bay nutrient and plankton dynamics. I. Bacterial biomass and production during spring tidal destratification in the York River, Virginia, Estuary. *Limnol. Oceanogr.* **27**, 651–659.

Ducklow H.W. (1984) Geographical ecology of marine bacteria: physical and biological variability at the mesoscale. In Klug M.J. & Reddy C.A. (eds) *Current Perspectives in Microbial Ecology*, pp. 22–31. American Society for Microbiology, Washington DC.

Ellis R.J. & Hemmingson S.M. (1989) Molecular chaperones: proteins essential for the biogenesis of some macromolecular structures. *Trends Biochem. Sci.* **14**, 339–342.

Enger Ö., Hoff K.A., Schei G.H. & Dundas I. (1990) Starvation survival of the fish pathogenic bacteria *Vibrio anguillarum* and *Vibrio salmonicida* in marine environments. *FEMS Microbiol. Ecol.* **74**, 215–220.

Faquin W.C. & Oliver J.D. (1984) Arginine uptake by a psychrophilic marine *Vibrio* sp. during starvation-induced morphogenesis. *J. Gen. Microbiol.* **130**, 1331–1335.

Flärdh K., Cohen P.S. & Kjelleberg S. (1992) Ribosomes exist in large excess over the apparent demand for protein synthesis during carbon starvation in the marine *Vibrio* sp. strain CCUG 15956. *J. Bacteriol.* **174**, (November issue).

Fuhrman J.A., Sleeter T.D., Carlson C.A. & Proctor L.M. (1989) Dominance of bacterial biomass in the Sargasso Sea and its ecological implications. *Mar. Ecol. Prog. Ser.* **57**, 207–217.

Geesey G.G. & Morita R.Y. (1979) Capture of arginine at low concentrations by a marine psychrophilic bacterium. *Appl. Environ. Microbiol.* **38**, 1092–1097.

Geesey G.G. & Morita R.Y. (1981) Relationship of cell envelope stability to substrate capture in a marine psychrophilic bacterium. *Appl. Environ. Microbiol.* **42**, 533–540.

Giovannoni S.J., Brischgi T.B., Moyer C.L. & Field K.G. (1990) Genetic diversity in Sargasso Sea bacterioplankton. *Nature* **345**, 60–63.

Gray T.R.G. (1976) Survival of vegetative microbes in soil. In Gray T.R.G. & Postgate J.R. (eds) *The Survival of Vegetative Microbes*, pp. 327–364. Cambridge University Press, Cambridge.

Groat R.G., Schultz J.E., Zychlinsky E., Bockman A. & Matin A. (1986) Starvation proteins in *Escherichia coli*: kinetics of synthesis and role in starvation survival. *J. Bacteriol.* **168**, 486–493.

Guckert J.B., Hood M.A. & White D.C. (1986) Phospholipid ester-linked fatty acid profile changes during nutrient deprivation of *Vibrio cholerae*: increases in the *trans/cis* ratio and proportions of cyclopropyl fatty acids. *Appl. Environ. Microbiol.* **52**, 794–801.

Heldal M. & Bratbak G. (1991) Production and decay of viruses in aquatic environments. *Mar. Ecol. Prog. Ser.* **72**, 205–212.

Hermansson M., Jones G.W. & Kjelleberg S. (1987) Frequency of antibiotic and heavy metal resistance, pigmentation and plasmids in bacteria of the marine air–water interface. *Appl. Environ. Microbiol.* **53**, 2338–2342.

Hood M.A., Guckert J.B., White D.C. & Deck F. (1986) Effect of nutrient deprivation on lipid, carbohydrate, DNA, RNA, and protein levels in *Vibrio cholerae. Appl. Environ. Microbiol.* **52**, 788–793.

Jenkins D.E., Auger E.A. & Matin A. (1991) Role of RpoH, a heat shock regulator protein in *Escherichia coli* carbon starvation protein synthesis and survival. *J. Bacteriol.* **173**, 1992–1996.

Joint I.R. & Pomroy A.J. (1987) Activity of heterotrophic bacteria in the euphotic zone of the Celtic Sea. *Mar. Ecol. Prog. Ser.* **41**, 155–165.

Jones K.L. & Rhodes-Roberts M.E. (1981) The survival of marine bacteria under starvation conditions. *J. Appl. Bacteriol.* **50**, 247–258.

Jouper-Jaan Å., Dahlöf B. & Kjelleberg S. (1986) Changes in the protein composition of three bacterial isolates from marine waters during short term energy and nutrient deprivation. *Appl. Environ. Microbiol.* **52**, 1419–1421.

Kirchman D. (1983) The production of bacteria attached to particles suspended in a freshwater pond. *Limnol. Oceanogr.* **28**, 858–872.

Kirchman D.L. (1990) Limitation of bacterial growth by dissolved organic matter in the subarctic Pacific. *Mar. Ecol. Prog. Ser.* **62**, 47–54.

Kirchman D.L., Suzuki S., Garside C. & Ducklow H.W. (1991) High turnover rates of dissolved organic carbon during a spring phytoplankton bloom. *Nature* **352**, 612–614.

Kjelleberg S. & Hermansson M. (1984) Starvation induced effects on bacterial surface characteristics. *Appl. Environ. Microbiol.* **48**, 497–503.

Kjelleberg S., Humphrey B.A. & Marshall K.C. (1983) Initial phases of starvation and activity of bacteria at surfaces. *Appl. Environ. Microbiol.* **46**, 978–984.

Kjelleberg S., Hermansson M., Mårdén P. & Jones G.W. (1987) The transient phase between growth and non-growth of heterotrophic bacteria with emphasis on the marine environment. *Ann. Rev. Microbiol.* **41**, 25–49.

Koch A.L. (1971) The adaptive response of *Escherichia coli* to a feast and famine existence. *Adv. Microbiol. Physiol.* **6**, 147–217.

Lancelot C. & Billen G. (1984) Activity of heterotrophic bacteria and its coupling to primary production during the spring phytoplankton bloom in the southern bight of the North Sea. *Limnol. Oceanogr.* **29**, 721–730.

Lange R. & Hengge-Aronis R. (1991) Identification of a central regulator of stationary-phase gene expression in *Escherichia coli. Mol. Microbiol.* **5**, 49–59.

Lee S. & Fuhrman J.A. (1990) DNA hybridization to compare species compositions of natural bacterioplankton assemblages. *Appl. Environ. Microbiol.* **56**, 739–746.

Lewis D.L. & Gattie D.K. (1990) Effects of cellular aggregation on the ecology of microorganisms. *ASM News* **56**, 263–268.

Losick R. & Kroos L. (1989) Dependence pathways for the expression of genes involved in endospore formation in *Bacillus subtilis.* In Smith I., Slepecky R.A. & Setlow P. (eds) *Regulation of Procaryotic Development*, pp. 223–241. American Society for Microbiology, Washington DC.

Mach H., Hecker M., Hill I., Schroeter A. & Mach F. (1989) Stringent control and starvation survival in *Escherichia coli. Z. Naturforsch.* **44**, 838–844.

Malmcrona-Friberg K., Tunlid A., Mårdén P., Kjelleberg S. & Odham G. (1986) Chemical changes in cell envelope and poly-β-hydroxybutyrate during short term starvation of a marine bacterial isolate. *Arch. Microbiol.* **144**, 340–345.

Malmcrona-Friberg K., Goodman A. & Kjelleberg S. (1990) Chemotactic responses of marine *Vibrio* sp. strain S14 (CCUG 15956) to low-molecular-weight substances under starvation and recovery conditions. *Appl. Environ. Microbiol.* **56**, 3699–3704.

Malmcrona-Friberg K., Blainey B.L. & Marshall K.C. (1991) Chemotactic response of a marine bacterium to breakdown products of an insoluble substrate. *FEMS Microbiol. Ecol.* **85**, 199–206.

Mårdén P., Tunlid A., Malmcrona-Friberg K., Odham G. & Kjelleberg S. (1985) Physiological and morphological changes during short term starvation of marine bacterial isolates. *Arch. Microbiol.* **142**, 326–332.

Mårdén P., Nyström T. & Kjelleberg S. (1987) Uptake of leucine by a marine

Gram-negative heterotrophic bacterium during exposure to starvation conditions. *FEMS Microbiol. Ecol.* **45**, 233–241.

Marshall N.B. (1979) *Developments in Deep-Sea Biology*. Blandford Press, Poole.

Marshall K.C. (1989) Growth of bacteria on surface-bound substrates: significance in biofilm development. In Hattori T., Ishida Y., Maruyama Y., Morita R.Y. & Uchida A. (eds) *Recent Advances in Microbial Ecology*, pp. 146–150. Japan Scientific Societies Press, Tokyo.

Matin A. (1990) Molecular analysis of the starvation stress in *Escherichia coli*. *FEMS Microbiol. Ecol.* **74**, 185–196.

Moriarty D.J.W. (1986) Measurement of bacterial growth rates in aquatic systems from rates of nucleic acid synthesis. *Adv. Microbiol. Ecol.* **9**, 245–292.

Moriarty D.J.W. (1989) Relationships of bacterial biomass and production to primary production in marine sediments. In Hattori T., Ishida Y., Maruyama Y., Morita R.Y. & Uchida A. (eds) *Recent Advances in Microbial Ecology*, pp. 349–354. Japan Scientific Societies Press, Tokyo.

Moriarty D.J.W. (1990) Interactions of microorganisms and aquatic animals, particularly the nutritional role of the gut flora. In Lésel R. (ed.) *Microbiology in Poecilotherms*, pp. 217–222. Elsevier, Amsterdam.

Moriarty D.J.W. & Pollard P.C. (1982) Diel variation of bacterial productivity in seagrass (*Zostera capricorni*) beds measured by rate of thymidine incorporation into DNA. *Mar. Biol.* **72**, 165–173.

Moriarty D.J.W. & Hansen J.A. (1990) Productivity and growth rates of coral reef bacteria on hard calcareous substrates and in sandy sediments in summer. *Aust. J. Mar. Freshwat. Res.* **41**, 785–794.

Moriarty D.J.W., Pollard P.C., Hunt W.G., Moriarty C.M. & Wassenberg T.J. (1985) Productivity of bacteria and microalgae and the effect of grazing by holothurians in sediments on a coral reef flat. *Mar. Biol.* **85**, 293–300.

Moriarty D.J.W., Roberts D.G. & Pollard P.C. (1990) Primary and bacterial productivity of tropical seagrass communities in the Gulf of Carpentaria, Australia. *Mar. Ecol. Prog. Ser.* **61**, 145–157.

Moriarty D.J.W., Syring G.W., O'Brien G.W. & Heggie D.T. (1991) Heterotrophic bacterial activity and growth rates in sediments of the continental margin of eastern Australia. *Deep-Sea Res.* **38**, 693–712.

Morita R.Y. (1982) Starvation-survival of heterotrophs in the marine environment. *Adv. Microbiol. Ecol.* **6**, 171–178.

Morita R.Y. (1985) Starvation and miniaturisation of heterotrophs, with special emphasis on maintenance of the starved viable state. In Fletcher M.M. & Floodgate G.D. (eds) *Bacteria in Their Natural Environments*, pp. 111–130. Academic Press, London.

Moyer C.L. & Morita R.Y. (1989a) Effect of growth rate and starvation-survival on the viability and stability of a psychrophilic marine bacterium. *Appl. Environ. Microbiol.* **55**, 1122–1127.

Moyer C.L. & Morita R.Y. (1989b) Effect of growth rate and starvation-survival on cellular DNA, RNA and protein of a psychrophilic marine bacterium. *Appl. Environ. Microbiol.* **55**, 2710–2716.

Neidhardt F.C., Ingraham J.L. & Schaechter M. (1990) *Physiology of the Bacterial Cell*. Sinauer Associates, Sunderland.

Nissen H. (1987) Long term starvation of a marine bacterium, *Alteromonas denitrificans*, isolated from a Norwegian fjord. *FEMS Microbiol. Ecol.* **45**, 173–184.

Novitsky J.A. & Morita R.Y. (1976) Morphological characterization of small cells resulting from nutrient starvation of a psychrophilic marine vibrio. *Appl. Environ. Microbiol.* **32**, 617–622.

Novitsky J.A. & Morita R.Y. (1977) Survival of a psychrophilic marine *Vibrio* under long-term nutrient starvation. *Appl. Environ. Microbiol.* **33**, 635–641.

Nyström T. & Kjelleberg S. (1989) Role of protein synthesis in the cell division and starvation induced resistance to autolysis of a marine *Vibrio* during the initial phase of starvation. *J. Gen. Microbiol.* **135**, 1599–1606.

Nyström T., Albertson N.H. & Kjelleberg S. (1988) Synthesis of membrane and periplasmic proteins during starvation of a marine *Vibrio* sp. *J. Gen. Microbiol.* **134**, 1645–1651.

Nyström T., Albertson N.H., Flärdh K. & Kjelleberg S. (1990a) Physiological and molecular adaptation to starvation and recovery from starvation by the marine *Vibrio* sp. S14. *FEMS Microbiol. Ecol.* **74**, 129–140.

Nyström T., Flärdh K. & Kjelleberg S. (1990b) Responses to multiple-nutrient starvation in marine *Vibrio* sp. strain CCUG 15956. *J. Bacteriol.* **172**, 7085–7097.

Nyström T., Olsson R-M. & Kjelleberg S. (1992) Survival, stress resistance, and alterations in protein expression in the marine *Vibrio* sp. strain S14 during starvation for different individual nutrients. *Appl. Environ. Microbiol.* **58**, 55–65.

Oliver J.D. & Stringer W.F. (1984) Lipid composition of a psychrophilic marine *Vibrio* sp. during starvation-induced morphogenesis. *Appl. Environ. Microbiol.* **47**, 461–466.

Östling J., Goodman A.E. & Kjelleberg S. (1991) Behaviour of IncP-1 plasmids and a miniMu transposon in a marine *Vibrio* sp.: isolation of starvation inducible *lac* operon fusions. *FEMS Microbiol. Ecol.* **86**: 83–94.

Painting S.J., Lucas M.I. & Muir D.G. (1989) Fluctuations in heterotrophic bacterial community structure, activity and production in response to development and decay of phytoplankton in a microcosm. *Mar. Ecol. Prog. Ser.* **53**, 129–141.

Phillips G.J. & Silhavy T.J. (1990) Heat-shock proteins DnaK and GroEL facilitate export of LacZ hybrid proteins in *E. coli*. *Nature* **344**, 882–884.

Pollard P.C. & Moriarty D.J.W. (1991) Organic carbon decomposition, primary and bacterial productivity, and sulphate reduction, in tropical seagrass beds of the Gulf of Carpentaria, Australia. *Mar. Ecol. Prog. Ser.* **69**, 149–159.

Preiss J. (1984) Bacterial glycogen synthesis and its regulation. *Ann. Rev. Microbiol.* **38**, 419–458.

Reeve C.A., Amy P.S. & Matin A. (1984a) Role of protein synthesis in the survival of carbon-starved *Escherichia coli* K-12. *J. Bacteriol.* **160**, 1041–1046.

Reeve C.A., Bockman A.T. & Matin A. (1984b) Role of protein degradation in the survival of carbon-starved *Escherichia coli* and *Salmonella typhimurium*. *J. Bacteriol.* **157**, 758–763.

Riemann B. & Bell R.T. (1990) Advances in estimating bacterial biomass and growth in aquatic systems. *Arch. Hydrobiol.* **118**, 385–402.

Roszak D.B. & Colwell R.R. (1987) Survival strategies of bacteria in the natural environment. *Microbiol. Rev.* **51**, 365–379.

Schmidt T.M., DeLong E.F. & Pace N.R. (1991) Analysis of a marine picoplankton community by 16S rRNA gene cloning and sequencing. *J. Bacteriol.* **173**, 4371–4378.

Sharp J.H. (1981) Inputs into microbial food chains. In Hobbie J.E. & Williams P.J. Le B. (eds) *Heterotrophic Activity in the Sea*, pp. 101–120. Plenum, New York.

Sieburth J.McN. (1979) *Sea Microbes*. Oxford University Press, New York.

Simon M. & Azam F. (1989) Protein content and protein synthesis rates of planktonic marine bacteria. *Mar. Ecol. Prog. Ser.* **51**, 201–213.

Simon M., Alldredge A.L. & Azam F. (1990) Bacterial carbon dynamics on marine snow. *Mar. Ecol. Prog. Ser.* **65**, 205–211.

Skowyra D., Georgopoulos C. & Zylicz M. (1990) The *E. coli* DnaK gene product, the hsp70 homologue, can reactivate heat-inactivated RNA polymerase in an ATP hydrolysis-dependent manner. *Cell* **62**, 939–944.

Smigielski A.J., Wallace B.J. & Marshall K.C. (1989) Changes in membrane functions during short term starvation of *Vibrio fluvialis* strain NCTC 11328. *Arch. Microbiol.* **151**, 336–341.

Smigielski A.J., Wallace B.J., Abrahams S. & Marshall K.C. (1990) Effects of respiratory activity on starvation survival of marine vibrios. *Arch. Microbiol.* **153**, 175–180.

Sochard M.R., Wilson D.F., Austin B. & Colwell R.R. (1979) Bacteria associated with the surface and gut of marine copepods. *Appl. Environ. Microbiol.* 37, 750–759.

Sorokin Y.I., Koppylov A.I. & Mamaeva N.V. (1985) Abundance and dynamics of microplankton in the central tropical Indian Ocean. *Mar. Ecol. Prog. Ser.* **24**, 27–41.

Spector M.P., Aliabadi Z., Gonzalez T & Foster J.W. (1986) Global control in *Salmonella typhimurium*: two-dimensional electrophoretic analysis of starvation-, anaerobiosis-, and heat shock-inducible proteins. *J. Bacteriol.* **168**, 420–424.

Spector M.P., Park Y.K., Tirgari S., Gonzalez T. & Foster J.W. (1988) Identification and characterization of starvation-regulated genetic loci in *Salmonella typhimurium* by using Mu d-directed *lacZ* operon fusions. *J. Bacteriol.* **170**, 345–351.

Spence J., Cegielska A. & Georgopoulos C. (1990) Role of *Escherichia coli* heat shock proteins DnaK and HtpG (C62.5) in response to nutritional deprivation. *J. Bacteriol.* **172**, 7157–7166.

Stouthamer A.H., Bulthuis B.A. & Van Verseveld H.W. (1990) Energetics of growth at low growth rates and its relevance for the maintenance concept. In Poole R.K., Bazin M.J. & Keevil C.W. (eds) *Microbial Growth Dynamics*, pp. 85–102. IRL Press, Oxford.

Tabor P.S., Ohwada K. & Colwell R.R. (1981) Filterable marine bacteria found in the deep sea: distribution, taxonomy, and response to starvation. *Microb. Ecol.* 7, 67–83.

Thomas T.D. & Batt R.D. (1969) Degradation of cell constituents by starved *Streptococcus lactis* in relation to survival. *J. Gen. Microbiol.* **58**, 347–362.

Torrella F. & Morita R.Y. (1981) Microcultural study of bacterial size changes and microcolony and ultramicrocolony formation by heterotrophic bacteria in seawater. *Appl. Environ. Microbiol.* **41**, 518–527.

Torrella F. & Morita R.Y. (1982) Starvation induced morphological changes, motility and chemotaxis patterns in a psychrophilic marine *Vibrio*. *Deuxieme Colloque de Microbiologie Marine. Pub. de Centre National pour l'exploitation des Oceans.* **13**, 45–60.

van Duyl F.C. & Kop A.J. (1988) Temporal and lateral fluctuations in production and biomass of bacterioplankton in the western Dutch Wadden Sea. *Neth. J. Sea Res.* **22**, 51–68.

van Duyl F.C., Bak R.P.M., Kop A.J. & Nieuwland G. (1990) Bacteria, auto- and heterotrophic nanoflagellates, and their relations in mixed, frontal and stratified waters of the North Sea. *Neth. J. Sea Res.* **26**, 97–109.

Weiss R. (1990) Making the most of MIPs. *Science News* **137**, 394–395.

Wheeler P.A. & Kirchman D.L. (1986) Utilization of inorganic and organic nitrogen by bacteria in marine systems. *Limnol. Oceanogr.* **31**, 998–1009.

Whittenbury R. & Dow C.S. (1977) Morphogenesis and differentiation in *Rhodomicrobium vannielii* and other budding and prosthecate bacteria. *Bacteriol. Rev.* **41**, 754–808.

Wrangstadh M., Szewzyk U., Östling J. & Kjelleberg S. (1990) Starvation-specific formation of a peripheral exopolysaccharide by a marine *Pseudomonas* sp., strain S9. *Appl. Environ. Microbiol.* **56**, 2065–2072.

Wright R.T. (1984) Dynamics of pools of dissolved organic carbon. In Hobbie J.E. & Williams P.J. Le B. (eds) *Heterotrophic Activity in the Sea*, pp. 121–154. Plenum, New York.

Wright R.T. (1988) Methods for evaluating the interaction of substrate and grazing as factors controlling planktonic bacteria. *Arch. Hydrobiol. Beih. Ergebn. Limnol.* **31**, 229–242.

Wright R.T. & Coffin R.B. (1984) Factors affecting bacterioplankton density and productivity in salt marsh estuaries. In Klug M.J. & Reddy C.A. (eds) *Current Perspectives in Microbial Ecology*, pp. 485–494. American Society for Microbiology, Washington DC.

13

David L. Kirchman

Particulate detritus and bacteria in marine environments

13.1 Introduction

Interactions between particles and bacteria have been examined since the beginning of marine microbiology. This interest in part has been and continues to be stimulated by practical problems associated with attached bacteria, such as biofouling and biocorrosion. In addition, until relatively recently, marine microbiologists thought that particulate surfaces were essential for the survival of bacteria in the oceans. It has been known for many years that the concentrations of dissolved organic matter (DOM) in the oceans are very low (about 100 μg or less of labile DOM l^{-1}) compared with concentrations typically used in laboratory cultures (milligrams per liter). Early studies suggested that unattached marine bacteria could not grow on these low DOM concentrations (Jannasch 1967). Consequently, marine microbiologists considered that many, if not most, bacteria in the oceans were dormant except perhaps those attached to particulate detritus (Stevenson 1978). Even today those marine microbiologists who work with pure cultures in the lab speak of "starving bacteria" (e.g., Moyer & Morita 1989; Nyström *et al.* 1990; see also Matin 1990 for work on nonmarine bacteria).

We now know that free-living bacteria in the oceans are not dormant and do not depend on particulate detritus for their survival. Microautoradiography has shown that a high, albeit variable, percentage of unattached bacteria are capable of assimilating several organic compounds (Meyer-Reil 1978); microautoradiographic results with thymidine suggest that these bacteria are actively dividing (Fuhrman & Azam 1982; Douglas *et al.* 1987). Also, free-living bacteria increase in abundance (and often cell size) following filtration (pore sizes as small as 0.6 μm) that removes all large particles (e.g., Fuhrman & Azam 1980). Clearly free-living bacteria do not depend on surfaces provided by particulate detritus for survival. Finally, several studies have found that free-living bacteria often greatly outnumber the attached cells (see below).

Nonetheless, attached bacteria and their impact on several

oceanographic processes may be as important as ever hypothesized by marine microbiologists. One goal of this review is to point out some of the new topics about attached bacteria that interest biological oceanographers. Although my focus is on salt water (especially the open ocean), most of the basic processes also occur in fresh water.

13.2 Number of attached bacteria

During the 1970s there was much discussion about whether bacteria in aquatic environments were predominately attached or free-living. The question then was somewhat difficult to answer because the methodology for enumerating bacteria by direct epifluorescence microscopy was just being developed (Zimmermann & Meyer-Reil 1974; Hobbie *et al.* 1977). Especially in oligotrophic aquatic environments, the free-living bacteria are very small (<0.5 μm in diameter) and difficult, if not impossible, to detect by other methodologies, although the estimates of Jannasch & Jones (1959) using other direct-count methods are not unreasonably low. Also, the hypothesized importance of particles for bacterial survival (see above) may have led to the premature conclusion that most bacteria should be attached.

It is true that attached bacteria can account for as much as 90% of total bacterial abundance in some lakes and estuaries. But usually particle-bound bacteria make up <10% of the total assemblage. Kirchman & Ducklow (1987) summarized the literature up to 1986, and their conclusions appear to be supported by more recent work (e.g., Cho & Azam 1988; Alldredge & Gotschalk 1990). Laanbroek & Verplanke (1986) and Albright *et al.* (1986), which were not included in the Kirchman & Ducklow (1987) review, did find fairly high percentages of attached bacteria, but still the free-living bacteria predominated.

The most important variable determining the numbers of attached bacteria appears to be the amount of particulate matter. Cammen & Walker (1982) showed that not only the numbers, but also the percentage of attached bacteria increased with the amount of suspended particulate matter. Why the percentage increases is not completely obvious. Perhaps it is because the number of free-living bacteria is set by other factors, e.g., primary production (Cole *et al.* 1988), and only the number of attached, not free-living bacteria, increases with particulate matter. Yamamoto & Lopez (1985) showed that bacterial abundance in sediments increased with specific surface area. The abundance of attached bacteria also increases with the quality of the particulate matter (Kirchman 1983).

Large particles, such as marine snow, can be intensively colonized by bacteria and other microorganisms at densities far exceeding that of surrounding water (Caron *et al.* 1982; Herndl 1988; Alldredge

1989). Also, these particles, especially delicate marine snow, are often excluded in normal sampling procedures. Nonetheless, the large particles are relatively rare. Alldredge & Silver (1988) found as few as $0.1\,l^{-1}$ particles compared with 10^6 for particles 1–10 µm, and even given their high densities of microorganisms, the attached assemblage is numerically unimportant compared with free-living bacteria (Alldredge *et al.* 1986; Alldredge & Gotschalk 1990).

13.3 Activity of attached bacteria

The effect of surfaces on metabolic activity of attached bacteria has always intrigued microbiologists since the beginnings of this field. The dogma still cited frequently is that surfaces can stimulate bacterial activity, especially in low nutrient concentrations. Recently, van Loosdrecht *et al.* (1990) critically reviewed some of the old literature that purported to support the dogma of surface-enhanced growth. These old studies (e.g., Heukelekian & Heller 1940) were relatively straightforward laboratory experiments with pure cultures, and so methodology, e.g., of counting bacteria, is not the problem. The review of van Loosdrecht *et al.* (1990) casts doubt on these previous studies and discusses how little we know about the influence of surfaces on bacterial activity.

What remains clear, however, is that attached bacteria differ from free-living bacteria in natural aquatic environments. Several studies have shown that the per cell uptake of organic compounds, such as glucose, is higher by attached bacteria than by free-living bacteria (see Kirchman & Ducklow 1987). In addition, bacteria attached to particles are much bigger than free-living bacteria (reviewed by Kirchman & Ducklow 1987; Alldredge 1989; Simon *et al.* 1990; although Herndl 1988 came to a different conclusion). An attached bacterium would have to assimilate more organic matter than a free-living bacterium to maintain the large cell size. In pure cultures, cell sizes increase with growth rate (Pierucci 1978; Larsson & Hagström 1982).

There are probably several reasons why attached bacteria are larger than free-living bacteria. For example, if attached to organic detritus, attached bacteria can degrade the detritus to support growth, as well as assimilate DOM. Another mechanism, which is discussed below, has been suggested by B.E. Logan and coworkers.

13.3.1 Effect of flow on attached bacteria

Logan & Hunt (1987) pointed out that bacteria associated with permeable aggregates, such as marine snow, are in a different hydrodynamic regime than free-living bacteria. As particles sink (or are mixed by turbulence), attached bacteria are affected by water flowing past the particle. This flow can increase the collision of DOM molecules with cell-surface transporters and thus theoretically in-

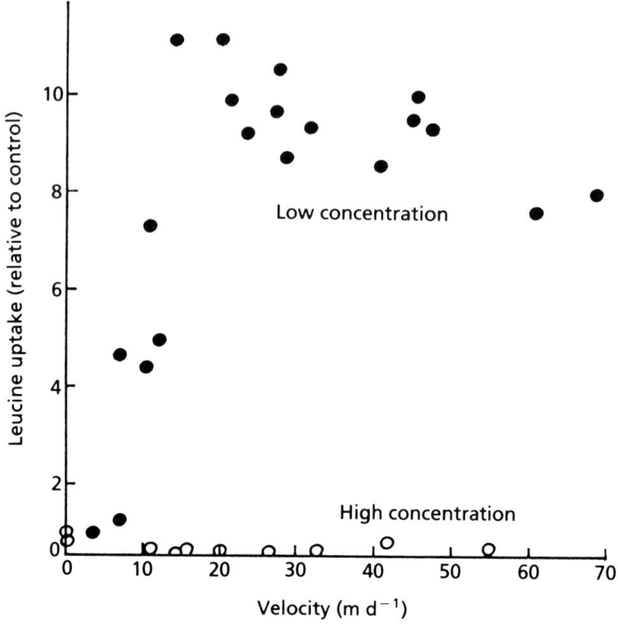

Figure 13.1 Effect of water flow on leucine uptake by marine bacterial assemblages held on a filter. Relative uptake of low ($1\,nmol\,l^{-1}$) and high ($10.8\,nmol\,l^{-1}$) added concentrations of ^{3}H-leucine. From Logan & Kirchman (1991).

crease DOM uptake. In contrast, free-living bacteria are substantially smaller than the Kolmogorov microscale of turbulence (Purcell 1977) and move with the water parcel. In low DOM concentrations, substrate transport to free-living bacteria may be limited by diffusion (see Pasciak & Gavis 1975). Logan & Dettmer (1989) first demonstrated that flow can increase uptake of leucine by *Zoogloea ramigera*.

Logan & Kirchman (1991) examined the effect of flow on natural marine bacterial assemblages. They found that flow past bacterial assemblages trapped on filters increased the uptake of leucine only when leucine concentrations were $1\,nmol\,l^{-1}$ (Figure 13.1). It was expected that flow would affect uptake only when uptake was not saturated (i.e., less than V_{max}). In contrast, flow had no effect on glucose uptake. Logan & Kirchman (1991) could not rule out the possibility that glucose uptake was saturated because glucose concentrations were not measured, although in the same sample leucine concentrations had decreased over time from 45 to $<1\,nmol\,l^{-1}$. They suggested another explanation. Flow should only affect uptake when the transporters are associated with the outer membrane of Gram-negative bacteria (the most common form in sea water), which is the case with amino acid transport. Glucose, on the other hand, diffuses through a general porin in the outer membrane and then is taken up by specific transporters in the inner membrane. Flow should not affect the uptake of compounds transported by this type of mechanism (Logan & Kirchman 1991).

In addition to the effect of flow on DOM uptake, the shear associated with flow may affect the biochemical composition of attached bacteria (Mittelman *et al.* 1990). This possibility has been examined only in pure culture but at least with a marine strain (*Pseudomonas atlantica*).

13.3.2 Growth rates of attached bacteria

In spite of the advantages (both purported and demonstrated) of association with particles, the growth rates of attached bacteria do not appear to differ substantially from the rates of free-living bacteria in natural aquatic environments (Ducklow & Kirchman 1983; Kirchman 1983; Pedrós-Alió & Brock 1983; Jacobsen & Azam 1984; Simon 1988; but see Newell & Christian 1981). Recently, Iriberri *et al.* (1990) confirmed that thymidine conversion factors and growth rates were similar for attached and free-living bacteria. Growth rates of bacteria associated with marine snow apparently vary greatly, and can be lower, the same, or higher than growth rates of free-living cells (Alldredge *et al.* 1986; Alldredge 1989; Alldredge & Gotschalk 1990). Rates may vary because of the age and decomposition state of the marine snow (Alldredge 1989; Alldredge & Gotschalk 1990).

Why growth rates of attached bacteria are not consistently higher than free-living bacteria is puzzling. It seems to contradict observations about higher uptake rates of organic compounds and larger cell sizes of attached bacteria relative to free-living cells. Perhaps attached cells are larger than free-living cells because of lower predation, which can select against large cells (Andersson *et al.* 1986). Caron (1987) did find that grazing by microflagellates was lower when bacteria were attached to particles or in aggregates.

13.3.3 Growth efficiency of attached bacteria

Kirchman (1983) suggested that attached bacteria have lower growth efficiencies than free-living bacteria. Thus attached bacteria would assimilate more DOM and perhaps be larger but not divide faster than free-living bacteria (see above). Indeed, R.C. Newell and coworkers (e.g., Newell *et al.* 1981; Linley & Newell 1984) have argued that growth efficiency of bacteria degrading particulate detritus is low (sometimes <10%) compared with utilization of DOM (>30%, e.g., Bjørnsen 1986). The molecular mechanism explaining these low growth efficiencies during detritus utilization is unclear. Van Loosdrecht *et al.* (1990) concluded that the pure culture studies offered no clearcut evidence that growth efficiency is affected by attachment to surfaces.

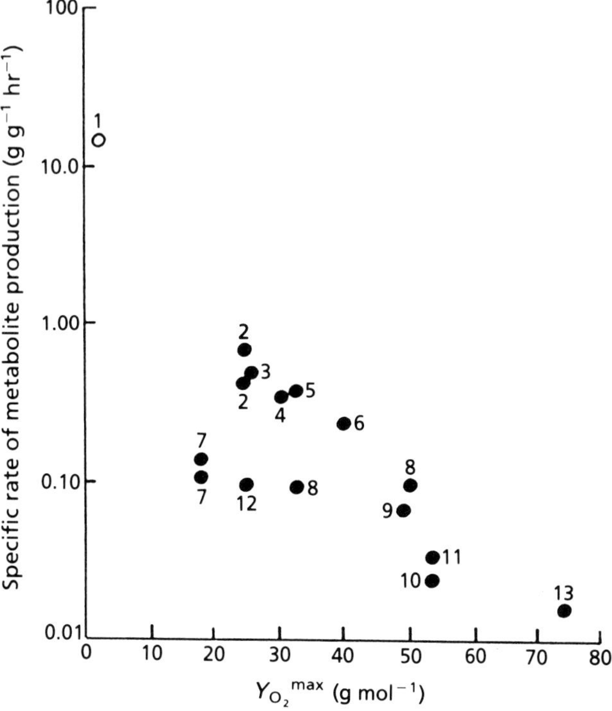

Figure 13.2 The highest specific rates of metabolite production reported as a function of the maximum growth ($Y_{O_2}^{max}$) of the producing organism: 1, total gluconate; 2, alginate; 3&5, xanthan; 4, succinogalactan; 6, succinoglucan; 7, glucomannan; 8, curdlan; 9, *Klebsiella pneumoniae* exopolysaccharide; 10, citric acid; 11, exoenzyme; 12, rhamnolipid; 13, polyhydroxybutyrate. From Linton (1990).

One mechanism that could explain low growth efficiencies is that attached bacteria may produce more exopolymers than free-living bacteria. This hypothesis was suggested by Kirchman (1983), although it has not been examined in any detail. As some support, Linton (1990) noted that growth efficiency of nonmarine bacterial strains decreases with increasing production of various extracellular metabolites, predominantly exopolymers (Figure 13.2). This inverse relationship between growth efficiency and exopolymer production may be very important for examining free-living as well as attached bacteria.

13.4 Impact of attached bacteria on total water column processes

Suspended particles and attached bacteria can have important, sometimes unique, roles in biogeochemical cycling of several elements. During the late 1970s and into the 1980s, several studies compared the uptake of simple DOM compounds, such as glucose, acetate, and amino acids, by attached and free-living bacteria (reviewed by Kirchman & Ducklow 1987). These studies demonstrated that attached bacteria assimilated more DOM than indicated by their abundance, although the problem of phytoplankton uptake in these size fractionation studies cannot be ignored. In any case, uptake by attached bacteria is usually <20% of total uptake (Kirchman & Ducklow 1987). Biomass production and accompanying respiration, i.e., oxygen consumption, is also usually low compared with the free-living bacteria. The important exceptions include waters with high amounts of detritus, such as some eutrophic lakes and estuaries.

13.4.1 Anoxic microenvironments

What is more interesting, and probably more important, is the role of particles as microenvironments. In particular, bacterial activity may reduce oxygen tensions within particles (including aggregates such as marine snow) to levels low enough to permit anaerobic metabolism. Alldredge & Cohen (1987) used microelectrodes to measure oxygen and pH gradients in marine snow and fecal pellets. Microelectrodes provide a powerful approach for examining the metabolism of attached bacteria (e.g., Nielsen *et al.* 1990), although small particles are difficult to study.

These anoxic microenvironments can be important because they would explain the presence, in oxic waters, of compounds produced by anaerobic metabolism, e.g., methane (Burke *et al.* 1983). The problem is that the existence of these anoxic particles does not necessarily explain quantitatively the anaerobic byproducts. Other explanations, including lateral transport from coastal benthos and excretion from anoxic digestive tracts of animals, are still possible.

13.4.2 Adsorption onto suspended particles

Another role of particles is as sites for adsorption of dissolved compounds. Adsorption can be an important step in the transport, transformation and even degradation of surface-active compounds. Some important examples include hydrophobic pollutants (van Loosdrecht *et al.* 1990), biological macromolecules such as proteins (Hollibaugh & Azam 1983; Kirchman *et al.* 1989) and DNA (Romanowski *et al.* 1991), and both natural and anthropogenic radionuclides (see review by Fowler & Knauer 1986).

Attached bacteria may affect adsorbed material by several mechanisms. Microbial activity may create reduced microenvironments (anoxic and low pε and pH) and thus change speciation of metals associated with particles. Perhaps the most topical example is iron (see Martin *et al.* 1990), which is reduced from Fe(III) in oxidized environments to Fe(II) in reduced environments (Stumm & Morgan 1981). This reduction may occur in anoxic sites in particles, and the change in iron speciation in an otherwise oxic environment may affect iron availability. Also, the exopolymers produced by attached microorganisms may adsorb metals (Mittelman & Geesey 1985) and other compounds. Of course attached bacteria may degrade adsorbed organic compounds (Samuelsson & Kirchman 1990), although adsorption can inhibit the degradation of some compounds such as DNA (Romanowski *et al.* 1991).

It should be pointed out that free-living bacteria and other microorganisms are themselves "particles." In fact, when modeling adsorption in the oceans, geochemists do not distinguish detrital surfaces from surfaces of living microorganisms. Obviously the two types differ greatly. I have avoided discussing the composition of detrital particles because it is difficult to determine composition in the oceans when living organisms are also present. Information about the chemical composition of detrital particles would be useful for examining degradation (see below), among many other processes. Kirchman *et al.* (1989), on the other hand, argued that adsorption of at least dissolved proteins depended on surface energy and not directly on the composition of the surface.

13.5 Formation of sinking particles

Theoretical models and direct observations indicate that most of the vertical flux of particulate matter is in the form of large particles (>100 μm; Fowler & Knauer 1986). In contrast, in the open ocean nearly all phytoplankton production and the direct grazers of the phytoplankton are much less than 100 μm; in fact, much of the biological production is <10 μm and even <1.0 μm (Azam *et al.* 1983). The question then arises: What are the mechanisms by which small particles are packaged into large, sinking particles?

Many of the attempts to answer this question do not involve attached bacteria and are outside the scope of this review. For example, the feeding and subsequent excretion of large fecal pellets by macrozooplankton can be important (e.g., Small *et al.* 1987). The aggregation of diatoms and other living phytoplankton into large flocs (Alldredge & Gotschalk 1989) is another potentially large contribution to the sinking particle flux. Jackson (1990) presented a model that combined coagulation theory and the kinetics of algal growth in order to examine the dynamics of algal floc formation.

Bacteria and cyanobacteria have been observed in deep-sea algal flocs (Lochte & Turley 1988), although it is unclear if these organisms are involved in floc formation or become associated with the floc after its formation. Even if not directly involved with initial algal floc formation, bacteria could be indirectly important by stressing the phytoplankton, e.g., direct attack or competition for limiting nutrients, which may increase exopolymer production by the algae. These exopolymers would affect floc "stickiness" (collision efficiency or the probability of sticking together after a collision between particles), which is an important parameter in floc formation (Jackson 1990). Bacteria associated with algal flocs could also produce exopolymers that affect the overall stickiness of the floc. Similarly, attached bacteria could affect, via exopolymer production, the aggregation of particles into large, sinking particles.

This discussion about floc formation points to the potential importance of exopolymers produced by bacteria, both attached to particles and free-living. In a comprehensive review, Decho (1990) pointed out the paucity of information about exopolymer production by natural marine bacteria. One problem is that these polymers are difficult to study; there are few methods available to measure even the quantities of exopolymers associated with bacterial cells. Progress in this area is hampered by the biochemical, biophysical, and physiological complexity of the exopolymers. Christensen (1989) has written a good review of extracellular polysaccharides produced by microorganisms in biofilms, which share characteristics of microorganisms associated with oceanic particles.

13.5.1 Coagulation of DOM and colloids into sinking particles

While the interest in flocs containing viable algae is relatively recent, biological oceanographers have examined the possible formation of marine snow from DOM since the 1960s (e.g., Baylor & Sutcliffe 1963). The main controversy at that time was the relative importance of bacteria versus physical mechanisms, such as coagulation of DOM during bubbling of sea water, in the formation of flocculent particles.

It seemed that Biddanda (1985) had solved the controversy. He showed that marine snow-like particles were produced only when bacteria were present. No particles were produced when sea water was poisoned with mercuric chloride or when all microorganisms (and other particles) were removed by filtration. Previous studies may have underestimated the importance of bacteria because epifluorescence microscopy was not used or was unavailable.

Johnson *et al.* (1986) then presented a model and experimental results to suggest that during bubbling of sea water, colloids and submicron particles (see below) coagulate into large particles. They argued that filtration would remove these colloids, in addition to eliminating bacteria. Poisons against the bacteria could disrupt coagulation chemistry. The data of Johnson *et al.* (1986) clearly showed that more particles were produced with bubbling than without bubbling of sea water. Furthermore, microscopic examination revealed that some particles (quantitative data were not reported) were devoid of attached bacteria, although they noted that all particles were quickly colonized by bacteria. They concluded that bacteria associated with the marine snow in their experiments were the result of, not the cause of, particle formation.

It seems premature to exclude bacterial effects during coagulation processes. The most convincing experiment to demonstrate the importance of totally abiotic mechanisms would be to eliminate or disable the bacteria without eliminating the colloids. Johnson *et al.* (1986) did not attempt this. It is difficult to believe that all antimicrobial agents would completely disrupt coagulation. The presence of marine snow particles devoid of bacteria in these experiments is some evidence for a totally abiotic mechanism but that observation does not exclude cell-free mechanisms involving bacteria, e.g., production of exopolymers. Also, bubbling may not be the most important mechanism of particle formation via coagulation. Although not tested by Johnson *et al.* (1986), Biddanda (1985) and Kepkay *et al.* (1990) found that mixing sea water without bubbling can lead to particle formation. Coagulation is certainly possible without bubbling.

Despite these criticisms, Johnson *et al.* (1986) is one paper among many in a growing literature about coagulation in marine environments (e.g., Jackson 1990; Alldredge & McGillivary 1991). Ultimately, the distinction between microbial and abiotic mechanisms, which Johnson *et al.* (1986) emphasizes, may not prove useful, given the ubiquity of microorganisms and the biological origin of colloids and DOM that may coagulate into particles. Although there are other studies on coagulation of relatively large particles (e.g., Alldredge & McGillivary 1991), more work is needed, like Johnson *et al.* (1986), that examines coagulation of submicron particles. I discuss below recent studies that have demonstrated the great abundance and mass of submicron particles in the oceans.

Coagulation theory is also being used to examine the formation of macroscopic particles from collision of particles $>100\,\mu m$. Alldredge & McGillivary (1991) found that 60–88% of collisions between particles ranging from 200 to $7600\,\mu m$ resulted in attachment and formation of larger particles. Although they did not examine attached bacteria, Alldredge & McGillivary (1991) discuss the importance of exopolymer production and microbial fibrils in affecting coagulation and particle formation. They hypothesized that the variation in collision efficiency that they observed may have been due to variation in the attached microbial assemblages.

A novel approach for examining marine snow formation via coagulation is fractal geometry. The basic measurement is to analyze particle numbers (N) composing an aggregate or aggregate mass versus size, e.g., $N \sim l^D$ where D_n is the dimension of the marine snow and l is the characteristic length scale (Logan & Wilkinson 1990). Although $D_3 = 3$ for a Euclidean sphere, aggregates formed by random collisions have D_3 significantly less than 3. Aggregates with noninteger dimensions (D_n) have fractal geometries. Logan & Wilkinson (1990) discuss how aggregates formed by different mech-

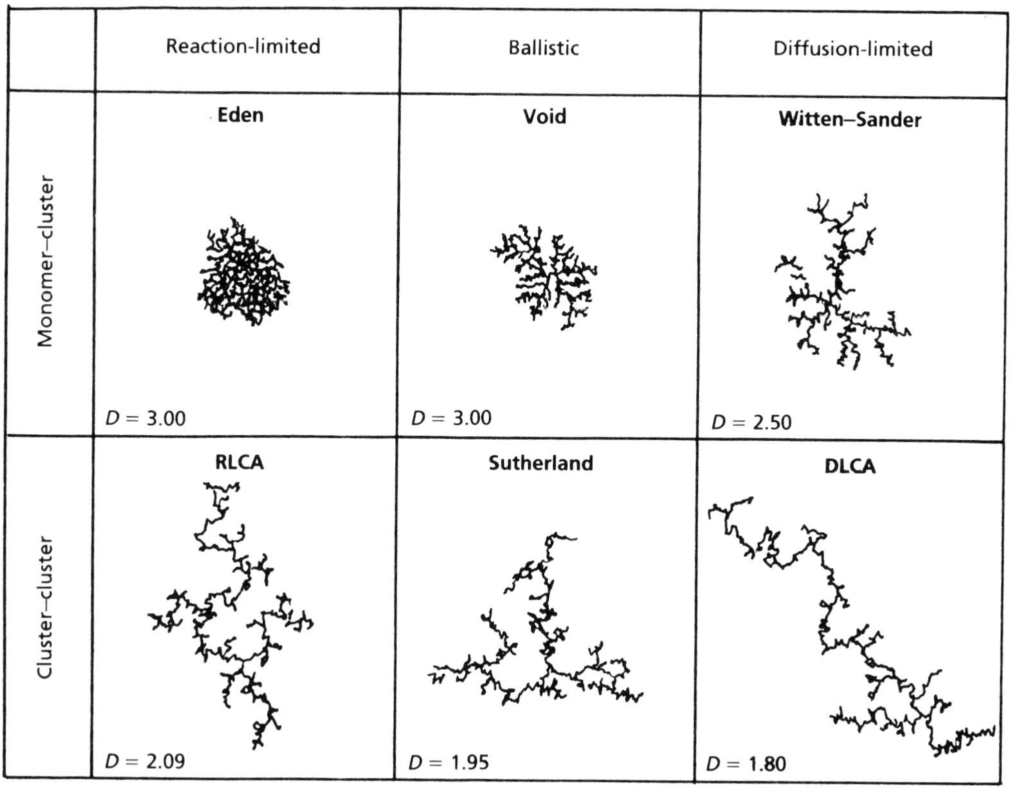

	Reaction-limited	Ballistic	Diffusion-limited
Monomer–cluster	**Eden** $D = 3.00$	**Void** $D = 3.00$	**Witten–Sander** $D = 2.50$
Cluster–cluster	**RLCA** $D = 2.09$	**Sutherland** $D = 1.95$	**DLCA** $D = 1.80$

Figure 13.3 Examples of particles formed by different coaggulation reactions where D is the fractal dimension of their three-dimensional analogs. From Schaefer (1989).

anisms have different fractal dimensions (Figure 13.3). In particular, they found that the fractal dimensions of marine snow were lower than those of aggregates formed from inorganic colloidal aggregation.

13.5.2 Formation of submicron detrital particles

Although the large amount of detritus in the oceans was well known, the report of Koike *et al.* (1990) surprised marine microbial ecologists. Koike *et al.* (1990) found a large number (roughly $10^7\,\mathrm{ml}^{-1}$ versus $10^6\,\mathrm{ml}^{-1}$ bacteria) of submicron (0.4–1 μm) detrital particles in the north Pacific. This was surprising because bacteria and other microorganisms by far dominate what is visible by standard epifluorescence microscopy (Fuhrman *et al.* 1989). In fact, the newly discovered submicron particles cannot be observed easily by epifluorescence microscopy and are stained only weakly (at most) by acridine orange or 4,6-diamidino-2-phenylindole (DAPI). Koike *et al.* (1990) found these particles with an electronic particle counter. Thin fibrils (3–10 nm diameter) and colloids have been observed with electron microscopy in lakes (Leppard 1986; Massalski & Leppard 1979).

Wheeler (personal communication, 1992) recently compared particulate nitrogen col lected on 0.2 μm Nuclepore polycarbonate filters versus that collected on standard glass fiber filters (Whatman® GF/F) which have a nominal pore size of 0.7 μm. Wheeler (personal communication, 1992) found that in the north-eastern Pacific a substantial amount of nitrogen was collected on the filter with the smaller pore size (as much as seven-fold greater than the GF/F nitrogen). Bacterial nitrogen accounted for <10% of this submicron particulate matter. Wheeler (personal communication, 1992) observed increases in this particulate matter at various depths, ranging from 100 to 1000 m, suggesting that there are several mechanisms that produce these small particles.

In surface waters, Koike *et al.* (1990) suggested that the submicron particles were produced from flagellates feeding on bacteria. In support of this hypothesis, they found that the number of submicron particles increased when flagellates were introduced into the bacterial size fraction, i.e., bacteria plus submicron particles, of coastal sea water. No particles were produced in formalin-killed controls.

Obviously we know very little about these submicron particles. Here we should be interested in their fate: What are the mechanisms by which bacteria "attach" to submicron particles in order to degrade them? Or can bacteria degrade them by excreted extracellular enzymes? Since some of the particles observed by Koike *et al.* (1990) are similar in size to bacteria, perhaps grazing by protozoa is significant. Another important issue is the possible coagulation of submicron detrital particles into large particles (see Section 13.5.1) that are capable of supporting attached microbial assemblages and of contributing to the sinking particle flux.

The discovery of submicron particles and the attention now being directed towards them illustrates a difficulty in separating DOM from particulate carbon and nitrogen. Traditionally biological oceanographers define DOM as the carbon and nitrogen that passes through a Whatman® GF/F and particulate matter is that which is retained on these filters. We now know that the traditionally defined "dissolved" pool contains bacteria and even eukaryotic microorganisms (Fuhrman & McManus 1984; Chisholm *et al.* 1988), viruses (Børsheim *et al.* 1990), and submicron detrital particles. This traditional DOM pool probably also contains many colloids that are not sampled by previous methods for measuring dissolved organic carbon (DOC) and nitrogen (DON) (Suzuki *et al.* 1985; Sugimura & Suzuki 1988). Figure 13.4 emphasizes the continuum of "particles," ranging from globular proteins to >1.0 μm microorganisms to macroscopic marine snow.

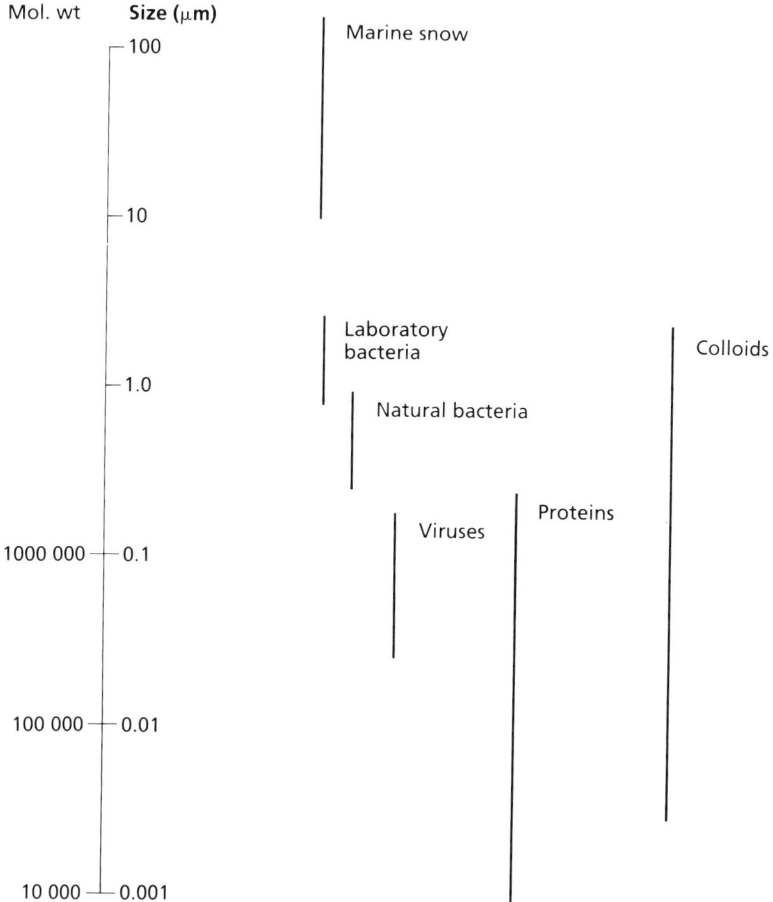

Figure 13.4 Size spectrum of some particles in sea water. The term "marine snow" is usually applied to macroscopic particles.

Nonetheless, it is useful to retain some distinctions among submicron particles. Obviously living microorganisms and viruses differ from detritus. It may be useful to call organic matter a "particle" only when it can be ingested by phagotrophic protozoa. This biological definition would include most of the traditionally defined oceanographic particles, plus the bacteria. It would exclude some submicron "particles," even some relatively large (>0.5 μm), if they are not ingested by protozoa. A more precise term for these particles is colloids.

Colloids are defined as any "particle, droplet, or gas bubble where the surface free energy dominates the bulk free energy" (Rees 1991). Colloids are being extensively examined in ground water because of their role in the transport of contaminants (Rees 1991). Perhaps marine microbial ecologists should look more closely at colloids and their possible roles in biogeochemical cycles.

13.6 Degradation of particles: the fate of sinking detritus in the oceans

Microbial ecologists have always been interested in the degradation of detritus and its role in the regeneration of inorganic nutrients, such as ammonium and phosphate. The problem takes on even greater importance as oceanographers attempt to understand the role of the oceans in buffering changes in atmospheric carbon dioxide. The vertical flux of particles, part of the "biological pump" (Longhurst & Harrison 1989), is one important mechanism by which carbon is transported to the deep oceans and sediments. The formation and degradation of this sinking particulate carbon are now being studied intensively.

The role of attached bacteria in the degradation of these sinking particles is unclear. One approach is to use estimates of bacterial biomass production, e.g., measured by thymidine incorporation, in samples collected by sediment traps (Ducklow *et al.* 1982). Based on this and other measures of microbial activity (and assuming a growth efficiency), it appears that direct decomposition of sinking particles by bacteria is relatively slow and insignificant (Ducklow *et al.* 1982; Ducklow *et al.* 1985; Karl *et al.* 1988).

Banse (1990) has criticized these studies because of several methodological problems. The most interesting problem is that previous experiments with sediment trap material did not mimic the flow regime of a particle sinking in the ocean. Once confined in a trap or an experimental container, particle degradation by attached bacteria may be impeded, for example, by the build-up of soluble by-products. Banse (1990) suggested some experimental approaches to solve this problem.

Cho & Azam (1988) hypothesized that attached bacteria helped solubilize particulate organic matter (POM) to DOM, which is subsequently used by free-living bacteria. Indeed, in the central-north Pacific gyre and in the Santa Monica basin, they measured bacterial production down to 1000 m and found it to nearly equal the estimated particulate carbon flux. Simon *et al.* (in press) used a different approach for estimating bacterial production, but found similar results in the subarctic Pacific. Banse (1990) pointed out that such high carbon consumption by free-living bacteria contradicts other estimates of carbon consumption by mesozooplankton (0.2–0.3 mm) and nekton. These organisms cannot live off free-living bacteria and must have some role in consuming sinking particles in mesopelagic waters.

There are many difficult problems in examining sinking particles in the ocean, including questions about the accuracy of sediment traps in collecting this material (e.g., Karl & Knauer 1989). If we look past these technical problems, several interesting questions in microbial ecology become apparent.

The solubilization of sinking POM to DOM is an important topic, even if free-living bacterial production in the deep ocean has been overestimated as Banse (1990) implies. It is quite likely that attached bacteria produce DOM as suggested by Cho & Azam (1988), but their results and those of Simon *et al.* (in press) imply that the amount of DOM released is very high compared with attached bacterial production. The turnover of POM by attached bacteria is at least one order of magnitude less than the degradation implied by the POM → DOM → free-living bacteria pathway. That is, if this hypothesis is correct, these attached bacteria would have to be very inefficient in assimilating the compounds released during hydrolysis of the POM. This seems unlikely, but we do not know much about exoenzymes of natural bacterial assemblages. Another possibility that has not been examined critically in marine systems is abiotic leaching of DOM from POM, which Cole *et al.* (1984) suggested to be important in lakes (although see Pett 1989). However, it is difficult to see how this mechanism could be important below the upper layers of the oceans. Finally, perhaps bacteria that are not directly attached to particles can help in POM degradation (Hermansson & Marshall 1985). If so, rates per cell of POM degradation or solubilization to DOM need not be large because sinking particles encounter large numbers of free-living bacteria in the oceans (Cho & Azam 1988).

Previous experiments have focused on the role of attached bacteria in POM degradation, even though other microorganisms, such as protozoa and even viable algae, can be present in sinking particles (Silver & Alldredge 1981). We may miss the importance of these other microorganisms if we only measure thymidine incorporation in sediment trap samples. Indeed, Banse (1990) pointed out that depletion of bulk POM, i.e., mineralization to carbon dioxide and solubilization to DOM or inorganic nutrients, in sediment trap samples can be very high. This degradation of POM, in contrast to specific measurements of bacterial activity, is the end result of many processes. There are probably many synergistic relationships among microorganisms associated with sinking particles that determine POM degradation.

13.7 Conclusions

Oceanographers are trying to understand the "biological pump" which transports carbon from the surface layer to the deep oceans and is one mechanism by which the oceans can influence the greenhouse effect. Although recent studies have pointed to the possible importance of DOM advection (Toggweiler 1989; Bacastow & Maier-Reimer 1991), the vertical flux of particles will remain an important component of this "biological pump." Consequently, biological oceanographers are interested in the formation, transport, and degradation of particles in the oceans. Some interests of microbial ecologists overlap with the oceanographers, some do not.

There is one unanswered question that interests microbial ecologists, but not oceanographers: Does the species composition of attached and free-living bacteria differ? We need not consider old studies because traditional pure culture techniques miss >99% of the bacteria found in natural environments. Only recently have RNA-based approaches, which do not require isolation of pure cultures from the oceans, been used to examine the taxonomy of natural marine bacterial assemblages (Giovannoni *et al.* 1990). Oceanographers, however, will not be interested in the species composition of attached versus free-living bacteria, unless that information helps them to understand processes, such as particle degradation. Perhaps information about some bacterial species could tell oceanographers that particular processes are occurring which may not be evident from other data. A good example is the possibility of anaerobic metabolism in oxic environments. Even so, oceanographers will still want quantitative information about rates, not just qualitative information about presence or absence of a metabolic type.

Other unanswered questions should interest both microbial ecologists and biological oceanographers. The important ones that emerge from this chapter, I suggest concern exopolymers. This material associated (perhaps even excreted occasionally) with the outer membrane of marine bacteria includes enzymes for degrading macromolecules, other proteins for specific attachment to surfaces, and polysaccharides for nonspecific attachment, protection, and possibly as a carbon reserve (Decho 1990). Even with this short list, it is clear that exopolymers are involved in the major unanswered questions about particles in the oceans: particle formation, e.g., via coagulation aided by "sticky" exopolymers; bacterial attachment; and the degradation of particles by both free-living and attached bacteria. By focusing on the bacteria as organisms rather than a "black box," microbial ecologists may help solve these important oceanographic problems.

Acknowledgments

Preparation of this chapter was supported by the Office of Naval research, contract N00014-87-0108. I thank Bruce Logan and John Hobbie for their constructive criticism.

References

Albright L.J., McCrae S.K. & May B.E. (1986) Attached and free-floating bacterioplankton in Howe Sound, British Columbia, a coastal marine fjord-embayment. *Appl. Environ. Microbiol.* 51, 614–621.

Alldredge A.L. (1989) The significance of suspended detrital aggregates of marine snow as microhabitats in the pelagic zone of the ocean. In Hattori T., Ishida Y., Maruyama Y., Morita R.Y. & Uchida A. (eds) *Recent Advances in Microbial Ecology*, pp. 108–112. Japan Scientific Societies Press, Tokyo.

Alldredge A.L. & Cohen Y. (1987) Can microscale chemical patches persist in the sea? Microelectrode study of marine snow, fecal pellets. *Science* **235**, 689–691.

Alldredge A.L. & Silver M.W. (1988) Characteristics, dynamics, and significance of marine snow. *Prog. Oceanogr.* **20**, 41–82.

Alldredge A.L. & Gotschalk C.C. (1989) Direct observations of the mass flocculation of diatom blooms: characteristics, settling velocities and formation of diatom aggregates. *Deep-Sea Res.* **36**, 159–171.

Alldredge A.L. & Gotschalk C.C. (1990) The relative contribution of marine snow of different origins to biological processes in coastal waters. *Cont. Shelf Res.* **10**, 41–58.

Alldredge A.L. & McGillivary P. (1991) The attachment probabilities of marine snow and their implications for particle coagulation in the ocean. *Deep-Sea Res.* **38**, 431–443.

Alldredge A.L., Cole J.J. & Caron D.A. (1986) Production of heterotrophic bacteria inhabiting macroscopic organic aggregates (marine snow) from surface waters. *Limnol. Oceanogr.* **31**, 68–78.

Andersson A., Larsson U. & Hagström A. (1986) Size-selective grazing by a microflagellate on pelagic bacteria. *Mar. Ecol. Prog. Ser.* **33**, 51–57.

Azam F., Fenchel T., Field J.G., Gray J.S., Meyer-Reil L.A. & Thingstad T. (1983) The ecological role of water-column microbes in the sea. *Mar. Ecol. Prog. Ser.* **10**, 257–263.

Bacastow R. & Maier-Reimer E. (1991) Dissolved organic carbon in modeling oceanic new production. *Global Biogeochem. Cycles* **5**, 71–85.

Banse K. (1990) New views on the degradation and disposition of organic particles as collected by sediment traps in the open sea. *Deep-Sea Res.* **37**, 1177–1195.

Baylor E.R. & Sutcliffe W.H.J. (1963) Dissolved organic matter in seawater as a source of particulate food. *Limnol. Oceanogr.* **4**, 369–371.

Biddanda B.A. (1985) Microbial synthesis of macroparticulate matter. *Mar. Ecol. Prog. Ser.* **20**, 241–251.

Bjørnsen P.K. (1986) Bacterioplankton growth yield in continuous seawater cultures. *Mar. Ecol. Prog. Ser.* **30**, 191–196.

Børsheim K.Y., Bratbak G. & Heldal M. (1990) Enumeration and biomass estimation of planktonic bacteria and viruses by transmission electron microscopy. *Appl. Environ. Microbiol.* **56**, 352–356.

Burke R.A. Jr, Reid D.F., Brooks J.M. & Lavoie D.M. (1983) Upper water column methane geochemistry in the eastern tropical North Pacific. *Limnol. Oceanogr.* **28**, 19–32.

Cammen L.M. & Walker J.A. (1982) Distribution and activity of attached and free-living suspended bacteria in the Bay of Fundy. *Can. J. Fish. Aquat. Sci.* **39**, 1655–1663.

Caron D.A. (1987) Grazing of attached bacteria by heterotrophic microflagellates. *Microb. Ecol.* **13**, 203–218.

Caron D.A., Davis P.G., Madin L.P. & Sieburth J.McN. (1982) Heterotrophic bacteria and bacterivorous protozoa in oceanic macroaggregates. *Science* **218**, 795–797.

Chisholm S.W., Olson R.J., Zettler E.R., Goericke R., Waterbury J.B. & Welschmeyer N.A. (1988) A novel free-living prochlorophyte abundant in the oceanic euphotic zone. *Nature* **334**, 340–343.

Cho B.C. & Azam F. (1988) Major role of bacteria in biogeochemical fluxes in the ocean's interior. *Nature* **332**, 441–443.

Christensen B.E. (1989) The role of extracellular polysaccharides in biofilms. *J. Biotech.* **10**, 181–202.

Cole J.J., Likens G.E. & Hobbie J.E. (1984) Decomposition of planktonic algae in an

oligotrophic lake. *Oikos* **42**, 257–266.
Cole J.J., Findlay S. & Pace M.L. (1988) Bacterial production in fresh and saltwater ecosystems: a cross-system overview. *Mar. Ecol. Prog. Ser.* **43**, 1–10.
Decho A.W. (1990) Microbial exopolymer secretions in ocean environments: their role(s) in food webs and marine processes. *Oceanogr. Mar. Biol. Annu. Rev.* **28**, 73–153.
Douglas D.J., Novitsky J.A. & Fournier R.O. (1987) Microautoradiography-based enumeration of bacteria with estimates of thymidine-specific growth and production rates. *Mar. Ecol. Prog. Ser.* **36**, 91–99.
Ducklow H.W. & Kirchman D.L. (1983) Bacterial dynamics and distribution during a spring diatom bloom in the Hudson River plume, USA. *J. Plank. Res.* **5**, 333–355.
Ducklow H.W., Kirchman D.L. & Rowe G.T. (1982) Production and vertical flux of attached bacteria in the Hudson River plume of the New York Bight as studied with floating sediment traps. *Appl. Environ. Microbiol.* **43**, 769–776.
Ducklow H.W., Hill S.M. & Gardner W.D. (1985) Bacterial growth and the decomposition of particulate organic carbon collected in sediment traps. *Cont. Shelf Res.* **4**, 445–464.
Fowler S.W. & Knauer G.A. (1986) Role of large particles in the transport of elements and organic compounds through the oceanic water column. *Prog. Oceanogr.* **16**, 147–194.
Fuhrman J.A. & Azam F. (1980) Bacterioplankton secondary production estimates for coastal waters of British Columbia, Antarctica, and California. *Appl. Environ. Microbiol.* **39**, 1085–1095.
Fuhrman J.A. & Azam F. (1982) Thymidine incorporation as a measure of heterotrophic bacterioplankton production in marine surface waters: evaluation and field results. *Mar. Biol.* **66**, 109–120.
Fuhrman J.A. & McManus G.B. (1984) Do bacteria-sized marine eukaryotes consume significant bacterial production? *Science* **224**, 1257–1260.
Fuhrman J.A., Sleeter T.D., Carlson C.A. & Proctor L.M. (1989) Dominance of bacterial biomass in the Sargasso Sea and its ecological implications. *Mar. Ecol. Prog. Ser.* **57**, 207–217.
Giovannoni S.J., Britschgi T.B., Moyer C.L. & Field K.G. (1990) Genetic diversity in Sargasso Sea bacterioplankton. *Nature* **345**, 60–63.
Hermansson M. & Marshall K.C. (1985) Utilization of surface localized substrate by non-adhesive marine bacteria. *Microb. Ecol.* **11**, 91–105.
Herndl G.J. (1988) Ecology of amorphous aggregations (marine snow) in the northern Adriatic Sea. II. Microbial density and activity in marine snow and its implication to overall pelagic processes. *Mar. Ecol. Prog. Ser.* **48**, 265–275.
Heukelekian H. & Heller A. (1940) Relation between food concentration and surface for bacterial growth. *J. Bacteriol.* **40**, 547–558.
Hobbie J.E., Daley R.J. & Jasper S. (1977) Use of Nuclepore filters for counting bacteria by fluorescence microscopy. *Appl. Environ. Microbiol.* **33**, 1225–1228.
Hollibaugh J.T. & Azam F. (1983) Microbial degradation of dissolved proteins in seawater. *Limnol. Oceanogr.* **28**, 1104–1106.
Iriberri J., Unanue M., Ayo B., Barcina I. & Egea L. (1990) Bacterial production and growth rate estimation from [^{3}H]thymidine incorporation for attached and free-living bacteria in aquatic systems. *Appl. Environ. Microbiol.* **56**, 483–487.
Jackson G.A. (1990) A model of the formation of marine algal flocs by physical coagulation processes. *Deep-Sea Res.* **37**, 1197–1211.
Jacobsen T.R. & Azam F. (1984) Role of bacteria in copepod fecal pellet decomposition: colonization, growth rates and mineralization. *Bull. Mar. Sci.* **35**, 492–502.
Jannasch H.W. (1967) Growth of marine bacteria at limiting concentrations of organic carbon in seawater. *Limnol. Oceanogr.* **12**, 264–271.
Jannasch H.W. & Jones G.E. (1959) Bacterial populations in sea water as determined by different methods of enumeration. *Limnol. Oceanogr.* **4**, 128–139.

Johnson B.D., Zhou X. & Wangersky P.J. (1986) Surface coagulation in sea water. *Neth. J. Sea Res.* **20**, 201–210.

Karl D.M. & Knauer G.A. (1989) Swimmers: a recapitulation of the problem and a potential solution. *Oceanogr. Mag.* **2**, 32–35.

Karl D.M., Knauer G.A. & Martin J.H. (1988) Downward flux of particulate organic matter in the ocean: a particle decomposition paradox. *Nature* **332**, 438–441.

Kepkay P.E., Muschenheim D.K. & Johnson B.D. (1990) Surface coagulation and microbial respiration at a tidal front on Georges Bank. *Cont. Shelf Res.* **10**, 573–588.

Kirchman D. (1983) The production of bacteria attached to particles suspended in a freshwater pond. *Limnol. Oceanogr.* **28**, 858–872.

Kirchman D.L. & Ducklow H.W. (1987) Trophic dynamics of particle-bound bacteria in pelagic ecosystems: a review. In Moriarty D.J.W. & Pullin R.S.V. (eds) *Detritus and Microbial Ecology in Aquaculture*, pp. 54–82. International Center for Living Aquatic Resource Management, Manila.

Kirchman D.L., Henry D.L. & Dexter S.C. (1989) Adsorption of proteins to surfaces in seawater. *Mar. Chem.* **27**, 201–217.

Koike I., Hara S., Terauchi K. & Kogure K. (1990) Role of sub-micrometre particles in the ocean. *Nature* **345**, 242–244.

Laanbroek H.J. & Verplanke J.C. (1986) Seasonal changes in percentages of attached bacteria enumerated in a tidal and a stagnant coastal basin: relation to bacterioplankton productivity. *FEMS Microb. Ecol.* **38**, 87–98.

Larsson U. & Hagström A. (1982) Fractionated phytoplankton primary production, exudate release and bacterial production in a Baltic eutrophication gradient. *Mar. Biol.* **67**, 57–70.

Leppard G.G. (1986) The fibrillar matrix component of lacustrine biofilms. *Wat. Res.* **20**, 697–702.

Linley E.A.S. & Newell R.C. (1984) Estimates of bacterial growth yields based on plant detritus. *Bull. Mar. Sci.* **35**, 409–425.

Linton J.D. (1990) The relationship between metabolite production and the growth efficiency of the producing organism. *FEMS Microbiol. Rev.* **75**, 1–18.

Lochte K. & Turley C.M. (1988) Bacteria and cyanobacteria associated with phytodetritus in the deep sea. *Nature* **333**, 67–69.

Logan B.E. & Hunt J.R. (1987) Advantages to microbes of growth in permeable aggregates in marine systems. *Limnol. Oceanogr.* **32**, 1034–1048.

Logan B.E. & Dettmer J.W. (1989) Increased mass transfer to microorganisms with fluid motion. *Biotech. Bioeng.* **35**, 1135–1144.

Logan B.E. & Wilkinson D.B. (1990) Fractal geometry of marine snow and other biological aggregates. *Limnol. Oceanogr.* **35**, 130–136.

Logan B.E. & Kirchman D.L. (1991) Uptake of dissolved organics by marine bacteria as a function of fluid motion. *Mar. Biol.* **111**, 175–181.

Longhurst A.R. & Harrison W.G. (1989) The biological pump: profiles of plankton production and consumption in the upper ocean. *Prog. Oceanogr.* **22**, 47–123.

Martin J.H., Fitzwater S.E. & Gordon R.M. (1990) Iron deficiency limits phytoplankton growth in antarctic waters. *Global Biogeochem. Cycles* **4**, 5–12.

Massalski A. & Leppard G.G. (1979) Morphological examination of fibrillar colloids associated with algae and bacteria in lakes. *J. Fish. Res. Board Can.* **36**, 922–938.

Matin A. (1990) Molecular analysis of the starvation stress in *Escherichia coli*. *FEMS Microb. Ecol.* **74**, 185–196.

Meyer-Reil L.A. (1978) Autoradiography and epifluorescence microscopy combined for the determination of number and spectrum of actively metabolizing bacteria in natural waters. *Appl. Environ. Microbiol.* **36**, 506–512.

Mittelman M.W. & Geesey G.G. (1985) Copper-binding characteristics of exopolymers from a freshwater-sediment bacterium. *Appl. Environ. Microbiol.* **49**, 846–851.

Mittelman M.W., Nivens D.E., Low C. & White D.C. (1990) Differential adhesion, activity, and carbohydrate: protein ratios of *Pseudomonas atlantica* monocultures attaching to stainless steel in a linear shear gradient. *Microb. Ecol.* **19**, 269–278.

Moyer C.L. & Morita R.Y. (1989) Effect of growth rate and starvation-survival on the viability and stability of a psychrophilic marine bacterium. *Appl. Environ. Microbiol.* 55, 1122–1127.

Newell R.C., Lucas M.I. & Linley E.A.S. (1981) Rate of degradation and efficiency of conversion of phytoplankton debris by marine micro-organisms. *Mar. Ecol. Prog. Ser.* **6**, 123–136.

Newell S.Y. & Christian R.R. (1981) Frequency of dividing cells as an estimator of bacterial productivity. *Appl. Environ. Microbiol.* **42**, 23–31.

Nielsen L.P., Christensen P.B., Revsbech N.P. & Sorensen J. (1990) Denitrification and oxygen respiration in biofilms studied with a microsensor for nitrous oxide and oxygen. *Microb. Ecol.* **19**, 63–72.

Nyström T., Flärdh K. & Kjelleberg S. (1990) Responses to multiple-nutrient starvation in marine *Vibrio* sp. strain CCUG 15956. *J. Bacteriol.* **172**, 7085–7097.

Pasciak W.J. & Gavis J. (1975) Transport limited nutrient uptake rates in *Ditylum brightwellii*. *Limnol. Oceanogr.* **20**, 604–617.

Pedrós-Alió C. & Brock T.D. (1983) The importance of attachment to particles for planktonic bacteria. *Arch. Hydrobiol.* **98**, 354–379.

Pett R.J. (1989) Kinetics of microbial mineralization of organic carbon from detrital *Skeletonema costatum* cells. *Mar. Ecol. Prog. Ser.* **52**, 123–128.

Pierucci O. (1978) Dimensions of *Escherichia coli* at various growth rates: model of envelope growth. *J. Bacteriol.* **135**, 559–574.

Purcell E.M. (1977) Life at low Reynolds number. *Am. J. Physics* **45**, 3–11.

Rees T.F. (1991) Transport of contaminants by colloid mediated processes. In Hutzinger O. (ed.) *The Handbook of Environmental Chemistry, vol. 2F: Reactions and Processes*, pp. 25–44. Springer Verlag, Heidelberg.

Romanowski G., Lorenz M.G. & Wackernagel W. (1991) Adsorption of plasmid DNA to mineral surfaces and protection against DNase I. *Appl. Environ. Microbiol.* 57, 1057–1061.

Samuelsson M.-O. & Kirchman D.L. (1990) Degradation of adsorbed protein by attached bacteria in relationship to surface hydrophobicity. *Appl. Environ. Microbiol.* **56**, 3643–3648.

Schaefer D.W. (1989) Polymers, fractals, and ceramic materials. *Science* **243**, 1023–1027.

Silver M.W. & Alldredge A.L. (1981) Bathypelagic marine snow: deep-sea algal and detrital community. *J. Mar. Res.* **39**, 501–530.

Simon M. (1988) Growth characteristics of small and large free-living and attached bacteria in Lake Constance. *Microb. Ecol.* **15**, 151–163.

Simon M., Alldredge A.L. & Azam F. (1990) Bacterial carbon dynamics on marine snow. *Mar. Ecol. Prog. Ser.* **65**, 205–211.

Simon M., Welschmeyer N.A. & Kirchman D.L. (In press) Bacterial production and the sinking flux of particulate organic matter in the subarctic Pacific. *Deep-Sea Res.*

Small L.F., Knauer G.A. & Tuel M.D. (1987) The role of sinking fecal pellets in stratified euphotic zones. *Deep-Sea Res.* **34**, 1705–1712.

Stevenson L.H. (1978) A case for bacterial dormancy in aquatic systems. *Microb. Ecol.* **4**, 127–133.

Stumm W. & Morgan J.J. (1981) *Aquatic Chemistry*. John Wiley & Sons, New York.

Sugimura Y. & Suzuki Y. (1988) A high-temperature catalytic oxidation method for the determination of non-volatile dissolved organic carbon in seawater by direct injection of a liquid sample. *Mar. Chem.* **24**, 105–131.

Suzuki Y., Sugimura Y. & Itoh T. (1985) A catalytic oxidation method for the determination of total nitrogen dissolved in seawater. *Mar. Chem.* **16**, 83–97.

Toggweiler J.R. (1989) Is the downward dissolved organic matter (DOM) flux important in carbon transport? In Berger W.H., Smetacek V.S. & Wefer G. (eds)

Productivity of the Ocean: Present and Past, pp. 65–84. J. Wiley & Sons, New York.
van Loosdrecht M.C.M., Lyklema J., Norde W. & Zehnder A.J.B. (1990) Influence of interfaces on microbial activity. *Microbiol. Rev.* **54**, 75–87.
Yamamoto N. & Lopez G. (1985) Bacterial abundance in relation to surface area and organic content of marine sediments. *J. Exp. Mar. Biol. Ecol.* **90**, 209–220.
Zimmermann R. & Meyer-Reil L.-A. (1974) A new method for fluorescence staining of bacterial populations on membrane filters. *Kieler Meeresforsch. Sonderh.* **30**, 24–27.

14

Hans W. Paerl

Interaction of nitrogen and carbon cycles in the marine environment

14.1 Introduction

The carbon and nitrogen cycles are of fundamental importance and inextricably linked to marine microbial production, mineralization, and sedimentation processes (Redfield 1958; Redfield *et al.* 1963; Dugdale 1967; Eppley *et al.* 1973; Carpenter & Capone 1983; Blackburn & Sorensen 1988). Aside from oxygen and hydrogen, carbon and nitrogen are the most abundant structural and functional constituents of organic macromolecules. In particular, those macromolecules mediating biochemical transformations (proteins–enzymes) and physiological regulatory, or genetic controls (hormones, growth regulators, DNA, RNA), exhibit a common reliance on specific ratios and configurations of carbon and nitrogen (Lehninger 1975). It is therefore not surprising that nitrogen and carbon limitations of microbial metabolic activity and growth have important consequences for production and mineralization of organic matter in the sea (Dugdale & Goering 1967; Eppley *et al.* 1973; Morris 1974; McCarthy 1980; Harrison *et al.* 1987; Eppley 1989).

Biogeochemically, carbon and nitrogen cycles share analogous traits, the most obvious being microbially-mediated transformations between gaseous and nongaseous states which are tightly controlled by specific oxygen tension and redox potential regimes (Woodwell *et al.* 1973; Delwiche 1981; Jørgensen 1983; Reeburgh 1983; Blackburn & Sorensen 1988; Capone 1991). Carbon and nitrogen also reveal remarkable degrees of interdependence in the face of nutrient-depleted conditions, which characterize many of the world's marine ecosystems. In this context, it has been well established that the availability of combined nitrogen is a key factor controlling primary and secondary production in geographically and trophically diverse waters, ranging from the ultraoligotrophic open ocean to heavily eutrophied estuaries (Dugdale 1967; Rhyther & Dunstan 1971; Nixon *et al.* 1976; Parsons *et al.* 1977; Mann 1982; Carpenter

& Capone 1983; Blackburn & Sorensen 1988; Codispoti 1989). In meeting biological demands in respective ecosystems, nitrogen transformations mediating availability of combined nitrogen often rely on oxidizable carbon supplies (Delwiche 1978; Sorensen 1978; Fenchel & Blackburn 1979; Jørgensen 1983; Paerl *et al.* 1987; Blackburn & Sorensen 1988; Smith & Hollibaugh 1989). The reverse also operates; that is, reduced carbon availability is frequently controlled by primary production, which in turn is controlled by nitrogen supplies (Redfield *et al.* 1963; Dugdale 1967; Parsons *et al.* 1977; Mann 1982; Blackburn & Sorensen 1988). This rather paradoxical and apparently circular dilemma concerning the interdependence of carbon and nitrogen availability can be illustrated by the case of nitrogen fixation and denitrification, as they are functionally coupled to the carbon cycle. As "sources and sinks" of biologically mediated nitrogen, nitrogen fixation and denitrification are invariably reliant on the availability of utilizable organic carbon, which is required as an energy and reductant source among heterotrophic nitrogen fixers and denitrifiers (Figures 14.1 and 14.2) (Stewart 1971; Delwiche 1978; 1981; Goering 1978; Payne 1981; Knowles 1982). In addition, localized oxygen depletion resulting from the oxidation of organic carbon creates microaerophilic/anaerobic conditions essential for both processes to function (Koike & Hattori 1978; Sorensen *et al.* 1979; Knowles 1982; Jørgensen 1983; Paerl 1990). Conversely, in unenriched (absence of anthropogenic inputs) ecosystems where primary production is frequently controlled by nitrogen availability, fixed nitrogen supply can tightly control photosynthetic carbon dioxide fixation rates (Harrison *et al.* 1987; Codispoti 1989; Eppley 1989). Open ocean, reef, coastal marsh, and benthic systems rely to varying extents on "within-system" nitrogen fixation to meet primary production nitrogen demands (Goering *et al.* 1973; Webb *et al.* 1975; Wiebe *et al.* 1975; Paerl *et al.* 1981; Carpenter & Capone 1983; Howarth *et al.* 1988; Blackburn & Sorensen 1988). Ironically, the rate at which autotrophically produced organic carbon is supplied to nitrogen fixers is often a strong determinant of the productive capacity and trophic states of such systems (Stewart 1971; Fogg 1982; Capone 1983; Paerl 1990).

These systems exhibit a strong mechanistically linked reliance on microbial interception and fixation (reduction) of the large pools of atmospheric (gaseous) carbon dioxide and nitrogen for maintenance of marine fertility, a characteristic not shared with other major nutrient cycles (P, S, Si, and most trace metals). One notable exception is iron, which may be intercepted in the form of aeolian dust, potentially ameliorating iron limitation in open ocean surface water

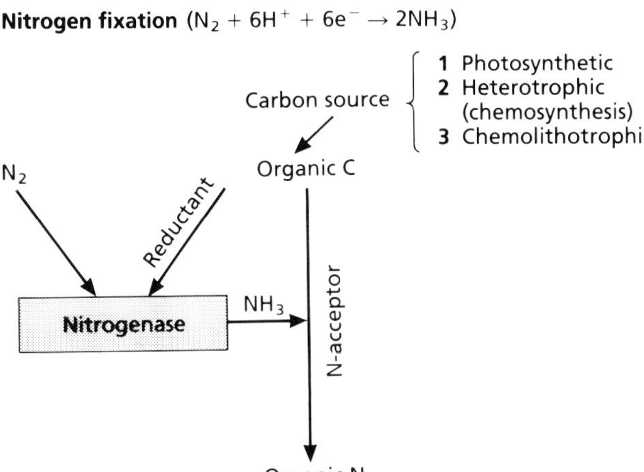

Microbial functional groups mediating marine N_2 fixation

1 Photoautotrophs
  Oxygenic photoautotrophs: cyanobacteria
    Nonheterocystous filamentous (*Trichodesmium, Lyngbya, Microcoleus, Oscillatoria*)
    Nonheterocystous, nonfilamentous (*Synechococcus, Gloeocapsa*)
    Heterocystous filamentous (*Calothrix, Scytonema, Nostoc*, and others)
  Anoxygenic photoautotrophs (*Chlorobium, Chromatium, Rhodobacter, Rhodospirillum, Ectothiorhodospira*)
2 Heterotrophs (chemosynthetic bacteria)
  Aerobes (*Azotobacter, Azospirillum, Corynebacterium, Methylococcus, Spirillum*)
  Facultative anaerobes (*Bacillus, Klebsiella, Vibrio*)
  Anaerobes (*Clostridium, Desulfovibrio*)
3 Chemolithotrophs (*Thiobacillus*)

Figure 14.1 Interactions of nitrogen fixation with carbon cycling. Enzyme-mediated steps and their coupling to carbon as reductant and/or energy sources are schematically shown. Specific microbial functional groups and representative genera mediating individual nitrogen transformations are also shown.

(Martin *et al.* 1990). With specific reference to nitrogen–carbon interactions and their impacts on microbial production, the equilibrium between carbon dioxide fixation, organic carbon mineralization, nitrogen fixation, intermediate nitrogen transformations, and denitrification determine the net fluxes of biologically utilizable forms of these elements between the hydrosphere and atmosphere (Goering *et al.* 1973; Nixon *et al.* 1976; Paul 1978; Fogg 1982; Blackburn & Sorensen 1988; Glibert 1988). An analogous dynamic equilibrium exists between the lithosphere and atmosphere (Delwiche 1978; 1981).

14.2 The evolutionary, physiological, and

The advent and proliferation of oxygenic photosynthesis on Earth has profoundly influenced the interaction of carbon and nitrogen cycling and resultant biological availability of these elements. From

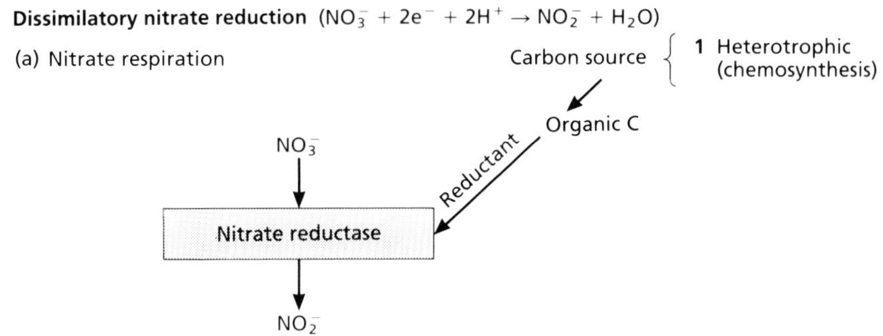

Microbial functional groups mediating marine nitrate respiration

1 Heterotrophs (chemosynthetic bacteria)
Aerobes (*Aeromonas*, *Corynebacterium*, *Pseudomonas*, *Halobacterium*, and others)
Facultative anaerobes (*Bacillus*, *Enterobacter*, *Klebsiella*, *Serratia*, *Vibrio*, and others)

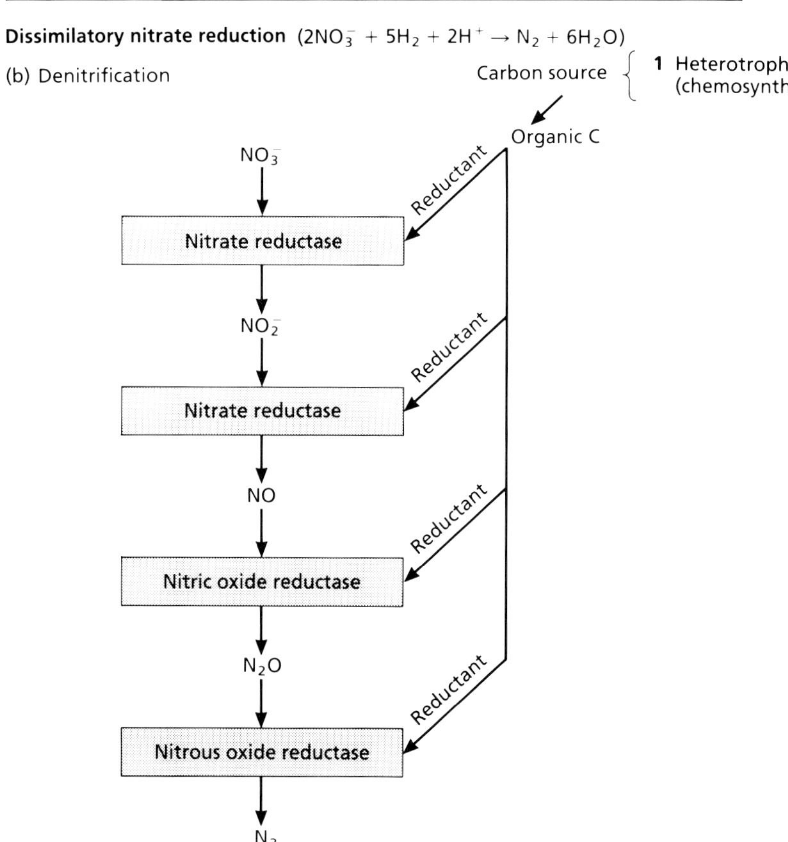

Microbial functional groups mediating marine denitrification

1 Heterotrophs (chemosynthetic bacteria)
Facultative anaerobes/obligate anaerobes (*Bacillus*, *Corynebacterium*, *Pseudomonas*, *Spirillum*, *Thiobacillus*, and others)

Figure 14.2 Interactions of dissimilatory nitrate reduction with carbon cycling. Schematically described as for Figure 14.1.

ecological controls of carbon and nitrogen transformations

a molecular perspective, nitrogen fixation and denitrification are highly conserved processes, being widespread among ancient anaerobic autotrophic and heterotrophic microbial genera, including methanogens, sulfur and sulfate reducers, photosynthetic bacteria, primitive anoxygenic cyanobacteria, and a wide range of anoxic heterotrophs (Cloud 1976; Knoll 1977; Walker *et al.* 1982). The development of cyanobacterial oxygenic photosynthesis, considered a key factor initiating the evolution of an oxygen-enriched biosphere, has led to environmental constraints on nitrogen fixation (Schopf & Walter 1982) and denitrification (Knowles 1982), to the extent that both are largely confined to oxygen-depleted microenvironments, or microzones, in the contemporary biosphere (Jørgensen 1983; Paerl 1990). Obligatory confinement to oxygen-depleted microzones has ushered in a variety of physiological, morphological and ecological adaptations among nitrogen-fixing cyanobacteria, including heterocyst differentiation (Wolk 1982), aggregate formation (Fogg *et al.* 1973), endosymbioses (Stewart *et al.* 1980; Ahmadjian & Paracer 1986), and consortial relationships with oxygen-consuming heterotrophs (Paerl 1978a). The problem of excluding oxygen is even more perplexing among nitrogen-fixing bacteria, which lack the ability to form heterocysts (thick-walled, biochemically differentiated cells in which oxygen evolution does not occur; this feature allows for nitrogen-fixing activity).

Most denitrifying bacteria are facultative aerobes, e.g., growth can occur in the presence of oxygen but must be anaerobic if NO_3^- is to serve as the terminal electron acceptor. Denitrification ($NO_3 \rightarrow N_2$) can therefore only occur in the absence of oxygen, making this process reliant on oxygen-depleted microenvironments. As such, denitrifyers are largely confined to oxygen-depleted surficial microzones or anoxic sediments (Fenchel & Blackburn 1979; Sorensen *et al.* 1979; Jørgensen & Revsbech 1989; Paerl 1990), thus excluding much of the oxic water column as a potential habitat (Koike & Hattori 1978; Sorensen *et al.* 1979; Payne 1981; Nishio *et al.* 1982; Seitzinger 1988).

Forced confinement of the "source and sink" of biologically utilizable nitrogen has, in all likelihood, led to severe constraints on the nitrogen fixation necessary for meeting ecosystem nitrogen demands in vast segments of the oceans. Furthermore, denitrification must be confined to oxygen-poor microzones, yet near sources of the oxidized substrates NO_3^- and NO_2^-. Over long time scales (thousands to millions of years), biogeochemical impacts of confinement of these processes may have ranged from alteration of the global nitrogen flux to enhancement of chronic nitrogen limitation (Codispoti 1989; Paerl 1990). Since it appears that both nitrogen

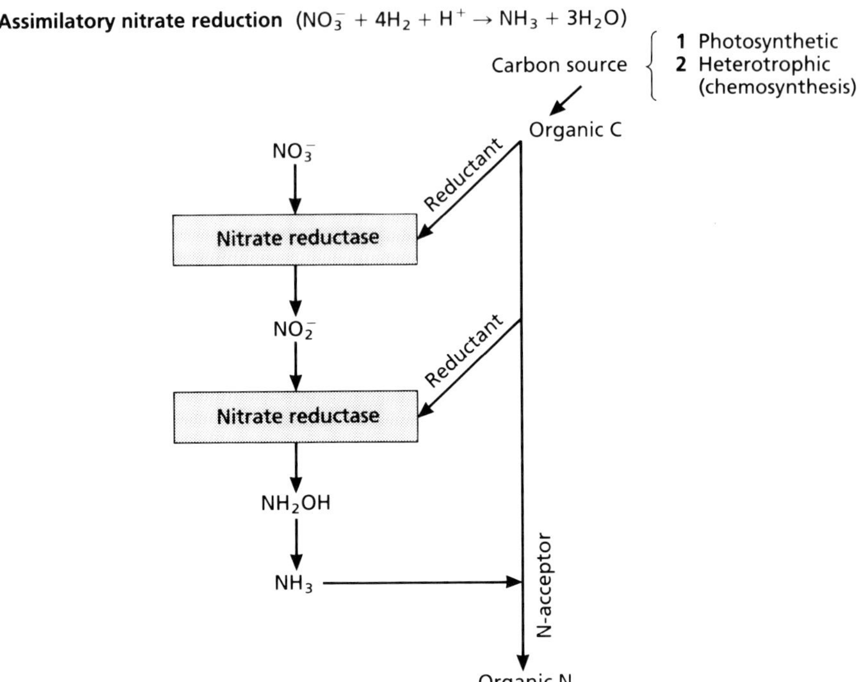

Functional groups mediating aquatic assimilatory nitrate reduction

1 Photoautotrophs
- Oxygenic photoautotrophs
  - Prokaryotes: cyanobacteria (most genera)
  - Eukaryotes
    - microalgae (virtually all genera)
    - macroalgae (virtually all genera)
    - macrophytes (virtually all genera)

2 Anoxygenic photoautotrophs
  - Prokaryotes: photosynthetic bacteria (*Rhodospirillum*)
- Aerobic heterotrophs
  - Prokaryotes: bacteria (most genera)
  - Eukaryotes: fungi (virtually all genera)

3 Facultative anaerobic/anaerobic heterotrophs
  - Prokaryotes: bacteria (most genera)

Figure 14.3 Interactions of assimilatory nitrate reduction with carbon cycling. Schematically described as for Figure 14.1.

fixation and denitrification have undergone some confinement, the severity, magnitude, and extent of marine nitrogen limitation may reflect the impact of environmental alteration of these processes. In this respect, altered long-term spatial and temporal patterns of atmospheric carbon dioxide concentration (and hence availability), climate, primary production, and resultant organic carbon availability and deposition, have undoubtedly played a key role in shaping the changing patterns of nitrogen limitation (Schopf *et al.* 1983; Berger & Keir 1987; Codispoti 1989).

Another product of the evolution of an oxygen-rich biosphere is

Ammonification (NH_3 regeneration)

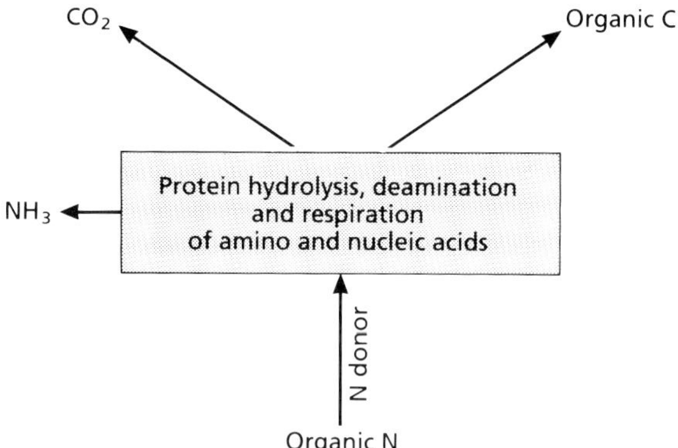

Microbial functional groups mediating marine ammonification

1 Photoautotrophs
   Oxygenic photoautotrophs
      Prokaryotes (cyanobacteria)
      Eukaryotes (all phyla and genera)
   Anoxygenic photoautotrophs (all genera)
2 Heterotrophs (chemosynthetic bacteria)
   Aerobes (most genera)
   Facultative anaerobes (most genera)
   Anaerobes (most genera)
3 Chemolithotrophs (most genera)

Figure 14.4 Interactions of ammonification with carbon cycling. Schematically described as for Figure 14.1.

the obligate spatial-temporal coupling of ammonification, nitrification, and assimilatory NO_3^-/NO_2^- reduction to nitrogen sources and sinks (Figures 14.3–5) (Alexander 1965; Koike & Hattori 1978; Sorensen *et al.* 1979; Delwiche 1981; Nishio *et al.* 1982; Klump & Martens 1983). Concomitant coupling to readily available carbon sources (as reductant and/or carbon skeleton sources) is also necessary (Suess 1976; Delwiche 1981; Blackburn & Sorensen 1988). Nitrifying bacteria are an example of adaptation to such multiple environmental constraints. This group mediates the aerobic oxidation of ammonia to nitrite and nitrate while chemoautotrophically reducing carbon dioxide (see Figure 14.5) (Alexander 1965). Nitrifyers are reliant on NH_4^+ as an electron and (to a lesser extent) nitrogen source. Since the biotic source of NH_4^+, ammonification (see Figure 14.4), is often concentrated near particles, surfaces, and benthic regions in NH_4^+-poor marine waters, nitrifyers are typically active at surficial interfaces and chemoclines (redoxclines), where NH_4^+ diffusion must overlap with oxic (oxidative) conditions (Figure 14.6) (Codispoti & Christensen 1985; Ward 1986; Blackburn & Sorensen 1988; Ward & Zafirou 1988).

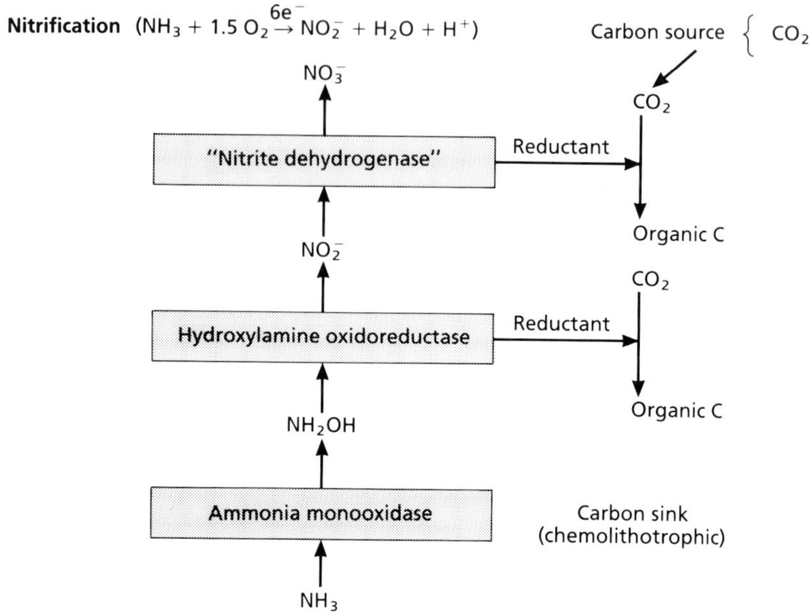

Microbial functional groups mediating marine nitrification

1 Chemolithotrophs Aerobes (*Nitrosomonas*: $NH_3 \rightarrow NH_2OH \rightarrow NO_2^-$) (*Nitrobacter*: $NO_2 \rightarrow NO_3^-$)

Figure 14.5 Interactions of nitrification with carbon cycling. Schematically described as for Figure 14.1.

Methane formation and its reoxidation are similarly reliant on appropriate overlap of adequate organic carbon supply, conducive oxygen gradients and appropriate physical–chemical conditions needed to support the microbial consortium involved in mediating these processes (Figure 14.7). Characteristically, the sediment–water interface or decomposing organic aggregates meet such environmental requirements (Zeikus 1977; Karl 1982; Martens 1984). These habitats provide a source of recently formed and accumulated organic matter (phytoplankton and macrophyte detritus) serving as a source of biopolymers (carbohydrates, proteins, lipids) which are subsequently hydrolyzed to monomers (sugars, amino acids, fatty acids). Oxidative breakdown (initially aerobic oxidation, then anaerobic oxidation and fermentation) leads to the formation of strong oxygen gradients (thus providing microzones for essential microbial consortia members) and provides the source of 2, 3, and 4 carbon acids, alcohols, and aldehydes serving as substrates for methanogenesis (Blackburn & Henricksen 1983; Claypool & Kvenvolden 1983; Heinrichs & Reeburgh 1987; Wolin & Miller

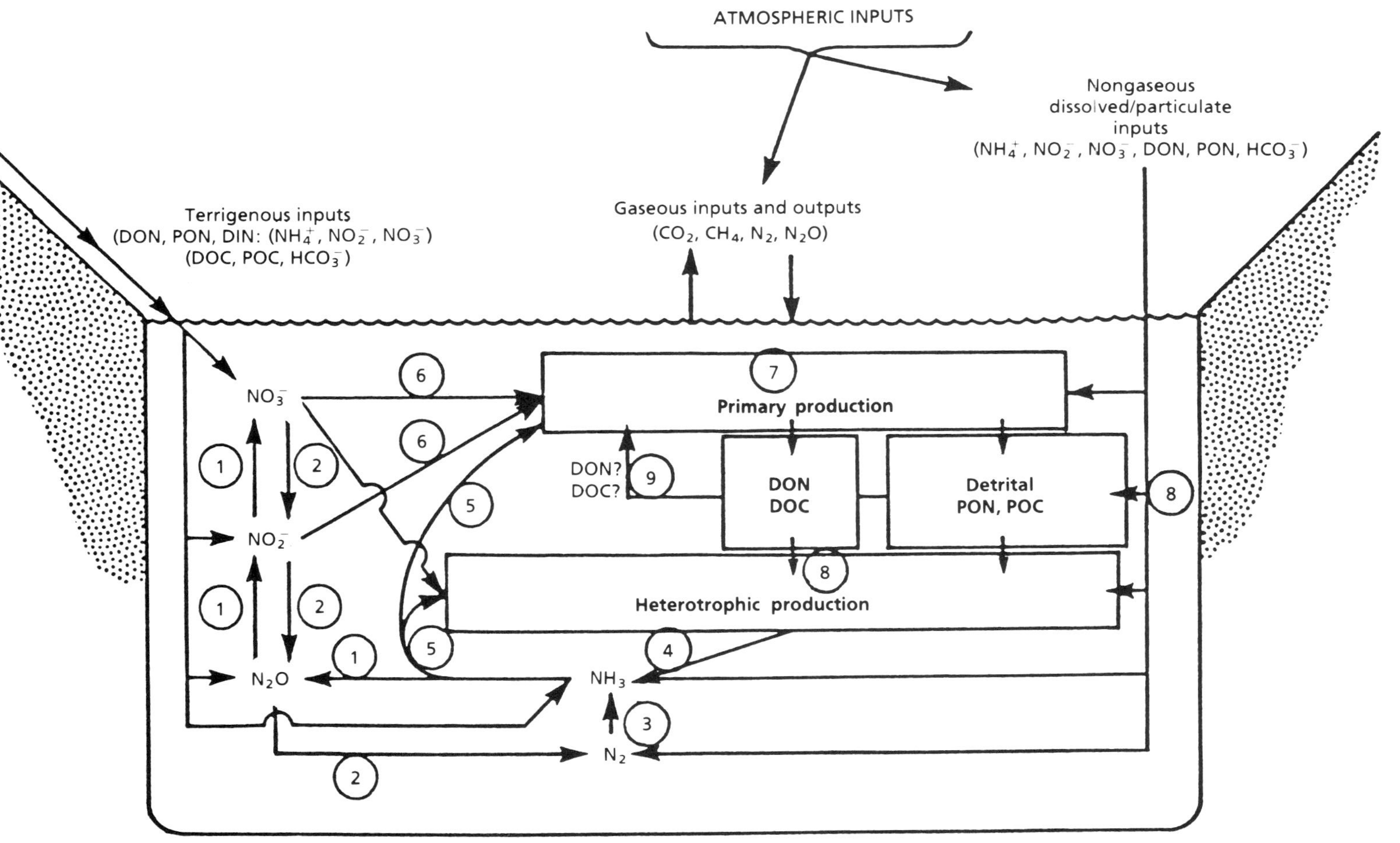

Figure 14.6 Major nitrogen and carbon inputs, within-system microbial transformations and outputs in a marine ecosystem. Planktonic and benthic habitats are jointly presented as sites of microbial transformations. Dominant carbon and nitrogen transformation processes are as follows: 1, nitrification; 2, denitrification; 3, nitrogen fixation; 4, ammonification; 5, ammonia assimilation; 6, nitrite and nitrate assimilation; 7, photosynthetic/chemolithotrophic carbon fixation; 8, heterotrophic (chemosynthetic) organic carbon and nitrogen assimilation; 9, photoheterotrophic organic carbon and nitrogen assimilation (this process has been observed but not quantified). Note; sizes of boxes and arrows are not indicative of pool size or flux rate.

Methane formation and oxidation

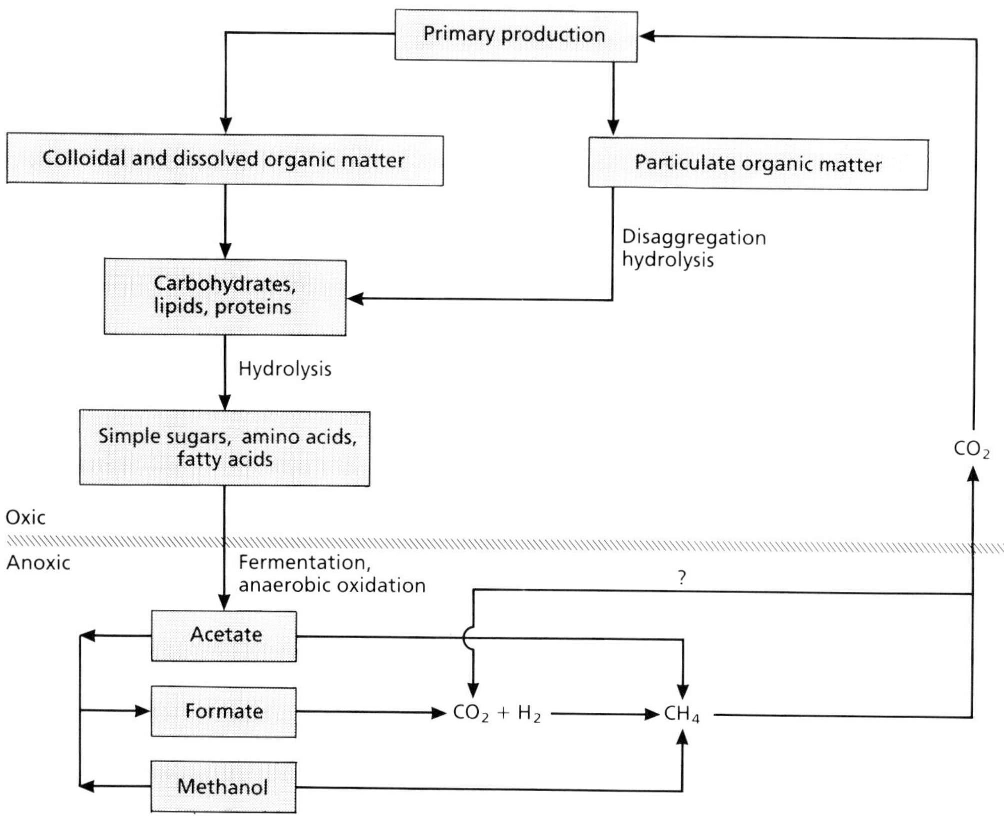

Microbial functional groups mediating marine methane formation and oxidation

1 Anaerobic fermentative bacterial methanogens (most common reactions below)

$4\,H_2 + CO_2 \rightarrow CH_4 + 2\,H_2O$

$4\,HCOO^- + 2\,H^+ \rightarrow CH_4 + CO_2 + 2\,HCO_3^-$

$4\,CH_3OH \rightarrow 3\,CH_4 + CO_2 + 2\,H_2O$

$4\,CH_3NH_2 + 2\,H_2O + 4\,H^+ \rightarrow 3\,CH_4 + CO_2 + NH_4^+$

Relevant genera: *Methanobacterium, Methanococcus, Methanomicrobium, Methanosarcina*

2 Aerobic methanotrophic methane oxidizing bacteria

$CH_4 + O_2 + NADH_2 \rightarrow CH_3OH + H_2O + NAD^+$

$CH_3OH + NAD^+ + H_2O \rightarrow HCOOH + NADH_2$

$HCOOH + NAD^+ \rightarrow CO_2 + NADH_2$

Relevant genera: *Methylococcus, Methylobacterium, Methylosinus*

3 Anaerobic methane oxidizing bacteria

Relevant microorganisms suspected but not yet isolated

Figure 14.7 Methane cycling as mediated by representative methanogenic and methane-oxidizing bacterial genera.

1987; Oremland 1988), and subsequent oxic and anoxic methane oxidation (see Figure 14.7) (Zehnder & Brock 1979; Devol 1983; Alperin & Reeburgh 1985; Blackburn & Sorensen 1988).

The microbial consortia mediating anaerobic mineralization of organic matter are strongly reliant on interspecific exchange of carbon, nitrogen, and hydrogen (H_2 as an electron donor; H^+ as an electron acceptor) along the redoxcline (Wolin & Miller 1987). Resident genera of these consortia include nitrogen fixers, e.g., *Clostridium* and *Methanococcus*, which not only assure adequate nitrogen supplies, but could participate in interspecies hydrogen transfer (nitrogenase produces hydrogen in addition to ammonia). In this manner methanogens (some of which may also be nitrogen fixers) readily benefit from hydrogen production by coupling oxidative steps in this fermentative pathway to reductive reactions of hydrogen-consuming species. This coupling avoids the accumulation of excess electrons resulting from the reoxidation of nicotinamide adenine dinucleotide hydrogen (NADH) among fermenters (Zehnder 1978; Wolin & Miller 1987; Oremland 1988).

Because marine nitrogen and carbon transformations yield intermediates spanning a wide range of oxidation states, interfaces supporting oxygen gradients and diffusive boundaries are geochemically important and biochemically active nitrogen transformation sites (Jørgensen 1983; Klump & Martens 1983; Martens 1984; Jørgensen & Revsbech 1989). Relevant examples include:

1 the air–water interface;
2 stratified water column thermoclines and oxyclines, and redoxclines (Reeburgh 1983; Jørgensen 1983; Ward 1990) and/or haloclines (Sharp *et al.* 1984);
3 submersed surfaces or biofilms (Paerl & Carlton 1988; Revsbech *et al.* 1989);
4 suspended detritus and sediments (Paerl & Prufert 1987);
5 benthic–water column interfaces (Suess 1976; Nixon *et al.* 1976; Capone 1983);
6 sediment oxyclines and redoxclines (Heinrichs & Reeburgh 1987; Blackburn & Sorensen 1988; Jørgensen & Revsbech 1989);
7 microbe–plant/microbe–animal/microbe–microbe symbioses and consortial associations (Ahmadjian & Paracer 1986).

Habitats supporting active microzones include:

1 stratified planktonic waters (McCarthy & Goldman 1979), especially those exhibiting microalgal blooms (Fogg 1982; Paerl 1988; 1990; Paerl & Bebout 1988);
2 suspended detrital and sediment communities, including "marine snow," (Alldredge & Cohen 1987; Paerl & Prufert 1987);
3 intertidal and subtidal reefs (Webb *et al.* 1975; Wiebe *et al.* 1975; Capone 1983);
4 intertidal mud and sandflats (Gotto *et al.* 1981; Bautista & Paerl 1985; Stal *et al.* 1985; Bebout *et al.* 1987);
5 subtidal benthic regions, including macrophyte (seagrass) beds (Capone 1983);

6 coastal lagoonal marsh and mangrove regions (Bohlool & Wiebe 1978; Kaplan *et al.* 1979; Valiela & Teal 1979; Paerl *et al.* 1981; Whitton & Potts 1982);
7 deep-water chlorophyll *a* and turbidity (particle) maxima (Parsons *et al.* 1977; Blackburn & Sorensen 1988);
8 aphotic sediments, including geothermal vent communities (Karl 1982; Jannasch 1985).

Quantitatively and qualitatively, dynamic nitrogen and carbon cycling can be expected at marine physical–chemical interfaces (Suess 1976; Marshall 1979; Reeburgh 1983; Paerl 1985a). The diversity of activities of microbial "life at the interface" is discussed below.

14.3 Microorganisms mediating marine carbon and nitrogen cycling

Metabolically diverse autotrophs and heterotrophs play vital roles in the cycling and balance of marine inorganic and organic carbon and nitrogen species. Daily turnover of carbon and nitrogen due to growth, reproduction, and death represents by far the bulk of biochemically mediated carbon and nitrogen cycling (Redfield *et al.* 1963; Eppley *et al.* 1973; Hargrave 1973; McCarthy 1980; Shaffer 1987; Glibert 1988). In terms of specific carbon and nitrogen transformation steps, a suite of microorganisms, varying in oxygen tolerance/requirements from obligate anaerobic to aerobic is responsible. Dominant microbial nitrogen transformations mediated by representative autotrophs and heterotrophs are shown in Figures 14.1–5. Figure 14.6 illustrates major nitrogen inputs, outputs, and transformations in the marine environment. The vertical distributions of gaseous and dissolved inorganic/organic nitrogen cycle intermediates are schematically shown in Figure 14.8. They are described in relation to microbially mediated and physicochemically controlled fluxes between a stratified water column and underlying sediments. The nature, sources, and exchanges of organic and inorganic carbon, as well as ambient oxic conditions, are of central importance in controlling the distributions and interactions of primary producers and mineralizers mediating the nitrogen cycle. Figures 14.9 and 14.10 depict carbon cycling. Figure 14.9 illustrates planktonic trophic interactions of primary producers and higher-ranked consumers, while Figure 14.10 broadly illustrates differences between oxic and anoxic carbon cycling as mediated by autotrophic and heterotrophic microorganisms. Examination of the relevant transformations and functional microbial groups reveals a substantial amount of similarity in mechanisms of carbon utilization and transformation among microorganisms with radically different roles in the nitrogen cycle. One example is the nitrogen fixing versus denitrifying microflora (see Figures 14.1 and 14.2). While nitrogen fixers reduce nitrogen to ammonia, denitrifyers reduce nitrate to nitrogen; yet

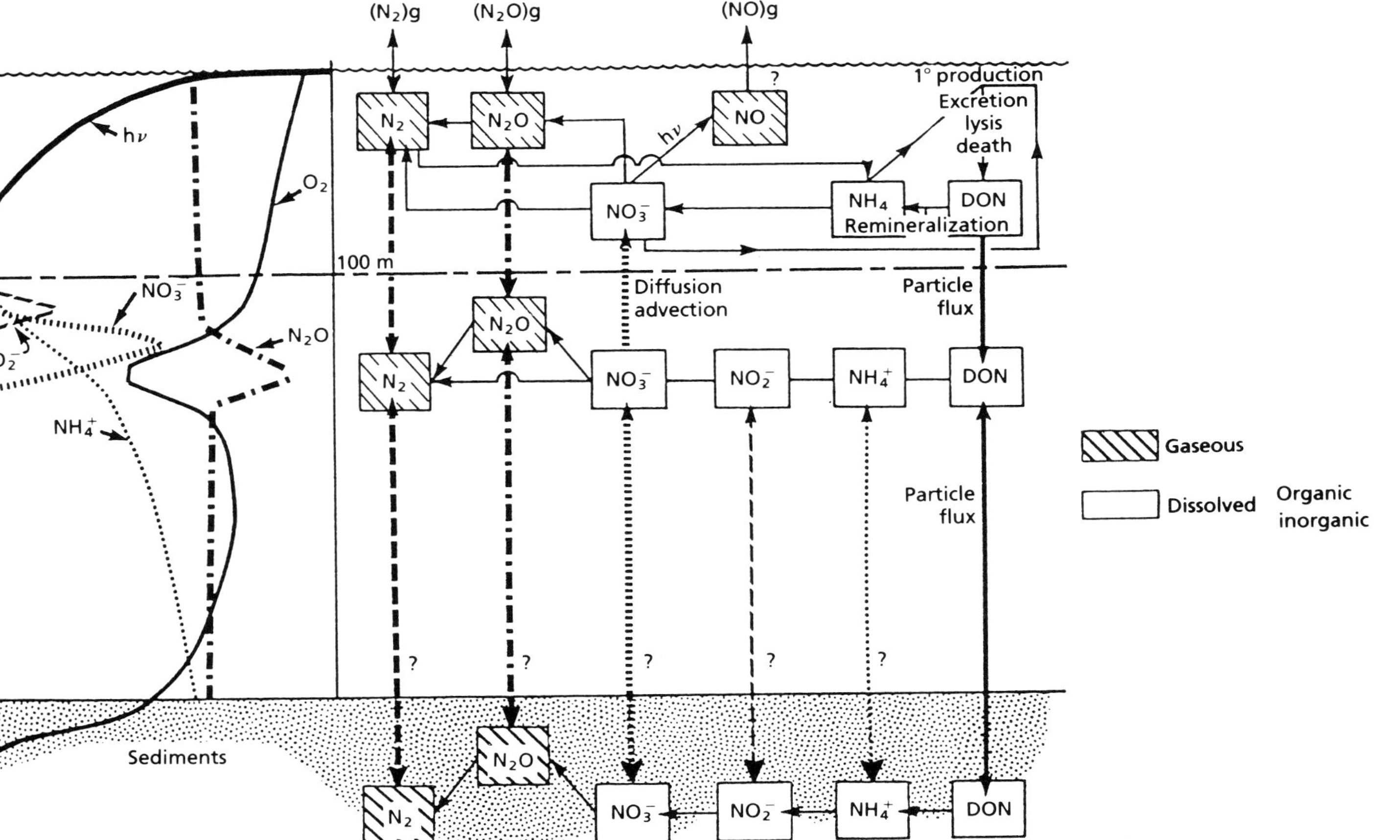

Figure 14.8 Vertical distributions of gaseous and dissolved forms of inorganic/organic nitrogen mediated by dominant transformation processes are shown on the left (see Figure 14.1). Fluxes and vertical transport of nitrogen cycle intermediates are shown for a representative vertically stratified water column. Exchanges between the water column and underlying sediments are also shown.

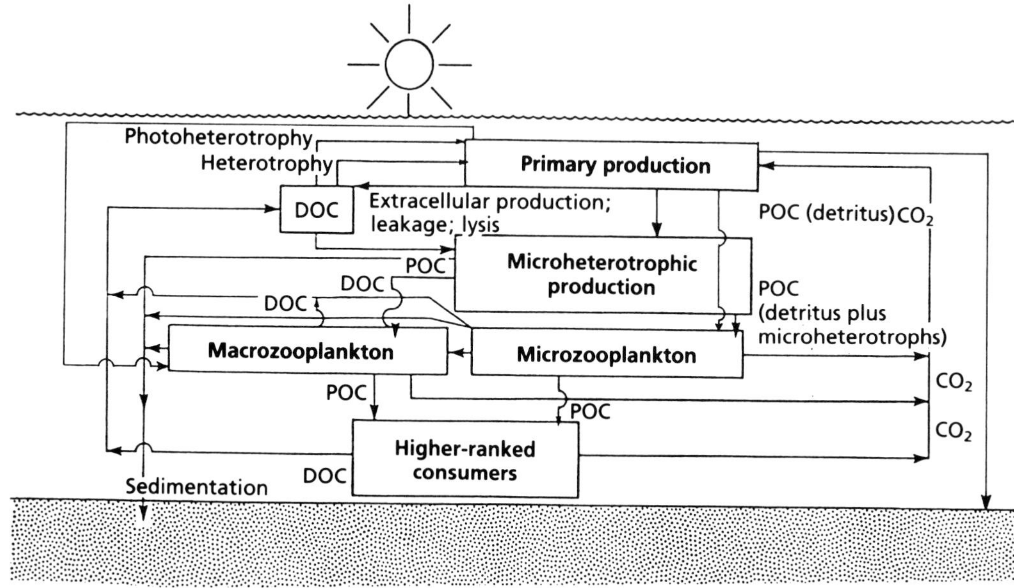

Figure 14.9 Planktonic trophic interactions between microbial primary producers, heterotrophs, and higher-ranked consumers. Resultant carbon fluxes under aerobic conditions are illustrated. Elements of the microbial organic and inorganic recycling component (microbial loop) of the microbial food web are also incorporated into this schematic diagram.

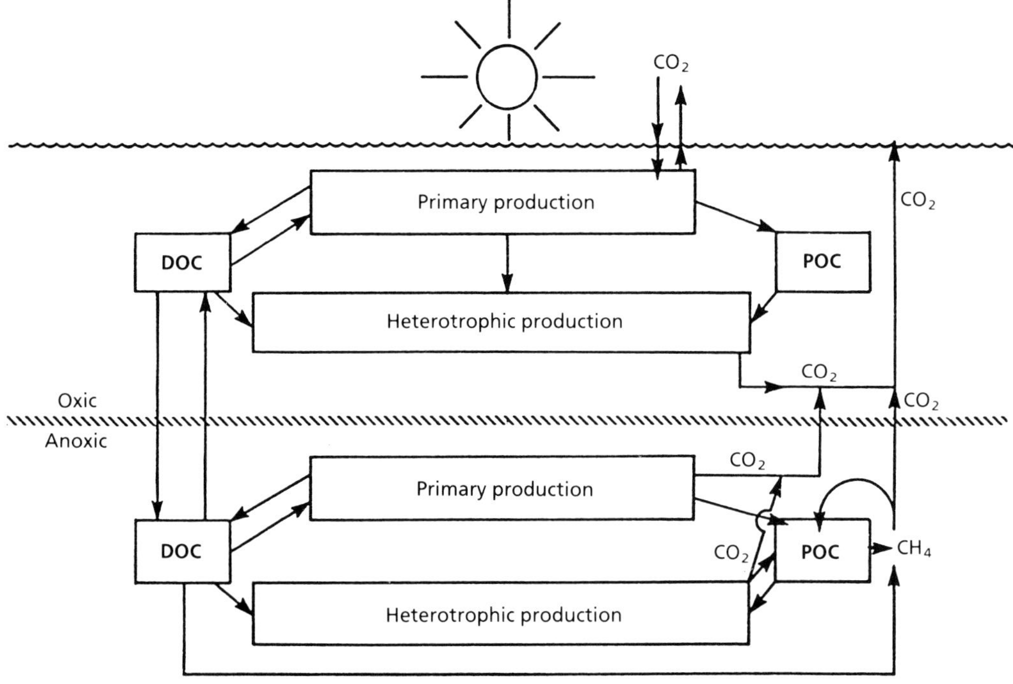

Figure 14.10 Carbon flux interactions between primary producers and microheterotrophs in representative oxic and anoxic waters. Soluble organic and inorganic carbon fluxes across the oxic–anoxic interface are also included.

both groups essentially rely on similar, and in some cases identical, organic carbon (energy) sources (Stewart 1971; Postgate 1978; Fenchel & Blackburn 1979; Payne 1981; Knowles 1982). Heterotrophic ammonifying bacteria can potentially make use of similar compounds as supplemental carbon sources (see Figure 14.4). These sources include a complex mix of carbohydrates, lipids, and proteinaceous compounds (Koike & Hattori 1978; Fenchel & Blackburn 1979; Fenchel 1980).

14.4 Recent conceptual and methodological considerations and evaluations of marine carbon/nitrogen cycling

The past two decades have witnessed some major revisions in our conceptual views of, and analytical approaches to, carbon and nitrogen cycling in planktonic and benthic habitats. Revisions are in response to recent findings that the bulk of planktonic and benthic carbon and nitrogen cycling involves microorganisms less than 2–3 μm in size (pico- and nanoplankton) (Johannes 1965; Fenchel 1980; Azam *et al.* 1983; Kirchman *et al.* 1989). Recently, a nutrient cycling role has been suggested for viral size organisms able to infect and lyze phytoplankton, thereby affecting species composition and possibly mediating bloom dynamics (Proctor & Fuhrman 1990; Bratbak *et al.* 1990; Suttle *et al.* 1990). These findings modify previous assumptions that micro- ($<$64 μm) and net ($>$64 μm) plankton were the dominant primary production and nutrient cycling organisms (Pomeroy 1984). The bulk of the pico- and nanoplankton are consumed by microorganisms not much larger in size, including ciliated protozoans (Fenchel 1980; Sieburth 1984; Porter *et al.* 1985; Jonsson 1987; Sherr & Sherr 1987; Sherr *et al.* 1989), and algal and bacterial flagellates (Gast 1985; Goldman & Caron 1985; Lessard & Swift 1985). Accordingly, a substantial portion of marine carbon and nitrogen fluxes and transformations appears confined to microbial size groups traditionally not considered in microscopic assessments of primary producers and microheterotrophs (Pomeroy 1984; Pace 1988; Kirchman *et al.* 1989). This has necessitated the application of novel identification, enumeration, and (habitat) characterization techniques. Included are: immunofluorescence, autofluorescence, both of which can now be coupled to detection and quantification via flow cytometry (Olson *et al.* 1990; Ward 1990); microautoradiography, coupled to cell staining and fluorescence microscopy techniques (Paerl 1974; Hoppe 1976; Meyer-Reil 1978); and the application of cell-specific nucleic acid and immunochemical probing techniques (Campbell & Iturriaga 1988; DeLong *et al.* 1989; Currin *et al.* 1990; Ward 1990). Production and carbon/nitrogen cycling dynamics associated with marine cyanobacterial picoplankton (*Synechococcus* spp., *Synechocystis* spp.) are of particular importance. These organisms are known to comprise a highly significant fraction of planktonic and benthic primary production in the oligotrophic ocean and certain coastal habitats (Waterbury *et al.* 1979; Iturriaga & Mitchell 1986; Olson *et al.* 1990). The contemporary study of picoplankton has necessitated consideration of appropriate and relevant scales applicable to environmental controls operating on such microorganisms (Platt & Li 1986; Waterbury *et al.* 1987). In addition, our perceptions of relevant nutrient uptake, transport, and assimilation kinetics and limits have been substantially altered (Raven 1986; Platt & Li 1987). Nutrient supply rates and concentrations previously thought to have been too low ("undetectable") to be important, are now receiving serious attention in light of findings that marginally detectable quantities and rates are highly relevant to the physiology and ecology of pico- and nanoplankton. Because of high surface/volume ratios and efficient uptake kinetics, nutrient utilization efficiencies of pico- and nanoplankton far exceed those for micro- and net plankton (Ducklow *et al.* 1986; Platt & Li 1986). Accordingly, analytical requirements have grown for determining "lower levels" of planktonically mediated nutrient flux and concentration. In addition, unique pico- and nanoplankton nutrient storage (nitrogen as the protein–pigments phycobilins; carbon as macromolecular polymers, e.g., PHBs) and consortial recycling capabilities have recently been emphasized. These are key ecophysiological strategies counteracting or compensating for exceedingly low, but periodically patchy, nutrient concentrations (McCarthy & Goldman 1979; Wyman *et al.* 1985).

Trophic roles of and interactions between photosynthetic pico- and nanoplankton, heterotrophic bacteria, and microzooplankton are far more complex than previously recognized. These complexities have a significant impact on the routes and amounts of oceanic carbon and nitrogen flux (Azam *et al.* 1983; Pomeroy 1984; Goldman & Caron 1985; Ducklow *et al.* 1986). For example, recently identified and frequently abundant microzooplankton, including ciliates and microflagellates, can under certain circumstances be the chief consumers of phototrophic picoplankton and microheterotrophs (Johannes 1965; Porter *et al.* 1985; Sherr & Sherr 1987). By doing so they function as an additional "link" in the planktonic food web, being situated between primary producers, heterotrophic decomposers, and crustacean macrozooplankton. Within the photosynthetic pico/nanoplankton grazed by micro- and macrozooplankton, heterotrophic assimilation of a variety of organic carbon and nitrogen substrates, including carbohydrates (simple sugars, organic acid), simple lipids (acetate), urea, and amino acids, has been demonstrated (McCarthy 1972; North 1975; Wheeler *et al.* 1977; Lewin & Hellebust 1978; Flynn & Butler 1987). Recent findings indicate that light-mediated

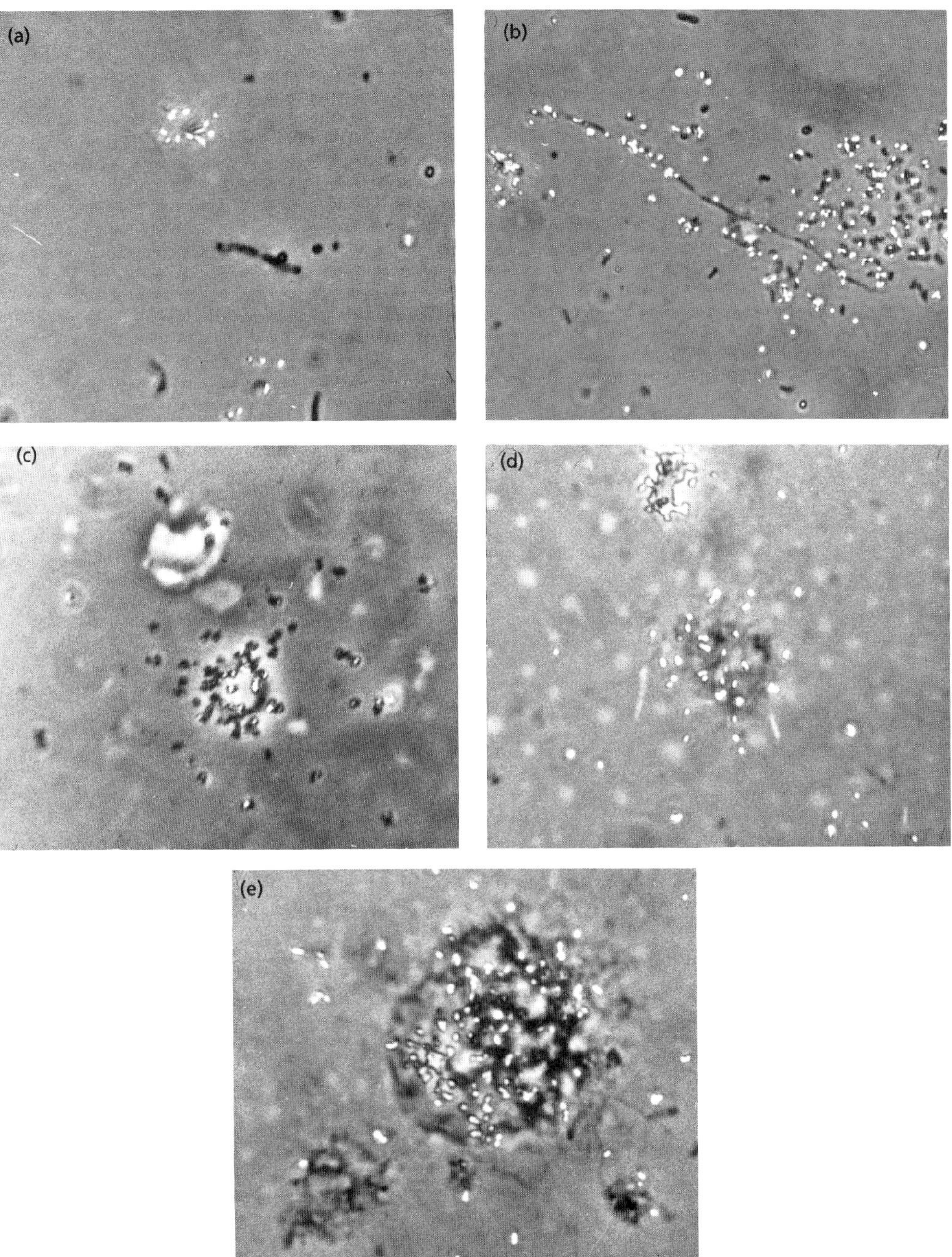

Figure 14.11 Microautoradiographs illustrating light-stimulated (photoheterotrophic) utilization of a ^{3}H-labeled amino acid mixture by picoplanktonic and microplanktonic phytoplankton in oligotrophic western Atlantic waters (sample taken near San Salvador Island, Bahamas). Both phytoplankton types revealed chlorophyll *a* autofluorescence when viewed with the epifluorescent microscope. Small white to gray dots covering and surrounding phytoplankton cells are nuclear track emulsion silver grains exposed by incorporated ^{3}H amino acids. Formalin-killed controls revealed no significant (above background) silver grain exposures, and hence no detectable ^{3}H amino acid incorporation. For details see Paerl (1991).

utilization of specific organic substrates (amino acids for example) also exists among picoplankton in nitrogen-depleted coastal and pelagic waters (Figure 14.11) (Paerl 1991). The overall biogeochemical and trophic ramifications of these additional microbial loop dissolved organic matter recycling mechanisms remain poorly understood (see Figures 14.6 and 14.9).

The commonly recognized biogeochemical role of microheterotrophs is as mineralizers of organic matter. However, it is rarely observed that more than 50% of assimilated organic carbon is respired as carbon dioxide (Williams 1970; Crawford *et al.* 1974; Lewin & Hellebust 1978; Hollibaugh *et al.* 1980); the remainder being incorporated as microbial biomass or excreted as extracellular mucous, capsular, and sheath material. Given such diverse fates of assimilated carbon, microheterotrophs should be considered in a dual fashion; as both mineralizers of organic matter and producers of particulate organic carbon (POC). Dissolved organic carbon (DOC) to POC conversion as mediated by heterotrophic bacteria (and perhaps fungi) therefore represents a microbial DOC retention or recovery system, in all likelihood optimizing trophic transfer of recently fixed organic carbon (Pomeroy 1984; Ducklow *et al.* 1986); the overall efficiencies and amounts of organic matter transferred from primary producers to microheterotrophs are subjects of some debate (Fuhrman & McManus 1984; Taylor & Joint 1990). It is safe to conclude however that microheterotrophs play a key functional role in DOC/DON (dissolved organic nitrogen) recovery and utilization by marine planktonic food webs (see Figure 14.9) (Paerl 1978b).

14.5 The roles and impacts of microscale versus macroscale interactions of carbon and nitrogen cycling

Contemporary experimental/empirical work on oceanic and estuarine carbon and nitrogen cycling has emphasized macroscale "big picture" measurements and conceptualizations (Woodwell *et al.* 1973; Paul 1978; Muller & Suess 1979; Premuzic *et al.* 1982; Lorenzen *et al.* 1983; Smith *et al.* 1983; Perry 1987; Falkowski & Kirshner 1990). Biogeochemical transformations, input/output, and flux determinations are invariably derived from measurements and calculations covering large volumes (>liters) and distances (>meters). The logistics of oceanographic and estuarine process and budgetary work justifiably dictate considerations on and extrapolations beyond such scales. However, underlying microbial transformations are mediated on far smaller scales. Environmental requirements and limitations relevant to microorganisms are frequently on the order of microliters to milliliters and microns to millimeters (Jørgensen 1983; Blackburn & Sorensen 1988; Jørgensen & Revsbech 1989). As such, controls on microbiological activities and production operate on scales far smaller than commonly measured or considered by biological and chemical oceanographers.

Recent research utilizing a variety of microscale electrochemical, immunological, molecular, and radiolabeling "probing" techniques, has demonstrated environmental heterogeneity and its control on a scale often distinct or independent from macroscale measurements. Perhaps most striking are contrasts in micro versus macroscale "patchiness" among such relevant parameters as dissolved and particulate organic matter (POM), pH, dissolved oxygen, inorganic nutrients (including carbon and nitrogen compounds), metallic, and sulfur compounds. The introduction and deployment of O_2, pH, Eh, S^{2-} and, recently, N_2O, NH_4^+, and NO_3^- microelectrodes has helped to evaluate scales relevant to biogeochemical transformations (Revsbech & Jørgensen 1986; Jørgensen & Revsbech 1989; Revsbech *et al.* 1989). Microhistological, microautoradiographic, and cell-specific molecular/immunological localization and probing techniques, have helped us realize and appreciate the fact that small-scale heterogeneity allows certain microbial processes to take place in microenvironments undetectable by larger scale sampling and analytical techniques. Examples include, particle- or surface-associated anoxic or microaerophilic processes, such as nitrogen fixation, denitrification, the coupling of ammonification to nitrification, methanogenesis, and sulfate and sulfur reduction in oxic bulk phase waters. In addition, differential (often enhanced) microbial assimilation and transformation of carbon- and nitrogen-containing inorganic and organic nutrients have been observed in association with particles and surfaces.

Appropriate time scales must also be considered, so as to accurately determine and characterize fluxes relevant to the ecosystem in question. Diel patterns in carbon and nitrogen transformations and net fluxes have been documented both on population and community levels. Among these, nitrogen transformations closely coupled to carbon transformation can be both interdependent and cross-inhibitory on a diel basis. An example is the closely coupled metabolic interplay of nitrogen fixation and oxygenic carbon dioxide fixation in individual cyanobacterial species, consortial (syntrophic) associations between cyanobacteria and eubacteria in microbial mats, detrital aggregates and surficial biofilm communities (Stal *et al.* 1985; Bebout *et al.* 1987; Paerl & Carlton 1988; Paerl *et al.* 1989). Because of strong inhibition of the oxygen-sensitive nitrogen-fixing complex nitrogenase by oxygenic photosynthesis, daytime nitrogen fixation rates are often suppressed; one result is the confinement and optimization of nitrogen fixation to night time, a period frequently under-

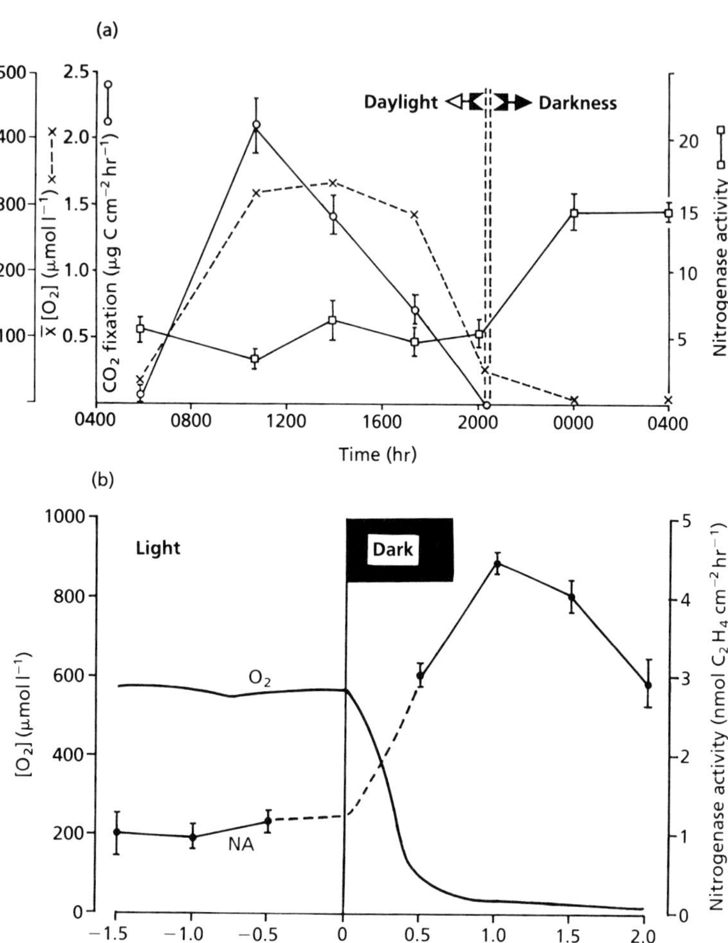

Figure 14.12 Examples of the temporal interaction of oxygenic photosynthesis and oxygen-sensitive nitrogen fixation in two representative marine habitats where these processes co-occur. (a) Illustrates the diel interactions of these processes in a marine intertidal microbial mat community dominated by filamentous cyanobacteria (*Lyngbya*, *Microcoleus*, *Oscillatoria*) and diatoms. Note that maximum rates of nitrogen fixation (acetylene reduction) characteristically occur in the evening hours after photosynthesis (carbon dioxide fixation) and concomitant oxygen evolution have ceased. Aerial carbon dioxide fixation was determined using the ^{14}C method, while oxygen saturation in the upper few millimeters of the mat was measured with microelectrodes. (b) Demonstrates a similar temporal interaction between these processes in a marine biofilm exposed to short-term (2 hr) alterations between illuminated and dark conditions. The microbial biofilm was established on Nitex (nylon) screens suspended in sea water. Screens were colonized by a diverse microbial community dominated by the coccoid cyanobacterium *Synechocystis* sp. Oxygen saturation was determined by maintaining a microelectrode near the biofilm surface during the light–dark cycles. Nitrogen fixation (NA, nitrogenase activity) was determined on pieces of screen, which were periodically examined using the acetylene reduction assay. (a) From Paerl (1990) and (b) from Paerl & Carlton (1988).

assessed or totally ignored in field determinations of "new" nitrogen inputs via this process (Figure 14.12).

Reductive processes, such as denitrification, sulfate reduction, and methanogenesis, experience similar inverse relationships with photosynthetic oxygen evolution on close-interval temporal and spatial scales (Sorensen *et al.* 1979; Nishio *et al.* 1982; Jørgensen 1983; Klump & Martens 1983; Reeburgh 1983). Clearly, relevant (both spatially and temporally) scalar assessments and their environmental controls will prove invaluable for defining, resolving, and quantifying the overall ecological and geochemical importance and interactions of microzone-mediated carbon and nitrogen transformations.

We are only now recognizing the ubiquity and potential geochemical implications of microscale carbon and nitrogen transformations on marine and freshwater fluxes and budgets extrapolated to ecosystem and global scales. Geochemical implications, in terms of revising and/or altering carbon or nitrogen fluxes, and their sources and sinks are manifold. A particularly relevant (and currently problematic) topic is that of "new" versus "regenerated" sources of nitrogen and carbon (Dugdale & Goering 1967; Eppley *et al.* 1973; Berner 1982; Emerson *et al.* 1985; Eppley 1989). From a biological perspective, nitrogen fixation represents a fundamentally important source of "new" nitrogen in aquatic ecosystems, potentially supporting new production in significant and widespread amounts (Paerl *et al.* 1981; Capone & Carpenter 1982). Recent recognition of the diverse genetic potentials of nitrogen fixers (as well as denitrifiers) leads to the following question. Are current estimations of new nitrogen inputs attributed to nitrogen fixation realistic and adequate (from geochemical nitrogen flux/budgetary modeling perspectives) or is this process underassessed?

14.6 Interaction of carbon and nitrogen cycles in supporting "new" versus "old" production: the paradigm versus consideration of new sources

It is generally agreed that the availability of biologically utilizable nitrogen controls microbial primary production in vast segments of the oceans and, to a lesser degree, freshwater ecosystems (Redfield *et al.* 1963; Dugdale 1967; Harrison *et al.* 1987; Hecky & Kilham 1988; Codispoti 1989). Because allochthonous nitrogen fixation appears to fall substantially (although the extent is still debatable) short of ecosystem demands, nitrogen remains chronically deficient and limiting in many pelagic environments (Capone & Carpenter 1982; Codispoti 1989). Pelagic waters under the influence of upwelling, and coastal waters receiving anthropogenic loading (from agricultural, domestic sewage, industrial, and other municipal sources), are closer to meeting ecosystem nitrogen demands (D'Elia *et al.* 1986; Shaffer 1987; Harrison *et al.* 1987). Not surprisingly, these nitrogen-enriched waters exhibit varying degrees of eutrophication, ranging from predictable phytoplankton blooms, i.e., upwelling and convergence zones, to nuisance algal blooms, including, most notably, toxic "red tide" dinoflagellates in euhaline waters and cyanobacteria in freshwater/estuarine habitats (McCarthy 1980; Mann 1982; Paerl 1988). In such nitrogen-sufficient waters other potentially growth-limiting nutrients such as phosphorus and iron may control primary production (Smith 1984; Doremus 1985; D'Elia *et al.* 1986; Codispoti 1989; LaPointe & O'Connell 1989; Martin *et al.* 1990).

Much of the marine environment, however, remains free of sizable and/or consistent (over time and space) nitrogen loading events. Under conditions of chronic nitrogen deficiency, nitrogen regeneration, including biochemical mineralization, reduction, and oxidation of preexisting "internal" organic nitrogen sources, is considered to be the major nitrogen supply for ecosystem maintenance (Dugdale & Goering 1967; Eppley & Peterson 1979; Harrison *et al.* 1987; Eppley 1989). Among pelagic waters, well away from continental margins, regeneration appears to be the chief nitrogen source supporting primary production (Harrison *et al.* 1987); on occasions being supplemented by ephemeral blooms of nitrogen fixing cyanobacteria (*Trichodesmium*; *Rhizoselenia-Richelia*) (Fogg 1982; Carpenter 1983) and possibly particle-associated nitrogen fixation (Paerl 1990). Coastal and marginal (continental shelf) waters offer a more complex interplay of new versus regenerated nitrogen inputs. In addition to well-documented terrigenous runoff and groundwater sources, atmospheric nitrogen inputs, in the form of wet and dry deposition, may represent an underestimated but potentially important source of new inorganic and organic nitrogen in such waters (Menzel & Spaeth 1962; Likens *et al.* 1974; Paerl 1985b; Duce 1986; Mopper & Zika 1987). Recent estimates of atmospheric nitrogen inputs are high (relative to other allochthonous (anthropogenic) nitrogen in coastal waters and estuaries), varying from 20 to 50% of nitrogen loading (Fisher *et al.* 1988; Prado-Fiedler 1990). This nitrogen source has been shown to significantly stimulate primary production in nitrogen-depleted coastal (Atlantic) North Carolina waters (Paerl *et al.* 1990). Given accelerating atmospheric emissions of biologically available nitrogen (as NO_x, NH_4^+, and DON) (Placit *et al.* 1986), the relative trophic importance of this source is likely to increase substantially as we enter the next century. This hitherto unaddressed source of nitrogen loading may not only change our perception and mechanistic understanding of new versus regenerated nitrogen loading, but, more seriously, alter the productive and trophic states of affected waters (Paerl *et al.* 1990; Prado-Fiedler 1990). As such, the relative roles and importance of regenerated versus new nitrogen inputs in supporting marine (near-shore) primary production are being reexamined with a critical eye on hitherto unrecognized and unassessed sources of nitrogen (Legendre & Gosselin 1989). These authors argued that regenerated nitrogen may play a less important (relative to atmospheric nitrogen) deterministic role in coastal ecosystems than previously proposed. The potential impacts of recent reevaluations of atmospheric N loading and N_2 fixation in coastal/neritic waters are shown in Tables 14.1–3.

This controversy has a potentially important bearing on the global carbon cycle. The connection between increased atmospheric nitrogen loading and enhanced carbon assimilation seems obvious, but requires verification from qualitative and quantitative experimental perspectives. Accelerating atmospheric nitrogen loading to nitrogen-depleted coastal surface waters enhances photosynthetic carbon dioxide fixation (Paerl *et al.* 1990). Resultant stimulated

Table 14.1 Estimates of new oceanic primary production supported by externally supplied (not regenerated in the euphotic zone) nitrogen inputs. Nonregenerated nitrogen inputs include: upwelling, eddy diffusion, terrestrial sources (runoff, sewage, river discharge), nitrogen fixation, and atmospheric deposition (rainfall plus dryfall). Previous estimates of Eppley & Peterson (1979) are compared with more recent, upward revisions of the relative contributions of atmospheric nitrogen deposition and nitrogen fixation from Legendre & Gosselin (1989). Legendre & Gosselin suggest that atmospheric deposition may be two- to three-times higher than previously considered in nitrogen budgets of nearshore, inshore, and neritic waters. Nitrogen fixation may also have been underestimated, in light of recent findings of surficial and microzonal inputs via this process. In the revised calculations nitrogen fixation inputs are doubled. Estimates are given as 10^9 tons C yr^{-1} total (P_T) or new (P_N) primary production

Oceanic region	Area (10^6 km^2)	(P_T)	(P_N)	P_N/P_T (%)
Oligotrophic open ocean	148	3.79 (EP)	0.24 (EP)	6
Transitional waters (between subtropical and subpolar zones)	83	4.22 (EP)	0.54 (EP)	13
Equatorial divergence waters/subpolar zones	86	6.31 (EP)	1.15 (EP)	18
Inshore waters	39	4.80 (EP) 5–6 (LG)	1.44 (EP) >2.75 (LG)	30 >50
Neritic waters	11	3.90 (EP) 5–6 (LG)	1.85 (EP) >3.30 (LG)	46 >60
Total	367	23.02 (EP) 25.32 (LG)	5.22 (EP) 7.98 (LG)	

EP, Eppley & Peterson; LG, Legendre & Gosselin.

Table 14.2 Calculated "new" nitrogen inputs to the Chesapeake Bay, USA, watershed, 1984. Note: organic nitrogen loading from atmospheric sources has not been included. Nitrogen fixation has also been omitted, but is considered to be relatively small (<5% of total inputs). Adapted from Fisher & Oppenheimer (1991)

Source	10^6 kg N yr^{-1}	Percentage of total
Atmospheric (precipitation)		
Nitrate	151	24
Ammonium	84	13
Animal waste	195	31
Fertilizer	158	25
Point sources (sewage and industrial discharge)	41	7
Total	629	100

Table 14.3 Estimates of the proportion of new nitrogen inputs attributable to atmospheric deposition in diverse estuarine, coastal, and pelagic seas. In some cases only wet deposition is considered. Where appropriate, wet plus dry deposition have been combined. Note: in all cases only inorganic nitrogen inputs have been considered

Water body	Percentage of new nitrogen inputs of atmospheric origin	Reference
Narragansett Bay, USA, 1983	5–10 (WD)	Nixon & Pilson (1983)
Chesapeake Bay, USA	39 (WD + DD)	Fisher & Oppenheimer (1991)
Neuse River–Pamlico Sound, USA	27 (WD + DD)	Paerl & Fogel (in press)
North Carolina Atlantic coastal waters (near Morehead City, USA)	35–60 (WD + DD)	Paerl & Fogel (in press)
Sargasso Sea (surface waters)	24 (WD)	Duce (1986)
North Pacific Gyre (surface waters)	4 (WD)	Duce (1986)
Western Mediterranean Sea, 1990	10–60 (WD)	Loye-Pilot *et al.* (1990)
Kiel Bight (Baltic Sea)	60 (WD + DD)	Prado-Fiedler (1990)
Baltic Sea	>25 (WD + DD)	Ambio (1990)

DD, dry desposition; WD, wet desposition.

carbon dioxide fixation should enhance the "drawdown" of atmospheric carbon dioxide, potentially affecting the air–sea equilibrium of this greenhouse gas. Quantitative data must be obtained before we can consider this scenario a biogeochemically significant interaction between the carbon and nitrogen cycles, with implications for both the regenerated versus new production concept and air–sea balances of carbon dioxide.

First and foremost is the need to quantify actual impacts of enhanced atmospheric nitrogen loading on large-scale (ecosystem-level) marine primary production. We need to assess the relative contribution of atmospheric versus other nitrogen input sources. In estuarine and coastal waters downwind from enriched atmospheric nitrogen sources, this contribution appears to be significant (Søderlund 1987; Fisher *et al.* 1988; Prado-Fiedler 1990). To what extent this remains significant in nearshore and more pelagic offshore coastal waters, or even open ocean environments, remains unknown (Paerl 1985b; Duce 1986; Paerl *et al.* 1990). Whether and how atmospheric inputs of new nitrogen and iron synergistically interact to control and enhance marine primary production also require experimental and quantitative assessments. The resultant amount of atmospheric-driven production which can be expected, as well as its trophic fate, remains unknown. However, this is critically important information because to exert measurable and significant impacts on global carbon budgets and the air–sea balance of carbon dioxide, intercepted atmospheric nitrogen (and iron) must be effectively transformed into plant biomass. Furthermore, this biomass must be transported to and buried as particulate carbon in the sediments (Suess 1980; Berner 1982; Toggweiler 1988). A bulk of the primary production resulting from atmospheric nitrogen loading occurs in shallow coastal shelf environments, where biomass and nonliving POM (detritus), formed by primary production, is often rapidly recycled (physically and biochemically) between the water column and underlying sediments, thus potentially minimizing permanent burial of recently fixed carbon dioxide (Muller & Suess 1979; Premuzic *et al.* 1982; Emerson *et al.* 1985). Ideally, recently fixed carbon should move seaward off the continental shelves to be buried in deeper oceanic sediments, as suggested by Walsh *et al.* (1981; 1985). Currently, insufficient information exists to allow for quantification of carbon fixation and transport mechanisms.

The manner by which carbon dioxide is fixed and its trophic transfer are important in predicting and quantifying its fate in estuarine, coastal, and pelagic ecosystems. Whether carbon dioxide is incorporated into planktonic or benthic plants, large versus small phytoplankton, toxic versus nontoxic micro- or macroalgal species, motile or sessile phytoplankton, all have bearing on location, physical dispersal or movement, edibility and hence transfer into, and transport by, higher-ranked trophic levels. Of additional concern is the linkage of accelerated nitrogen inputs to nuisance algal blooms (Paerl

1988; Paerl *et al.* 1990). While this concern is important to water quality and environmental health, it may also be a strong determinant in the geochemical fate of fixed carbon dioxide. Nuisance euhaline dinoflagellate (and freshwater cyanobacterial) blooms, are sometimes inedible to macrozooplankton and hence are utilized and regenerated (with respect to carbon and nitrogen) via trophic routes quite distinct from the more conventional zooplankton grazer → planktivorous fish → carnivorous fish food chain common to planktonic ecosystems (Steele 1974; Parsons *et al.* 1977; Mann 1982). Remineralization via a microbial loop in which a bulk of recently fixed carbon and nitrogen are transferred by bacterial heterotrophs to microzooplankton (Hagstrom *et al.* 1988), may represent a less effective means of burial of recently intercepted atmospheric carbon and nitrogen as opposed to the conventional macrozooplankton utilization route. Macrozooplankton often "package" unincorporated or nonrespired carbon as fecal pellets, which are known to enhance the downward transport of particulates in aquatic environments (Pomeroy *et al.* 1984; Fowler & Knauer 1986). Lastly, production and sinking of amorphous aggregates (marine snow), partially mediated by microorganisms and microzooplankton, have considerable impacts on vertical transport of carbon and nitrogen (Alldredge & Silver 1988).

If off-shelf transport of POC and PON is a significant component of carbon flux, then coastal interception of carbon dioxide may significantly mediate the air–sea carbon dioxide balance (Walsh *et al.* 1981; 1985). In addition, if atmospheric nitrogen deposition forms a significant fraction of new nitrogen inputs in waters responsible for a bulk of coastal and oceanic carbon flux, then a compelling argument can be made that atmospheric nitrogen influx in surface waters may play a significant role in mediating global air–sea carbon dioxide dynamics.

In considering how carbon and nitrogen inputs and their utilization via primary production may moderate or control each other in terms of global budgets of greenhouse gases such as carbon dioxide, we must not lose sight of other transformations of the carbon and nitrogen cycle which may concomitantly be affected by production and trophic changes, and whose altered activities may in turn affect the air–sea balance of these elements. One such transformation is denitrification. It has been shown that enhanced primary production, leading to increased inputs of oxidizable organic matter, stimulates denitrification (Goering 1978; Knowles 1982; Hattori 1983; Seitzinger 1988). Both the formation of oxygen-depleted microhabitats and organic matter as an energy/carbon source are generally promoted under enhanced "internal" (primary production) carbon loading. Thus, the rate of reduction of oxidized forms of inorganic N (such as NO_x) to N_2 via denitrification might be expected to increase. What does this mean for the oceanic nitrogen and carbon balance? Will enhanced production due to NO_x enrichment in surface waters be offset by nitrogen losses via stimulated denitrification? Clearly, many questions persist when we consider all facets of the linkage of carbon and nitrogen cycles as we try to characterize, quantify, and predict the biogeochemical effects of anthropogenic modifications of the biosphere.

14.7 The current DOC and DON controversy: What does it mean for microbial heterotrophic production and mineralization?

DOC and DON constitute vast and dynamic pools of combined elements (Lancelot & Billen 1985). In the case of nitrogen, DON is by far the largest constituent of the combined nitrogen pool (Sharp 1983). However, its reactivity in terms of biological utilization and oxidation is poorly understood (Jackson & Williams 1985). In open ocean waters, DON can account for nearly 80–90% of total combined nitrogen (DON + PON + DIN). Recent analytical modifications indicate that the DON pool may be greater than previously assumed (Sugamura & Suzuki 1988; Williams & Druffel 1989). Since extensive areas of the oceans exhibit nitrogen-limited conditions, the microbial utilization and consequent trophic/geochemical fate of biologically reactive DON and DOC warrant close evaluation (Fuhrman & Bell 1985; Toggweiler 1989). This is particularly relevant information in light of the fact that certain constituents of the DON/DOC pool (including amino acids, urea, organic acids, sugars, and sugar alcohols) are rapidly assimilated and recycled by heterotrophic bacteria (Hobbie *et al.* 1968; Williams 1970; McCarthy 1972; Crawford *et al.* 1974; Dawson & Goeke 1978; Hollibaugh *et al.* 1980; Paul 1983; Wheeler & Kirchman 1986). An unknown fraction of the DON and DOC pools is apparently also utilized via dark- and light-mediated mechanisms in prokaryotic (cyanobacteria) and eukaryotic (chlorophyceans, dinophyceans, pyrrhophyceans, bacillariophyceans) pico- and microplanktonic algae (Neilson & Lewin 1974; Wheeler *et al.* 1977; Lewin & Hellebust 1978; Paerl 1991) and benthic microalgae (Paerl *et al.* submitted). Whether such light-stimulated utilization of DON/DOC constitutes true photoheterotrophy remains to be determined. In a strict sense, photoheterotrophy implies that the participant is capable of growth on DON/DOC as the sole carbon source, a criterion that has neither been tested nor proven among microorganisms exhibiting light-stimulated organic matter incorporation (Paerl 1991). Using the photosynthetic (photosystem II) inhibitor 3(3,4-dichloro-phenyl)-1,1-dimethylurea (DCMU) however, it can be demonstrated that at least one criterion of photoheterotrophy, namely partial insensitivity to the action of this inhibitor (Neilson & Lewin 1974; Paerl 1991), can be met.

The light-stimulated reincorporation of DON and DOC constituents by phototrophic microorganisms, which are responsible for producing such compounds in the first place, via excretion, lysis, and "sloppy feeding" by consumer organisms (Sharp 1977), brings up several questions relevant to our conceptual and functional understanding of carbon and nitrogen cycling in the sea. These questions include:

1 What are the physiological and ecological rationales for such routes of internal carbon and nitrogen recycling among primary producers?

2 Does this mechanism confer a selective advantage to the microorganisms, in terms of obtaining organic nitrogen compounds in the face of severe DIN limitation?

3 Does this form of DON/DOC incorporation involve chemical signaling (specific DON/DOC compounds acting as chemical messages)?

4 How does light-mediated DON/DOC incorporation effect our current understanding of the microbial loop?

The last question merits some attention, for the conceptual and functional design of the microbial loop has been largely based on heterotrophic bacteria (rather than photoheterotrophs) as being the chief recyclers of organic matter in planktonic food chains (Azam & Hodson 1977; Azam *et al.* 1983; Ducklow *et al.* 1986). Much of the experimental/field verification of picoplanktonic DON/DOC utilization has been based on dark-incubated uptake of isotopically labeled amino acids, sugars, and organic acids; an approach which would assess utilization by heterotrophs (Azam & Hodson 1977; Fuhrman & Bell 1985), but not necessarily photoheterotrophs. Furthermore, size-fractionation experiments (using various pore size Nuclepore filters) have largely been based on the assumption that isotopically labeled picoplankton passing a 2 μm filter are exclusively bacterial. However, microautoradiographic examinations of this size-fraction reveal the presence and periodic preponderance of autofluorescent (chlorophyll *a* containing) picoplankton. This picoplankton component reveals both dark-mediated and light-stimulated incorporation of a wide range of organics (Paerl 1991).

These results implicate photosynthetic picoplankton (and microplankton) as playing a dual role as primary producers and heterotrophs. This dual role suggests that DOC/DON may be recycled among functionally similar or different microorganisms in a syntrophic or consortial manner. The conventional microbial loop in which dark-mediated microheterotrophs "recover" DOC/DON otherwise "lost" from the microbial food web, needs revision to accommodate this additional photoheterotrophic route. Lastly, this route could in part help explain the seemingly unusually high amount of "heterotrophic bacterial" (in fact some of this "bacterial" is algal) biomass periodically accompanying primary producer biomass.

14.8 Concluding remarks

Because they are key nutrients limiting and controlling primary production and heterotrophic decomposition, respectively, nitrogen and carbon are both indispensable and inseparable in considerations and evaluations of marine fertility, trophic structure, and material flux. Both elements are critically important structural and functional constituents of biochemically important molecules. The "meat and potatoes" components of microbial biomass, namely proteins, nucleic acids, carbohydrates/polysaccharides, and, to a lesser extent, lipids, largely owe their structural uniqueness and functional effectiveness to specific molecular carbon–nitrogen configurations. Many of the regulatory molecules controlling nutrient assimilation, photosynthesis, respiration, growth, and reproduction are reliant on specific orientations of carbon–nitrogen moieties.

Geochemically, the redox-mediated transformations and exchanges between gaseous and dissolved and particulate forms of carbon and nitrogen reveal both a remarkable parallel among respective cycles and interdependence in fulfilling cellular energetic, structural and growth needs. Close interactions of carbon and nitrogen cycling can be traced from the Archean beginnings of autotrophic and heterotrophic microbial "life styles" to the modern-day oxic biosphere. The evolution from anoxygenic (hydrogen sulfide-based) to oxygenic (water-based) photosynthesis as mediated by photosynthetic bacteria, cyanobacteria, and eukaryotic algae serves as an excellent example of biotic adaptations revolving around the common need to ensure adequate carbon and nitrogen sufficiency for growth and reproduction. Respiratory, nutrient, and energy storing mechanisms reveal a similar reliance on a high degree of coordination and synergism between efficient carbon and nitrogen cycling.

Accordingly, diverse microbial strategies have evolved and endured to meet and maintain cellular carbon and nitrogen demands in the ocean, which by and large is a desert-like, oligotrophic environment often depleted in biologically available forms of these elements. I have provided only a few familiar examples of the close interplay of parallel, complementary, biochemical carbon and nitrogen transformations and their adaptation to oligotrophy in representative marine habitats. These adaptations are critical in meeting microbial population and community, as well as ecosystem-level production and nutrient-cycling demands. Lastly, major biogeochemical fluxes, such as air–sea exchanges of gaseous forms of carbon (CO_2, CO,

CH_4) and nitrogen (NO_3, NO_2, NH_3, N_2O, N_2), remineralization, sedimentation, and burial of particulate forms of carbon and nitrogen, are joint products of nitrogen/carbon limited primary production and mineralization processes in planktonic and benthic marine environments.

Given such close coupling of carbon and nitrogen cycling in marine production and mineralization processes, it seems intuitively obvious that our alteration of, and interference with, natural cycling of these elements has and will continue to yield significant impacts on marine fertility and resultant water quality. It is not surprising, for example, that the best documented cases of marine anthropogenic or "cultural" eutrophication (i.e., human induced acceleration of microalgal primary production due to increased nutrient loading) have implicated enhanced combined nitrogen (NO_3^-, NH_4^+, DON) or organic carbon (as POC and/or DOC) loading, either individually or jointly. While in many such cases the primary production and biomass (as chlorophyll *a* or turbidity) responses to such loading events have been obvious and, at times, detrimental (toxic or nuisance algal blooms), acute and chronic impacts at higher trophic levels have yet to be realized, examined, or evaluated in most cases. Similarly, the abiotic fate of primary production in terms of sedimentation and burial is in most instances unknown. It must therefore be concluded that the overall impacts on ecosystem-level production, nutrient regeneration, and resultant carbon and nitrogen fluxes and budgets, as well as carbon and nitrogen-driven marine eutrophication, still require clarification and quantification. Superimposed on this is the realization that we have only begun to differentiate or filter out anthropogenic from naturally driven perturbations on marine carbon and nitrogen cycling.

A formidable and compelling set of research questions face marine microbiologists, biogeochemists, and biological oceanographers as we enter the twenty-first century. There exists increasing evidence that previously pristine coastal and pelagic oceanic waters are experiencing incipient, and, in some coastal-estuarine cases, more advanced symptoms of eutrophication. To what extent this constitutes "cultural eutrophication" and the biogeochemical and ecological ramifications of this process, are largely unknown. Clearly, a diverse set of critically important research topics lies ahead as we attempt to elucidate the dynamics of marine carbon and nitrogen cycling within the context of global change, and the implications for water quality and our resources.

Acknowledgments

I appreciate the assistance of L. White, B. Bright, H. Barnes, B. Bebout, and S. Joye with the preparation and documentation of this chapter. Research in this area and relevant experimental results discussed were supported by the National Science Foundation (Projects OCE 8820036; OCE 9012496; DEB 9210495; BSR 89-18482). The University of North Carolina Water Resources Research Institute–US Geological Survey (Project 5-46295) and the North Carolina Sea Grant Program (Project REHP-1).

References

Ahmadjian V. & Paracer S. (1986) *Symbiosis: An Introduction to Biological Associations*, pp. 28–56. University Press of New England, Hanover.

Alexander M. (1965) Nitrification. *Agronomy* **10**, 307–343.

Alldredge A.L. & Cohen Y. (1987) Can microscale chemical patches persist in the sea? Microelectrode study of marine snow, fecal pellets. *Science* **235**, 689–691.

Alldredge A.L. & Silver M.W. (1988) Characteristics, dynamics, and significance of marine snow. *Prog. Oceanogr.* **20**, 41–82.

Alperin M.J. & Reeburgh W.S. (1985) Inhibition experiments on anaerobic methane oxidation. *Appl. Environ. Microbiol.* **50**, 940–945.

Ambio (1990) Marine Eutrophication (whole issue). *Ambio* **19**, 101–176.

Azam F. & Hodson R.E. (1977) Size distribution and activity of marine microheterotrophs. *Limnol. Oceanogr.* **22**, 492–501.

Azam F., Fenchel T., Field J.G., Gray J.S., Meyer-Reil L.A. & Thingstad F. (1983) The ecological role of water-column microbes in the sea. *Mar. Ecol. Prog. Ser.* **10**, 257–263.

Bautista M.F. & Paerl H.W. (1985) Diel N_2 fixation in an intertidal marine cyanobacterial mat community. *Mar. Chem.* **16**, 369–377.

Bebout B.M., Paerl H.W., Crocker K.M. & Prufert L.E. (1987) Diel interactions of oxygenic photosynthesis and N_2 fixation (acetylene reduction) in a marine microbial mat community. *Appl. Environ. Microbiol.* **53**, 2353–2362.

Berger W.H. & Keir R.S. (1987) Glacial-holocene changes in atmospheric CO_2 and the deep sea record. In Hansen J.E. & Takahashi T. (eds) *Climate Processes and Climate Sensitivity*, Americ. Geophys. Union Geophys. Monogr. **29**, 337–351.

Berner R.A. (1982) Burial of organic carbon and pyrite in the modern ocean: its geochemical and environmental significance. *Am. J. Sci.* **82**, 451–475.

Blackburn T.H. & Henricksen K. (1983) Nitrogen cycling in different types of sediments from Danish waters. *Limnol. Oceanogr.* **28**, 477–493.

Blackburn T.H. & Sorensen J. (1988) *Nitrogen Cycling in Coastal Marine Environments*. John Wiley & Sons, New York.

Bohlool B.B. & Wiebe W.J. (1978) Nitrogen-fixing communities in an intertidal ecosystem. *Can. J. Microbiol.* **24**, 932–938.

Bratbak G., Heldal M., Norland S. & Thingstad T.F. (1990) Viruses as partners in spring bloom microbial trophodynamics. *Appl. Environ. Microbiol.* **56**, 1400–1405.

Campbell L. & Iturriaga R. (1988) Identification of *Synechococcus* spp. in the Sargasso Sea by immunofluorescence and fluorescence excitation spectroscopy performed on individual cells. *Limnol. Oceanogr.* **33**, 1196–1201.

Capone D.G. (1983) Benthic nitrogen fixation. In Carpenter E.J. & Capone D.G. (eds) *Nitrogen in the Marine Environment*, pp. 105–137. Academic Press, New York.

Capone D.G. (1991) Aspects of the marine nitrogen cycle with relevance to the dynamics of nitrous and nitric oxide. In Whitman W.B. & Rogers J.E. (eds) *Microbial Production and Consumption of Radiatively Important Trace Gases*, pp. 473–501. American Society for Microbiology, Washington DC.

Capone D.G. & Carpenter E.J. (1982) Nitrogen fixation in the marine environment. *Science* **217**, 1140–1142.

Carpenter E.J. (1983) Nitrogen fixation by marine *Oscillatoria* (*Trichodesmium*) in the world's oceans. In Carpenter E.J. & Capone D.G. (eds) *Nitrogen in the Marine*

Environment, pp. 65–104. Academic Press, New York.

Carpenter E.J. & Capone D.G. (1983) *Nitrogen in the Marine Environment.* Academic Press, New York.

Claypool G.E. & Kvenvolden K.A. (1983) Methane and other hydrocarbon gases in marine sediments. *Ann. Rev. Earth Planet. Sci.* **11**, 299–327.

Cloud P. (1976) Beginnings of biospheric evolution and their biogeochemical consequences. *Paleobiology* **2**, 351–387.

Codispoti L.A. (1989) Phosphorus vs. nitrogen limitations of new and export production. In Berger W.H., Smetacek V.S. & Wefer G. (eds) *Productivity of the Ocean: Present and Past*, pp. 377–394. John Wiley & Sons, New York.

Codispoti L.A. & Christensen J.P. (1985) Nitrification, denitrification and nitrous oxide cycling in the eastern tropical Pacific. *Mar. Chem.* **16**, 277–300.

Crawford C.C., Hobbie J.E. & Webb K.L. (1974) The utilization of dissolved free amino acids by estuarine microorganisms. *Ecology* **55**, 551–563.

Currin C.A., Paerl H.W., Suba G. & Alberte R.S. (1990) Immunofluorescent detection and characterization of N_2-fixing microorganisms from aquatic environments. *Limnol. Oceanogr.* **35**, 59–71.

Dawson R. & Goeke K. (1978) Heterotrophic activity in comparison to the free amino acid concentration in Baltic Sea water samples. *Oceanol. Acta* **1**, 45–54.

D'Elia C.F., Sanders J.G. & Boynton W.R. (1986) Nutrient enrichment studies in a coastal plain estuary: phytoplankton growth in large scale, continuous cultures. *Can. J. Fish. Aquat. Sci.* **43**, 397–406.

DeLong E.F., Wickham G.S. & Pace N.R. (1989) Phylogenetic strains: ribosomal RNA-based probes for the identification of single cells. *Science* **243**, 1360–1363.

Delwiche C.C. (1978) Biological production and utilization of N_2. *Pure Appl. Geophys.* **116**, 414–422.

Delwiche C.C. (1981) The nitrogen cycle and nitrous oxide. In Delwiche C.C. (ed.) *Denitrification, Nitrification and Atmospheric Nitrous Oxide*, pp. 1–15. Wiley Interscience, New York.

Devol A.H. (1983) Methane oxidation rates in the anaerobic sediments of Saanich Inlet. *Limnol. Oceanogr.* **28**, 739–742.

Doremus C. (1985) Geochemical control of dinitrogen fixation in the open ocean. *Biol. Oceanogr.* **1**, 429–436.

Duce R.A. (1986) The impact of atmospheric nitrogen, phosphorus, and iron species on marine biological productivity. In Menard P. (ed.) *The Role of Air–Sea Exchange in Geochemical Cycling*, pp. 497–529. D. Reidel, Berlin.

Ducklow H.W., Purdie D.A., Williams P.J. Le B. & Davies J.M. (1986) Bacterioplankton: A sink for carbon in a coastal marine plankton community. *Science* **232**, 865–867.

Dugdale R.C. (1967) Nutrient limitation in the sea: dynamics, identification and significance. *Limnol. Oceanogr.* **12**, 685–695.

Dugdale R.C. & Goering J.J. (1967) Uptake of new and regenerated nitrogen forms in primary productivity. *Limnol. Oceanogr.* **12**, 196–206.

Emerson S., Fischer K., Reimers C. & Heggie D. (1985) Organic carbon dynamics and preservation in deep-sea sediments. *J. Mar. Res.* **32**, 1–21.

Eppley R.W. (1989) New production: history, methods, and problems. In Berger W.H., Smetacek V.S. & Wefer G. (eds) *Productivity of the Ocean: Present and Past*, pp. 85–97. John Wiley & Sons, New York.

Eppley R.W. & Peterson B.J. (1979) Particulate organic matter flux and planktonic new production in the deep ocean. *Nature* **282**, 677–680.

Eppley R.W., Renger E.H., Venrick E.L. & Mullen M.M. (1973) A study of plankton dynamics and nutrient cycling in the Central Gyre of the North Pacific Ocean. *Limnol. Oceanogr.* **18**, 534–551.

Falkowski P.G. & Kirshner L.M. (1990) *Ocean Biology and Global Climate. Natural Science: The World and I*, pp. 284–291. News World Communications, New York.

Fenchel T. (1980) Suspension feeding in ciliated protozoa: feeding rates and their ecological significance. *Microb. Ecol.* **6**, 13–26.

Fenchel T. & Blackburn T.H. (1979) *Bacteria and Mineral Cycling*. Academic Press, New York.

Fisher D., Ceraso J. & Oppenheimer M. (1988) *Polluted Coastal Waters: The Role of Acid Rain*. Environmental Defense Fund, New York.

Fisher D.C. & Oppenheimer M.J. (1991) Atmospheric nitrogen deposition and the Chesapeake Bay Estuary. *Ambio* **20**, 102–108.

Flynn K.J. & Butler I. (1987) Nitrogen sources for the growth of marine microalgae: role of dissolved free amino acids. *Mar. Ecol. Prog. Ser.* **34**, 281–304.

Fogg G.E. (1982) Nitrogen cycling in sea water. *Philos. Trans. R. Soc. Lond. Ser.* **296**, 511–520.

Fogg G.E., Stewart W.D.P., Fay P. & Walsby A.E. (1973) *The Blue-Green Algae*. Academic Press, New York.

Fowler S.W. & Knauer G.A. (1986) Role of large particles in the transport of elements and organic compounds through the water column. *Prog. Oceanogr.* **16**, 147–194.

Fuhrman J.A. & McManus G.B. (1984) Do bacteria-sized marine eukaryotes consume significant bacterial production. *Science* **224**, 1257–1260.

Fuhrman J.A. & Bell T.M. (1985) Biological considerations in the measurement of dissolved free amino acids in seawater and implications for chemical and microbiological studies. *Mar. Ecol. Prog. Ser.* **25**, 13–21.

Gast V. (1985) Bacteria as a food source for microzooplankton in the Schlei Fjord and Baltic Sea with special reference to ciliates. *Mar. Ecol. Prog. Ser.* **22**, 107–120.

Glibert P.M. (1988) Primary productivity and pelagic nitrogen cycling. In Blackburn T.H. & Sorensen J. (eds) *Nitrogen Cycling in Coastal Marine Environments*, pp. 3–31. John Wiley & Sons, New York.

Goering J.J. (1978) Denitrification in marine systems. In Schlessinger D. (ed.) *Microbiology*, pp. 357–361. American Society for Microbiology, Washington DC.

Goering J.J., Richards F.A., Codispoti L.A. & Dugdale R.C. (1973) Nitrogen fixation and denitrification in the ocean: Biogeochemical budgets. *Proc. Symp. Hydrogeochem. Biogeochem.* **2**, 12–27.

Goldman J.C. & Caron D.A. (1985) Experimental studies on an omnivorous microflagellate: implications for grazing and nutrient regeneration in the marine microbial food web. *Deep-Sea Res.* **32**, 889–915.

Gotto J.W., Tabita F.R. & Van Baalen C. (1981) Nitrogen fixation in intertidal environments of the Texas Gulf Coast. *Estuarine, Coastal and Shelf Science* **12**, 231–235.

Hagstrom A., Azam F., Anderson A., Wilkner J. & Rassoulzadegan J. (1988) Microbial loop in an oligotrophic pelagic marine ecosystem: possible roles of cyanobacteria and nanoflagellates in the organic fluxes. *Mar. Ecol. Prog. Ser.* **49**, 171–178.

Hargrave B.T. (1973) Coupling carbon flow through some pelagic and benthic communities. *J. Fish. Res. Board. Can.* **30**, 1317–1326.

Harrison W.G., Platt T. & Lewis M.R. (1987) F-Ratio and its relationship to ambient nitrate concentration in coastal waters. *J. Plank. Res.* **9**, 235–248.

Hattori A. (1983) Denitrification and dissimilatory nitrate reduction. In Carpenter E.J. & Capone D.G. (eds) *Nitrogen in the Marine Environment*, pp. 191–232. Academic Press, New York.

Hecky R.E. & Kilham P. (1988) Nutrient limitation of phytoplankton in freshwater and marine environments: a review of recent evidence on the effects of enrichment. *Limnol. Oceanogr.* **33**, 796–822.

Heinrichs S.M. & Reeburgh W.S. (1987) Anaerobic mineralization of marine sediment organic matter: rates and the role of anaerobic processes in the oceanic carbon economy. *Geomicrobiol. J.* **5**, 191–237.

Hobbie J.E., Crawford C.C. & Webb K.L. (1968) Amino acid flux in an estuary. *Science* **159**, 1463–1464.

Hollibaugh J.T., Carruthers A.B., Fuhrman J.A. & Azam F. (1980) Cycling of organic nitrogen in marine plankton communities studied in enclosed water columns. *Mar. Biol.* **59**, 15–21.

Hoppe H.G. (1976) Determination and properties of actively metabolizing heterotrophic bacteria in the sea, investigated by means of microautoradiography. *Mar. Biol.* **36**, 291–302.

Howarth R.W., Marino R., Lane J. & Cole J.J. (1988) Nitrogen fixation in freshwater, estuarine and marine ecosystems. 1. Rates and importance. *Limnol. Oceanogr.* **33**, 619–687.

Iturriaga R. & Mitchell B.G. (1986) Chrococcoid cyanobacteria: a significant component in the food web dynamics of the open sea. *Mar. Ecol. Prog. Ser.* **28**, 291–297.

Jackson A.J. & Williams P.M. (1985) Importance of dissolved organic nitrogen and phosphorus to biological nutrient cycling. *Deep-Sea Res.* **32**, 223–235.

Jannasch H.W. (1985) The chemosynthetic support of life and the microbial diversity at deep-sea hydrothermal vents. *Proc. R. Soc. Lond.* **225**, 277–297.

Johannes R.E. (1965) Influence of marine protozoa in nutrient regeneration. *Limnol. Oceanogr.* **10**, 434–442.

Jonsson P.R. (1987) Photosynthetic assimilation of inorganic carbon in marine oligotrich ciliates (Ciliophora, Oligotrichina). *Mar. Microb. Food Webs* **2**, 55–68.

Jørgensen B.B. (1983) Processes at the sediment–water interface. In Bolin B. & Cook R. (eds) *The Major Biogeochemical Cycles and their Interactions. Scope 21*, pp. 477–509. John Wiley & Sons, New York.

Jørgensen B.B. & Revsbech N.P. (1989) Oxygen uptake, bacterial distribution, and carbon–nitrogen–sulfur cycling in sediments from the Baltic Sea–North Sea transition. *Ophelia* **31**, 51–72.

Kaplan W., Valiela I. & Teal J.M. (1979) Denitrification in a salt marsh ecosystem. *Limnol. Oceanogr.* **24**, 726–734.

Karl D.M. (1982) Microbial transformations of organic matter at oceanic interfaces: a review and prospectus. *EOS* **63**, 138–140.

Kirchman D.L., Keil R.G. & Wheeler P.A. (1989) The effect of amino acids on ammonium utilization and regeneration by heterotrophic bacteria in the subarctic Pacific Ocean. *Deep-Sea Res.* **36**, 1763–1776.

Klump J.V. & Martens C.S. (1983) Benthic nitrogen regeneration. In Carpenter E.J. & Capone D.G. (eds) *Nitrogen in the Marine Environment*, pp. 411–457. Academic Press, New York.

Knoll A.H. (1977) Paleomicrobiology. In Laskins A.I. & LeChevalier H.A. (eds) *Handbook of Microbiology*, pp. 8–29. CRC Press, Boca Raton.

Knowles R. (1982) Denitrification. *Microbiol. Rev.* **46**, 43–70.

Koike I. & Hattori A. (1978) Denitrification and ammonia formation in anaerobic coastal sediments. *Appl. Environ. Microbiol.* **35**, 278–282.

Lancelot C. & Billen G. (1985) Carbon–nitrogen relationships in nutrient metabolism of coastal marine ecosystems. In Jannasch H.W. & Williams P.J. Le B. (eds) *Advances in Aquatic Microbiology*, pp. 263–321. Academic Press, London.

LaPointe E. & O'Connell J. (1989) Nutrient-enhanced growth of *Cladophora prolifera* in Harrington Sound, Bermuda: eutrophication of a confined, phosphorus-limited marine ecosystem. *Estuarine, Coastal and Shelf Science* **28**, 347–360.

Legendre L. & Gosselin M. (1989) New production and export of organic matter to the deep ocean: consequences of some recent discoveries. *Limnol. Oceanogr.* **34**, 1374–1380.

Lehninger A.L. (1975) *Biochemistry*. Worth, New York.

Lessard E.J. & Swift E. (1985) Species-specific grazing rates of heterotrophic dinoflagellates in oceanic waters, measured with a dual-label radioisotope technique. *Mar. Biol.* **87**, 289–296.

Lewin J. & Hellebust J.A. (1978) Utilization of glutamate and glucose for heterotrophic growth by the marine pennate diatom *Nitzchia laevis*. *Mar. Biol.* **47**, 1–7.

Likens G.E., Bormann F.H. & Johnson N.M. (1974) Acid rain. *Environment* **14**, 33–40.

Lorenzen C.J., Welschmeyer N.J. & Copping A.E. (1983) Particulate organic flux in the subarctic Pacific. *Deep-Sea Res.* **30**, 639–643.

Loye-Pilot M.D., Martin J.M. & Morelli J. (1990) Atmospheric input of inorganic nitrogen to the Western Mediterranean. *Biogeochemistry* **9**, 117–134.

Mann K.H. (1982) *Ecology of Coastal Waters*. University of California Press, Berkeley.

Marshall K.C. (1979) Growth at interfaces. In Shilo M. (ed.) *Proceedings of the Dahlem Workshop on Strategy of Life in Extreme Environments*, pp. 281–290. Chemie Verlag, Berlin.

Martens C.S. (1984) Recycling of organic carbon near the sediment–water interface in coastal environments. *Bull. Mar. Sci.* **35**, 566–575.

Martin J., Fitzwater S.E. & Gordon R.M. (1990) Iron deficiency limits phytoplankton growth in Antarctic waters. *Global Geochem. Cycles* **4**, 5–12.

McCarthy J.J. (1972) The uptake of urea by natural populations of marine phytoplankton. *Limnol. Oceanogr.* **17**, 738–748.

McCarthy J.J. (1980) Nitrogen and phytoplankton ecology. In Morris I. (ed.) *The Physiological Ecology of Phytoplankton*, pp. 191–233. Blackwell Scientific Publications, Oxford.

McCarthy J.J. & Goldman J.C. (1979) Nitrogenous nutrition of marine phytoplankton in nutrient-depleted waters. *Science* **203**, 670–672.

Menzel D.W. & Spaeth J.P. (1962) Occurrence of ammonia in Sargasso Sea waters and in rain water at Bermuda. *Limnol. Oceanogr.* **7**, 159–162.

Meyer-Reil L.A. (1978) Autoradiography and epifluorescence microscopy combined for the determination of number and spectrum of actively metabolizing bacteria in natural waters. *Appl. Environ. Microbiol.* **36**, 506–512.

Mopper K. & Zika R.G. (1987) Free amino acids in marine rains: evidence for oxidation and potential role in nitrogen cycling. *Nature* **325**, 246–249.

Morris I. (1974) Nitrogen assimilation and protein synthesis. In Stewart W.D.P. (ed.) *Algal Phycology and Biochemistry*, pp. 583–609. Blackwell Scientific Publications, Oxford.

Muller P.J. & Suess E. (1979) Productivity, sedimentation, and sedimentary organic carbon in the oceans. I. Organic carbon preservation. *Deep-Sea Res.* **26**, 1347–1362.

Neilson A.H. & Lewin R.A. (1974) The uptake and utilization of organic carbon by algae: an essay in comparative biochemistry. *Phycologia* **13**, 227–264.

Nishio T., Koike I. & Hattori A. (1982) Denitrification, nitrate reduction, and oxygen consumption in coastal and estuarine sediments. *Appl. Environ. Microbiol.* **43**, 648–653.

Nixon S.W. & Pilson M.E.Q. (1983) Nitrogen in estuarine and coastal marine ecosystems. In Carpenter E.J. & Capone D.G. (eds) *Nitrogen in the Marine Environment*, pp. 565–648. Academic Press, New York.

Nixon S.W., Oviatt C.A. & Hale S.S. (1976) Nitrogen regeneration and the metabolism of coastal marine bottom communities. In Anderson J.M. & Macfayden A. (eds) *The Role of Terrestrial and Aquatic Organisms in Decomposition Processes*, pp. 269–283. Blackwell Scientific Publications, Oxford.

North B.B. (1975) Primary amines in California coastal waters: utilization by phytoplankton. *Limnol. Oceanogr.* **20**, 20–27.

Olson R.J., Chisholm S.W., Zettler E.R. & Armbrust E.V. (1990) Pigments, size and distributions of *Synechococcus* in North Atlantic and Pacific Oceans. *Limnol. Oceanogr.* **35**, 45–58.

Oremland R.S. (1988) Biogeochemisty of methanogenic bacteria. In Zehnder A.J.B. (ed.) *Biology of Anaerobic Microorganisms*, pp. 641–705. John Wiley & Sons, New York.

Pace M.L. (1988) Bacterial mortality and the fate of bacterial production.

Hydrobiologia **159**, 41–50.
Paerl H.W. (1974) Bacterial uptake of dissolved organic matter in relation to detrital aggregation in marine and freshwater systems. *Limnol. Oceanogr.* **19**, 966–972.
Paerl H.W. (1978a) Role of heterotrophic bacteria in promoting N_2 fixation in bacteria by *Anabaena* in aquatic habitats. *Microb. Ecol.* **40**, 215–231.
Paerl H.W. (1978b) Microbial carbon recovery in aquatic ecosystems. *Limnol. Oceanogr.* **23**, 927–935.
Paerl H.W. (1985a) Influence of attachment on microbial metabolism and growth in aquatic ecosystems. In Savage D.C. & Fletcher M. (eds) *Bacterial Adhesion: Mechanisms and Physiological Significance*, pp. 363–400. Plenum, New York.
Paerl H.W. (1985b) Enhancement of marine primary production by nitrogen-enriched acid rain. *Nature* **316**, 747–749.
Paerl H.W. (1988) Nuisance phytoplankton blooms in coastal, estuarine, and inland waters. *Limnol. Oceanogr.* **33**, 823–847.
Paerl H.W. (1990) Physiological ecology and regulation of N_2 fixation in natural waters. *Adv. Microbiol. Ecol.* **11**, 305–344.
Paerl H.W. (1991) Ecophysiological and trophic implications of light stimulated amino acid utilization in marine picoplankton. *Appl. Environ. Microbiol.* **57**, 473–479.
Paerl H.W. & Prufert L.E. (1987) Oxygen-poor microzones as potential sites of microbial N_2 fixation in nitrogen-depleted aerobic marine waters. *Appl. Environ. Microbiol.* **53**, 1078–1087.
Paerl H.W. & Bebout B.M. (1988) Direct measurement of O_2-depleted microzones in marine *Oscillatoria*: relation to N_2 fixation. *Science* **241**, 442–445.
Paerl H.W. & Carlton R.G. (1988) Control of nitrogen fixation by oxygen depletion in surface-associated microzones. *Nature* **332**, 260–262.
Paerl H.W. & Fogel M.L. (In press) Atmospheric deposition to estuarine and coastal waters: is there a link to eutrophication? In Dyer K.R. (ed.) *Changes in Fluxes in Estuaries: Implications for Science and Management.* Olsen and Olsen, Copenhagen.
Paerl H.W., Webb K.L. & Wiebe W.J. (1981) Nitrogen fixation in waters. In Broughton W.J. (ed.) *Nitrogen Fixation. Volume 1: Ecology*, pp. 193–240. Oxford Science Publications, Oxford.
Paerl H.W., Crocker K.M. & Prufert L.E. (1987) Limitation of N_2 fixation in coastal marine waters: relative importance of molybdenum, iron, phosphorus, and organic matter availability. *Limnol. Oceanogr.* **32**, 525–536.
Paerl H.W., Bebout B.M., Prufert L.E. & Currin C.A. (1989) Mediation of aquatic N_2 fixation potentials by O_2-depleted microzones. In Hattori T., Ishida Y., Maruyama Y., Morita R.Y. & Uchida A. (eds) *Recent Advances in Microbial Ecology*, pp. 113–117. Japan Scientific Society Press, Tokyo.
Paerl H.W., Rudek J. & Mallin M.A. (1990) Stimulation of phytoplankton production in coastal waters by natural rainfall inputs: nutritional and trophic implications. *Mar. Biol.* **107**, 247–254.
Parsons T.R., Takahashi M. & Hargrave B. (1977) *Biological and Oceanographic Processes.* Pergamon, Oxford.
Paul E.A. (1978) Contribution of nitrogen fixation to ecosystem functioning and nitrogen fluxes on a global basis. *Ecol. Bull.* **26**, 282–293.
Paul J.H. (1983) Uptake of organic nitrogen. In Carpenter E.J. & Capone D.G. (eds) *Nitrogen in the Marine Environment*, pp. 275–308. Academic Press, New York.
Payne W.J. (1981) *Denitrification.* John Wiley & Sons, New York.
Perry M.J. (1987) Assessing marine primary production from space. *Bioscience* **36**, 461–467.
Placit M., Streets D.G. & Williams E.R. (1986) *Environmental Trends Associated with the 5th National Energy Policy Plan. Prepared for Argonne National Laboratory. ANL-EES-TM-323.* National Acid Precipitation Assessment Program, Washington DC.
Platt T. & Li W.K.W. (eds) (1986) *Photosynthetic Picoplankton. Can. Bull. Fish.*

Aquat. Sci. **214** (special issue).

Pomeroy L.R. (1984) Significance of microorganisms in carbon and energy flow in marine ecosystems. In Klug M.J. & Reddy C.A. (eds) *Current Perspectives in Microbial Ecology*, pp. 405–411. American Society for Microbiology, Washington DC.

Pomeroy L.R., Hanson R.B., McGillivary P.A., Sherr B.F., Kirchman D.L. & Deidel D. (1984) Microbiology and chemistry of fecal products of pelagic tunicates: rates and fates. *Bull. Mar. Sci.* **35**, 426–439.

Porter K.G., Sherr E.B., Sherr B.F., Pace M. & Sanders R.W. (1985) Protozoa in planktonic food webs. *J. Protozool.* **32**, 409–415.

Postgate J.R. (1978) *Nitrogen Fixation. Studies in Biology No. 92.* Edward Arnold, London.

Prado-Fiedler R. (1990) Atmospheric input of inorganic nitrogen species to the Kiel Bight. *Helgol. Meeresunters.* **44**, 21–30.

Premuzic E.T., Benkovitz C.M., Gaffney J.S. & Walsh J.J. (1982) The nature and distribution of organic matter in the surface sediments of world oceans and seas. *Org. Geochem.* **4**, 63–77.

Proctor L.M. & Fuhrman J.A. (1990) Viral mortality of marine bacteria and cyanobacteria. *Nature* **343**, 60–62.

Raven J.A. (1986) Physiological consequences of extremely small size for autotrophic organisms in the sea. In Platt T. & Li W.K.W. (eds) *Photosynthetic Picoplankton. Can. Bull. Fish. Aquat. Sci.* **214** (special issue), 1–70.

Redfield A.C. (1958) The biological control of chemical factors in the environment. *Am. Sci.* **46**, 206–226

Redfield A.C., Ketchum B.H. & Richards F.A. (1963) The influence of organisms on the composition of sea water. In Hill M.N. (ed.) *The Sea*, vol. 2, pp. 26–77. Wiley Interscience, New York.

Reeburgh W.S. (1983) Rates of biogeochemical processes in anoxic sediment. *Ann. Rev. Earth Planet. Sci.* **11**, 269–294.

Revsbech N.P. & Jørgensen B.B. (1986) Microelectrodes: their use in microbial ecology. *Adv. Microbiol. Ecol.* **9**, 293–352.

Revsbech N.P., Christensen P.B., Nielsen L.P. & Sorensen J. (1989) Denitrification in a trickling filters biofilm studied by a microsensor for oxygen and nitrous oxide. *Wat. Res.* **23**, 1–5.

Ryther J.H. & Dunstan W.M. (1971) Nitrogen, phosphorus and eutrophication in the coastal marine environment. *Science* **171**, 1008–1012.

Schopf J.W. & Walter M.R. (1982) Origin and early evolution of cyanobacteria: the geological evidence. In Carr N.C. & Whitton B.A. (eds) *The Biology of Cyanobacteria*, pp. 543–564. Blackwell Scientific Publications, Oxford.

Schopf J.W., Hayes J.M. & Walter M.R. (1983) Evolution of Earth's earliest ecosystems: recent progress and unsolved problems. In Schopf J.W. (ed.) *Earth's Earliest Biosphere: Its Origin and Evolution*, pp. 361–384. Princeton University Press, Princeton.

Seitzinger S. (1988) Denitrification in freshwater and coastal marine ecosystems: ecological and geochemical significance. *Limnol. Oceanogr.* **33**, 702–724.

Shaffer G. (1987) Redfield ratios, primary production, and organic carbon burial in the Baltic Sea. *Deep-Sea Res.* **34**, 769–784.

Sharp J.H. (1977) Excretion of organic matter by marine phytoplankton. Do healthy cells do it?. *Limnol. Oceanogr.* **22**, 381–399.

Sharp J.H. (1983) The distributions of inorganic nitrogen and dissolved and particulate organic nitrogen in the sea. In Carpenter E.J. & Capone D.G. (eds) *Nitrogen in the Marine Environment*, pp. 1–35. Academic Press, New York.

Sharp H.J., Pennock J.R., Church T.M., Tramatono J.K. & Cifuentes L.A. (1984) The estuarine interaction of nutrients, organics and metals: a case study in the Delaware Estuary. In Kennedy V.S. (ed.) *The Estuary as a Filter*, pp. 135–159. Academic Press, New York.

Sherr B.F., Sherr E.B. & Pedrós-Alió C. (1989) Simultaneous measurement of bacterioplankton production and protozoan bacterivory in estuarine water. *Mar. Ecol. Prog. Ser.* **54**, 209–219.

Sherr E.B. & Sherr B.F. (1987) High rates of consumption of bacteria by pelagic ciliates. *Nature* **325**, 710–711.

Sieburth J.McN. (1984) Protozoan bactivory in pelagic marine waters. In Hobbie J.E. & Williams P.J. Le B. (eds) *Heterotrophic Activity in the Sea*, pp. 405–444. Plenum, New York.

Smith R.C., Campbell J.W., Esaias N.E. & McCarthy J.J. (1983) *Primary Production of the World Ocean*, pp. 26–83. Scripps Institute of Oceanography, La Jolla.

Smith S.V. (1984) Phosphorus vs. nitrogen limitation in the marine environment. *Limnol. Oceanogr.* **29**, 1149–1160.

Smith S.V. & Hollibaugh J.T. (1989) Carbon-controlled nitrogen cycling in a marine "macrocosm": an ecosystem-scale model for managing cultural eutrophication. *Mar. Ecol. Prog. Ser.* **52**, 103–109.

Søderlund R. (1987) *Deposition estimates to the Baltic Sea area based on reported data for 1983/85*, pp. 1–13. Baltic Marine Environment Protection Commission, HELCOM-Rep. EGAP 4/4a. Stockholm, Sweden.

Sorensen J. (1978) Denitrification rates in marine sediment as measured by the acetylene inhibition technique. *Appl. Environ. Microbiol.* **36**, 139–143.

Sorensen J., Jørgensen B.B. & Revsbech N.P. (1979) A comparison of oxygen, nitrate, and sulfate respiration in marine sediments. *Mar. Ecol.* **5**, 105–115.

Stal L.J., Van Gemerden H. & Krumbein W.E. (1985) Structure and development of a benthic marine microbial mat. *FEMS Microbiol. Ecol.* **31**, 111–125.

Steele J.M. (1974) *The Structure of Marine Ecosystems*. Harvard University Press, Cambridge.

Stewart W.D.P. (1971) Nitrogen fixation in the sea. In Costlow J.D. Jr (ed.) *Fertility of the Sea*, vol. 2, pp. 537–564. Gordon & Breach, New York.

Stewart W.D.P., Rowell P. & Rai A.N. (1980) Symbiotic nitrogen-fixing cyanobacteria. In Stewart W.D.P. & Gallon J.R. (eds) *Proceedings of an International Symposium on Nitrogen Fixation*, pp. 239–277. Oxford University Press, Oxford.

Suess E. (1976) Nutrients near the depositional interface. In McCave I.N. (ed.) *The Benthic Boundary Layer*, pp. 57–59. Plenum, New York.

Suess E. (1980) Particulate organic carbon flux in the oceans: surface productivity and oxygen utilization. *Nature* **288**, 260–263.

Sugamura Y. & Suzuki Y. (1988) A high temperature catalytic oxidation method of non-volatile dissolved organic carbon in seawater by direct injection of liquid samples. *Mar. Chem.* **24**, 105–111.

Suttle C.A., Chen A.M. & Cotrell M.T. (1990) Infection of phytoplankton by viruses and reduction of primary productivity. *Nature* **347**, 467–469.

Taylor A.H. & Joint I. (1990) A steady state analysis of the "microbial loop" in stratified systems. *Mar. Ecol. Prog. Ser.* **59**, 1–17.

Toggweiler J.R. (1988) Deep-sea carbon, a burning issue. *Nature* **344**, 468.

Toggweiler J.R. (1989) Is the downward dissolved organic matter (DOM) flux important in carbon transport? In Berger W.H., Smetacek V.S. & Wefer G. (eds) *Productivity of the Ocean: Present and Past*, pp. 65–83. John Wiley & Sons, New York.

Valiela I. & Teal J.M. (1979) The nitrogen budget of a salt marsh ecosystem. *Nature* **280**, 652–656.

Walker J.C.G., Klein C., Schidlowski M., Schopf J.W., Stevenson D.J. & Walter M.R. (1982) Environmental evolution of the Archean-early Proterozoic Earth. In Schopf J.W. (ed.) *Origin and Evolution of Earth's Earliest Biosphere*, pp. 50–79. Princeton University Press, Princeton.

Walsh J.J., Rowe G.T., Iverson R.L. & McRoy C.P. (1981) Biological export of shelf carbon is a sink of the global CO_2 cycle. *Nature* **291**, 196–201.

Walsh J.J., Premuzic E.T., Gaffney J.S., Rowe G.T., Harbottle G., Stoenner R.W., Balsam W.L, Betzer P.R. & Macke S.A. (1985) Organic storage of CO_2 on the continental slope of the mid-Atlantic Bight, the south-eastern Bering Sea and the Peru coast. *Deep-Sea Res.* **32**, 853–883.

Ward B.B. (1986) Nitrification in marine environments. In Prosser J.I. (ed.) *Nitrification*, pp. 157–184. IRL Press, Oxford.

Ward B.B. (1990) Application of immunology in oceanography. *Oceanography* **2**, 65–72.

Ward B.B. & Zafirou O.C. (1988) Nitrification and nitric oxide in the oxygen minimum of the eastern tropical North Pacific. *Deep-Sea Res.* **35**, 1127–1142.

Waterbury J.B., Watson S.W., Guillard R.R. & Brand L.E. (1979) Widespread occurrence of a unicellular, marine planktonic cyanobacterium. *Nature* **277**, 293–294.

Waterbury J.B., Watson S.W., Valois F.W. & Franks D.G. (1987) Biological and ecological characterization of the marine unicellular cyanobacterium *Synechococcus*. In Platt T. & Li W.K.W. (eds) *Photosynthetic Picoplankton. Can. Bull. Fish. Aquat. Sci.* **214** (special issue), 71–120.

Webb K.C., Dupaul W.D., Wiebe W., Scottile W. & Johannes R.E. (1975) Enewetak (Eniwetok) Atoll: aspects of the nitrogen cycle on a coral reef. *Limnol. Oceanogr.* **20**, 198–210.

Wheeler P., North B., Littler M. & Stevens G. (1977) Amino acid uptake by marine phytoplankters. *Limnol. Oceanogr.* **19**, 249–259.

Wheeler P.A. & Kirchman D.L. (1986) Utilization of inorganic and organic nitrogen by bacteria in marine systems. *Limnol. Oceanogr.* **19**, 249–259.

Whitton B.A. & Potts M. (1982) Marine littoral. In Carr N.G. & Whitton B.A. (eds) *The Biology of Cyanobacteria*, pp. 515–542. Blackwell Scientific Publications, Oxford.

Wiebe W.J., Johannes R.E. & Webb K.L. (1975) Nitrogen fixation in a coral reef community. *Science* **188**, 257–259.

Williams P.J. (1970) Heterotrophic utilization of dissolved organic compounds in the sea. I. Size distribution of population and relationship between respiration and incorporation of growth substrates. *J. Mar. Biol. Assoc. UK* **50**, 859–870.

Williams P.M. & Druffel E.R.M. (1989) Dissolved organic matter in the ocean: comments on a controversy. *Oceanography* **1**, 14–18.

Wolin M.J. & Miller T.L. (1987) Bioconversion of organic carbon to CH_4 and CO_2. *Geomicrobiol. J.* **5**, 239–259.

Wolk C.P. (1982) Heterocysts. In Carr N.G. & Whitton B.A. (eds) *The Biology of Cyanobacteria*, pp. 359–386. Blackwell Scientific Publications, Oxford.

Woodwell G., Rich P.H. & Hall C. (1973) Carbon in estuaries. In Woodwell G. & Pecan E. (eds) *Carbon and the Biosphere. Natl. Tech. Inf. Serv. CONF-72051*, pp. 221–240. US Government Printing Office, Washington DC.

Wyman M., Gregory R.D.F. & Carr N.G. (1985) Novel role for phycoerythrin in a marine cyanobacterium *Synechococcus* strain DC2. *Science* **230**, 818–820.

Zehnder A.J.B. (1978) Ecology of methane formation. In Mitchell R. (ed.) *Water Pollution Microbiology*, pp. 349–376. John Wiley & Sons, New York.

Zehnder A.J.B. & Brock T.D. (1979) Methane formation and methane oxidation by methanogenic bacteria. *J. Bacteriol.* **137**, 420–432.

Zeikus J.G. (1977) The biology of methanogenic bacteria. *Bacteriol. Rev.* **41**, 514–541.

15

Donn J. Kushner

Microbial life in extreme environments

15.1 Introduction

What is an extreme environment? A definition made by humans is likely to be anthropocentric: an environment that we, and microorganisms closely associated with us, find extreme. A narrow definition could include environments that would preclude the growth of *Escherichia coli*: temperatures >40°C and <10°C; pH values >9.5 or <4 (Thimann 1963); very low nutrient conditions; and toxic concentrations (usually in the millimolar range) of heavy metals and other inhibitory conditions.

We could also, in a more fanciful mood, define such environments in terms of how they would "look" to different creatures. Thus, marine and freshwater organisms might consider each other as living in extreme environments. The same would be true for acidophiles and alkalophiles, for barophiles from the deep sea trenches and surface marine organisms, for extreme halophiles and freshwater bacteria, mesophiles and extreme thermophiles, and so on. Human blood, with its antibodies and phagocytes, is an extreme environment for pathogenic bacteria. Every environment will be extreme to some organisms.

The classic extreme environments (high temperatures, extremes of pH, high concentrations of solutes and toxic substances) are usually so restrictive that only a few types of organisms can grow in them. These are usually, though not always, prokaryotic cells; very often, in recent years, the latter have proved to be members of the archaebacteria.

Many environments generally considered extreme are characterized by low species diversity (Brock 1969). This is certainly true of hot and salty environments and, to a lesser extent, of dry environments. It is not true of cold, high-pressure environments such as the bulk of the oceans.

Another characteristic of true extreme environments may be termed "difficulty of access by outsiders." One or a few mutations are not sufficient to overcome the environmental barrier. There have been few reports of mutations to growth in higher or lower temperatures or salt concentrations, and the changes reported have been moderate and sometimes unstable. The ability to grow in such environments must be determined by a number of different genes.

Overviews of extreme environments, and of more specialized areas, have appeared in a number of books and chapters (Heinrich 1976; Ponnamperuma 1976; Kushner 1978a; Shilo 1979; Gould & Corry 1980; Herbert & Codd 1986; Edwards 1990a; Horikoshi & Grant 1990; and others cited below). Some of the inhabitants of such environments will be considered, together with how they adapt to, or why they require, such harsh and unusual conditions.

15.2 High-temperature environments

We have long been fascinated by the ability of microorganisms to survive and grow at high temperatures, an ability that has become more impressive the more it is studied. Many high-temperature environments have been studied (Kushner 1980; Brock 1986a), including solar-heated soils and rocks, coal, trash and compost heaps warmed by microbial action to 70°C or higher, volcanic fumaroles, and terrestrial and submarine hot springs (some of which can reach temperatures well over 100°C). Certain submarine hydrothermal vents, the "black smokers," are hotter than 300°C. Their existence has raised fundamental questions about the upper temperature limits of life (see below).

Table 15.1 lists the upper temperature limits for growth of a number of microorganisms and a few macroorganisms. It describes growth rather than survival. For the latter, bacterial endospores remain the most heat tolerant form of cells. Many of the vegetative cells of spore formers show no unusual heat resistance. Others do, including the most studied moderate thermophiles, *Bacillus stearothermophilus*, and the food spoiler, *B. coagulans*.

Table 15.1 also shows that as the temperature increases simpler forms of life predominate. The highest temperature at which animals and higher plants can live and reproduce is 50°C. The upper limit for eukaryotic microorganisms is about 60°C. Some cyanobacteria and anoxygenic phototrophic bacteria can grow at 70°C or slightly higher, but life at higher temperatures is limited to heterotrophic and chemoautotrophic bacteria, especially archaebacteria.

Table 15.1 shows a continuous range of upper temperature limits of growth, and any classification based on temperature limit is necessarily arbitrary. Since hardly any eukaryotic organisms grow above 60°C, and natural habitats with sustained temperatures above 55–60°C are largely limited to geothermal environments, it is convenient to consider 55–60°C as the thermophile boundary (Brock 1986b). "Extreme thermophiles" (hyperthermophiles) have been

Table 15.1 Upper temperature limits (°C) for growth

	Temperature (°C)	Reference
Animals		
Fish	38	Brock (1986b)
Insects	45–50	
Plants		
Vascular plants	45	
Mosses	50	
Eukaryotic microorganisms		
Protozoa	56	
Algae	55–60	
Fungi	60–62	
Photosynthetic bacteria		
Cyanobacteria	55–70	
Purple bacteria	55–60	
Green bacteria	70–73	
Other eubacteria		
Gram-positive bacteria (including *Bacillus, Clostridium*, lactic acid bacteria, and actinomycetes)	50–75	
Gram-negative aerobes and anaerobes	50–75	
Thermus	85	Tansey & Brock (1978)
Thermotoga maritima	90	Huber *et al.* (1986)
Archaebacteria		
Thermoplasma	65	Tansey & Brock (1978)
Methanobacterium	75	Stetter (1986)
Methanothermus	97	
Sulfolobus	87	Brock & Madigan (1989)
Acidianus	95	
Pyrococcus	103	
Pyrodictium	110	

considered to be those with temperature optima ≥80°C (Stetter 1986). All the extreme thermophiles shown in Table 15.1 are archaebacteria, with the exception of the interesting organism, *Thermotoga maritima* (Huber *et al.* 1986). The latter, however, is probably the most closely related eubacterium to the thermophilic archaebacteria (Woese 1987).

Optimal and lower temperature limits for thermophiles reflect the upper ones rather closely so that, in general, the actual temperature range in which each species can grow is only about 30°C. Thus, highly thermotolerant microorganisms are also truly thermophilic, i.e., lovers of high temperatures.

Brock's classic studies in the hot springs in Yellowstone Park (Brock & Madigan 1989; 1991) revealed biological gradients that followed the temperature gradients in the springs' outflow. The positions of the different kinds of microorganisms were determined by the temperature of the cooling springs.

In more recent years submarine hot springs and other thermal environments have yielded extremely interesting thermophilic archaebacteria including the heterotrophic moderate thermophile, *Thermoplasma* (first isolated from self-heating coal piles), and sulfur-dependent extreme thermophiles, such as *Sulfolobus*, *Acidianus*, and *Pyrodictium*. The latter derive their energy from the oxidation or reduction of elemental sulfur. *Pyrodictium* species, disc-shaped bacteria, often connected by short stalks, have temperature optima of 105°C. At this temperature, their growth must be studied under pressure.

Acidianus, which can both oxidize and reduce elemental sulfur, and *Sulfolobus*, which oxidizes sulfur to sulfuric acid, have pH optima of 2–3 (Brock & Madigan 1989). These and other similar bacteria have been called "thermoacidophiles," although the expression "sulfur-dependent archaebacteria" is now more generally used.

Do the temperatures of growth of the bacteria listed in Table 15.1 reflect the true *in situ* temperature limits of growth? A good deal of excitement arose a few years ago from the apparent demonstration in deep hydrothermal vents ("black smokers") of bacteria able to grow at 250°C at 300 bar pressure (Baross & Deming 1983). Later work, however suggested that these apparent "bacteria" were artifacts of the culture conditions and that proteins, and even amino acids, would be unstable at such temperatures (Trent *et al.* 1984; White 1984). It seems likely that the upper temperature limit of life is only between 110 and 150°C, although bacteria able to grow in the upper ranges of this limit have so far not been isolated (Stetter 1986).

15.2.1 Mechanisms of adaptation to high temperatures

Much work has been done on mechanisms whereby bacteria are able to withstand, and indeed require, high temperatures for growth (reviewed by Amelunxen & Murdock 1978; Langworthy & Pond 1986; Sundaram 1986; Edwards 1990a). This property now seems to be mainly due to changes in their macromolecules, which are extraordinarily stable at high temperatures and, in the case of some enzymes, function best at such temperatures. Very high degrees of thermostability have been observed with different proteins from thermophiles. As an extreme example, the ferridoxin of *Pyrococcus furiosus* was stable for 24 hours at 95°C (Aono *et al.* 1989).

The molecular determinants of such thermostability are thought to lie in noncovalent linkages in the proteins. Few characteristic differences between thermophilic and nonthermophilic proteins have been found, and a change in as little as one amino acid may make a great difference in an enzyme's thermal response (Sundaram 1986; Aono *et al.* 1989). Many thermophilic enzymes show low activity at temperatures <40 or 50°C, and it has been suggested that low enzyme activity and/or improper assembly of ribosomes at the suboptimum temperature could determine temperature minima for growth (Sundaram 1986).

Other cell components that could determine thermostability have been considered and then discounted. For example, the "melting" temperatures of double-stranded DNA increase with the guanine+ cytosine (G+C) content, but there is no correlation between G+C content of DNA and temperature maxima. Indeed, some of the most thermophilic sulfur-dependent archaebacteria have G+C contents as low as 31 mol%. Pure DNA of this composition would certainly not exist in the double-stranded form at the growth temperatures of these bacteria, and binding proteins seem essential for keeping the strands together (Stetter *et al.* 1986).

Thermine and other special polyamines seem essential for stabilizing protein synthesis systems in several thermophiles (Oshima 1986), and presumably add some protection to the natural thermostability of the proteins.

15.2.2 Membrane lipids of thermophiles

Thermophilic bacteria possess unusual membrane lipids, though it is not certain how these relate to life at high temperatures (Langworthy 1985; Langworthy & Pond 1986). Structures of some of these lipids are shown in Figure 15.1.

Thermophilic eubacteria have high amounts of branched-chain fatty acids, and some lipids with fatty acids connected to glycerol by ether rather than ester linkages. Some species have long-chain diols instead of glycerol-linked lipids as well as other unusual lipids (Langworthy & Pond 1986).

All archaebacteria, including nonthermophilic forms, possess di- or tetraethers of phytanyl groups linked to glycerol. The tetraethers are found in large proportions in certain thermophilic archaebacteria where they form monolayers with glycerol groups on the inside and outside of the membrane rather than the more usual bilayers. Membranes containing them may retain a liquid crystalline structure up to 100°C. It has been suggested that such lipids may be responsible for the membrane stability of extreme thermophiles (Langworthy 1985; Langworthy & Pond 1986).

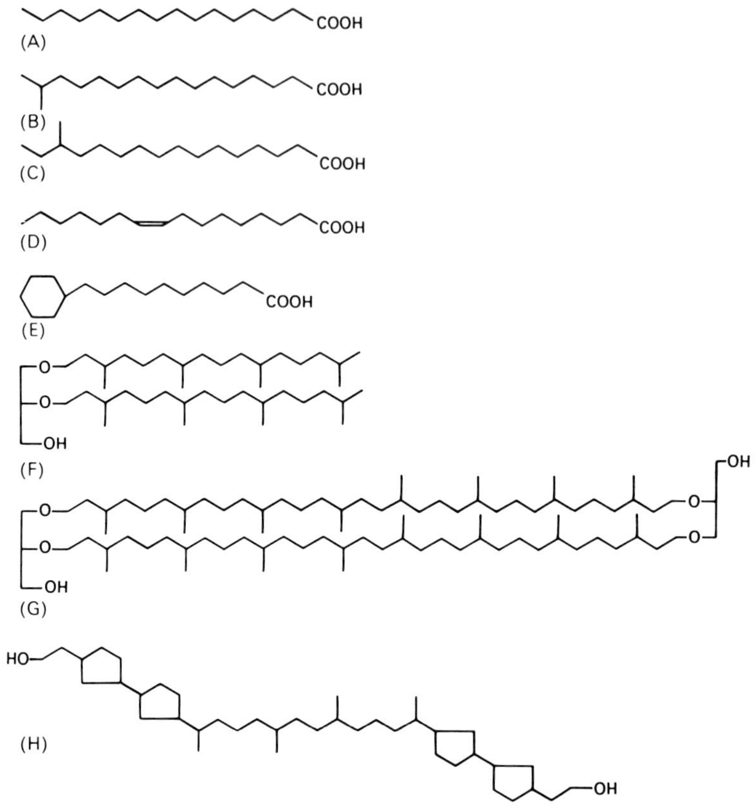

Figure 15.1 Constituents of lipids of thermophilic bacteria. (A) Saturated straight chain; (B) *iso*-branched; (C) *ante iso*-branched; (D) unsaturated chain; (E) alicyclic cyclohexyl fatty acid; (F) phytanyl glycerol diether; (G) dibiphytanyl diglycerol tetraether; (H) C_{40} biphytanyl diol containing four pentacyclic rings. (A), (B) and (C) are found in thermophilic eubacteria, such as *Bacillus stearothermophilus*; (E) appears in *B. acidocaldarius*, which is both thermophilic and acidophilic; (F) and (G) are found in mesophilic and thermophilic archaebacteria, including *Methanobacterium thermoautotrophicum*, *Sulfolobus*, and *Thermoplasma*. In the last two, tetraethers (G) predominate, some with pentacyclic rings. Halophilic archaebacteria have C_{20} and C_{25} phytyl groups (Kushner 1985). From Edwards (1990b).

15.3 Life in high salt and solute concentrations

Food microbiology played an essential role in early studies of this subject. Although high salt and sugar concentrations have long been used to preserve foods, such foods may still be spoiled by microbial growth. Manufacture of soy sauce and other products depends on the action of microorganisms that can grow in high salt concentrations. Many studies of halophilic bacteria, "osmophilic" yeasts, and "xerophilic" fungi have developed through concerns of the food industries. These organisms, especially the bacteria, are reviewed in detail in Brown (1976), Kushner (1978a; 1985), Vreeland (1987), Rodriguez-Valera (1988), Gilmour (1990), and Grant (1990).

The reddening of salted fish is a characteristic manifestation of the growth of the so-called red halophilic bacteria, whose responsibility was determined early in this century (Harrison & Kennedy 1922). Much earlier manifestations of such bacteria were noted as the reddening of solar salterns (ponds near the sea where sea water is evaporated into salt). The red color of such ponds could be used as an index of the degree of saturation, since the red halophiles only grow and predominate at near-saturating conditions (Baas-Becking 1931).

Halophilic archaebacteria are not the only red-pigmented organisms in hypersaline environments. The flagellated eukaryotic algae, *Dunaliella viridis* and *D. salina*, also grow in high salt concentrations, though they grow optimally in somewhat lower salinity than the most halophilic archaebacteria. They contribute to the color in such environments, though probably not to the final color of saturated solar salterns as they make up a relatively small proportion of the microbiota (Kushner 1978b; Rodriguez-Valera 1988; 1991; Javor 1989).

15.3.1 A survey of halophiles and their environments

The salt responses of different kinds of bacteria are summarized in Table 15.2. Truly halophilic bacteria are those that require substantial amounts of salt for growth (since all bacteria need some salt). These bacteria come from different sources, often as contaminants of salted food products or culture media. Some of the extremely halophilic archaebacteria (including species of *Halobacterium*, and *Halococcus*) were first isolated from salted foods and hides, undoubtedly as contaminants of solar salt prepared from sea water. These and other halophilic archaebacteria, including *Haloferax*, *Natronobacterium*, *Natronococcus*, and others, as well as halophilic eubacteria, were isolated from hypersaline lakes. The chemistry and biology of such lakes has been described in detail (Javor 1989). As Table 15.3 shows, all hypersaline environments do not have the same salt composition. Thus, in marine solar salterns Na^+ is the predominant cation, whereas in the Dead Sea (which has supplied some of the most intensively studied halophiles) the concentration of Mg^{2+} is higher than that of Na^+. In contrast to both of these, Lake Magadi, like many carbonate-based alkaline salt lakes, has virtually no Mg^{2+}. The Mg^{2+} requirement and tolerance of halophilic archaebacteria isolated from these lakes may reflect the Mg^{2+} concentration in which they live (Kushner 1985).

It has been pointed out that the salt range for growth may depend on environmental factors (Kushner & Kamekura 1988). Generally, bacteria are more salt dependent at higher than at lower temperatures; sometimes they are more specifically Na^+ dependent. This may be true of the "osmophilic" yeasts, which may require salts or other

Table 15.2 Salt responses of different microorganisms. From Kushner (1992)

Category	Salt (NaCl) concentration (mol l^{-1})		
	Minimal	Optimal	Maximal
Nonhalophiles[1]	–	0.2	0.3–1.0
Slight halophiles[2]	0.2	0.2–0.5	2.0
Moderate halophiles[3]	0.4	0.5–2.0	3.0–3.5
Borderline extreme halophiles[4]	1.0	2.0–3.0	4.0
Extreme halophiles[5]	2.0	3.0–4.0	5.2 (saturated)
Halotolerant[6]	Nonhalophiles, which can tolerate high salt concentrations. Extremely halotolerant cells can grow in higher than 2.5 mmol l^{-1} NaCl		
Facultative halophiles[7]	Require about 0.5 mol l^{-1} NaCl only under certain environmental conditions		

Examples:
[1] Most terrestrial eubacteria and most freshwater microorganisms.
[2] Many marine microorganisms.
[3] *Vibrio costicola, Paracoccus halodenitrificans*, many other eubacteria and algae.
[4] *Ectothiorhodospira halophila, Actinopolyspora halophila, Haloferax* spp.
[5] *Halobacterium, Natronobacterium, Halococcus, Natronococcus* spp. and other "red halophiles."
[6] *Staphylococcus* and *Halomonas* spp., some of the latter being "extremely halotolerant."
[7] *Planococcus halophila.*

Table 15.3 Ionic contents of different hypersaline waters (data in g 100 ml^{-1})

Ions	Great Salt Lake	Dead Sea	Lake Magadi	Marine saltern
Na^+	10.5	3.97	10.6	9.84
K^+	0.67	0.76	0.16	0.49
Mg^{2+}	1.11	4.24	0.0024	1.45
Ca^{2+}	0.03	1.72	0.24	0.04
Cl^-	18.1	21.9	2.28	18.7
SO_4^{2-}	2.7	0.04	–	1.93
CO_3^{2-}	–	–	11.7	–
pH	–	6.1	11.03	ca. 8

Figures for the Great Salt Lake (North Arm), Dead Sea (lower water mass before overturn in 1978), and marine saltern (at onset of NaCl saturation) from Javor (1989). Figures for Lake Magadi (analyses made in 1976) from Grant & Tindall (1986).

solutes only at higher temperatures. Halophilic bacteria may also grow over wider ranges of salt concentrations in nutritionally rich media than in poor media (Kushner 1978b; Kauri *et al.* 1990).

The halophilic archaebacteria mentioned above usually have

temperature optima for growth of about 50°C (Hochstein 1988). Franzmann *et al.* (1987; 1988) identified psychrotrophic halophilic eubacteria able to grow at 4°C and lower and no higher than 37°C in hypersaline Antarctic lakes.

In addition to the mainly aerobic halophiles mentioned above, a number of strictly anaerobic archaebacteria, halophilic methanogens, have been found in hypersaline environments (Mathrani & Boone 1985; Paterek & Smith 1985; Zhilina 1986). *Halomethanococcus doii* needs at least 1.8 mol l^{-1} NaCl for growth and can be considered an extreme halophile (Yu & Kawamura 1987).

15.3.2 Limits of water availability

In considering effects of solutes, we must remember that all of them reduce the amount of free water. This is often calculated as water activity;

$$a_w = p/p_o,$$

where p is the vapour pressure of the solution at a particular temperature and p_o that of the solvent. Water availability is also expressed as water potential, Ψ, where (in the absence of hydrostatic pressure)

$$\Psi = \mathrm{R}T \log_e a_w / V,$$

where R is the gas constant, T the absolute temperature, and V the partial molar volume (Brown 1990; Jennings 1990). The last references should be consulted for detailed discussion of different kinds of water potential, including the "matric" potential that plays an important role in water availability in soil. Water activity has been especially useful in studying the limits of growth of food-spoilage microorganisms (Kushner 1978b; Brown 1990) The a_w of pure water is 1.000, that of 0.3 mol l^{-1} NaCl 0.99, and that of saturated NaCl about 0.75; corresponding Ψ values are 0, −1.38, and −39.1 MPa (Jennings 1990). The inhibitory effects of different solutes are sometimes due primarily to nonspecific lowering of water activity or potential. However, in general, halophilic bacteria have specific requirements for high Na^+ concentrations rather than for low a_w (Kushner 1978a).

A number of microorganisms, especially yeasts and "xerophilic" fungi, can grow at considerably lower a_w than that provided by saturated NaCl solutions. These water activities have been measured by equilibrating agar cultures on slides against different H_2SO_4 concentrations or mixtures of certain salts. However, the lowest a_w at which growth of any microorganism seems possible is about 0.60 (Kushner 1978b; Brown 1990).

Table 15.4 Internal ionic contents of bacteria growing in different concentrations of NaCl. Figures from Kushner & Kamekura (1988)

	Ion concentration ($mol\,l^{-1}$)					
	In medium			Cell associated		
Bacteria	Na^+	K^+	Cl^-	Na^+	K^+	Cl^-
Salmonella oranienburg	0.15	0.025	0.18	0.13	0.24	0.005
Vibrio costicola	1.00	0.14	1.00	0.68	0.22	0.14
Vibrio alginolyticus	0.40	0.046	0.40	0.30	0.40	0.10
Vibrio alginolyticus	0.80	0.046	0.80	0.50	0.38	0.22
Haloanaerobium prevalens	2.22	0.013	2.30	1.52	1.59	2.24
Halobacterium salinarium	4.00	0.032	4.00	1.37	4.57	3.61

Some microorganisms can overcome water limitations by finding protected niches. Thus, in extremely arid environments, such as the Antarctic dry valleys, the outside air is too dry for microbial growth. However, cryptoendolithic microbial communities, composed mainly of lichens, may live just below the surface of rocks using small amounts of moisture retained within the rock (Palmer & Friedmann 1990).

15.3.3 Osmoregulation and compatible solutes

Halophilic and salt-tolerant microorganisms do not adapt to high external solute concentrations by actively maintaining much lower internal concentrations. Rather, they maintain internal osmolar conditions equivalent to the external ones, but use solutes that permit intracellular processes to function. Such "compatible solutes" vary widely (Brown 1976). As the figures in Table 15.4 show, certain halophilic archaebacteria and halophilic anaerobic eubacteria contain cell-associated monovalent cations and anions roughly equivalent to those in the medium. All halophilic archaebacteria studied contain high concentrations of KCl, which may be regarded as the main compatible solute. Polyols are common in yeasts and algae; for example, *Dunaliella* spp. contain very high glycerol concentrations, which vary directly with the external salinity. Cyanobacteria use sucrose, trehalose, glycine betaine, and glutamate betaine. Nonhalophilic eubacteria (other than cyanobacteria) use proline, glutamate, glycine betaine, and the cyclic amino acid ectoine when growing at higher external salt concentrations (Kushner & Kamekura 1988; Grant 1990; Wohlfarth *et al.* 1990).

15.3.4 Physiological determinants of halophilic life

Most detailed work has been carried out on the halophilic archaebacteria, with special reference to the effects of salts on enzyme activity and stability. Many of the enzymes of these bacteria need, or can tolerate, high salt (NaCl or KCl) concentrations for activity. A

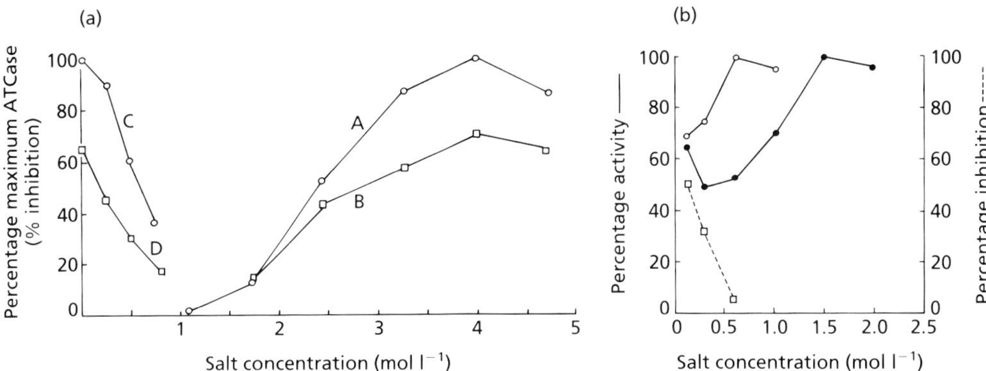

Figure 15.2 Effects of salts on activity and regulation of halophilic and nonhalophilic aspartate transcarbamylase (ATCase). (a) ATCase of yeast, *Saccharomyces cerevisiae*, and *Halobacterium cutirubrum*, an extremely halophilic archaebacterium. Lines A and C show enzyme activity; lines B and D show feedback inhibition by 2 mmol l^{-1} uridine triphosphate (for yeast) or cytidine triphosphate (CTP) (for the halophile). Increasing salt (NaCl) inhibits activity and regulation of the yeast enzyme. On the contrary, the halophile enzyme requires high salt for both activity and regulation. This enzyme is quickly and irreversibly inactivated in the absence of salt. KCl has approximately the same effect on the halophilic enzyme as NaCl. Note that other enzymes of halophilic archaebacteria can be active in the absence of salts, but most or all are unstable under these conditions (Kushner 1978b; 1985). From Liebl *et al.* (1969). (b) Effects of salts on activity and regulation of ATCase of a moderate halophile, *Vibrio costicola* in the presence of NaCl (solid circles) and KCl (clear circles). Regulation (feedback inhibition by 2 mmol l^{-1} CTP) was studied in the presence of NaCl. Note that although salt stimulated enzyme activity, the highest regulation is found in the absence of salt (clear squares). Other enzymes of these bacteria are inhibited by all added salt concentrations (as chlorides; see text). From Ahonkhai *et al.* (1989).

comparison of the aspartate transcarbamylase from an extremely halophilic archaebacterium, a moderate halophile and a nonhalophile (Figure 15.2) shows that salt can affect both activity and regulation (feedback inhibition) in different ways. In the absence of salts most enzymes of extreme halophiles are reversibly or, more often, irreversibly inactivated. Many of the difficulties of protein isolation at very high salt concentrations have now been overcome by hydrophobic chromatography and other techniques, so that highly purified enzymes and other proteins of the extremely halophilic archaebacteria have now been studied in detail (Kushner 1985; Eisenberg & Wachtel 1987).

It has been known for some time that most proteins of halophilic archaebacteria are acidic, and that they contain relatively low proportions of nonpolar amino acids (Lanyi 1974). High salt concentrations are needed partly to neutralize the mutual repulsion of negatively charged proteins (a function of the cations of the salts) and, especially, to promote the formation of hydrophobic bonds needed to stabilize the proteins (Lanyi 1974; Eisenberg & Wachtel 1987).

A more detailed study of a few purified enzymes of an extreme halophile, *Halobacterium marismortui*, has shown that the negatively charged carboxylate groups from glutamic acid are situated on the outside of the molecule. With such an arrangement the protein has an extraordinary ability to bind Na^+, K^+, and water eight- to 10-fold higher than bovine serum albumin. These proteins have been considered to be "anionic sponges" (Eisenberg & Wachtel 1987).

Proteins of halophilic eubacteria have not attracted the same level of study, although about a dozen have been isolated (Kushner & Kamekura 1988). Some of the effects of salts on the eubacterial enzymes are still difficult to interpret. Analyses of the Na^+ and K^+ contents of several aerobic halophilic eubacteria suggest that the sum of these cations is approximately equal to the sum of the extracellular ions, over a range of external salt concentrations. However, many of the intracellular enzymes are most active in the absence of salt and are strongly inhibited by such salt concentrations *in vitro*. Such findings are partly due to the common use of NaCl and KCl as the "salts" to be tested, whereas many of these cells exclude Cl^- ions (see Table 15.4) as do, indeed, many eukaryotic cells (Kamekura & Kushner 1984). It has been shown (Kamekura & Kushner 1984; Choquet *et al.* 1989; Choquet & Kushner 1990) that Cl^- ions are toxic to several steps in protein synthesis. Thus, *in vitro* protein synthesis in *Vibrio costicola* is completely inhibited by $0.6\,mol\,l^{-1}$ NaCl, but is stimulated by $0.6\,mol\,l^{-1}$ Na-glutamate. However, these cells contain only about $0.1\,mol\,l^{-1}$ glutamate (Kamekura & Kushner 1984).

Some aspects of the salt response of enzymes of *V. costicola* are still difficult to explain. When these cells grow in the highest salt concentration possible ($3-3.5\,mol\,l^{-1}$), the internal Cl^- concentration (about $1.5\,mol\,l^{-1}$) seems sufficient to inhibit most processes studied. Certainly the *in vitro* protein synthesizing systems of such bacteria are as sensitive to NaCl and KCl as the systems of cells grown in $1\,mol\,l^{-1}$ NaCl, whose Cl^- concentration is only about $0.1\,mol\,l^{-1}$, but the cells do grow (Kamekura & Kushner 1984; M. Kamekura & D.J. Kushner unpublished). Glycine betaine was able partly to counteract the toxic action of Cl^- ions, but not sufficiently to account for growth at all external salt concentrations (Kamekura & Kushner 1984; Choquet *et al.* 1989; Choquet & Kushner 1990). Effects of other organic solutes of halophilic eubacteria on the salt response of internal enzymes should certainly be studied.

Of course, the measurements of Cl^- ions are bulk measurements, and may not apply to all cells. Furthermore, the aqueous environment of all intracellular enzymes must be considered uncertain (Kushner 1988). We must even consider the possibility that ionic contents of distinct regions of the cytoplasm can vary widely (Wiggins 1990), which adds to the difficulties of understanding the true internal environment of halophilic microorganisms.

The physiology of certain halophilic anaerobic eubacteria may be easier to understand. These, in common with halophilic archaebacteria, maintain internal Cl^- ion concentrations approximately equal to that of the external medium (see Table 15.4). The enzymes of the halophilic anaerobic eubacteria studied thus far resemble those of the halophilic archaebacteria in their salt responses (Rengipipat *et al.* 1988).

15.4 Microbial life at extreme pH values

Many acidic and alkaline environments are well populated by microorganisms, and it has long been recognized that many microorganisms can grow at pH values that are toxic for animal cells. Some can withstand, and are limited to, acid or alkaline conditions. Others can grow over wide pH ranges, from 2 to 11. Tabel 15.5 gives a representative, but by no means exhaustive, listing of such microorganisms.

The acidophiles include the acetic acid bacteria (*Acetobacter* species), the thiobacilli which by oxidation of sulfides, sulfur, or Fe^{2+}, are responsible for the leaching of ores, and can live in high concentrations of H_2SO_4 (Brock & Madigan 1989), and a number of thermophilic archaebacteria and eubacteria. The alkalophilic bacteria include several *Bacillus* species, salt-tolerant and halophilic phototrophic bacteria of the genus *Ectothiorhodospira*, archaebacteria, such as *Natronobacterium* and *Natronococcus* species, and others which grow in the very alkaline carbonate lakes mentioned in Section 15.3.

Organisms that grow in alkaline environments have been separated into two main groups: alkali-tolerant organisms that grow best at pH 7.0–9.0 and cannot grow above pH 9.5, and alkalophiles which grow best between pH 10.0 and 12.0. In the latter group, facultative alkalophiles can also grow at neutral pH values, while obligate alkalophiles cannot grow below pH 8.5–9.0 (Krulwich & Guffanti 1989a).

The biochemical adaptations necessary for such abilities usually involve the cell envelope, especially the cytoplasmic membrane which controls ion transport, rather than changes within the cytoplasm. It was pointed out some years ago (Gale 1943) that internal enzymes of bacteria that grow at extremes of pH are themselves quite susceptible to such extremes and usually have pH optima near 7, suggesting that the internal pH of such bacteria is also near neutrality.

This suggestion has since been supported by measurements of internal pH (pH_i) through the distribution of weak acids or bases.

Table 15.5 pH ranges for microbial growth

Organisms	pH range for growth	Reference
Bacteria		
Thiobacillus thioooxidans	0.9–4.5	Langworthy (1978)
Thiobacillus ferrooxidans	1.5–4.0	
Thermoplasma acidophilum	1.0–4.0	
Bacillus acidocaldarius	2.0–6.0	
Sulfolobus acidocaldarius	0.9–5.8	
Bacillus alcalophilus	8.5–11.6	Krulwich & Guffanti (1983)
Actinomycete sp.	8.0–11.5	
Flavobacterium sp.	7.0–11.2	
Exiguobacterium sp.	7.0–11.5	
Plectonema nostocorum	to 13	Langworthy (1978)
Spirulina sp.	8.0–11.0	Krulwich & Guffanti (1983)
Synechococcus sp.	6.5–10.0	
Ectothiorhodospira sp.	8.0–10.0	
Natronobacterium and *Natronococcus* spp.	opt. 9.0–10.0	
Fungi		
Aspergillus, *Penicillium*, and *Fusarium* spp.	2–10	Langworthy (1978)
Phycomyces blakesleeanus, *Acontium velatium*, and *Cephalosporium*	2.0–7.0 (opt. 3) (grow in 2.5 $mol\,l^{-1}$ H_2SO_4 and 4% $CuSO_4$)	
Algae		
Chlorella pyrenoidosum	2–10	
Cyanidium caldarium	<2–5	
Protozoa		
Polytomella caeca	1.4–9.6	
Flagellates and amoebae	grow at pH 2	

There is no "magic pH_i" for all bacteria. The values for acidophiles are in the range 6.5–7.0, those for neutrophiles are 7.5–8.0, and those for alkalophiles are 8.4–9.0 (Booth 1985). The values found are generally much narrower than those in which such microorganisms actually grow. However, some of the alkali-tolerant bacilli (*Bacillus circulans* and *B. firmis*) have pH_i values quite close to the external values, up to 9.0–9.5, and it has been suggested that such internal values are the highest compatable with life. At much higher values, RNA itself would be unstable (Krulwich & Guffanti 1983).

Some of the most fascinating microorganisms in Table 15.5 are the eukaryotic cells, especially fungi and protozoa, that can grow over very wide ranges of pH and, according to the original reports, up to values as high as 13. The physiology and internal pH of these fungi and protozoa have not been studied extensively.

The maintenance of pH_i values near neutrality is mainly or entirely due to the movement of protons in and out of the cell, often coupled with the movement of other ions (Krulwich & Guffanti 1983; 1989a,b; Booth 1985). Na^+/H^+ antiport activity seems especially important in adjusting pH_i of alkalophiles. Most, or all, such organisms need Na^+ ions for growth, apparently because of a special Na^+ cycle or cycles involved in pH_i homeostasis (Krulwich & Guffanti 1989a,b).

The existence of microorganisms that grow in extremes of pH has also posed fascinating problems in bioenergetics. Normally, protonmotive force (pmf) is made of two components, ΔpH, with acid outside, and $\Delta\Psi$ or membrane potential, with (+) outside. Garland (1977) pointed out that because of the great increase in ΔpH of acidophiles and the decrease to a negative value in alkalophiles, values and orientation of $\Delta\Psi$ might change in such organisms. In fact, the $\Delta\Psi$ for several acidophiles is (+) inside, while alkalophiles have very high $\Delta\Psi$ values, (+) outside. These changes serve to balance the different ΔpH values of acidophiles and alkalophiles (Krulwich & Guffanti 1983).

An alkaline external pH also makes it difficult to generate adenosine triphosphate (ATP) through a normal F_1F_0 ATPase. It has been suggested that alkalophiles have special channels for conserving the H^+ gradient generated through respiration, so that the protons do not have to mix with the alkaline bulk medium (Krulwich & Guffanti 1989a,b).

Many organisms that grow at extremes of pH possess unusual surface components, whose physiological role is still uncertain. These components include the ether-linked lipids of acidophilic and alkalophilic archaebacteria, which would be resistant to hydrolysis under alkaline, acid, or high temperature conditions (Langworthy 1978; 1985; Langworthy & Pond 1986). Alkalophilic and acidophilic bacteria may also possess some unusual glycoproteins (Krulwich & Guffanti 1983). Some of the external structures, such as heat- and acid-stable pili of *Sulfolobus*, are adapted to extreme pH (Krulwich & Guffanti 1983).

Exoenzymes of alkalophiles are of special interest because of their actual or potential commercial importance. Such enzymes from alkalophilic *Bacillus* species include cellulases and xylanases for waste treatment and possibly energy production; and proteases for detergent additives and for use in food and tanning industries. Such enzymes tend to be thermostable and their ability to act in an alkaline environment may increase their activity (Horikoshi & Akiba 1982; Horikoshi 1986; Krulwich & Guffanti 1989a; Horikoshi & Grant 1990).

15.5 High concentrations of toxic metals

Metals such as Hg, Cd, Cu, Cr, As, and others are generally toxic to most forms of life. Certain bacteria, such as *E. coli* and *Pseudomonas* species, are inhibited by 10^{-5}–10^{-4} mol l^{-1} concentrations or less of these metals. Bacterial resistance to such levels is usually plasmid directed and involves the prevention of entry of these ions, facilitation of their exit, or their conversion to a less toxic form. For example, Hg^{2+} may be reduced to Hg^0 (Silver & Misra 1988). Many other studies have been made of mechanisms of metal resistance in bacteria, algae, and fungi (Gadd 1986; 1990; Collins & Stotzky 1989). There is no single mechanism of metal resistance. Cells may withstand toxic metals by precipitating them as sulfides or phosphates, by binding them to extracellular polymers or smaller molecules such as oxalate or siderophores (which mainly serve to bring essential Fe ions into the cell). Resistant cells may bind metals to their walls, and, in the cytoplasm, to polyphosphate inclusions or to sulfur-containing proteins such as metallothionines and phytochelatins (Gadd 1990).

The highest levels of metal tolerance are found under acid conditions, presumably because the cell envelope is less negatively charged under such conditions and less attractive to metal cations. Thus, some *Thiobacillus* species can grow below pH 4 in the presence of 0.1 mol l^{-1} or higher concentrations of Al^{3+}, Zn^{2+}, Co^{2+}, and Cu^{2+}, and over 15 mmol l^{-1} Cr^{3+} (Ehrlich 1978; Lundgren & Silver 1980; Wong *et al.* 1982).

Fungi seem especially tolerant of heavy metals and may be the predominant microorganisms in heavy metal-contaminated soil near smelters (Gadd 1986). *Cephalosporium*, *Scytalidium*, and other fungi have been reported to grow on saturated $CuSO_4$, but only under acid conditions (Starkey & Waksman 1943; Gadd 1986; 1990). At neutral pH values, the fungi are inhibited by millimole concentrations of Cu^{2+} (Gadd 1990).

The alga *Euglena mutabilis* is often found in very acid and metal-polluted sites. One strain of these algae showed exceptional tolerance (10–100-fold more than previously found for algae) to Al, Fe, and Ni, at low pH (down to 2), especially if grown in coculture with a yeast, *Cryptococcus* sp., with which it was normally found associated in nature (Nakatsu & Hutchinson 1988).

15.6 Extreme environments with wide biological diversity

15.6.1 Low temperatures

The Earth contains many low-temperature environments, the oceans being the most important since they cover the greater part of the Earth's surface, and since 90% of their volume is at 5°C or less (Herbert 1986; Gounot 1991) the bulk of the biosphere may be considered cold. Many other cold environments have been studied, including polar soils, the upper atmosphere, caves, glaciers, and refrigerated foods. Different microorganisms, including bacteria (mainly Gram-negative), fungi, yeasts, and algae can grow in all of them; indeed, some microbial growth seems possible as long as liquid water is present, though in practice little microbial growth or activity has been observed below −5 to −10°C (Baross & Morita 1978; Herbert 1986).

By a generally accepted definition, psychrophilic (cold-loving) microorganisms' growth can range from <0 to 20°C, with an optimum at 15°C or less. Psychrotrophic (cold-tolerant) microorganisms can grow at 0–5°C, with maximum temperatures >25°C (Baross & Morita 1978; Herbert 1986). Because true psychrophiles die at ambient temperatures, care is needed in isolating and maintaining them, which has limited the study of this group of microorganisms, so important in terms of global microbial ecology.

Many mesophilic bacteria, such as *E. coli*, grow in the temperature range about 10–40°C. Although the upper temperature limit of growth is presumably determined by heat sensitivity of one or more key proteins, there is usually no obvious reason why enzyme activity would cease at low temperatures (i.e., <10°C). Apparently, other cellular changes can define the lower temperature limit of life. These may include transport processes, which fail when the membrane lipids solidify, too strict metabolic regulation, improper assembly, and function of ribosomes, and doubtless other failures in cellular metabolism (Inniss & Ingraham 1978; Herbert 1986).

The upper temperature limit of growth of psychrophiles seems determined by temperature sensitivity of key enzymes or of protein synthesis, by activation of autolytic enzymes (Inniss & Ingraham 1978; Gounot 1991), and doubtless by other failings.

15.6.2 Life under pressure in the cold dark sea

Interesting high-pressure environments are found in deep wells, freshwater lakes, and underground deposits, but the greatest pressures are reached in the ocean depths, where pressure increases about 1 bar (atmosphere) with every 10 m (Marquis & Matsumura 1978; Morita 1986).

Living microorganisms are found in the greatest depths of the sea, at over 1000 bar pressure. The first isolates were exposed to normal atmospheric pressure during collection; evidence was obtained, however, that some of these grew better at high pressures (ZoBell & Morita 1957; Marquis & Matsumura 1978). More recently, techniques have been devised for collecting microorganisms and studying their growth at *in situ* pressures. Such techniques have revealed the presence of truly barophilic bacteria, strongly adapted to life under high pressure. Several hundred strains of such bacteria have been isolated, often associated with animals or animal remains from the same depths (Yayanos 1986).

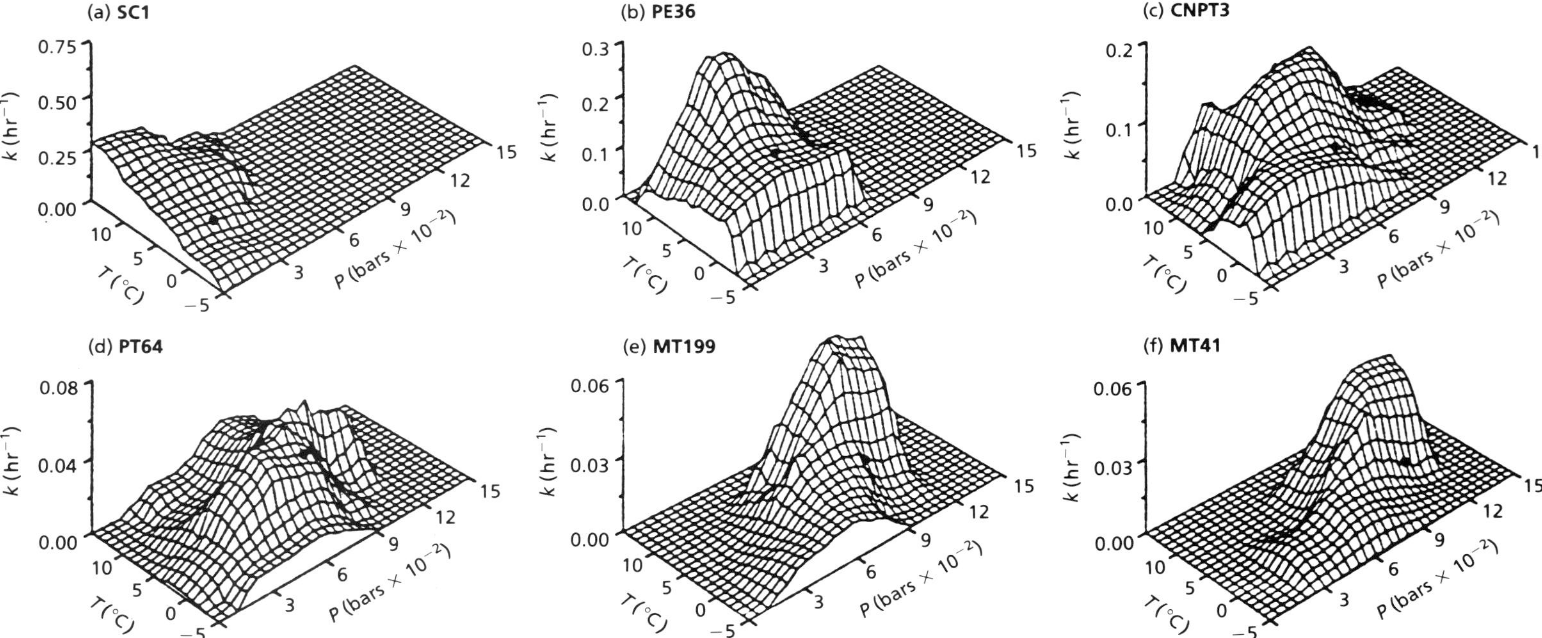

Figure 15.3 Three-dimensional views of the temperature–pressure dependence of exponential growth rate constants (k) of six deep-sea bacterial isolates from different North Pacific Ocean depths. The black dot is at the temperature and pressure of the presumed habitat of the isolate. (a) Strain SC1, from 1975 m was only marginally barophilic. (b) Strain PE36, from 3584 m was barophilic at its habitat temperature. (c) Strain CNPT3, from 5782 m was obligatorily barophilic at temperatures above 6°C. (d) Strain PT64, from 7100 m was nearly obligatorily barophilic at 2°C and barophilic at higher temperatures. (e) Strain MT199 from 8961 m was obligately barophilic at all temperatures. (f) Strain MT41, from 10 476 m was the most extreme barophile isolated. Adapted from Yayanos (1986).

Adaptation to pressure increased with the depth of isolation. Barophily, that is increased growth with higher pressure occurred in bacteria isolated at 2000 m depths (about 200 bar) or deeper. Strict barophily, the inability to grow at atmospheric pressure and the requirement of a few hundred bar for growth, occurred in cells isolated from depths of 6370 m and deeper. The pressure–temperature relations of growth for some of these isolates are shown in Figure 15.3.

This figure shows that the degree of barophily was dependent on temperature. Obligate, or strict barophily occurred for all isolates at temperatures above those permitting growth at atmospheric pressure. This latter temperature was always low – as low as about 5°C for the isolate from the greatest depth. Thus, these are among the most psychrophilic bacteria known.

In all cases, increased pressure increased the maximum growth temperature, a response not observed with mesophilic or thermophilic bacteria from other habitats (Yayanos 1986).

Pressure raises the melting points of lipids so that they gel or crystallize. In order to maintain membrane fluidity, bacteria that live at high pressures produce long-chain polyunsaturated fatty acids (C20:5), which are able to maintain a liquid state even at high pressures. The proportions of such lipids increase with increasing pressure and lower temperature for several barophilic and non-barophilic psychrophilic bacteria (DeLong & Yayanos 1985; 1986; Wirsen *et al.* 1987).

Pressure may also affect the protein composition of cells. This was shown for a thermophilic methanogen (Jaenicke *et al.* 1988) and for a deep-sea bacterium, in which transcription of the *ompH* gene is controlled by hydrostatic pressure (Bartlett *et al.* 1989).

15.7 Brief mention of other extreme environments

15.7.1 Low-nutrient conditions

Very low concentrations of organic carbon (usually less than 10 mg l^{-1}) are found in the open seas, in many bodies of fresh water, and in laboratory distilled water. Microbial inhabitants of such environments have been considered by Dow & Whittenbury (1980), Morgan & Dow (1986), and Fry (1990). In distilled water, gases from the air provide enough nutrients for growth, not only of *Caulobacter* species, which are often found in oligotrophic environments, but also such bacteria as *Pseudomonas fluorescens*, which are normally grown in rich media. Free-living bacteria in natural low-nutrient environments often have the ability to use many carbon sources, and have very active transport systems for such sources. Others develop very small cellular forms (filterable bacteria or ultramicrobacteria), whose increased surface/volume ratios facilitate transport of the low concentrations of nutrients available. Other bacteria form stalks (prosthecae) which provide increased surface area, and which may also bind cells to surfaces which are well known to be zones of higher nutrient concentration. Yet other bacteria form biofilms on surfaces, which may provide an excellent means of concentrating nutrients by adsorption/binding to extracellular fibers or capsular material (Marshall 1980).

15.7.2 Bacteria in organic solvents

These are synthetic rather than natural extreme environments, found in the laboratory and in certain industries. Recently, soil samples have provided bacteria that can grow in the presence of pure toluene (which is normally toxic to bacteria at a 0.1% concentration). One strain, IH-2000 (a pseudomonad), can also grow in the presence of *p*-xylene, ethylbenzene (but not benzene), cyclohexane, and other solvents that are toxic to *E. coli* and other bacteria. The mechanism of tolerance, still not established, is thought to be due to a change in the cell surface (Inoue 1990).

15.8 Concluding remarks

This brief survey shows that microbial life can occur in quite unlikely conditions, ones that we, and many of our fellow creatures, would certainly find extreme. Some microorganisms, indeed, withstand more than one extreme environment, such as those that grow in hot acid (sometimes with high concentrations of toxic metals as well), or in hypersaline, highly alkaline lakes.

Since my first review of the subject of this chapter (Kushner 1964), I have been pleased to see it develop from a microbiological curiosity to a major enterprise. The study of different "extremophiles" has been fully justified through the contributions made to biotechnology, to possibilities of life on other planets and, especially, to knowledge of all forms of life. To take only a few examples, the bacteriorhodopsin system of the halobacteria helped establish the chemiosmotic hypothesis (Kushner 1985). Acidophiles and alkalophiles have increased our insight into cell membrane physiology. We are beginning to understand some aspects of life in the greatest volume of our biosphere, the cold deep seas.

We have also learned that some conditions do preclude life. It seems likely that none can exist at temperatures higher than 150°C, if that high.

Curiously, the most limiting condition still seems to be the availability of water. As far as is known nothing can grow within solid ice, nor in steam. In solutions or on surfaces a substantial amount of water is needed. The limiting a_w value of 0.6 still seems to hold, so that only about 40% of available water can be removed before growth of even the most desiccation resistant microorganism stops. The reasons for this are not obvious. Future workers on extreme environments must look further into the structure and functions of water itself, especially intracellular water. We may find that some microorganisms can grow in almost any environment, provided that the temperature is not high enough to disrupt essential molecules and that the cells are able to maintain a "comfortable" form of intracellular water.

Acknowledgments

Expenses involved in preparing this manuscript were supported by a grant from the Natural Sciences and Engineering Research Council of Canada. I have to thank an anonymous reviewer for highly critical comments.

References

Ahonkhai I., Kamekura M. & Kushner D.J. (1989) Effects of salts on the aspartate transcarbamylase of a halophilic eubacterium, *Vibrio costicola*. *Biochem. Cell Biol.* **67**, 666–669.

Amelunxen R.E. & Murdock A.L. (1978) Microbial life at high temperatures: mechanisms and molecular aspects. In Kushner D.J. (ed.) *Microbial Life in Extreme Environments*, pp. 217–278. Academic Press, London.

Aono S., Bryant F.O. & Adams M.W.W. (1989) A novel and remarkably thermostable ferredoxin from the hyperthermophilic archaebacterium *Pyrococcus furiosus*. *J. Bacteriol.* **171**, 3433–3439.

Baas-Becking L.G.M. (1931) Historical notes on salt and salt manufacture. *Sci. Month.* **32**, 434–446.

Baross J.A. & Morita R.Y. (1978) Microbial life at low temperatures: ecological aspects. In Kushner D.J. (ed.) *Microbial Life in Extreme Environments*, pp. 9–71. Academic Press, London.

Baross J.A. & Deming J.W. (1983) Growth of "black smoker" bacteria at temperatures of at least 250°C. *Nature* **303**, 423–426.

Bartlett D., Wright M., Yayanos A.A. & Silverman M. (1989) Isolation of a gene regulated by hydrostatic pressure in a deep-sea bacterium. *Nature* **342**, 572–574.

Booth I.R. (1985) Regulation of cytoplasmic pH in bacteria. *Microbiol. Rev.* **49**, 359–378.

Brock T.D. (1969) Microbial growth under extreme conditions. *Symp. Soc. Gen. Microbiol.* **19**, 15–42.

Brock T.D. (ed.) (1986a) *Thermophiles: General, Molecular, and Applied Microbiology*. John Wiley & Sons, New York.

Brock T.D. (1986b) Introduction: an overview of the thermophiles. In Brock T.D. (ed.) *Thermophiles: General, Molecular, and Applied Microbiology*, pp. 1–16. John Wiley & Sons, New York.

Brock T.D. & Madigan M.T. (1989; 1991) *Biology of Microorganisms*, 5th and 6th edn. Prentice Hall, Englewood Cliffs.

Brown A.D. (1976) Microbial water stress. *Bacteriol. Rev.* **40**, 803–846.

Brown A.D. (1990) *Microbial Water Stress Physiology: Principles and Perspectives*. John Wiley & Sons, New York.

Choquet C. & Kushner D.J. (1990) Use of natural messenger RNAs in cell-free protein synthesizing systems of the moderate halophile *Vibrio costicola*. *J. Bacteriol.* **172**, 3462–3468.

Choquet C., Kamekura M. & Kushner D.J. (1989) *In vitro* protein synthesis by the moderate halophile, *Vibrio costicola*: the site of action of Cl^- ions. *J. Bacteriol.* **171**, 880–886.

Collins Y.E. & Stotzky G. (1989) Factors affecting the toxicity of heavy metals to microbes. In Beveridge T.J. & Doyle R.J. (eds) *Metal Ions & Bacteria*, pp. 31–90. John Wiley & Sons, New York.

DeLong E.F. & Yayanos A.A. (1985) Adaptation of the membrane lipids of a deep-sea bacterium to changes in hydrostatic pressure. *Science* **228**, 1101–1102.

DeLong E.F. & Yayanos A.A. (1986) Biochemical function and ecological significance of novel bacterial lipids in deep-sea procaryotes. *Appl. Env. Microbiol.* **51**, 730–737.

Dow C.S. & Whittenbury R. (1980) Procaryotic form and function. In Ellwood D.C., Hedger J.N., Latham M.J., Lynch J.M. & Slater J.H. (eds) *Contemporary Microbial Ecology*, pp. 391–417. Academic Press, London.

Edwards C. (ed.) (1990a) *Microbiology of Extreme Environments*. Open University Press, Milton Keynes.

Edwards C. (1990b) Thermophiles. In Edwards C. (ed.) *Microbiology of Extreme Environments*, pp. 1–32. Open University Press, Milton Keynes.

Ehrlich H.L. (1978) How microbes cope with heavy metals, arsenic and antimony in their environment. In Kushner D.J. (ed.) *Microbial Life in Extreme Environments*, pp. 381–408. Academic Press, London.

Eisenberg H. & Wachtel E.J. (1987) Structural studies of halophilic proteins, ribosomes, and organelles of bacteria adapted to extreme salt concentrations. *Ann. Rev. Biophys. Chem.* **16**, 69–92.

Franzmann P.D., Burton H.R. & McMeekin T.A. (1987) *Halomonas subglaciescola*, a new species of halotolerant bacteria isolated from Antarctica. *Int. J. System. Bacteriol.* **37**, 27–34.

Franzmann P.D., Stackenbrandt E., Sanderson K., Volkman J.K., Cameron D.E., Stevenson P.L., McMeekin T.A. & Burton H.R. (1988) *Halobacterium lacusprofundii* sp. nov., a halophilic bacterium isolated from Deep Lake, Antarctica. *System. Appl. Microbiol.* **11**, 20–27.

Fry J.C. (1990) Oligotrophs. In Edwards C. (ed.) *Microbiology of Extreme Environments*, pp. 93–116. Open University Press, Milton Keynes.

Gadd G.M. (1986) Fungal responses towards heavy metals. In Herbert R.A. & Codd G.A. (eds) *Microbes in Extreme Environments* (Special Publications of the Society for General Microbiology; no. 17), pp. 83–110. Academic Press, London.

Gadd G.M. (1990). Metal tolerance. In Edwards C. (ed.) *Microbiology of Extreme Environments*, pp. 178–210. Open University Press, Milton Keynes.

Gale E.F. (1943) Factors influencing the enzymatic activities of bacteria. *Bacteriol. Rev.* **7**, 139–173.

Garland P.B. (1977) Energy transduction and transmission in microbial systems. *Symp. Soc. Gen. Microbiol.* **27**, 1–22.

Gilmour D. (1990) Halotolerant and halophilic microorganisms. In Edwards C. (ed.) *Microbiology of Extreme Environments*, pp. 147–177. Open University Press, Milton Keynes.

Gould G.W. & Corry J.E.L. (eds) (1980) *Microbial Growth and Survival in Extremes of Environments*. Academic Press, London.

Gounot A.M. (1991) Bacterial life at low temperature: physiological aspects and biotechnological implications. *J. Appl. Bacteriol.* **71**, 386–397.

Grant W.D. (1990) General view of halophiles. In Horikoshi K. & Grant W.D. (eds) *Superbugs: Microorganisms in Extreme Environments*, pp. 15–27. Japan Scientific Societies Press, Tokyo.

Grant W.D. & Tindall B.J. (1986) The alkaline saline environment. In Herbert R.A. & Codd G.A. (eds) *Microbes in Extreme Environments* (Special publications of the Society for General Microbiology; no. 17), pp. 55–82. Academic Press, London.

Harrison F.C. & Kennedy M.E. (1922) The red discolouration of cured codfish. *Trans. R. Soc. Canada Section V*, pp. 101–151.

Heinrich M.R. (ed.) (1976) *Extreme Environments: Mechanisms of Microbial Adaptation*. Academic Press, New York.

Herbert R.A. (1986) The ecology and physiology of psychrophilic microorganisms. In Herbert R.A. & Codd G.A. (eds) *Microbes in Extreme Environments* (Special publications of the Society for General Microbiology; no. 17), pp. 1–23. Academic Press, London.

Herbert R.A. & Codd G.A. (eds) (1986) *Microbes in Extreme Environments* (Special publications of the Society for General Microbiology; no. 17). Academic Press, London.

Hochstein L. (1988) The physiology and metabolism of the extremely halophilic bacteria. In Rodriguez-Valera F. (ed.) *Halophilic Bacteria*, vol. 2, pp. 67–84. CRC Press, Boca Raton.

Horikoshi K. (1986) Genetic applications of alkalophilic microorganisms. In Herbert R.A. & Codd G.A. (eds) *Microbes in Extreme Environments* (Special publications of the Society for General Microbiology; no. 17), pp. 297–315. Academic Press, London.

Horikoshi K. & Akiba T. (1982) *Alkalophilic Microorganisms*. Springer Verlag, Heidelberg.

Horikoshi K. & Grant W.D. (eds) (1990) *Superbugs: Microorganisms in Extreme Environments*. Japan Scientific Societies Press, Tokyo.

Huber R., Langworthy T.A., König H., Thomm M., Woese C.R., Sleytr U.B. & Stetter K.O. (1986) *Thermotoga maritima* sp. nov. represents a new genus of unique extremely thermophilic eubacteria growing up to 90°C. *Arch. Microbiol.* **144**, 324–333.

Inniss W.E. & Ingraham J.L. (1978) Microbial life at low temperatures: mechanisms and molecular aspects. In Kushner D.J. (ed.) *Microbial Life in Extreme Environments*, pp. 73–104. Academic Press, London.

Inoue A. (1990) Solvent tolerant microorganisms. In Horikoshi K. & Grant W.D. (eds) *Superbugs: Microorganisms in Extreme Environments*, pp. 43–53. Japan Scientific Societies Press, Tokyo.

Jaenicke R., Bernhardt G., Ludemann H.-D. & Stetter K.O. (1988) Pressure-induced alterations in the protein pattern of the thermophilic archaebacterium *Methanococcus thermolithotrophicus*. *Appl. Env. Microbiol.* **54**, 2375–2380.

Javor B. (1989) *Hypersaline Environments: Microbiology and Biogeochemistry*. Springer Verlag, New York.

Jennings D.H. (1990) Osmophiles. In Edwards C. (ed.) *Microbiology of Extreme Environments*, pp. 117–146. Open University Press, Milton Keynes.

Kamekura M. & Kushner D.J. (1984) Effect of chloride and glutamate ions on *in vitro* protein synthesis by the moderate halophile *Vibrio costicola*. *J. Bacteriol.* **160**, 385–390.

Kauri T., Wallace R. & Kushner D.J. (1990) Nutrition of the halophilic archaebacterium, *Haloferax volcanii*. *System. Appl. Microbiol.* **13**, 14–18.

Krulwich T.A. & Guffanti A.A. (1983) Physiology of acidophilic and alkalophilic bacteria. *Adv. Microbial Physiology* **24**, 173–214.

Krulwich T.A. & Guffanti A.A. (1989a) Alkalophilic bacteria. *Ann. Rev. Microbiol.* **43**, 435–463.

Krulwich T.A. & Guffanti A.A. (1989b) The Na^+ cycle of extreme alkalophiles: a secondary Na^+/H^+ antiporter and Na^+ solute symporters. *J. Bioenerget. Biomembr.* **21**, 663–677.

Kushner D.J. (1964) Microbial resistance to harsh and destructive environmental conditions. In Schnitzer R.J. & Hawking F. (eds) *Experimental Chemotherapy. Vol. II. Chemotherapy of Bacterial Infections, Part I*, pp. 114–169. Academic Press, New York.

Kushner D.J. (ed.) (1978a) *Microbial Life in Extreme Environments*. Academic Press, London.

Kushner D.J. (1978b) Life in high salt and solute concentrations: halophilic bacteria. In Kushner D.J. (ed.) *Microbial Life in Extreme Environments*, pp. 317–368. Academic Press, London.

Kushner D.J. (1980) Extreme environments. In Ellwood D.C., Hedger J.N., Latham M.J., Lynch J.M. & Slater J.H. (eds) *Contemporary Microbial Ecology*, pp. 29–54. Academic Press, London.

Kushner D.J. (1985) The Halobacteriaceae. In Woese C.R. & Wolfe R.S. (eds) *Archaebacteria*. In Sokatch J.R. & Ornston L.N. (eds) *The Bacteria: A Treatise on*

Structure and Function, vol. 8, pp. 171–214. Academic Press, New York.

Kushner D.J. (1988) What is the "true" internal salt concentration of halophilic and other bacteria? *Can. J. Microbiol.* **34**, 482–486.

Kushner D.J. (1992) Growth and nutrition of halophilic bacteria. In Vreeland R. & Hochstein L. (eds) *Microbiology of Extreme and Unusual Environments.* Telford Press (in press).

Kushner D.J. & Kamekura M. (1988) Physiology of halophilic eubacteria. In Rodriguez-Valera (ed.) *Halophilic Bacteria*, vol. 1, pp. 109–138. CRC Press, Boca Raton.

Langworthy T.A. (1978) Microbial life in extreme pH values. In Kushner D.J. (ed.) *Microbial Life in Extreme Environments*, pp. 279–315. Academic Press, London.

Langworthy T.A. (1985) Lipids of Archaebacteria. In Woese C.R. & Wolfe R.S. (eds) *Archaebacteria.* In Sokatch J.R. & Ornston L.W. (eds) *The Bacteria: A Treatise on Structure and Function*, vol. 8, pp. 459–497. Academic Press, New York.

Langworthy T.A. & Pond J.L. (1986) Membranes and lipids of thermophiles. In Brock T.D. (ed.) *Thermophiles: General, Molecular, and Applied Microbiology*, pp. 107–135. John Wiley & Sons, New York.

Lanyi J.K. (1974) Salt-dependent properties of proteins from extremely halophilic bacteria. *Bacteriol. Rev.* **38**, 272–290.

Liebl V., Kaplan J.G. & Kushner D.J. (1969) Regulation of a salt-dependent enzyme; the aspartate transcarbamylase of an extreme halophile. *Can. J. Biochem.* **47**, 1095–1097.

Lundgren D.G. & Silver M. (1980) Ore leaching by bacteria. *Ann. Rev. Microbiol.* **34**, 263–283.

Marquis R.E. & Matsumura P. (1978) In Kushner D.J. (ed.) *Microbial Life in Extreme Environments*, pp. 105–158. Academic Press, London.

Marshall K.C. (1980) Reactions of microorganisms, ions and macromolecules at interfaces. In Ellwood D.C., Hedger J.N., Latham M.J., Lynch J.M. & Slater J.H. (eds) *Contemporary Microbial Ecology*, pp. 107–136. Academic Press, London.

Mathrani I.M. & Boone D.R. (1985) Isolation and characterization of a moderately halophilic methanogen from a solar saltern. *Appl. Env. Microbiol.* **50**, 140–143.

Morgan P. & Dow C.S. (1986) Bacterial adaptations for growth in low nutrient environments. In Herbert R.A. & Codd G.A. (eds) *Microbes in Extreme Environments* (Special publications of the Society for General Microbiology; no. 17), pp. 187–214. Academic Press, London.

Morita R.Y. (1986) Pressure as an extreme environment. In Herbert R.A. & Codd G.A. (eds) (1986) *Microbes in Extreme Environments* (Special publications of the Society for General Microbiology; no. 17), pp. 171–185. Academic Press, London.

Nakatsu C. & Hutchinson T.C. (1988) Extreme metal and acid tolerance of *Euglena mutabilis* and an associated yeast from Smoking Hills, Northwest Territories, and their apparent mutualism. *Microb. Ecol.* **16**, 213–231.

Oshima T. (1986) Genes and genetic apparatus of extreme thermophiles. In Brock T.D. (ed.) *Thermophiles: General, Molecular, and Applied Microbiology*, pp. 137–157. John Wiley & Sons, New York.

Palmer R.J. Jr & Friedmann E.I. (1990) Water relations and photosynthesis in the cryptoendolithic microbial habitat of hot and cold deserts. *Microb. Ecol.* **19**, 111–118.

Paterek J.R. & Smith P. (1985) Isolation and characterization of a halophilic methanogen from Great Salt Lake. *Appl. Env. Microbiol.* **50**, 877–881.

Ponnamperuma C. (ed.) (1976) *Chemical Evolution of the Giant Planets.* Academic Press, New York.

Rengipipat S., Lowe S.E. & Zeikus J.G. (1988) Effect of extreme salt concentrations on the physiology and biochemistry of *Halobacteroides acetoethylicus. J. Bacteriol.* **170**, 3065–3071.

Rodriguez-Valera F. (ed.) (1988) *Halophilic Bacteria.* CRC Press, Boca Raton.

Rodriguez-Valera F. (ed.) (1991) *General and Applied Aspects of Halophilic Microorganisms.* Plenum, New York.

Shilo M. (ed.) (1979) *Strategies of Microbial Life in Extreme Environments*. Dahlem Konferenzen. Verlag Chemie, Berlin.

Silver S. & Misra T.K. (1988) Plasmid-mediated heavy metal resistances. *Ann. Rev. Microbiol.* **42**, 717–743.

Starkey R.L. & Waksman S.A. (1943) Fungi tolerant to extreme acidity and high concentrations of copper sulfate. *J. Bacteriol.* **45**, 509–519.

Stetter K.O. (1986) Diversity of extremely thermophilic archaebacteria. In Brock T.D. (ed.) (1986a) *Thermophiles: General, Molecular, and Applied Microbiology*, pp. 39–74. John Wiley & Sons, New York.

Stetter K.O., Segerer A., Zillig W., Huber G., Fiala G., Huber R. & König H. (1986) Extremely thermophilic sulfur-metabolizing archaebacteria. *System. Appl. Microbiol.* **7**, 393–397.

Sundaram T.K. (1986) Physiology and growth of thermophilic bacteria. In Brock T.D. (ed.) *Thermophiles: General, Molecular, and Applied Microbiology*, pp. 75–106. John Wiley & Sons, New York.

Tansey M.R. & Brock T.D. (1978) Microbial life at high temperatures: ecological aspects. In Kushner D.J. (ed.) *Microbial Life in Extreme Environments*, pp. 159–216. Academic Press, London.

Thimann K.V. (1963) *The Life of Bacteria*, 2nd edn. Macmillan, New York.

Trent J.D., Chastain R. & Yayanos A.A. (1984) Possible artifactual basis for apparent bacterial growth at 250°C. *Nature* **307**, 737–740.

Vreeland R.H. (1987) Mechanisms of halotolerance in microorganisms. *CRC Critical Reviews in Microbiology* **14**, 311–356.

White R.H. (1984) Hydrolytic stability of biomolecules at high temperatures and its implication for life at 250°C. *Nature* **310**, 430–432.

Wiggins P.M. (1990) Role of water in some biological processes. *Microbiol. Rev.* **54**, 432–449.

Wirsen C.O., Jannasch H.W., Wakeham S.G. & Canuel E.A. (1987) Membrane lipids of a psychrophilic and barophilic deep-sea bacterium. *Curr. Microbiol.* **14**, 319–322.

Woese C.R. (1987) Bacterial evolution. *Microbiol. Rev.* **51**, 221–271.

Wohlfarth A., Severin J. & Galinski E.A. (1990) The spectrum of compatible solutes in heterotrophic halophilic eubacteria of the family Halomonadaceae. *J. Gen. Microbiol.* **136**, 705–712.

Wong C., Silver M. & Kushner D.J. (1982) Effects of chromium and manganese on *Thiobacillus ferrooxidans*. *Can. J. Microbiol.* **29**, 536–544.

Yayanos A.A. (1986) Evolutional and ecological implication of the properties of deep-sea barophilic bacteria. *Proc. Natl Acad. Sci. USA* **83**, 9542–9546.

Yu I.K. & Kawamura F. (1987) *Halomethanococcus doii* gen. nov., sp. nov.: an obligately halophilic methanogenic bacterium from solar salt ponds. *J. Gen. Appl. Microbiol.* **33**, 303–310.

Zhilina T.N. (1986) Methanogenic bacteria from hypersaline environments. *System. Appl. Microbiol.* **7**, 216–222.

ZoBell C.E. & Morita R.Y. (1957) Barophilic bacteria in some deep sea sediments. *J. Bacteriol.* **73**, 563–568.

16

James S. Maki

The air–water interface as an extreme environment

16.1 Introduction

The air–water interface and associated surface microlayers of marine and inland waters represent a unique habitat for microorganisms compared to the respective bulk waters in either situation. Collectively, these microorganisms (Figure 16.1) are known as the neuston (Naumann 1917). Continued interest in the air–water interface is demonstrated by the number of review articles over the past decade or so covering various aspects of its biology, chemistry, and physical nature (Norkrans 1980; Hunter & Liss 1981; Lion & Leckie 1981; Hardy 1982; Sieburth 1983; Kjelleberg 1985; Gladyshev 1986; Romano 1989; Hermansson 1990).

The neuston has often been characterized as being more numerous than the plankton (see Norkrans 1980; Lion & Leckie 1981; Hardy 1982; Kjelleberg 1985; Hermansson 1990). However, despite these large numbers of microorganisms, some investigators have reported that measurements of neuston viability and metabolic activity are not as great as corresponding measurements in the plankton (e.g., Marumo *et al.* 1971; Dietz *et al.* 1976; Bell & Albright 1982). This low neuston activity and/or viability has been suggested to be due to several types of environmental stress (see reviews by Norkrans 1980; Hardy 1982). The major forms of this stress include solar radiation, accumulated toxic substances (e.g., heavy metals, pesticides), temperature fluctuations, and salinity changes (particularly in the marine environment).

The question that arises is: Does the sum total of these environmental stresses constitute the basis for consideration of the air–water interface as an extreme environment for microorganisms? Unfortunately, the answer is not completely straightforward. Although there is no absolute definition of the term "extreme" (see Chapter 15) (Brock 1969; Alexander 1976; Kushner 1978; Smith 1982; Edwards 1990), extreme environments have usually been described in terms of environmental parameters or the organisms found there. In the first case, extreme has been suggested to be the ends of

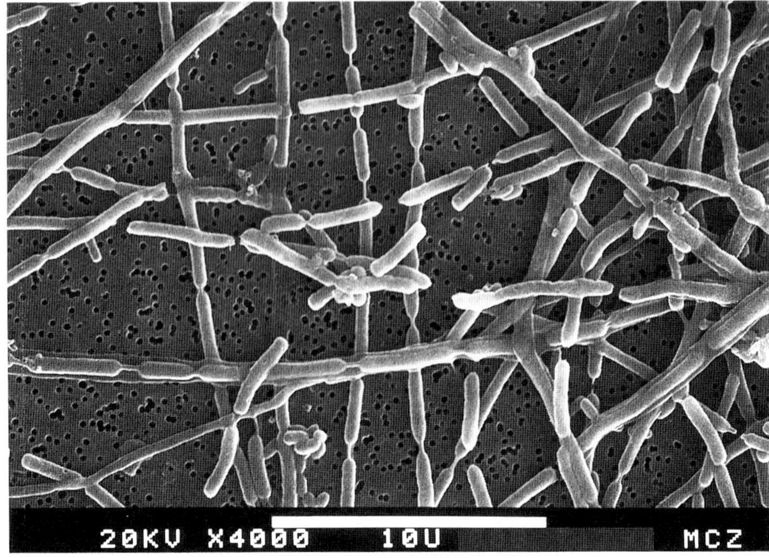

Figure 16.1 Assemblage of neuston from the surface of a freshwater pond as prepared (Maki & Remsen 1983) for scanning electron microscopy. From J.S. Maki (unpublished micrograph), courtesy of R. Pinto, Electron Microscope Facility, Museum of Comparative Zoology, Harvard University.

environmental gradients, e.g., temperature, salinity, pH, nutrients, and pressure (Vestal & Hobbie 1988). In the second case, extreme has been suggested to be associated with a decrease in organismal diversity (Brock 1969; Alexander 1976; Kushner 1978; Edwards 1990). There are, of course, some problems with both of these definitions (see Chapter 15) (Smith 1982), e.g., what is a "normal" or an "average" condition of an environmental parameter, and/or what constitutes a "lower" diversity?

In another examination of extreme environments, Shilo (1979) pointed out that there exist ubiquitous ecosystems that have both sporadic and regular fluctuations in extreme conditions, and therefore merit consideration as extreme environments. Furthermore, the organisms from environments with fluctuating conditions would require great plasticity in order to accommodate these changes (Shilo 1979). Thus, in spite of the fact that the air–water interface could be considered to be at the end of an environmental gradient (i.e., solar radiation) and that there is some indication that a lower diversity of microorganisms can be found in the neuston compared to the plankton (see discussions by Lion & Leckie 1981; Romano 1989), this habitat appears to fall best into the "fluctuating environment" category. Intense solar radiation and temperature can have both "regular" (i.e., diel and seasonal) and "irregular" fluctuations (i.e., periodic cloud cover), in addition to fluctuations of factors such as concentrations of heavy metals, pesticides, and salinity. In some respects, by this definition, all aquatic environments could be considered extreme. However, the air–water interface and the neuston, at the very top of the water column, bear the brunt of these fluctuations and are affected when the rest of the water column may not be. Therefore, it is in light of the comments by Shilo (1979) that the air–water interface will be considered as an extreme environment in this chapter, and that the characteristics of the neuston that contribute to their successful adaptation to this habitat will be examined.

In order to provide a better understanding of the habitat and what is known of its chemistry and biology, a brief discussion of sampling techniques, models of the structure of water layers, and the chemistry important to the neuston will be presented. There are some limitations on the scope of this chapter. First, although the neuston is composed of many types of microorganisms, the primary focus will be on bacteria (i.e., bacterioneuston). Secondly, although an air–water interface also exists when gas bubbles rise through a water column (see review by Kjelleberg 1985), only the interface, associated microlayers, and foams at the surface of marine and inland waters will be covered. For those requiring further information, the reader is encouraged to examine the previously mentioned reviews and the extensive report presented by Williams *et al.* (1986).

16.2 Sampling

What is already known and what will be learned in the future, regarding the neuston and the chemistry of the air–water interface, has been, and will be, somewhat influenced by the sampling procedures employed. As will be discussed below, the top organic molecular layer of the water column will be referred to as the surface microlayer, and all surface strata will be referred to as the surface film following the definitions of Norkrans (1980) and Hermansson (1990).

The ultimate aim of sampling the air–water interface, for the most part, is to collect only the organic molecules and organisms at the interface and leave the subsurface water behind (Kjelleberg 1985). In most, if not all, of the investigations on this zone, the surface film is operationally defined based upon the depth of the sample layer collected, which is, in turn, dependent upon the sampler employed (Wangersky 1976; Garrett & Duce 1980). A number of different sampling devices and/or methodologies are presented in Table 16.1.

Several factors are important in the choice of a sampler and consequently in the evaluation of surface film data. These include the needs of the investigator, i.e., biologists and chemists may have

Table 16.1 Examples of surface film sampling devices and some of their modifications

Sampling method	Depth sampled	Reference
Paper/membrane filters	8–100 μm	Amies (1956), Crow *et al.* (1975), Kjelleberg *et al.* (1979), Sewell *et al.* (1981), Maki & Remsen (1983)
Calcium alginate gauze	–	Amies (1956)
Screen sampler	300–400 μm	Garrett (1965), Sieburth (1965)
Rotating drum	34–100 μm	Harvey (1966); Brockmann *et al.* (1976), Daumas *et al.* (1976), Södergren (1984), Carlson *et al.* (1988), Hardy *et al.* (1988)
Bubble microtome	0.5–10 μm	MacIntyre (1968), Fasching *et al.* (1974)
Germanium prisms	30 nm	Baier (1970)
Glass plate	20–100 μm	Harvey & Burzell (1972), Hardy *et al.* (1985a)
Teflon plate	50–100 μm	Larsson *et al.* (1974)
Teflon disc	–	Miget *et al.* (1974)
Electron microscope grids	–	Young (1978), Fuerst *et al.* (1987)
Freezing probe	1000 μm	Hamilton & Clifton (1979)
Polyvinylchloride film	1 μm	Hamilton & Clifton (1979)

different requirements for the type of sampler employed (Wangersky 1976); the volume of sample required to complete all the desired analyses; and the type of material the sampler is constructed from. This last factor is particularly important for those sampling devices which function through adsorption of the surface film because different materials will adsorb different components (Garrett & Duce 1980). Comparisons of samplers and sampling methodologies both in the field (Daumas *et al.* 1976; Kjelleberg *et al.* 1979; Carlson 1982a; Crawford *et al.* 1982; Estep *et al.* 1985; Hardy *et al.* 1988) and in the laboratory (Hatcher & Parker 1974a; Van Vleet & Williams 1980) have indicated that samplers can have both qualitative and quantitative differences in their collection of various biological and chemical parameters. Their results indicate that direct comparisons between sampling devices are at best difficult (e.g., Carlson 1982a).

To compare components from samples collected from the air–water interface with corresponding samples from the bulk phase, the term enrichment is used. Simply, enrichment is the ratio between the concentration or density of a component in the surface sample and in the bulk phase (i.e., $E = X_{surface}/X_{bulk}$ or $E = [X_{surface}/X_{bulk}] - 1$). In the first case an $E > 1$ indicates a positive enrichment, while in the second case it is indicated by $E > 0$. These enrichment factors are often given in terms of the thickness of the surface layer collected by the sampler. However, some authors have corrected their sampling enrichment to the proposed thickness of the organic microlayer (e.g., Sieburth *et al.* 1976).

16.3 Surface film structure

The models that have been proposed for the structure of the air–water interface and associated surface microlayers are based upon the presence of certain major organic chemical components found there (i.e., proteins, polysaccharides, and lipids). This organic layer will be referred to as the surface microlayer, following the definition used by Norkrans (1980) and Hermansson (1990).

Some of the early investigations of the chemistry of surface films concentrated upon lipids and fatty acids (Garrett 1967; Jarvis *et al.* 1967; Larsson *et al.* 1974; Kattner & Brockmann 1978). Other investigations demonstrated the importance of proteins and polysaccharides in the surface film (Baier *et al.* 1974; Sieburth *et al.* 1976). Initial models separated these components into two layers (Figure 16.2a) (see Norkrans 1980):

1 a dry surfactant layer at the very surface composed of lipids, about 1–2 nm thick when they are in a monolayer and about 16 nm thick when they are multilayered;

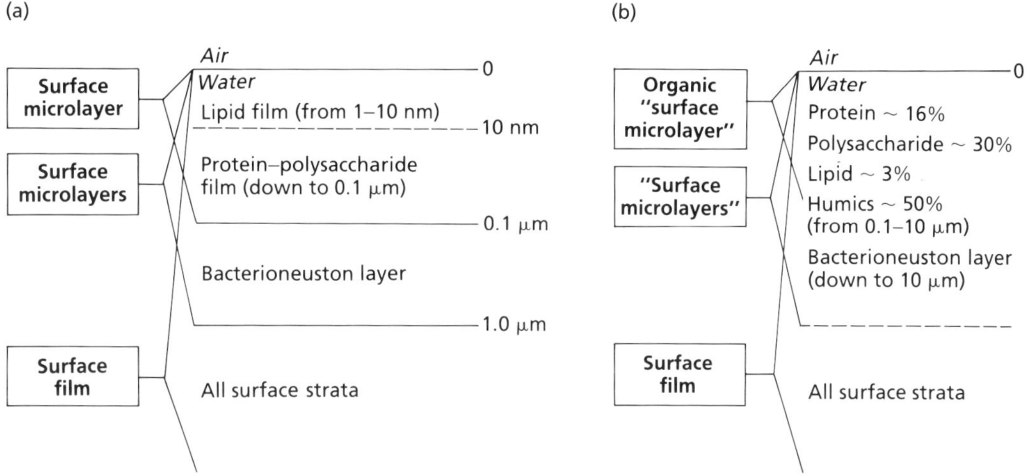

Figure 16.2 Models of the structure of water surface strata at the air–water interface. (a) Model of dry surfactant and wet surfactant layers composed of lipids and proteins–polysaccharides, respectively. The bacterioneuston are concentrated in a layer approximately 1 μm thick. (b) Model of a gel-like matrix composed of proteins, polysaccharides, lipids, and humic substances. The bacterioneuston are found in layers up to 10 μm thick. In each case surface microlayer refers to the organic layer, surface microlayers refers to the organic layer plus the bacterioneuston layer, and surface film refers to these layers plus all additional surface strata (Norkrans 1980; Hermansson 1990). (a) Adapted from Norkrans 1980; (b) adapted from Sieburth (1983).

2 a wet surfactant layer below it consisting of the proteins and polysaccharides.

However, other data has indicated that the importance of lipids to surface microlayers was overstated, and that lipids were not in high enough concentrations to form a coherent layer (see discussions by Hunter & Liss 1981; Sieburth 1983; Williams *et al.* 1986), although small concentrations could influence the surface potential and film pressure (Van Vleet & Williams 1983). An alternative model proposes a matrix of lipids, proteins, and polysaccharides, some of which are colloidal, forming one hydrated gel-like layer in combination with humic substances (Figure 16.2b) (Sieburth 1983). This organic layer of lipids, proteins, polysaccharides, and humic substances, originally suggested to be on the order of about 0.1 μm thick (Sieburth *et al.* 1976; Norkrans 1980), may actually be on the order of micrometers thick in its hydrated form (Sieburth 1983). The quantity of these components and, thus, the thickness of these films can vary both spatially and temporally (e.g., Carlson 1983; Williams *et al.* 1986).

Compression of film material through wind action and internal waves can form visible slicks on the water surface (Dietz & LaFond 1950; Ewing 1950; Barger *et al.* 1974) which are more viscous than nonslick areas (Carlson 1987). However, it is important to note that surface films can be found on all waters whether a visible slick is observed or not (Sieburth 1983). When wind speed increases, slicks can become compressed to the point that they collapse and form fibers and particulate material (Goldacre 1958; Wheeler 1975) which may then be carried into the bulk phase. Rain may also work to remove surface-film material (Green & Houk 1979). Although surface-film material can be disrupted by wind and rain, experimental data has indicated that it is renewed rather quickly (Jarvis 1967; Dragcevic & Pravdic 1981; Van Vleet & Williams 1983). However, the accumulation processes of material within various portions of a slick and between two separate slicks has been observed to be different (Romano & Laborde 1987). Methods for demonstrating the presence of surface films include determining a change in surface tension (e.g., Adam 1937; Hardman 1941; Goldacre 1949) and measuring ultraviolet absorbance of constituent materials (Carlson 1982b).

Below the organic surface microlayer is the bacterioneuston layer, originally suggested to be on the order of 1 μm thick (Sieburth *et al.* 1976; Norkrans 1980; Hermansson 1990). However, Sieburth (1983) has indicated that microcolonies of bacteria may form layers that are several micrometers thick at the surface. The thickness of the bacterioneuston layer may also vary due to the surface-active properties of the bacteria involved (Bungay & Masak 1981). Additional information on the bacterioneuston interaction with the air–water interface and surface films will appear in following sections. Other components of the neuston, including algae and protozoa, as well as particulate material, are found in thicker layers (Norkrans 1980; Hardy 1982; Sieburth 1983). Following the definition of Norkrans (1980) and Hermansson (1990), the surface microlayer and all additional surface strata will be referred to as the surface film. It is apparent that most of the samplers listed in Table 16.1 will collect a much thicker layer of water than either the surface microlayer or even the bacterioneuston layer (see Figure 16.2).

16.4 Chemical factors

As discussed above, the more recent model proposes a surface microlayer consisting of proteins, carbohydrates, lipids, and humic substances, in a gel-like matrix (Sieburth 1983). In addition, other organic and inorganic compounds are present which can influence the biology of the air–water interface. A number of processes are involved in the transport of dissolved and particulate material into and out of surface films as well as their transformation within these films (Table 16.2). Additional details regarding these processes and the chemistry of surface films can be found in Baier *et al.* (1974), Liss (1975), Hunter (1980), Hunter & Liss (1981), Lion & Leckie (1981), Armstrong & Elzerman (1982), Hardy (1982), Williams *et al.* (1986), and Hermansson (1990).

For the bacterioneuston, the advantage of being associated with the air–water interface, particularly in oligotrophic bulk waters, is the presence of nutrients. Enrichments of either dissolved organic carbon (including the proteins, polysaccharides, and lipids discussed above), phosphate, ammonia, nitrate, or nitrite have been reported

Table 16.2 Factors involved in the input and output and transformation of dissolved and particulate material in surface films. Compiled from Liss (1975) and Hunter (1980)

Inputs	Transformations	Outputs
From the atmosphere	In the microlayer	To the atmosphere
Dry deposition	Biological	Wind-generated aerosol
Rain	Photochemical	Bursting bubbles
		Evaporation
From the bulk phase		To the bulk phase
Upwelling		Wave action
Convection		Dissolution
Brownian diffusion		Sedimentation
Eddy diffusion		Brownian diffusion
Bubbles		Eddy diffusion

from surface films and foams in both marine (Goering & Menzel 1965; Williams 1967; Sieburth *et al.* 1976; Carlson 1983; De Souza Lima 1985; Henrichs & Williams 1985; Kattner *et al.* 1985; Williams *et al.* 1986; Marty *et al.* 1988) and freshwater environments (Hatcher & Parker 1974b; Saijo *et al.* 1974; Eisenreich *et al.* 1978; Danos *et al.* 1983; Maki *et al.* 1984; Estep & Remsen 1985; Södergren 1987; Maki & Remsen 1989). Sieburth *et al.* (1976) suggested that nutrient concentrations in samples collected from surface films and adjusted to a microlayer thickness of 0.1 μm, approach those concentrations of nutrients found in laboratory cultures. In support of this idea, laboratory studies have shown that starving copiotrophic bacteria resume growth when they come into contact with nutrients at an air–water interface (Kjelleberg *et al.* 1982). Furthermore, greater enrichments of bacteria in surface films have been observed in oligotrophic lakes than in eutrophic ones (Crawford *et al.* 1982).

Table 16.3 Examples of petroleum, polyaromatic (PAHs), and chlorinated hydrocarbons measured in marine, estuarine, and freshwater surface films

Ecosystem	Sampler	Compounds	Reference
Marine	Bottle	DDT, DDE, dieldrin, aldrin	Seba & Corcoran (1969)
Marine	Metal screen	Hydrocarbons, PCBs	Duce *et al.* (1972)
Marine	Metal screen	PCBs, DDT	Bidleman & Olney (1974)
Marine and freshwater	Teflon plate	DDT, DDD, DDE, PCBs	Larsson *et al.* (1974)
Marine	Metal screen	Hydrocarbons	Wade & Quinn (1975)
Marine	Metal screen	Normal alkanes	Marty & Saliot (1976)
Marine	Metal screen	Normal alkanes	Hardy *et al.* (1977)
Freshwater (foam)	Plastic cylinder	DDT, DDD, DDE, dieldrin	Eisenreich *et al.* (1978)
Marine and freshwater	Teflon sheets	PCBs	Ofstad *et al.* (1979)
Marine	Metal screen	Hydrocarbons, PAHs	Boehm (1980)
Freshwater	Glass plate	Hydrocarbons	Meyers & Kawka (1982)
Freshwater	Metal screen	PCBs	Rice *et al.* (1982)
Marine	Teflon drum	PCBs, DDT, DDD, DDE, PAHs	Cross *et al.* (1987)
Marine	Glass plate and Teflon drum	Hydrocarbons, PAHs, Pesticides, PCBs	Hardy *et al.* (1987)
Marine	Teflon drum	Alkanes, PAHs	Hardy *et al.* (1990)

DDD, dichlorodiphenyldichloroethane; DDE, dichlorodiphenyldichloroethylene; DDT, dichlorodiphenyltrichloroethane; PAH, polyaromatic hydrocarbon; PCB, polychlorinated biphenyl.

In addition to the nutrients, other types of compounds that can be sources of stress to the bacterioneuston have also been demonstrated to be enriched in surface films: pesticides, polyaromatic chlorinated, and other hydrocarbons, as well as heavy metals.

A number of investigators have reported petroleum hydrocarbons, polyaromatic hydrocarbons (PAHs), polychlorinated biphenyls (PCBs), and pesticides in the surface film (Table 16.3). In addition, dissolved phenolic compounds have also been demonstrated to be enriched in surface films (Carlson & Mayer 1980; Carlson 1982b). Petroleum hydrocarbons consist of a range of different products, some of which are easily broken down, while others are much more recalcitrant (Bartha 1986). These hydrocarbons in aquatic environments spread on the surface over large areas (Bartha 1986). However, they have a nonuniform distribution at the air–water interface due to surface chemical considerations (Garrett 1974). The more volatile compounds disappear first, some through photodegradation (e.g., Hansen 1977), leaving the less volatile compounds (Garrett 1974). The microlayer is often enriched with hydrocarbons although their nature can be quite different from hydrocarbons in the bulk phase (Boehm 1980). PCBs can become enriched in surface films from atmospheric deposition (Södergren *et al.* 1990) and both they and pesticides can become concentrated in slicks (Platford 1982).

The presence of a number of different metals has been examined in both freshwater and marine surface films (Table 16.4). Metal accumulation in surface films at any given time is controlled by a number of factors including the composition of microlayer material, the physical and chemical form of the metal, the processes controlling transport, longevity of the film, and the residence time of the metal (see Hunter 1980; Armstrong & Elzerman 1982). High metal enrichments have been reported to be associated with particulates in the surface film (e.g., Duce *et al.* 1972; Piotrowicz *et al.* 1972; Elzerman & Armstrong 1979; Lion *et al.* 1979; Hardy *et al.* 1990), although the degree of association will vary with each metal (e.g., Lion & Leckie 1982). Cross *et al.* (1987) reported higher metal concentrations in the dissolved fraction in offshore waters and higher concentrations associated in the particulate fraction in inshore waters. This may reflect the amount of particulate material in either location. Metals also appear to be complexed with organic matter (e.g., Piotrowicz *et al.* 1972; Pojasek & Zajicek 1978; Pellenbarg 1981; Hardy *et al.* 1985a) and concentrated in slicks (Piotrowicz *et al.* 1972; Elzerman & Armstrong 1979; Hardy *et al.* 1985a). Sources of metals to the surface film include both atmospheric deposition (Hunter 1980; Eisenreich 1982; Hardy *et al.* 1985b) and bubble scavenging (Eisenreich 1982). Residence times for metals

Table 16.4 Examples of metals measured in surface film samples from marine, estuarine, and fresh waters

Ecosystem	Sampler	Metals	Reference
Marine	Plastic screen	Cu, Fe, Ni, Pb	Duce *et al.* (1972)
Marine	Plastic screen	Al, Cu, Fe, Mn, Ni, Pb, V	Piotrowicz *et al.* (1972)
Freshwater	Drum, metal screen, tray	Fe, Mn, Zn	Hatcher & Parker (1974b)
Marine and freshwater	Teflon plate	Ni, Zn	Larsson *et al.* (1974)
Freshwater (foam)	Plastic cylinder	Cd, Cu, Fe, Pb, Zn	Eisenreich *et al.* (1978)
Freshwater	Plastic screen	Cd, Cu, Pb, Zn	Elzerman & Armstrong (1979)
Salt-marsh	Plastic screen	Cd, Pb	Lion *et al.* (1979)
Salt-marsh	Plastic boom	Cu, Fe, Zn	Pellenbarg & Church (1979)
Marine	Nylon screen	Cd, Cu, Fe, Mn, Ni, Pb, Zn	Hunter (1980)
Salt-marsh	Plastic boom	Cd, Cu, Fe, Pb, Zn	Pellenbarg (1981)
Freshwater	Nylon screen	Al, Cd, Cr, Cu, Fe, Mn, Ni, Pb, Zn	Eisenreich (1982)
Salt-marsh	Teflon screen	Pb	Harvey *et al.* (1982)
Salt-marsh	Teflon screen	Cd, Cu, Pb	Lion & Leckie (1982)
Marine	Glass plate	Cd, Cu, Fe, Pb, Zn	Hardy *et al.* (1985a)
Marine	Metal screen	Organotins	Cleary & Stebbing (1987)
Marine	Teflon drum	Ag, Cd, Cr, Cu, Fe, Mn, Ni, Pb, Zn	Cross *et al.* (1987)
Marine	Glass plate and Teflon drum	Ag, Cd, Cu, Pb, Zn	Hardy *et al.* (1987)
Estuarine	Teflon drum	Ag, Cd, Cu, Cr, Hg, Ni, Pb, Se, Zn	Hardy *et al.* (1990)

in surface films have been estimated to be from minutes to hours (Eisenreich 1982; Hardy *et al.* 1985b). Toxicity of these metals to the bacterioneuston may be moderated by metal binding to organics and/or particulates. Sequestering of the metals by organic compounds at the air–water interface may limit the exposure of the bacterioneuston to their toxic ionic forms.

16.5 Bacterioneuston

Although there is considerable literature regarding the numbers and activity of the neuston, there is not much known about what constitutes a "successful" bacterioneuston. For example, currently there have only been two reports of examinations of bacterioneuston for the presence of plasmids (Hermansson *et al.* 1987; Jones *et al.* 1991), even though in an exposed environment like the air–water interface, the information found on plasmids may be important to survival (i.e., resistance to ultraviolet (UV) light and heavy metal toxicity, catabolism of pesticides and other hydrocarbons). Therefore, the following discussion of bacterioneuston survival is limited in the sense that many characteristics that could be important to their survival have not necessarily been demonstrated or even looked for in bacterioneuston isolates. Thus, this discussion will draw upon both reported characteristics of bacterioneuston as well as other characteristics that may be applicable to their selection to and continued survival in this habitat.

16.5.1 *Firm and reversible adhesion of bacteria to the interface*

The air–water interface with its adsorbed macromolecules has been suggested to "appear" to a bacterium as a solid substratum (Sieburth 1983; Kjelleberg 1985). As such, it is not surprising that cell surface characteristics like hydrophobicity and charge, that are important in the adhesion of bacteria to solid substrata (Marshall 1985), appear to play a similar role in bacterial adhesion to the air–water interface (Kjelleberg & Stenström 1980; Dahlbäck *et al.* 1981; Hermansson *et al.* 1982; Kjelleberg 1985). Furthermore, the presence of extracellular polymers, fibers, and fimbriae observed associated with bacterioneuston (Young 1978; Maki & Remsen 1983; Fuerst *et al.* 1987) may mediate firm adhesion of the bacteria to the air–water interface as they do to solid substrata (Marshall 1985).

Laboratory experiments using artificial surface films composed of mono- or multilayers of lipids and fatty acids have indicated that some cells actually penetrate these films while other bacteria form layers underneath (Kjelleberg *et al.* 1976). The thickness of bacterial layers can be influenced by the surface active properties of the cell (Bungay & Masak 1981). Multilayered films can have a higher enrichment of bacterial cells than monolayers (Norkrans & Sörensson 1977) and, in addition, the architecture and charge of the film play roles in bacterial adhesion (Kjelleberg & Stenström 1980).

Bacteria exhibit two types of adhesion to surfaces: firm, mediated by extracellular polymeric material, and reversible, where bacteria are attracted to the surface but do not attach firmly (Marshall 1985). Bacterioneuston have been suggested to be preconditioned for rapid attachment to any substratum to which they may become exposed (DiSalvo 1973) and significant proportions have been reported to be associated with particulate material in surface films (e.g., Harvey & Young 1980a). However, in some situations the bacterioneuston may only reversibly attach or be nonadhesive to other substrata they may come in contact with (Maki *et al.* 1990). Reversibly attached cells are able to scavenge the adsorbed nutrients on a surface and then move away (Hermansson & Marshall 1985; Power & Marshall 1988). Bacterioneuston may be capable of an exchange between the interface and the bulk phase (Hermansson & Dahlbäck 1983). This suggests that a motile portion of the bacterioneuston may undergo a diel cycle, moving into the surface film at opportune moments, scavenging nutrients, and then moving back to the bulk phase. Changes in the number of bacterioneuston over a diel period have been reported in freshwater environments (Södergren 1984; Maki & Herwig 1991) but in some cases these changes may be masked by the number of cells trapped in thick surface films (e.g. Maki & Remsen 1989).

Other mechanisms of transport could account for changes in the number of bacterial cells reported above. One such mechanism is bubble scavenging. Bacteria from the bulk phase can be captured by rising bubbles and transported to the surface film or transferred to the air (e.g., Carlucci & Williams 1965; Blanchard & Syzdek 1978). These bubbles capture and transport some types of bacteria from the bulk phase more favorably than others (Carlucci & Williams 1965; Bezdek & Carlucci 1972; Blanchard & Syzdek 1978). Bacteria that are pigmented are one type that are more easily captured and transported, possibly because of differences in their cell surface characteristics (Blanchard & Syzdek 1978). These same characteristics may also contribute to their enrichment in surface films (Hermansson *et al.* 1979). Thus, this differential transport of bacteria by bubbles is an initial selection mechanism for potential bacterioneuston. Furthermore, bubbles represent a mechanism for microorganisms to return to the air–water interface after the surface film is disrupted by wind, waves and/or rain.

16.5.2 Neuston activity

As pointed out in Section 16.1, in spite of high numbers of cells, viability and/or activity of the neuston may be low in comparison with the plankton. Many studies indicated that viable bacteria, as determined by colony forming units on solid media, were more numerous in the bacterioneuston than in the plankton (e.g., Dratchev *et al.* 1957; Hatcher & Parker 1974b; Carty & Colwell 1975; Crow *et al.* 1975; Dietz *et al.* 1976; Fehon & Oliver 1979; Danos *et al.* 1983). Further studies have demonstrated that the bacterioneuston isolates expressed proteolytic and/or lipolytic activity (Babenzien & Schwartz 1970; Sieburth 1971a; Tsyban 1971; Crow *et al.* 1976; Dahlbäck *et al.* 1982). However, in some cases higher percentages of bacteria with these abilities were found in the bulk water (Kjelleberg & Håkansson 1977). Studies using direct microscopic methods to enumerate total bacteria in combination with either formazan dyes to determine respiring bacteria, or microautoradiography to determine active bacteria, showed that respiring or active bacteria were enriched in salt marsh, marine, and freshwater surface films (Harvey & Young 1980b; Hermansson & Dahlbäck 1983; Maki & Remsen 1989).

Most of the respiring bacteria were associated with surface film particulates in the salt marsh (Harvey & Young 1980b). Although active and respiring bacteria were enriched in the marine and freshwater surface films, the percentage of active or respiring cells was similar to that in the plankton (Hermansson & Dahlbäck 1983; Maki & Remsen 1989). Although some investigators have reported higher concentrations in the bulk phase (Dietz *et al.* 1976; Passman *et al.* 1979), in most cases measurements of adenosine triphosphate (ATP), an estimate of microbial biomass, have been found to be enriched in surface films (e.g., Daumas *et al.* 1976; Sieburth *et al.* 1976; Bell & Albright 1982; De Souza Lima & Romano 1983; De Souza Lima & Chretiennot-Dinet 1984; Williams *et al.* 1986; Mimura *et al.* 1988).

Several important points are indicated by the results of the above studies. These include:

1 ATP enrichment indicates that there is a living microbial biomass in surface films;

2 measurements of colony forming units and direct counts of respiring bacteria indicate that a portion of the bacteria are viable;

3 these bacteria apparently possess the enzymatic capability (e.g., proteolytic and lipolytic) to utilize surface film organic material.

These points suggest that a portion of the bacterioneuston is capable of active metabolism in spite of any inhibitory environmental stress. Further support for this is provided by studies examining the utilization (turnover times) or uptake of radiolabeled compounds in the neuston and the plankton. Some investigators have reported that turnover times of different substrates in the neuston were similar to those in the plankton (De Souza Lima 1985; Carlucci *et al.* 1985; 1986; 1991; Williams *et al.* 1986). Higher neuston activity has been observed in the presence of slicks (De Souza Lima & Chretiennot-Dinet 1984) and utilization rates in surface films in oligotrophic waters have been shown to be higher than in corresponding bulk waters, although in eutrophic waters rates were similar (Carlucci *et al.* 1985; Williams *et al.* 1986). Even studies that have indicated that the utilization or uptake of labeled compounds was lower in the neuston than the plankton have shown neuston activity (Dietz *et al.* 1976; Passman *et al.* 1979; Bell & Albright 1982; Maki & Herwig 1991). The results of all these investigations point to an active heterotrophic bacterial component in the neuston. Thus, in terms of the air–water interface as an extreme environment, the reports of activity indicate that the cells in many cases appear to be able to compensate for any stressful conditions quite well. Factors that may contribute to the survival of these active bacteria will be discussed in the following sections.

16.5.3 Solar radiation

Because of its location, the air–water interface is subject to intense solar radiation (see reviews by Norkrans 1980; Gladyshev 1986). Of the solar radiation reaching the air–water interface, the most biologically important in terms of their detrimental effects are the mid-UV (UV-B, 290–320 nm) and the near-UV (UV-A, 320–400 nm) (Thomas 1977; Nasim & James 1978; Jagger 1983), although harmful effects can be caused by wavelengths up to 700 nm (Nasim & James 1978). In general, these wavelengths could have both lethal or nonlethal effects on bacteria. The lethal effects are primarily caused by lesions in the DNA. Nonlethal effects include growth delay, growth inhibition, inhibition of induced enzyme synthesis, reduced active transport, and mutagenesis (Thomas 1977; Jagger 1983). Therefore, it is not surprising that solar radiation has been indicated to be one of the major environmental factors to have a deleterious effect on both neuston viability and activity (e.g., Marumo *et al.* 1971; Horrigan *et al.* 1981; Bailey *et al.* 1983 – the last paper deals with bacteria in the top 6 mm of the water column). As such, it would be an important selection mechanism for the surviving bacterioneuston.

Despite the above reports on its inhibitory effects, additional studies have reported little or no difference in neuston activity between intense sunlight or cloudy conditions (Hermansson & Dahlbäck 1983) and little effect of UV and visible light on the total activity (Mitamura & Matsumoto 1981; Carlucci *et al.* 1985; 1986; Williams *et al.* 1986; Garabetian 1991). In addition, Dahlbäck (1983) indicated that a higher proportion of surface film than bulk-water bacteria were able to grow on agar plates after irradiation by UV light. These reports indicate that a portion of the neuston that are metabolically active are relatively resistant to the effects of solar radiation. Obviously, one method for dealing with the intense solar radiation at the air–water interface would be to avoid it by only being a resident of surface films during periods of low light intensity. For the bacterioneuston unable to avoid the solar radiation, tolerance could be accomplished through at least three mechanisms (Jagger 1983):

1 physical screening by UV-absorbing molecules in the cell wall or membrane that either reflect the light or degrade it into heat;

2 interference or protection of hazardous photoproducts like singlet oxygen by carotenoid pigments;

3 systems for the repair of DNA damage.

Although reports of molecules that would perform mechanism 1 are so far unreported in the bacterioneuston, mechanisms 2 and 3 will be discussed below.

Many investigators have found that a considerable portion of heterotrophic bacteria isolated from surface films are pigmented

(Babenzien & Schwartz 1970; Tsyban 1971; DiSalvo 1973; Fehon & Oliver 1979; Hermannson *et al.* 1987; Guerin 1989; Jones *et al.* 1991). However, they are not always present in the highest frequency (e.g., Crow *et al.* 1975; Brockmann *et al.* 1976). Exposure to UV-B has been shown to cause an increase in the proportion of pigmented bacteria (Thomson *et al.* 1980). These pigments can be important to survival. For example, carotenoid pigments protect against the harmful effects of visible light by quenching the products of photochemically induced oxidations (Krinsky 1979). However, carotenoid pigments are not effective against DNA damage caused by UV light (Thomas 1977; Nasim & James 1978) so additional protective mechanisms would need to be employed.

The harmful effect of UV light centers on the formation of thymine dimers on strands of DNA, although other types of damage can occur (Nasim & James 1978; Jagger 1983). One possible protection mechanism is the base ratio of the bacteria involved. Singer & Ames (1970) proposed that bacteria that were exposed to sunlight would evolve a high guanosine+cytosine (G+C) content to avoid thymine damage from UV radiation in sunlight. If this were true one could expect that the genera of bacterioneuston that were adapted to existence at the air–water interface would have high mol% G+C. Examination of the mol% G+C in common bacterial genera isolated from surface films (Table 16.5) reveals that many of them (10 out the 16 genera listed) have mol% G+C in excess of 50, including the potentially pathogenic genera *Klebsiella* and *Salmonella*. The most commonly reported genera, *Alcaligenes*, *Micrococcus*, and *Pseudomonas*, all have mol% G+C between 56 and 75. This cursory examination of genera of bacterioneuston does lend some support to the hypothesis of Singer & Ames (1970).

However, in spite of the mol% G+C, repair systems (photoreactivation, excision repair, recombination repair, and SOS repair) for damaged DNA would be important to bacterioneuston adapting to life at the air–water interface. Induction of the SOS DNA repair system in bacteria occurs after exposure to UV light (Ossanna *et al.* 1987). This system involves activating approximately 20 operons, the products of which repair UV caused DNA damage (Ossanna *et al.* 1987). Recently, Hermansson *et al.* (1987) and Jones *et al.* (1991) have reported the presence of plasmids in bacteria isolated from marine and freshwater surface films. Bacterial resistance to UV may also be encoded on plasmids (Strike & Lodwick 1987) suggesting that acquisition of plasmids by bacterioneuston may be an important step in survival. However, the SOS repair systems for UV damage of DNA may be error-prone due to incorrect bases being placed in the DNA strand, leading to an increase in mutation in surviving bacteria

Table 16.5 Examples of some common bacterial genera isolated from marine and freshwater surface films and the mol% G+C content of their DNA

Genus	Mol% G+C	Surface film reference
Acinetobacter	38–47[1]	Fehon & Oliver (1979)
Aeromonas	57–63[1]	Carty & Colwell (1975), Cook *et al.* (1978)
Alcaligenes	56–70[1]	Babenzien & Schwartz (1970), Sieburth (1971a), Goutx *et al.* (1987)
Bdellovibrio	33.4–51.5[1]	Williams (1987)
Caulobacter	62–67[2]	Babenzien & Schwartz (1970)
Enterobacter	52–60[1]	Sieburth (1971a), Fehon & Oliver (1979)
Flavobacterium	31–42[1]	Fehon & Oliver (1979)
Klebsiella	53–58[1]	Cook *et al.* (1978)
Leptothrix	69.5–71[2]	Babenzien & Schwartz (1970), Frølund (1977)
Micrococcus	64–75[3]	Babenzien & Schwartz (1970), Tsyban (1971), Carty & Colwell (1975)
Moraxella	40–47.5[1]	Fehon & Oliver (1979)
Nevskia	60.4[1]	Babenzien & Schwartz (1970)
Pseudomonas	58–70[1]	Babenzien & Schwartz (1970), Sieburth (1971a), Tsyban (1971), Carty & Colwell (1975), Fehon & Oliver (1979)
Salmonella	50–53[1]	Cook *et al.* (1978)
Staphylococcus	30–39[3]	Cook *et al.* (1978)
Vibrio	38–51[1]	Carty & Colwell (1975)

[1] From Krieg (1984).
[2] From Staley *et al.* (1989).
[3] From Sneath *et al.* (1986).

(Ossanna *et al.* 1987; Strike & Lodwick 1987). Error-prone DNA repair can account for about half of the spontaneous mutagenesis in bacteria (Smith & Sargentini 1985). Thus, although repair systems may be essential for the survival of bacterioneuston during exposure to intense solar radiation, they may also contribute to an increase in mutation rate.

16.5.4 Temperature effects

Bacterioneuston are exposed to heat from both the water and from direct absorption (Norkrans 1980). Thus, in addition to the "regular" diel and seasonal fluctuations in temperature, they are also exposed to rapid sporadic changes due to factors like cloud cover (Norkrans 1980). As pointed out by Norkrans (1980) there appears to be no fluctuation data for these surfaces in nature. However, Maki & Herwig (1991) found that the number of total bacteria and colony forming units in the neuston of a freshwater Antarctic pond correlated positively to air temperature and surface water temperature, respectively. Short-term temperature fluctuations have been shown to increase bacterial diversity in chemostat enrichments of sediment bacteria, indicating that separate temperature niches may be utilized by different bacteria (Upton *et al.* 1990). Thus, these temperature fluctuations may affect single types of bacteria but not adversely affect the bacterioneuston population as a whole.

One seasonal temperature effect that bacterioneuston may face is ice formation. Nocturnal ice formation was suggested to have affected the neuston population size and activity in an Antarctic freshwater pond (Maki & Herwig 1991). These authors suggested that a period of time was required for the neuston to "recover" after the nocturnal freezing. Surface film water from Lake Ontario was found to freeze at significantly higher temperatures than the bulk water and it was suggested that these freezing characteristics may be due to a biotic component (i.e., ice-nucleating bacteria) (Simon & Makarewicz 1990). Bacteria capable of ice-nucleation are normally reported associated with plants (Lindow 1985) and if present in the neuston may have arrived at the air–water interface through atmospheric deposition or in association with leaf litter.

16.5.5 Pesticides and hydrocarbons

Pesticides and other chlorinated hydrocarbons in surface films could have an inhibitory effect on neuston growth. They can affect the cell membranes, the synthesis of nucleic acids, amino acids and proteins, enzyme activity, and, in addition, can cause mutations (see reviews by Lal & Saxena 1982; Tripathi 1987). Pesticides have been shown to inhibit hydrocarbon degradation by bacterioneuston isolates (Crow *et al.* 1976). However, these compounds are subject to biodegradation, and the mechanisms have been recently reviewed (Leahy & Colwell 1990; Chaudhry & Chapalamadugu 1991). The genes carrying the information for these biodegradation processes are often plasmid associated (Leahy & Colwell 1990; Chaudhry & Chapalamadugu 1991).

Petroleum hydrocarbons can also affect microorganisms, e.g., chemotaxis (Chet & Mitchell 1976). In terms of the bacterioneuston, petroleum hydrocarbon-degrading bacteria have been demonstrated in both surface films and surface foams (Gunkel 1973; Krasilnikov *et al.* 1973; Fehon & Oliver 1977; Rambeloarisoa *et al.* 1984). However, their population size varies between locations (Gunkel 1973). In addition, the utilization of hydrocarbons and PAHs may be less in the neuston than in the plankton (Fehon & Oliver 1977; Guerin 1989). Controlled ecosystem experiments have also shown no increase in biodegradation of hydrocarbons in the surface film

(Gearing & Gearing 1982). Examinations of hydrocarbon-degrading bacterioneuston have generally shown the presence of extracellular polymers (Wyndham & Costerton 1982; Goutx *et al.* 1987). These polymers appear to have an emulsifying action upon the petroleum hydrocarbons (Rambeloarisoa *et al.* 1984; Goutx *et al.* 1987), thus, aiding in their degradation. For further information regarding the degradation of petroleum hydrocarbons see the review by Leahy & Colwell (1990).

16.5.6 Heavy metals

Some metals are essential for growth, reproduction, and survival of microorganisms (Duxbury 1985). But even high concentrations of some of these essential elements can have a toxic effect on cells (Duxbury 1985). The toxicity manifests itself by affecting the growth, morphology, and metabolic activity of microorganisms (Collins & Stotzky 1989). On an ecosystem level heavy metals can alter microbial community structure both quantitatively and qualitatively (Duxbury 1985). The various factors involved in metal toxicity have been recently reviewed (Collins & Stotzky 1989).

The toxicity of metals is caused by the ionic form. As noted previously (see Section 16.4), in surface films most metals are associated with organic molecules and/or particulates. Although this may reduce the toxic effect, ionic forms of the metals may still be in high enough concentration to cause problems to microorganisms (e.g., Hardy *et al.* 1985a). Furthermore, the high numbers of bacterioneuston associated with particulates in surface films may be readily exposed to the particulate bound metals (e.g., Lion *et al.* 1979; Harvey *et al.* 1982). The bacterioneuston may contribute to the sequestering of metals. They may accomplish this through the production of extracellular polymers (e.g., Young 1978) which in turn may protect the cells by binding the toxic metals (Geesey & Jang 1989).

Bacteria isolated from the air–water interface have been demonstrated to be more resistant/tolerant to some metals than bacteria from the bulk water (Kim 1985; Hermansson *et al.* 1987; Jones *et al.* 1991). Resistance to metal toxicity is often associated with plasmids (Silver & Misra 1988; Summers & Barkay 1989). Some of the genetically determined resistances to metals reported in surface films (see Table 16.4) are presented in Table 16.6. As pointed out above, surface-film bacteria in at least one study have been reported to have a higher incidence of plasmids than bacteria in corresponding bulk water samples (Hermansson *et al.* 1987). Although a direct link between these plasmids and specific metal resistance has not been established, future research using specific genetic probes may show their importance to metal resistance in the bacterioneuston.

Table 16.6 Genetic metal resistance for heavy metals found in surface films. Compiled from Silver & Misra (1988) and Summers & Barkay (1989)

Surface-film metal	Genomic location	Mechanism of resistance
Ag	Plasmid(?)	Dependent on presence of Ag complexing components (halide ions)
	Chromosome	Change in outer membrane protein
Cd	Plasmid	Efflux system for Cd (may also include Zn and Co)
Cu	Plasmid	Efflux system for Cu (dependent upon chromosomal loci)[1]
Cr	Plasmid	Unknown
Hg	Plasmid	Transport system captures Hg for reduction to less toxic form
Ni	Plasmid	Unknown
Pb	Plasmid	Unknown
Zn	Plasmid	See Cd

[1] Only characterized for Gram-negative organisms.

16.5.7 *Salinity changes and rain effects*

Not much is known about the effects of salinity changes on the marine bacterioneuston, although it is known that some marine bacteria are sensitive to changes in salinity (see discussion by Norkrans 1980). Both Crow *et al.* (1976) and Fehon & Oliver (1979) reported that the majority of the bacterial isolates they collected from coastal surface films did not require added salt for growth, suggesting that they may be from terrestrial or freshwater origin. It is interesting to note that Carlucci *et al.* (1991) found that although rain caused a decrease in neuston population densities, their activities were stimulated. What changes occurred in the salinity of the surface film after the rain and whether this had any affect on the increase in activity is unknown. Acid rain is another type of problem that occurs with freshwater systems (Schindler 1988). Lower pH has been shown to affect the physiology of bacteria from lakes receiving acid precipitation (e.g., Rao *et al.* 1984). Although the neuston may be the first community of microorganisms to be affected by acid rain, no studies have appeared to examine this directly.

16.5.8 *Biological factors*

Some additional biological factors may also affect bacterioneuston activity and viability. Cases of bacterioneuston inhibition have been observed in marine waters (Sieburth 1971b; Sieburth *et al.* 1976). Although this inhibition may have been caused by anthropogenic materials it is also possible that the adverse effect may have been due to natural toxic substances from marine algae (see discussion by Sieburth 1983).

Viruses have recently been reported to be present in high numbers in aquatic environments (Bergh *et al.* 1989). It is possible that they also may play a role in decreasing microbial viability and activity in surface films. Because viruses can adsorb to rising bubbles (Baylor *et al.* 1977; Blanchard 1978) they can be transferred to the air in jet drops when the bubble bursts at the surface. However, bubble scavenging could also conceivably act as a mechanism to enrich viruses in surface films at the air–water interface, particularly because viruses do adsorb to surfaces (Bitton 1980) and as discussed above, the air–water interface may "appear" as a solid surface. In addition, UV light and mutagenic/carcinogenic chemicals may inactivate the phage repressors in bacteria at the air–water interface which maintain the lysogenic state (Schmieger 1990), thus causing a release of virions due to lysis of the bacteria. However, exposure to air–liquid interfaces (Trouwborst *et al.* 1974) and solar radiation (Bitton 1980) can lead to virus inactivation or removal. Bacterioneuston may resist the viruses through alteration of surface structures involved in virus recognition or require an assortment of restriction enzymes to protect themselves after a virus attack. Bdellovibrios have also been shown to be enriched in some surface films (Williams 1987) and would also be potential predators of bacterioneuston. However, the presence of heavy metals and pesticides in surface films could also affect their predatory behavior (Varon & Shilo 1981). Protozoans are also potential predators of bacterioneuston (Zaitsev 1971; Sieburth *et al.* 1976; Davis *et al.* 1978; Foissner 1979; Ricci *et al.* 1991). In fact, the bacterioneuston have been considered to be the first link in the whole community of organisms associated with the air–water interface (Zaitsev 1971).

16.6 Conclusion

The above discussions indicate that despite several types of potentially inhibitory environmental stress, some bacterioneuston survive and actively metabolize in surface films at the air–water interface. The fact that not all cells arriving at this habitat are able to survive and grow indicates that not all are able to overcome the environmental stresses encountered. To counter various forms of stress in a fluctuating extreme environment, Shilo (1979) pointed out that a certain amount of plasticity would be required. This would appear to be the case for bacterioneuston. A number of different mechanisms (Table 16.7) could aid in their continued survival in this "extreme" habitat. Variability displayed by bacteria can arise from either intracellular rearrangement or intercellular exchange of genes (Terzaghi & O'Hara 1990). Certainly, the acquisition of genetic information

Table 16.7 Compilation of the various forms of environmental stress at the air–water interface and possible adaptive mechanisms for the survival of bacterioneuston

Stress	Adaptive mechanisms
Solar radiation/UV light	Pigmentation, high percentage G+C content, effective DNA repair system, possibly plasmid associated
Toxic organic compounds	Plasmids containing genes for catabolism of toxic compounds (e.g., pesticides, other chlorinated hydrocarbons) and emusifying extracellular polymers to aid in degradation of petroleum hydrocarbons
Metals	Extracellular polymers for binding of metal ions, plasmids containing genes for resistance
Temperature fluctuations	Unknown
Salinity fluctuations	Unknown
Acid rain	Unknown
Natural algal antibiotics	Resistance factors(?)
Viruses	Changes in surface structures to prevent virus recognition, restriction enzymes

from other bacteria (see DeFlaun & Levy 1989; and Chapter 19, this book) might be beneficial in countering the various forms of inhibitory stress. Recently, in a laboratory investigation, Jones *et al.* (1991) found that conjugal transfer frequencies per donor between two strains of *Pseudomonas aeruginosa* were much higher at the air–water interface than in the bulk phase. The accumulation of high densities of cells at the interface was suggested as an explanation of the results. The presence of nutrients at the air–water interface also appeared to contribute in a positive manner (possibly by maintaining growth) to this plasmid transfer (Jones *et al.* 1991). For bacterioneuston in the environment, factors like UV light could inhibit transconjugant production. However, UV light and other forms of stress appear to be able to increase the movement of resistance determining transposable elements to conjugative plasmids (Viljanen & Boratynski 1991). They may also inhibit the restriction of DNA, and thus, cause a higher efficiency of gene transfer between cells (Viljanen & Boratynski 1991). If viruses are enriched at the air–water interface they may also contribute to gene exchange through transduction (see Chapter 19) (Schmieger 1990).

One question that arises is: Is there a specific successful bacterioneuston? Again, the answer to this question does not appear to be straightforward. Conditions in different films are likely to vary so much that it is difficult to make generalizations on what is a successful bacterioneuston. Instead, there appear to be a number of selection mechanisms, from the ability to be transported and attach to the air–water interface to the ability to survive there, that would contribute to the definition of a successful bacterioneuston organism. However, currently there is not enough information on how these bacteria specifically adapt. The use of molecular probes in the future to look for specific genes involved in resistance and catabolism in the bacterioneuston both in the laboratory and in the field may provide important information on the mechanisms of their survival.

References

Adam N.K. (1937) A rapid method for determining the lowering of tension of exposed water surfaces, with some observations on the surface tension of the sea and inland waters. *Proc. R. Soc. Lon. Biol.* **122**, 134–139.

Alexander M. (1976) Natural selection and the ecology of microbial adaptation in a biosphere. In Heinrich M.R. (ed.) *Extreme Environments*, pp. 3–25. University Park Press, Baltimore.

Amies C.R. (1956) Surface films on swimming pools. *Can. J. Publ. Health* **47**, 93–103.

Armstrong D.E. & Elzerman A.W. (1982) Trace metal accumulation in surface microlayers. *J. Great Lakes Res.* **8**, 282–287.

Babenzien H.-D. & Schwartz W. (1970) Studien zur Mikrobiologie des Neustons. *Limnologica* **7**, 247–272.

Baier R.E. (1970) Surface quality assessment of natural bodies of water. *Proc. 13th Conf. Great Lakes Res.*, pp. 114–127.

Baier R.E., Goupil D.W., Perlmutter S. & King R. (1974) Dominant chemical composition of sea surface films, natural slicks, and foams. *J. Rech. Atmos.* **8**, 571–600.

Bailey C.A., Neihof R.A. & Tabor P.S. (1983) Inhibitory effect of solar radiation on amino acid uptake in Chesapeake Bay bacteria. *Appl. Environ. Microbiol.* **46**, 44–49.

Barger W.R., Daniel W.H. & Garrett W.D. (1974) Surface chemical properties of banded sea slicks. *Deep-Sea Res.* **21**, 83–89.

Bartha R. (1986) Biotechnology of petroleum pollutant biodegradation. *Microb. Ecol.* **12**, 155–172.

Baylor E.R., Peters V. & Baylor M.B. (1977) Water-to-air transfer of virus. *Science* **197**, 763–764.

Bell C.R. & Albright L.J. (1982) Bacteriological investigation of the neuston and plankton in the Fraser River estuary, British Columbia. *Estuarine, Costal and Shelf Science* **15**, 385–394.

Bergh Ø., Børsheim K.Y., Bratbak G. & Heldal M. (1989) High abundance of viruses found in aquatic environments. *Nature* **340**, 467–468.

Bezdek H.F. & Carlucci A.F. (1972) Surface concentration of marine bacteria. *Limnol. Oceanogr.* **17**, 566–569.

Bidleman T.F. & Olney C.E. (1974) Chlorinated hydrocarbons in the Sargasso Sea atmosphere and surface water. *Science* **183**, 516–518.

Bitton G. (1980) Adsorption of viruses to surfaces: technological and ecological implications. In Bitton G. & Marshall K.C. (eds) *Adsorption of Microorganisms to Surfaces*, pp. 331–374. John Wiley & Sons, New York.

Blanchard D.C. (1978) Jet drop enrichment of bacteria, virus, and dissolved organic material. *Pageoph* **116**, 302–308.

Blanchard D.C. & Syzdek L.D. (1978) Seven problems in bubble and jet drop researches. *Limnol. Oceanogr.* **23**, 389–400.

Boehm P.D. (1980) Evidence for the decoupling of dissolved, particulate and surface microlayer hydrocarbons in northwestern Atlantic continental shelf waters. *Mar. Chem.* **9**, 255–281.

Brock T.D. (1969) Microbial growth under extreme conditions. *Symp. Soc. Gen. Microbiol.* **19**, 15–41.

Brockmann U.H., Kattner G., Hentzschel G., Wandschneider K., Junge H.D. & Huhnerfuss H. (1976) Naturliche Oberflächenfilme im Seegebiet vor Sylt. *Mar. Biol.* **36**, 135–146.

Bungay H.R. & Masak R.D. (1981) Estimation of thickness of bacterial films at an air–water interface. *Biotech. Bioengin.* **23**, 1155–1157.

Carlson D.J. (1982a) A field evaluation of plate and screen microlayer sampling techniques. *Mar. Chem.* **11**, 189–208.

Carlson D.J. (1982b) Surface microlayer phenolic enrichments indicate sea surface slicks. *Nature* **296**, 426–429.

Carlson D.J. (1983) Dissolved organic materials in surface microlayers: temporal and spatial variability and relation to sea state. *Limnol. Oceanogr.* **28**, 415–431.

Carlson D.J. (1987) Viscosity of sea-surface slicks. *Nature* **329**, 823–825.

Carlson D.J. & Mayer L.M. (1980) Enrichment of dissolved phenolic material in the surface microlayer of coastal waters. *Nature* **286**, 482–483.

Carlson D.J., Cantey J.L. & Cullen J.J. (1988) Description of and results from a new surface microlayer sampling device. *Deep-Sea Res.* **35**, 1205–1213.

Carlucci A.F. & Williams A.F. (1965) Concentration of bacteria from sea water by bubble scavenging. *J. Cons. Perm. Int. Explor. Mer.* **30**, 28–33.

Carlucci A.F., Craven D.B. & Henrichs S.M. (1985) Surface-film microheterotrophs: amino acid metabolism and solar radiation effects on their activities. *Mar. Biol.* **85**, 13–22.

Carlucci A.F., Craven D.B., Robertson K.J. & Williams P.M. (1986) Surface-film microbial populations: diel amino acid metabolism, carbon utilization, and growth rates. *Mar. Biol.* **92**, 289–297.

Carlucci A.F., Craven D.B. & Wolgast D.M. (1991) Microbial populations in surface films and subsurface waters: amino acid metabolism and growth. *Mar. Biol.* **108**, 329–339.

Carty C. & Colwell R.R. (1975) A microbiological study of air and surface water microlayers in the open ocean. *J. Wash. Acad. Sci.* **65**, 148–153.

Chaudhry G.R. & Chapalamadugu S. (1991) Biodegradation of halogenated organic compounds. *Microbiol. Rev.* **55**, 59–79.

Chet I. & Mitchell R. (1976) Ecological aspects of microbial chemotactic behavior. *Ann. Rev. Microbiol.* **30**, 221–239.

Cleary J.J. & Stebbing A.R.D. (1987) Organotin in the surface microlayer and subsurface waters of southwest England. *Mar. Pollut. Bull.* **18**, 238–246.

Collins Y.E. & Stotzky G. (1989) Factors affecting the toxicity of heavy metals to microbes. In Beveridge T.J. & Doyle R.J. (eds) *Metal Ions and Bacteria*, pp. 31–90. John Wiley & Sons, New York.

Cook W.L., Michaels G.B., Crow S.A. & Rector D.D. (1978) Indicator bacteria in surface microlayers of Lake Sidney Lanier. *Dev. Ind. Microbiol.* **19**, 473–476.

Crawford R.L., Johnson L. & Martinson M. (1982) Bacterial enrichments in surface films of freshwater lakes. *J. Great Lakes Res.* **8**, 323–325.

Cross J.N., Hardy J.T., Hose J.E., Hershelman G.P., Antrim L.D., Gossett R.W. & Crecelius E.A. (1987) Contaminant concentrations and toxicity of sea-surface microlayer near Los Angeles, California. *Mar. Environ. Res.* **23**, 307–323.

Crow S.A., Ahearn D.G., Cook W.L. & Bourquin A.W. (1975) Densities of bacteria and fungi in coastal surface films as determined by a membrane-adsorption procedure. *Limnol. Oceanogr.* **20**, 644–646.

Crow S.A., Cook W.L., Ahearn D.G. & Bourquin A.W. (1976) Microbial populations in coastal surface slicks. In Sharpley J.M. & Kaplan A.M. (eds) *Proc. Third Internat. Biodegrad. Symp.* pp. 93–98. Applied Sciences, London.

Dahlbäck B. (1983) *Marine Microorganisms and Interfaces.* PhD dissertation, University of Göteborg.

Dahlbäck B., Hermansson M., Kjelleberg S. & Norkrans B. (1981) The hydrophobicity of bacteria – an important factor in their initial adhesion at the air–water interface. *Arch. Microbiol.* **128**, 267–270.

Dahlbäck B., Gunnarsson L.A.H., Hermansson M. & Kjelleberg S. (1982) Microbial investigations of surface microlayers, water column, ice and sediment in the Arctic Ocean. *Mar. Ecol. Prog. Ser.* 9, 101–109.

Danos S.C., Maki J.S. & Remsen C.C. (1983) Stratification of microorganisms and nutrients in the surface microlayer of small freshwater ponds. *Hydrobiologia* **98**, 193–202.

Daumas R.A., Laborde P.L., Marty J.-C. & Saliot A. (1976) Influence of sampling method on the chemical composition of water surface film. *Limnol. Oceanogr.* **21**, 319–326.

Davis P.G., Caron D.A. & Sieburth J.McN. (1978) Oceanic amoebae from the North Atlantic: culture, distribution, and taxonomy. *Trans. Amer. Micros. Soc.* **97**, 73–88.

DeFlaun M.F. & Levy S.B. (1989) Genes and their varied hosts. In Levy S.B. & Miller R.V. (eds) *Gene Transfer in the Environment*, pp. 1–32. McGraw-Hill, New York.

De Souza Lima Y. (1985) Accumulation de sels nutritifs dans la microcouche de surface: influence possible des facteurs abiotiques et biotiques du milieu. *Oceanol. Acta* **8**, 47–58.

De Souza Lima Y. & Romano J.-C. (1983) Ecological aspects of the surface microlayer. I. ATP, ADP, AMP contents, and energy charge ratios of microplanktonic communities. *J. Exp. Mar. Biol. Ecol.* **70**, 107–122.

De Souza Lima Y. & Chretiennot-Dinet M.J. (1984) Measurements of biomass and activity of neustonic microorganisms. *Estuarine, Coastal and Shelf Science* **19**, 167–180.

Dietz A.S., Albright L.J. & Tuominen T. (1976) Heterotrophic activities of bacterioneuston and bacterioplankton. *Can. J. Microbiol.* **22**, 1699–1709.

Dietz R.S. & LaFond E.C. (1950) Natural slicks on the ocean. *J. Mar. Res.* **9**, 69–76.

DiSalvo L.H. (1973) Contamination of surfaces by bacterial neuston. *Limnol. Oceanogr.* **18**, 165–168.

Dragcevic Dj. & Pravdic V. (1981) Properties of the seawater–air interface. 2. Rates of surface film formation under steady state conditions. *Limnol. Oceanogr.* **26**, 492–499.

Dratchev S.M., Korsh L.E. & Mityagina O.V. (1957) On the microflora of the water surfaces. *J. Hygiene Epidemiol. Microbiol. Immunol.* **1**, 431–440.

Duce R.A., Quinn J.G., Olney C.E., Piotrowicz S.R., Ray B.J. & Wade T.L. (1972) Enrichment of heavy metals and organic compounds in the surface microlayer of Narragansett Bay, Rhode Island. *Science* **176**, 161–163.

Duxbury T. (1985) Ecological aspects of heavy metal responses in microorganisms. *Adv. Microb. Ecol.* **8**, 185–235.

Edwards C. (1990) Introduction. In Edwards C. (ed.) *Microbiology of Extreme Environments*, pp. ix–xi. McGraw-Hill, New York.

Eisenreich S.J. (1982) Atmospheric role in trace metal exchange at the air–water interface. *J. Great Lakes Res.* **8**, 243–256.

Eisenreich S.J., Elzerman A.W. & Armstrong D.E. (1978) Enrichment of micronutrients, heavy metals, and chlorinated hydrocarbons in wind-generated lake foam. *Environ. Sci. Tech.* **12**, 413–417.

Elzerman A.W. & Armstrong D.E. (1979) Enrichment of Zn, Cd, Pb, and Cu in the surface microlayer of Lakes Michigan, Ontario, and Mendota. *Limnol. Oceanogr.* **24**, 133–144.

Estep K.W. & Remsen C.C. (1985) Influence of the surface microlayer on nutrients, chlorophyll and algal diversity of a small eutrophic bog pond. *Hydrobiologia* **121**, 203–213.

Estep K.W., Maki J.S., Danos S.C. & Remsen C.C. (1985) The retrieval of material from the surface microlayer with screen and plate samplers and its implications for partitioning of material within the microlayer. *Freshwat. Biol.* **15**, 15–19.

Ewing G. (1950) Slicks, surface films and internal waves. *J. Mar. Res.* **9**, 161–187.

Fasching J.L., Courant R.A., Duce R.A. & Piotrowicz S.R. (1974) A new surface microlayer sampler utilizing the bubble microtome. *J. Rech. Atmos.* 8, 649–652.

Fehon W.C. & Oliver J.D. (1977) Degradation of crude oil by mixed populations of bacteria from the surface microlayer in an estuarine system. *J. Elisha Mitchell Soc.* **93**, 72–73.

Fehon W.C. & Oliver J.D. (1979) Taxonomy and distribution of surface microlayer bacteria from two estuarine sites. *Estuaries* **2**, 194–197.

Foissner W. (1979) Ökologische und systematische Studien über das Neuston alpiner Kleingewässer, mit besonderer Berücksichtigung der Ciliaten. *Int. Revue Ges. Hydrobiol.* **64**, 99–140.

Frølund A. (1977) The seasonal variation of the neuston of a small pond. *Bot. Tidsskrift* 72, 45–56.

Fuerst J.A., McGregor A. & Dickson M.R. (1987) Negative staining of freshwater bacterioneuston sampled directly with electron microscope specimen support grids. *Microb. Ecol.* **13**, 219–228.

Garabetian F. (1991) ^{14}C-glucose uptake and ^{14}C-CO_2 production in surface microlayer and surface-water samples: influence of UV and visible radiation. *Mar. Ecol. Prog. Ser.* 77, 21–26.

Garrett W.D. (1965) Collection of slick-forming materials from the sea surface. *Limnol. Oceanogr.* **10**, 602–605.

Garrett W.D. (1967) The organic chemical composition of the ocean surface. *Deep-Sea Res.* **14**, 221–227.

Garrett W.D. (1974) The surface activity of petroleum and its influence on the spreading and weathering of oil films at sea. *J. Rech. Atmosph.* **8**, 555–562.

Garrett W.D. & Duce R.A. (1980) Surface microlayer samplers. In Dobson F., Hasse L. & Davis R. (eds) *Air–Sea Interaction*, pp. 471–490. Plenum, New York.

Gearing P.J. & Gearing J.N. (1982) Transport of no. 2 fuel oil between water column, surface microlayer and atmosphere in controlled ecosystems. *Mar. Environ. Res.*, 6, 133–143.

Geesey G.G. & Jang L. (1989) Interactions between metal ions and capsular polymers. In Beveridge T.J. & Doyle R.J. (eds) *Metal Ions and Bacteria*, pp. 325–357. John Wiley & Sons, New York.

Gladyshev M.I. (1986) Neuston of inland waters (a review). *Gidrobiol. Zh.* **24**, 12–19.

Goering J.J. & Menzel D.W. (1965) The nutrient chemistry of the sea surface. *Deep-Sea Res.* **12**, 839–843.

Goldacre R.J. (1949) Surface films on natural bodies of water. *J. Anim. Ecol.* **18**, 36–39.

Goldacre R.J. (1958) Surface films, their collapse on compression, the shapes and sizes of cells and the origin of life. In Danielli J.F., Pankhurst K.G.A. & Riddiford A.C. (eds) *Surface Phenomena in Chemistry and Biology*, pp. 278–298. Pergamon, New York.

Goutx M., Mutaftshiev S. & Bertrand J.-C. (1987) Lipid and exopolysaccharide production during hydrocarbon growth of a marine bacterium from the sea surface. *Mar. Ecol. Prog. Ser.* **40**, 259–265.

Green T. & Houk D.F. (1979) The removal of organic surface films by rain. *Limnol. Oceanogr.* **24**, 966–970.

Guerin W.F. (1989) Phenanthrene degradation by estuarine surface microlayer and bulk water microbial populations. *Microb. Ecol.* 17, 89–104.

Gunkel W. (1973) Distribution and abundance of oil-oxidizing bacteria in the North Sea. In Ahearn D.G. & Meyers S.P. (eds) *The Microbial Degradation of Oil Pollutants*, pp. 127–139. Louisiana State University, Baton Rouge.

Hamilton E.I. & Clifton R.J. (1979) Techniques for sampling the air–sea interface for estuarine and coastal waters. *Limnol. Oceanogr.* **24**, 188–193.

Hansen H.P. (1977) Photodegradation of hydrocarbon surface films. *Rapp. P.-v. Reun. Cons. Int. Explor. Mer* **171**, 101–106.

Hardman Y. (1941) The surface tension of Wisconsin lake waters. *Trans. Wis. Acad. Sci. Arts Lett.* **33**, 395–404.

Hardy J.T. (1982) The sea surface microlayer: biology, chemistry and anthropogenic enrichment. *Prog. Oceanogr.* **11**, 307–328.

Hardy J.T., Apts C.W., Crecelius E.A. & Bloom N.S. (1985a) Sea-surface microlayer metals enrichments in an urban and rural bay. *Estuarine, Coastal and Shelf Science* **20**, 299–312.

Hardy J.T., Apts C.W., Crecelius E.A. & Fellingham G.W. (1985b) The sea-surface microlayer: fate and residence times of atmospheric metals. *Limnol. Oceanogr.* **30**, 93–101.

Hardy J.T., Crecelius E.A., Antrim L.D., Broadhurst V.L., Apts C.W., Gurtisen J.M. & Fortman T.J. (1987) The sea-surface microlayer of Puget Sound: Part II. Concentrations of contaminants and relation to toxicity. *Mar. Environ. Res.* **23**, 251–271.

Hardy J.T., Coley J.A., Antrim L.D. & Kiesser S.L. (1988) A hydrophobic large-volume sampler for collecting aquatic surface microlayers: characterization and comparison with the glass plate method. *Can. J. Fish. Aquat. Sci.* **45**, 822–826.

Hardy J.T., Crecelius E.A., Antrim L.D., Kiesser S.L., Broadhurst V.L., Boehm P.D., Steinhauer W.G. & Coogan T.H. (1990) Aquatic surface microlayer contamination in Chesapeake Bay. *Mar. Chem.* **28**, 333–351.

Hardy R., Mackie P.R., Whittle K.J., McIntyre A.D. & Blackman R.A.A. (1977) Occurrence of hydrocarbons in the surface film, sub-surface water and sediment in the waters around the United Kingdom. *Rapp. P.-v. Reun. Cons. Int. Explor. Mer* **171**, 61–65.

Harvey G.W. (1966) Microlayer collection from the sea surface: a new method and initial results. *Limnol. Oceanogr.* **11**, 608–613.

Harvey G.W. & Burzell L.A. (1972) A simple microlayer method for small samples. *Limnol. Oceanogr.* **17**, 156–157.

Harvey R.W. & Young L.Y. (1980a) Enrichment and association of bacteria and particulates in salt marsh surface water. *Appl. Environ. Microbiol.* **39**, 894–899.

Harvey R.W. & Young L.Y. (1980b) Enumeration of particle-bound and unattached respiring bacteria in the salt marsh environment. *Appl. Environ. Microbiol.* **40**, 156–160.

Harvey R.W., Lion L.W., Young L.Y. & Leckie J.O. (1982) Enrichment and association of lead and bacteria at particulate surfaces in a salt-marsh surface layer. *J. Mar. Res.* **40**, 1201–1211.

Hatcher R.F. & Parker B.C. (1974a) Laboratory comparisons of four surface microlayer samplers. *Limnol. Oceanogr.* **19**, 162–165.

Hatcher R.F. & Parker B.C. (1974b) Microbiological and chemical enrichment of freshwater-surface microlayers relative to the bulk-subsurface water. *Can. J. Microbiol.* **20**, 1051–1057.

Henrichs S.M. & Williams P.M. (1985) Dissolved and particulate amino acids and carbohydrates in the sea surface microlayer. *Mar. Chem.* **17**, 141–163.

Hermansson M. (1990) The dynamics of dissolved and particulate organic material in surface microlayers. In Wotton R.S. (ed.) *The Biology of Particles in Aquatic Systems*, pp. 145–159. CRC Press, Boca Raton.

Hermansson M. & Dahlbäck B. (1983) Bacterial activity at the air/water interface. *Microb. Ecol.* **9**, 317–328.

Hermansson M. & Marshall K.C. (1985) Utilization of surface localized substrate by non-adhesive bacteria. *Microb. Ecol.* **11**, 91–105.

Hermansson M., Kjelleberg S. & Norkrans B. (1979) Interactions of pigmented wildtype and pigmentless mutant of *Serratia marcescens* with lipid surface film. *FEMS Microbiol. Lett.* **6**, 129–132.

Hermansson M., Kjelleberg S., Korhonen T.K. & Stenström T.-A. (1982) Hydrophobic and electrostatic characterization of surface structures of bacteria and its relationship to adhesion to an air–water interface. *Arch. Microbiol.* **131**, 308–312.

Hermansson M., Jones G.W. & Kjelleberg S. (1987) Frequency of antibiotic and heavy metal resistance, pigmentation, and plasmids in bacteria of the marine air–water interface. *Appl. Environ. Microbiol.* 53, 2338–2342.

Horrigan S.G., Carlucci A.F. & Williams P.M. (1981) Light inhibition of nitrification in sea-surface films. *J. Mar. Res.* **39**, 557–565.

Hunter K.A. (1980) Processes affecting particulate trace metals in the sea surface microlayer. *Mar. Chem.* **9**, 49–70.

Hunter K.A. & Liss P.S. (1981) Organic sea surface films. In Duursma E. & Dawson R. (eds) *Marine Organic Chemistry*, pp. 259–298. Elsevier, Amsterdam.

Jagger J. (1983) Physiological effects of near-ultraviolet radiation on bacteria. *Photochem. Photobiol. Rev.* 7, 1–75.

Jarvis N.L. (1967) Adsorption of surface-active material at the sea–air interface. *Limnol. Oceanogr.* **12**, 213–221.

Jarvis N.L., Garrett W.D., Scheiman M.A. & Timmons C.O. (1967) Surface chemical characterization of surface-active material in seawater. *Limnol. Oceanogr.* **12**, 88–96.

Jones G.W., Baines L. & Genthner F.J. (1991) Heterotrophic bacteria of the freshwater neuston and their ability to act as plasmid recipients under nutrient deprived conditions. *Microb. Ecol.* **22**, 15–25.

Kattner G.G. & Brockmann U.H. (1978) Fatty-acid composition of dissolved and particulate matter in surface films. *Mar. Chem.* **6**, 233–241.

Kattner G., Nagel K., Eberlein K. & Hammer K.-D. (1985) Components of natural surface microlayers and subsurface water. *Oceanol. Acta* **8**, 175–183.

Kim S.-J. (1985) Effect of heavy metals on natural populations of bacteria from surface microlayers and subsurface water. *Mar. Ecol. Prog. Ser.* **26**, 203–206.

Kjelleberg S. (1985) Mechanisms of bacterial adhesion at gas–liquid interfaces. In Savage D.C. & Fletcher M. (eds) *Bacterial Adhesion*, pp. 163–194. Plenum, New York.

Kjelleberg S. & Håkansson N. (1977) Distribution of lipolytic, proteolytic, and amylolytic marine bacteria between the lipid film and the subsurface. *Mar. Biol.* **39**, 103–109.

Kjelleberg S. & Stenström T.-A. (1980) Lipid surface films: interaction of bacteria with free fatty acids and phospholipids at the air/water interface. *J. Gen. Microbiol.* **116**, 417–423.

Kjelleberg S., Norkrans B., Löfgren H. & Larsson K. (1976) Surface balance study of the interaction between microorganisms and lipid monolayer at the air/water interface. *Appl. Environ. Microbiol.* **31**, 609–611.

Kjelleberg S., Stenström T.A. & Odham G. (1979) Comparative study of different hydrophobic devices for sampling lipid surface films and adherent microorganisms. *Mar. Biol.* **53**, 21–25.

Kjelleberg S., Humphrey B.A. & Marshall K.C. (1982) Effect of interfaces on small, starved marine bacteria. *Appl. Environ. Microbiol.* **43**, 1166–1172.

Krasilnikov N.A., Tsyban A.V. & Koronelli T.V. (1973) Assimilation of normal alkanes and crude oil by marine bacteria. *Okeanologiya* **13**, 877–882.

Krieg N.R. (ed.) (1984) *Bergey's Manual of Systematic Bacteriology*, vol. 1. Williams & Wilkins, Baltimore.

Krinsky N.I. (1979) Carotenoid pigments: multiple mechanisms for coping with the stress of photosensitized oxidations. In Shilo M. (ed.) *Strategies of Microbial Life in Extreme Environments*, pp. 163–177. Verlag Chemie, Berlin.

Kushner D.J. (1978) Introduction: a brief overview. In Kushner D.J. (ed.) *Microbial Life in Extreme Environments*, pp. 1–7. Academic Press, London.

Lal R. & Saxena D.M. (1982) Accumulation, metabolism, and effects of organochlorine insecticides on microorganisms. *Microbiol. Rev.* **46**, 95–127.

Larsson K., Odham G. & Södergren A. (1974) On lipid surface films on the sea. I. A simple method for sampling and studies of composition. *Mar. Chem.* **2**, 49–57.

Leahy J.G. & Colwell R.R. (1990) Microbial degradation of hydrocarbons in the environment. *Microbiol. Rev.* **54**, 305–315.

Lindow S.E. (1985) Ecology of *Pseudomonas syringae* relevant to the field use of ice deletion mutants constructed *in vitro* for plant frost control. In Halvorson H.O., Pramer D. & Rogul M. (eds) *Engineered Organisms in the Environment: Scientific Issues*, pp. 23–35. American Society for Microbiology, Washington DC.

Lion L.W. & Leckie J.O. (1981) The biogeochemistry of the air–sea interface. *Ann. Rev. Earth Planet. Sci.* **9**, 449–486.

Lion L.W. & Leckie J.O. (1982) Accumulation and transport of Cd, Cu, and Pb in an estuarine salt marsh surface microlayer. *Limnol. Oceanogr.* **27**, 111–125.

Lion L.W., Harvey R.W., Young L.Y. & Leckie J.O. (1979) Particulate matter. Its association with microorganisms and trace metals in an estuarine salt marsh microlayer. *Environ. Sci. Tech.* **13**, 1522–1525.

Liss P.S. (1975) Chemistry of the sea surface microlayer. In Riley J.P. & Skirrow G. (eds) *Chemical Oceanography*, vol. 2, pp. 193–243. Academic Press, New York.

MacIntyre F. (1968) Bubbles: a boundary-layer "microtome" for micro-thick samples of a liquid surface. *J. Phys. Chem.* **72**, 589–592.

Maki J.S. & Remsen C.C. (1983) A membrane adsorption-SEM technique for observing neuston organisms. *Microb. Ecol.* **9**, 177–183.

Maki J.S. & Remsen C.C. (1989) Examination of a freshwater surface microlayer for diel changes in the bacterioneuston. *Hydrobiologia* **182**, 25–34.

Maki J.S. & Herwig R.P. (1991) A diel study of the neuston and plankton bacteria in an Antarctic pond. *Antarctic Sci.* **3**, 47–51.

Maki J.S., Danos S.C. & Remsen C.C. (1984) Quantitative changes in fungal colony-forming units in the surface microlayers of two freshwater ponds. *Can. J. Microbiol.* **30**, 578–586.

Maki J.S., Little B.J., Wagner P. & Mitchell R. (1990) Biofilm formation on metal surfaces in Antarctic waters. *Biofouling* **2**, 27–38.

Marshall K.C. (1985) Mechanisms of bacterial adhesion at solid-water interfaces. In Savage D.C. & Fletcher M. (eds) *Bacterial Adhesion*, pp. 133–161. Plenum, New York.

Marty J.C. & Saliot A. (1976) Hydrocarbons (normal alkanes) in the surface microlayer of seawater. *Deep-Sea Res.* **23**, 863–873.

Marty J.C., Zutic V., Precali R., Saliot A., Cosovic B., Smodlaka N. & Cauwet G. (1988) Organic matter characterization in the northern Adriatic Sea with special reference to the sea surface microlayer. *Mar. Chem.* **25**, 243–263.

Marumo R., Taga N. & Nakai T. (1971) Neustonic bacteria and phytoplankton in surface microlayers of the equatorial waters. *Bull. Plankt. Soc. Japan* **18**, 36–41.

Meyers P.A. & Kawka O.E. (1982) Fractionation of hydrophobic organic materials in surface microlayers. *J. Great Lakes Res.* **8**, 288–298.

Miget R., Kator H., Oppenheimer C., Laseter J.L. & Ledet E.J. (1974) New sampling device for the recovery of petroleum hydrocarbons and fatty acids from aqueous surface films. *Analyt. Chem.* **46**, 1154–1157.

Mimura T., Romano J.-C. & De Souza Lima Y. (1988) Microbiomass structure and respiratory activity of microneuston and microplankton in the northwestern Mediterranean Sea influenced by Rhone river water. *Mar. Ecol. Prog. Ser.* **49**, 151–162.

Mitamura O. & Matsumoto K. (1981) Uptake rate of urea nitrogen and decomposition rate of urea carbon at the surface microlayer in Lake Biwa. *Verh. Int. Verein. Theor. Angew. Limnol.* **21**, 556–564.

Nasim A. & James A.P. (1978) Life under conditions of high irradiation. In Kushner D.J. (ed.) *Microbial Life in Extreme Environments*, pp. 409–439. Academic Press, London.

Naumann E. (1917) Beiträge zur Kenntnis des Teichnannoplanktons. II. Über das Neustons des Susswasser. *Biol. Zentrabl.* **37**, 98–106.

Norkrans B. (1980) Surface microlayers in aquatic environments. *Adv. Microb. Ecol.* **4**, 51–85.
Norkrans B. & Sörensson F. (1977) On the marine lipid surface microlayer – bacterial accumulation in model systems. *Botanica Marina* **20**, 473–478.
Ofstad E.B., Lunde G. & Drangsholt H. (1979) Chlorinated organic compounds in the fatty surface film on water. *Int. J. Environ. Anal. Chem.* **6**, 119–131.
Ossanna N., Peterson K.R. & Mount D.W. (1987) UV-inducible SOS response in *Escherichia coli*. *Photochem. Photobiol.* **45**, 905–908.
Passman F.J., Novitsky T.J. & Watson S.W. (1979) Surface microlayers of the North Atlantic: microbial populations, heterotrophic and hydrocarbonoclastic activities. In Bourquin A.W. & Pritchard P.H. (eds) *Proc. Workshop: Microbial Degradation of Pollutants in Marine Environments*, pp. 214–226. US EPA, Gulf Breeze, Florida.
Pellenbarg R. (1981) Trace metal partitioning in the aqueous surface microlayer of a salt marsh. *Estuarine, Coastal and Shelf Science* **13**, 113–117.
Pellenbarg R.E. & Church T.M. (1979) The estuarine surface microlayer and trace metal cycling in a salt marsh. *Science* **203**, 1010–1012.
Piotrowicz S.R., Ray B.J., Hoffman G.L. & Duce R.A. (1972) Trace metal enrichment in the sea-surface microlayer. *J. Geophys. Res.* **77**, 5243–5254.
Platford R.E. (1982) Pesticide partitioning in artificial surface films. *J. Great Lakes Res.* **8**, 307–309.
Pojasek R.B. & Zajicek O.T. (1978) Surface microlayers and foams – source and metal transport in aquatic systems. *Wat. Res.* **12**, 7–10.
Power K. & Marshall K.C. (1988) Cellular growth and reproduction of marine bacteria on surface-bound substrate. *Biofouling* **1**, 163–174.
Rambeloarisoa E., Rontani J.F., Giusti G., Duvnjak Z. & Bertrand J.-C. (1984) Degradation of crude oil by a mixed population of bacteria isolated from sea-surface foams. *Mar. Biol.* **83**, 69–81.
Rao S.S., Paolini D. & Leppard G.G. (1984) Effects of low-pH stress on the morphology and activity of bacteria from lakes receiving acid precipitation. *Hydrobiologia* **114**, 115–121.
Ricci N., Erra F., Russo A. & Banchetti R. (1991) The air–water interface: a microhabitat for hypotrichous settlers (Protista, Ciliata). *Limnol. Oceanogr.* **36**, 1178–1188.
Rice C.P., Eadie B.J. & Erstfeld K.M. (1982) Enrichment of PCBs in Lake Michigan surface films. *J. Great Lakes Res.* **8**, 265–270.
Romano J.-C. (1989) Les populations bacteriennes a l'interface air–mer. In Bianchi M., Marty D., Bertrand J.-C., Caumette P. & Gauthier M. (eds) *Micro-Organismes dans les Ecosystèmes Océaniques*, pp. 153–184. Masson, Paris.
Romano J.-C. & Laborde P. (1987) Hétérogénéité spatiale a petite echelle des structures d'accumulation du matériel particulaire a l'interface air–mer. *Oceanol. Acta* **10**, 63–71.
Saijo Y., Mitamura O. & Ogiyama K. (1974) Chemical composition of surface film in the coastal sea area and in freshwater lakes. *Japan. J. Limnol.* **35**, 110–116.
Schindler D.W. (1988) Effects of acid rain on freshwater ecosystems. *Science* **239**, 149–157.
Schmieger H. (1990) Phage genetics and ecology. In Fry J.C. & Day M.J. (eds) *Bacterial Genetics in Natural Environments*, pp. 41–51. Chapman and Hall, London.
Seba D.B. & Corcoran E.F. (1969) Surface slicks as concentrators of pesticides in the marine environment. *Pest. Monit. J.* **3**, 190–193.
Sewell L.M., Bitton G. & Bays J.S. (1981) Evaluation of membrane adsorption-epifluorescence microscopy for the enumeration of bacteria in coastal surface films. *Microb. Ecol.* **7**, 365–369.
Shilo M. (1979) Introduction. In Shilo M. (ed.) *Strategies of Microbial Life in Extreme Environments*, pp. 11–13. Verlag Chemie, Berlin.
Sieburth J.McN. (1965) Bacteriological samplers for air–water and water–sediment

interfaces. In *Trans. Joint Conf. Ocean Sci. Ocean Engin*, pp. 1064–1068. American Soc. Limnol. Oceanogr., Washington DC.

Sieburth J.McN. (1971a) Distribution and activity of oceanic bacteria. *Deep-Sea Res.* **18**, 1111–1121.

Sieburth J.McN. (1971b) An instance of bacterial inhibition in oceanic surface water. *Mar. Biol.* **11**, 98–100.

Sieburth J.McN. (1983) Microbiological and organic-chemical processes in surface and mixed layers. In Liss P.S. & Slinn W.G.N. (eds) *Air–Sea Exchange of Gases and Particles*, pp. 121–172. D. Reidel, Hingham.

Sieburth J. McN., Willis P.-J., Johnson K.M., Burney C.M., Lavoie D.M., Hinga K.R., Caron D.A., French III F.W., Johnson P.W. & Davis P.G. (1976) Dissolved organic matter and heterotrophic microneuston in the surface microlayers of the North Atlantic. *Science* **194**, 1415–1418.

Silver S. & Misra T.K. (1988) Plasmid-mediated heavy metal resistances. *Ann. Rev. Microbiol.* **42**, 717–743.

Simon R.D. & Makarewicz J.C. (1990) Microbial ecology of the air/water interface in Lake Ontario. *Abst. 90th Ann. Meet. Amer. Soc. Microbiol., Anaheim, Calif.*, N65. American Society for Microbiology, Washington DC.

Singer C.E. & Ames B.N. (1970) Sunlight ultraviolet and bacterial DNA base ratios. *Science* **170**, 822–826.

Smith D.W. (1982) Extreme natural environments. In Burns R.G. & Slater J.H. (eds) *Experimental Microbial Ecology*, pp. 555–574. Blackwell Scientific Publications, Oxford.

Smith K.C. & Sargentini N.J. (1985) Metabolically-produced "UV-like" DNA damage and its role in spontaneous mutagenesis. *Photochem. Photobiol.* **42**, 801–803.

Sneath P.H.A., Mair N.S. & Sharpe M.E. (eds) (1986) *Bergey's Manual of Systematic Bacteriology*, vol. 2. Williams & Wilkins, Baltimore.

Södergren A. (1984) Small-scale temporal changes in the biological and chemical composition of surface microlayers in a eutrophic lake. *Verh. Int. Verein. Theor. Angew. Limnol.* **22**, 765–771.

Södergren A. (1987) Origin and composition of surface slicks in lakes of differing trophic status. *Limnol. Oceanogr.* **32**, 1307–1316.

Södergren A., Larsson P., Knulst J. & Bergqvist C. (1990) Transport of incinerated organochlorine compounds to air, water, microlayer, and organisms. *Mar. Pollut. Bull.* **21**, 18–24.

Staley J.T., Bryant M.P. & Pfennig N. (eds) (1989) *Bergey's Manual of Systematic Bacteriology*, vol. 3. Williams & Wilkins, Baltimore.

Strike P. & Lodwick D. (1987) Plasmid genes affecting DNA repair and mutation. *J. Cell Sci. Suppl.* **6**, 303–321.

Summers A.O. & Barkay T. (1989) Metal resistance genes in the environment. In Levy S.B. & Miller R.V. (eds) *Gene Transfer in the Environment*, pp. 287–308. McGraw-Hill, New York.

Terzaghi E. & O'Hara M. (1990) Microbial plasticity. The relevance to microbial ecology. *Adv. Microb. Ecol.* **11**, 431–460.

Thomas G. (1977) Effects of near ultraviolet light on microorganisms. *Photochem. Photobiol.* **26**, 669–673.

Thomson B.E., Van Dyke H. & Worrest R.C. (1980) Impact of UV-B radiation (290–320 nm) upon estuarine bacteria. *Oecologia* **47**, 56–60.

Tripathi A.K. (1987) The cytology and biochemistry of pesticide microbiology. *CRC Crit. Rev. Microbiol.* **15**, 223–246.

Trouwborst T., Kuyper S., de Jong J.C. & Plantinga A.D. (1974) Inactivation of some bacterial and animal viruses by exposure to liquid air interfaces. *J. Gen. Virol.* **24**, 155–165.

Tsyban A.V. (1971) Marine bacterioneuston. *J. Oceanogr. Soc. Japan* **27**, 56–66.

Upton A.C., Nedwell D.B. & Wynn-Williams D.D. (1990) The selection of microbial communities by constant or fluctuating temperatures. *FEMS Microbiol. Ecol.* **74**, 243–252.

Van Vleet E.S. & Williams P.M. (1980) Sampling sea surface films: a laboratory evaluation of techniques and collecting materials. *Limnol. Oceanogr.* **25**, 764–770.

Van Vleet E.S. & Williams P.M. (1983) Surface potential and film pressure measurements in seawater systems. *Limnol. Oceanogr.* **28**, 401–414.

Varon M. & Shilo M. (1981) Inhibition of the predatory activity of *Bdellovibrio* by various environmental pollutants. *Microb. Ecol.* **7**, 107–111.

Vestal J.R. & Hobbie J.E. (1988) Microbial adaptations to extreme environments. In Lynch J.M. & Hobbie J.E. (eds) *Micro-organisms in Action: Concepts and Applications in Microbial Ecology*, pp. 193–206. Blackwell Scientific Publications, Oxford.

Viljanen P. & Boratynski J. (1991) The susceptibility of conjugative resistance transfer in Gram-negative bacteria to physicochemical and biochemical agents. *FEMS Microbiol. Rev.* **88**, 43–54.

Wade T.L. & Quinn J.G. (1975) Hydrocarbons in the Sargasso Sea surface microlayer. *Mar. Pollut. Bull.* **6**, 54–57.

Wangersky P.J. (1976) The surface film as a physical environment. *Ann. Rev. Ecol. Syst.* **7**, 161–176.

Wheeler J.R. (1975) Formation and collapse of surface films. *Limnol. Oceanogr.* **20**, 338–342.

Williams H.N. (1987) The recovery of high numbers of bdellovibrios from the surface water microlayer. *Can. J. Microbiol.* **33**, 572–575.

Williams P.M. (1967) Sea surface chemistry: organic carbon and organic and inorganic nitrogen and phosphorus in surface films and subsurface waters. *Deep-Sea Res.* **14**, 791–800.

Williams P.M., Carlucci A.F., Henrichs S.M., Van Vleet E.S., Horrigan S.G., Reid F.M.H. & Robertson K.J. (1986) Chemical and microbiological studies of sea-surface films in the southern Gulf of California and off the west coast of Baja California. *Mar. Chem.* **19**, 17–98.

Wyndham R.C. & Costerton J.W. (1982) Bacterioneuston involved in the oxidation of hydrocarbons at the air–water interface. *J. Great Lakes Res.* **8**, 316–322.

Young L.Y. (1978) Bacterioneuston examined with critical point drying and transmission electron microscopy. *Microb. Ecol.* **4**, 267–277.

Zaitsev Y.P. (1971) *Marine Neustonology*. Israel Prog. Scientific Transl., Jerusalem.

17

Michael Boyle

Microbial ecology of sewage treatment

17.1 Introduction

The food we eat nourishes not only us, but eventually, microorganisms. One route for the cycling of our wastes is through a sewage treatment facility which accelerates natural microbial processes. The operational parameters for wastewater systems are based on many years of empirical optimization. But before we can enhance the decomposition process, either by designing better treatment plants or "engineering" more efficient microbes, the limits of microbial degradation must be identified. This chapter will briefly review the concepts of microbial incorporation, separation, and decomposition occurring in the largest bioremediation industry – the treatment of sewage waste water and sludge.

17.2 Aerobic treatment

The goal of aerobic treatment is the microbial oxidation of organic matter, often measured in $mg\,l^{-1}$ of O_2 consumed over a 5-day incubation period which is referred to as biological oxygen demand (BOD). Oxygen is consumed as a terminal electron acceptor in the decomposition of this dilute carbon-limited substrate which has $0.7–1.0\,g\,l^{-1}$ of total solids and approximately $300\,mg\,l^{-1}$ BOD (Schroeder 1977). The organic portion, usually 70% of the solids present, consists roughly of carbohydrates (30%), volatile and non-volatile acids (20%), amino acids, peptides and proteins (11%), while the rest is composed of fats, grease, and surfactants (Sterritt & Lester 1988).

Prior to biological treatment, waste water usually passes through a primary treatment settling tank which removes most of the grit (40–60% suspended solids and 25–35% of the BOD) and where floating scum is skimmed from the surface. After biological treatment but before the effluent is discharged, some form of partial disinfection, such as chlorination, is employed to reduce the pathogenic bacteria (e.g., *Shigella*, *Salmonella*, toxigenic *Escherichia coli*, *Vibrio*, *Yersinia*, *Mycobacterium*), protozoans (e.g., *Giardia lamblia*, *Entamoeba histolytica*) and viruses (e.g., hepatitis A, Norwalk, coxsackie, polio) as measured by indicator coliform counts. Coliform counts specifications depend on effluent use (e.g., agricultural or landscape irrigation, stream, lake, or ocean discharge).

Oxygen ponds and lagoons are a treatment option where land availability and climate are favorable. Waste water is retained in a series of ponds for extended periods (weeks to months) where algae are encouraged to grow to facilitate oxygen transfer. If some of the ponds are deep enough, anaerobic conditions can prevail in the lower regions (Atlas & Bartha 1987).

Nearly 4000 facilities use trickle filtration in the USA, a simple form of aerobic treatment in which waste water flows over microorganisms that are attached to gravel of approximately 5–10 cm in size. The attached microbial community is a complex mixture of algae, bacteria, and fungi. Protozoans (e.g., *Opercularia*, *Vorticella*, *Aspidisca*, *Carchesium*, *Chilodonella*, *Epistylis*), nematodes, and insect larvae are also present. These biofilms (optimally 0.1–0.3 mm), composed of living and dead microbial cells and extracellular polymers rich in polysaccharides, can compose up to 50% of the pore volume. After the waste water passes over the attached microbial community it moves into a settling tank to improve effluent clarity (Figure 17.1a). The effluent can be recycled, processed through a series of filters, or used in conjunction with a suspended growth treatment to increase the hydraulic residence time and improve oxidation. Trickle or percolating filters are capable of removing 80–85% of the BOD. Because oxygen is introduced passively, the 1–2 m deep filter bed can become stratified if the pore spaces within the lower sections become saturated. Anaerobic conditions can also occur if ponding develops due to an excess growth of fungi and filamentous bacteria.

In contrast to the percolating filters and oxygenation ponds, the more common activated sludge treatment is an energy-intensive process. Oxygen, substrate, and microorganisms are mixed in aeration basins for 5–10 hr to encourage the incorporation of soluble organic matter into cellular tissue (see Figure 17.1b) (Jewell 1987). Microorganisms grow and aggregate into 0.2–1 mm clusters (flocs) that settle out from the aerated fluid. Soluble and colloidal BOD is also removed by adsorption to the slime matrix of the flocs. These flocs are composed of organic matter, fungi, protozoa, and filamentous and nonfilamentous bacteria. The microbial community found in the flocs ($10^8\,ml^{-1}$ or $10^{10}\,g^{-1}$) has similar Gram-negative bacterial populations (e.g., *Enterobacter*, *Achromobacter*, *Flavobacter*, *Bacillus*, *Pseudomonas*, etc.) but is probably less complex than that found on biological filters. One of the more common heterotrophic species reported is *Zoogloea ramigera*, which produces copious

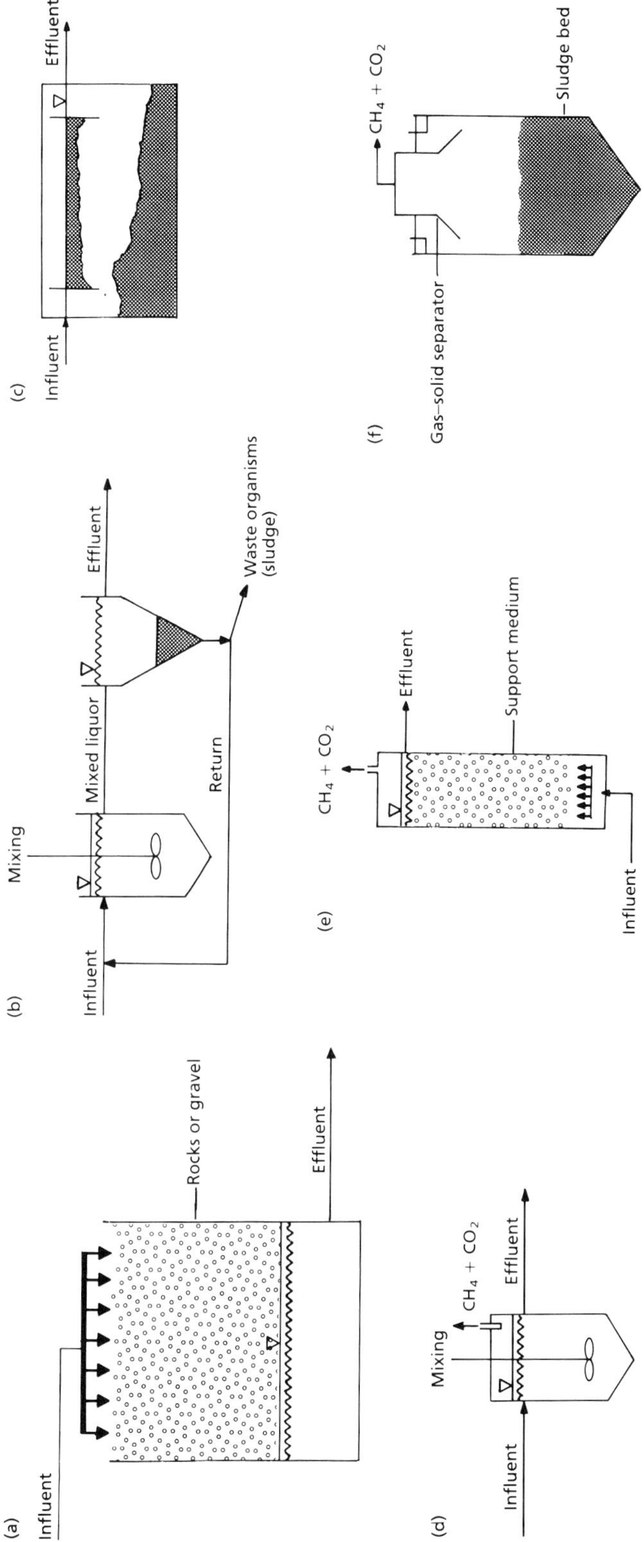

Figure 17.1 Wastewater treatment designs: (a) trickle filtration (aerobic); (b) activated sludge (aerobic); (c) septic tank (anaerobic); (d) completely mixed (anaerobic); (e) fluidized bed (anaerobic); (f) digester (anaerobic).

amounts of extracellular polysaccharide. The grazing protozoa population (10^4 protozoa ml^{-1}) helps reduce pathogenic organisms and nonsettling flocs (Lawler 1986; Rittmann 1987).

The overproduction of nonsettling flocs is the most common source of poor solid–liquid separation (i.e., sludge bulking). It is often caused by certain filamentous bacteria that are distinguished by their growth characteristics:

1 rapid growth rates and a high affinity for low concentration of dissolved oxygen (e.g., *Sphaerotilus natans*);
2 slow growth and slow decay rates;
3 able to oxidize reduced sulfur compounds (e.g., *Thiothrix*).

Poor separation is also caused by recalcitrant "pin point" flocs and by excessive production of gas (from denitrification) which increases the buoyancy of the flocs.

Because most pollutants are removed from aerobic waste water as particles or attached to particles, various compounds (e.g., polymers, Fe^{3+}, Cu^{2+}, powdered activated carbon) are often added to improve the efficiency of floc settling. Alum is also used but mostly to remove soluble inorganic phosphorus. In chemical separation, a form of enhanced primary treatment, these cohesive compounds are added in lieu of biological treatment (Lawler 1986).

For biological treatment to be effective, both soluble carbon and nitrogen need to be removed. Ammonification is the heterotrophic conversion of organic nitrogen into inorganic ammonium. Nitrification proceeds through the oxidation of ammonium to nitrate through the combined activity of the autotrophic bacteria typified by *Nitrosomonas* and *Nitrobacter*. In biological treatment, the operational parameters (e.g., cell retention time, volume of suspended solids) must be controlled to prevent heterotrophs from outcompeting the slower growing nitrifying bacteria. Nitrification utilizes oxygen; about 4 g of oxygen are required for the complete oxidation of 1 g ammonium to nitrate, which can contribute to 40% of the BOD. After nitrate is formed, heterotrophic denitrifying bacteria (*Pseudomonas*) use NO_3^- as their electron acceptor, in the absence of oxygen, converting it to nitrogen gas. If there is no denitrification step, 50% of the original nitrogen content of the sewage (approx. $40\,mg\,N\,l^{-1}$) will remain in the effluent after secondary treatment.

17.3 Anaerobic treatment

Anaerobic treatment of waste water has the potential to greatly reduce the amount of sludge produced from certain waste streams. This sludge reduction results from the much lower production of microbial cells than activated sludge treatment. (The large growing biomass in activated sludge treatment not only causes disposal problems but also contributes to effluent BOD through the production of soluble organic metabolites.) Anaerobic treatment requires the optimization of bacterial growth through proper biomass retention time, adequate mixing, proper pH (6.5–7.6), adequate concentration of micro- and macronutrients, and absence of toxic materials. The effectiveness of the fermentation process is usually determined by daily monitoring of CH_4, CO_2, pH, volatile solids, and total alkalinity. The maintenance of constant and uniform temperature is also critical. Optimal temperature for gas production is either 30–38°C or the thermophilic range of 50–60°C. The higher temperature conditions can double the production rate of biogas – a mixture composed of methane (60–75%) and carbon dioxide. Chemical energy is conserved in the formation of methane, which results in a lower adenosine triphosphate (ATP) yield than from aerobic conversion to carbon dioxide and water (McCarty & Smith 1986).

The production of methane from low molecular weight organics requires the combined activity of at least three different microbial populations (Figure 17.2). The hydrolysis and fermentation step performed by facultative and strict anaerobes converts low molecular

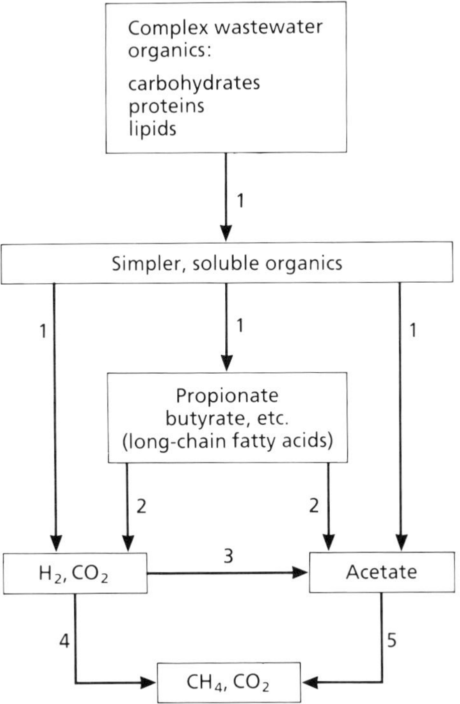

Figure 17.2 The process of methanogenesis. Bacterial groups: 1, fermentative and hydrolytic; 2, hydrogen-producing acetogenic; 3, hydrogen-consuming acetogenic; 4, carbon dioxide-reducing methanogens; 5, aceticlastic methanogens. From Parkin & Owen (1986).

weight hydrocarbons (e.g., ethanol, propionate, and butyrate) to hydrogen and acetate. The second species, a methanogen, combines the hydrogen and carbon dioxide to form methane (Equation 17.1). The third species is also a methanogen which splits acetate into carbon dioxide and methane (Equation 17.2).

$$CO_2 + 4H_2 \rightarrow CH_4 + 2H_2O \quad (17.1)$$

$$\underset{\text{(acetate)}}{CH_3COOH} \rightarrow CH_4 + CO_2 \quad (17.2)$$

About one-third of the methane is generated from hydrogen oxidation, while two-thirds comes from acetate cleavage (Figure 17.2, pathway 4 and 5, respectively). The carbon dioxide produced is either reduced to methane, volatilized, or converted to bicarbonate alkalinity, which is the dominant buffering agent in the system.

To make methanogenesis thermodynamically feasible, the products, acetate and especially hydrogen, must be removed. If hydrogen concentration is higher than 10^{-4} atm., then methane production is inhibited and butyric and propionic acids accumulate. Methanogens provide the sink for hydrogen and acetate while producing the most reduced and oxidized form of carbon. Because diffusion is the primary method of hydrogen transport between species, a close packing of different microorganisms is necessary for high hydrogen transfer and substrate conversion rates (McCarty & Smith 1986).

17.4 Bioreactor design

The simplest anaerobic reactor is a septic tank, which separates domestic waste into a settling solid phase, a liquid effluent and floating scum layer (see Figure 17.1c). In a completely mixed reactor (see Figure 17.1d), the minimum allowable time that solids must remain in a system is determined by the growth rate of methanogens which have a minimum doubling time of 3.3 days at 35°C. This cellular biomass has to be continually regenerated to prevent the flow of water from carrying the microorganisms out of the reactor ("washout").

To maintain a large resident biomass of slow-growing organisms, anaerobic filters, sludge blankets (floating flocs), and fluidized (expanded) beds are often used (see Figure 17.1e) (Jewell 1987). The microbial biomass is attached to granular sludge particles, organic polymers, zeolites, or activated carbon that are fluidized by the upward flow of dilute waste water. This allows the retention time of wastes to be dependent on the quality of wastes, rather than on the generation time of the slow-growing and toxic-sensitive methanogens.

Although mainly associated with anaerobic decomposition, the fluidized bed process can be adapted to aerobic treatment as well (Rittmann 1987). Plastic biological support particles and activated

Table 17.1 The sequence of events in the formation of a biofilm. From Boyle *et al.* 1990

1 Adsorption of macromolecules to surfaces
2 Transport of microorganisms to surfaces
3 Physicochemical and biological attraction of microbes for a surface in close proximity
4 Reversible adhesion of microbes by long range forces
5 Irreversible adhesion by polymeric bridging, and short range forces

carbon provide ideal surfaces for biofilm attachment. Activated carbon can also buffer the system by adsorbing metals and by partitioning hydrophobic organics within its pores (more than 90% of the pores are less than 0.1 μm diameter). The new plastics particles can be designed for specific surface characteristics, internal porosity, size, and density (Bryers & Characklis 1990).

As we have seen, biofilms can be attached to stationary (trickle filters), or mobile particles (fluidized bed). In addition, biofilms can be grown on biological contactors which are partially submerged plastic discs (or lattice) that rotate through a wastewater bath. The initial formation of a biofilm, which for nitrifiers can take weeks, is outlined in Table 17.1. These films promote redox gradients through microbial utilization of oxygen and decreased molecular diffusion. The incorporation of both aerobic and anaerobic degradation in the same bioreactor would require that the thickness of the biofilm be maintained such that the internal surfaces are anaerobic while the outer layer remains aerobic. Biofilms could facilitate nitrogen removal by promoting denitrification within the anaerobic zone and nitrification on the aerobic surface. A microhabitat that possesses a range of redox potentials may also facilitate the breakdown of certain halogenated xenobiotics (e.g., DDT and methoxychlor). The thickness of the biofilm is a difficult parameter to control because not only is it dependent on the nutrient load of the waste water, the adsorptive surface, and the type of microbial community attached but also on fluid temperature and velocity. In steady-state conditions, biofilm growth is equal to loss caused by both detachment from abrasion and shear stress, and by substrate utilization caused by grazing and microbial maintenance requirements (Rittmann 1987).

17.5 Sludge

Sludge is not only composed of living and dead cells, but also serves as microbial habitat and substrate. Sludge composition depends on the type of treatment and the quality of the influent wastewater. Sludge variability within one plant over time can be greater than between different treatment facilities. Generally, the level of organic carbon in sludge tends to be about 20–30%, while nitrogen content composes approximately 3% (Sommers 1977).

The inorganic portion of sludge affects microbial activity by providing sites for physical attachment, which increase the sedimentation rate of cells. Inorganic surfaces also contribute to the pH and nutrient buffering capacity of sludge. The nonorganic fraction of sludge includes sand, silt, clay, precipitates of calcium, phosphate, and iron as well as inorganic nutrients. Municipal sludge can also contain high concentrations of heavy metals (e.g., zinc, nickel, and copper) from industrial discharge and residential pipes (Sommers 1977). The metals in soil–sludge mixtures can take a variety of chemical forms (Lake *et al.* 1984):

1 ions in solution;
2 instantaneously exchangeable ions;
3 bound to internal surfaces of organic colloids;
4 (co-) precipitated with oxides, carbonates, phosphates, or clays;
5 ions in crystal lattices of primary minerals.

Soluble and exchangeable metals generally represent less than 10% of the total metals found in sludge. The precipitates of carbonate, phosphate, hydroxide, and sulfide dominate the soluble concentration of most metals (Gould & Genetelli 1975; Gadd & Griffiths 1978). However, the organometal complex (i.e., chelation) can strongly influence metal migration, bioavailability, and toxicity.

In such a complex mixture as sludge, there will be a range of decay rates associated with the various organic components. The labile fraction will be made up of the easily degradable carbohydrates and proteins with turnover rates of days, while the stable fraction (30–80% of municipal sludge depending on treatment) contains more recalcitrant components of microbial cell walls and humic material. The relative abundance of the labile (fast decay rates) and the stable fraction (slower decay rates) will determine overall sludge decomposition behavior.

The process of sludge stabilization is important because the more sludge resembles stable organic matter, the less likely it will contain pathogenic bacteria, protozoans, viruses, and parasites (e.g., *Taenia saginata*). Sludge stability is typically reported by engineers in terms of volatile solids (VS) concentration or the percentage of VS destroyed. Volatile solids content of less than $500\,\mathrm{mg\,l^{-1}}$ and a VS reduction of more than 40% usually indicates stabilization. But environmental health concerns about pathogens, bioavailable heavy metals, malodors, toxic organics, and methane production demonstrate that activated sludge is not a stable form of organic matter. A more process-oriented definition of sludge stabilization focuses on the degree in which the readily decomposable portion becomes mineralized to carbon dioxide, or converted to humic material (Hartenstein 1981).

Chemical separation has been the classical method to define stable humic matter. Humic acids (HA) are a mixture of compounds that precipitate out from an alkaline extraction when acid is added. The organic fraction that remains in the acidified solution (pH 2) is defined as fulvic acids (FA). Fulvic acids have a higher proportion of carboxyl and hydroxyl functional groups and a higher percentage of oxygen than HA. Humic acids contain a higher percentage of carbon and usually range between 2000 and 100 000 MW (Stevenson 1982).

Ideally, stabilized sludge should mimic the decomposition characteristics of humus, which has a mean residence time of hundreds of years (Waksman 1938; Kononova 1966). The humification process stabilizes organic matter through physical, chemical, and microbial actions. During humification, sludge undergoes extensive decarboxylation, dehydration, and demethylation. There is also a reduction in cellulose, fat, and hemicellulose complexes along with a relative increase in the proportion of aromatic and lignin-like compounds. Beginning with human dietary intake and continuing through the anaerobic digestion process, the carbon content of the organic material decreases (Table 17.2) and the oxygen/carbon ratio increases. With the onset of the humification process, the carbon content of the material increases and both oxygen/carbon and hydrogen/carbon ratios decrease (Hartenstein 1981).

Biochemically stabilized organic matter has been suggested to be formed under the following conditions (Martin & Haider 1986):

1 when *p*-hydroxyphenylpropane units, along with microbially synthesized polyhydroxyphenols, are oxidized to quinones by extracellular phenoloxidases;

2 when quinones react with amino acids, peptides, and alkanes to form a complex three-dimensional organic mesh.

This stable organic matter, with its high cation exchange capacity (up to 500 meq $100\,g^{-1}$) is composed of hydrophobic aliphatic

Table 17.2 Approximate percentage composition of organic materials (dry weight). Adapted from Hartenstein (1981)

Element	Carbohydrates	Proteins	Fats	Primary sludge[1]	Activated sludge	Digested sludge	Fulvic acids	Humic acids
Carbon	44	58	75	60	32	27	47	59
Hydrogen	6	7	12	9	4	4	4	5
Nitrogen		11		7	3[2]	3[2]	2	3
Oxygen	49	23	12	24	37	42	46	34

[1] Parkin & Owen 1986.
[2] Sommers 1977.

($-CH_2-$) chains as well as hydrophilic phenolic and carboxylic functional groups that form polyelectrolytic complexes with peptides, Ca^{2+}, metals, hydrous oxides, and clays. The random nature of the many bond types characteristic of this "polycondensate" suggests a large degree of entropy, and the large free radical concentration implies a high level of reactivity which could denature degrading enzymes.

17.6 Sludge treatment

The main objective of anaerobic digestion of sludge is to further stabilize the organic residue. Wastes that are relatively rich in organic solids (BOD 2–20 $g\,l^{-1}$) such as primary and/or secondary treated sludge are good candidates for anoxic digestion. (Although BOD was developed to gauge aerobic treatment, it is often used to describe anaerobic degradation efficiency.) During the 2- to 5-week anaerobic process, the volatile solids content can be reduced by as much as 75% and the BOD by 90% (Grant & Long 1981). Initially the large solids content is solubilized by hydrolytic, protolytic, and cellulytic bacteria (e.g., *Clostridium*, *Bacteroides*) and fermenters (e.g., *Butyribacterium*, *Propionibacterium*) before methanogenesis can be carried out by archaebacteria (e.g., *Methanobacter*, *Methanococcus*, *Methanothrix*) as outlined earlier (see Figure 17.2). The digestion process uses organic acids and carbon dioxide as terminal electron acceptors in the absence of NO_3^- or SO_4^{2-}. Biogas production can reach 10 m^3 of biogas per m^3 of bioreactor volume per day (see Figure 17.1f). The presence of toxins tend to increase microbial generation time which necessitates longer digestion and lower efficiency, but microbial adaptation to metals and xenobiotic stress is not uncommon within a treatment plant.

The residue organic material consists of a mixture of microbial tissue, cellulose, lipids, organic nitrogen compounds, and humic-like materials. Certain compounds are recalcitrant under anaerobic conditions. Saturated hydrocarbons and condensed aryl structures appear to need molecular oxygen for further enzymatic degradation. In addition, oxygen radicals (e.g., superoxides, peroxides, and hydroxyl radicals) that are toxic to anaerobes are required for the nonspecific oxidation of complex compounds such as lignin (Schink 1988).

The degradation process of many recalcitrant organics including xenobiotics, often involves cometabolism, a process in which the partial degradation of a substance is not accompanied by carbon incorporation or energy utilization by the degrading microorganism (Bull 1980; Alexander 1981). The microorganisms gain their energy for growth from another source (i.e., primary substrate). The detection of anaerobic cometabolism is difficult because anaerobes have such low growth yields which even under optimal conditions incor-

Table 17.3 A summary of the sludge composting process

1 An aerobic process carried out by mesophilic and thermophilic bacteria (e.g., *Bacillus*, *Thermonospora*, *Thermoactinomyces*) and fungi (e.g., *Geotrichum*, *Aspergillus*, *Thermoascus*)
2 Optimal temperature for degradation is 60–70°C
3 Bulking agents (e.g., wood chips, bark) are used to increase air flow
4 Forced aeration is used to remove water, heat, and carbon dioxide
5 The percentage of carbohydrate content in the residue decreases while carboxyl, alkyl, and aromatic increase

porate less than 10% of substrate into their cell biomass (Schink 1988).

Although soil has been traditionally used as a filter, barrier, and medium for the transformation of organic wastes, it is in the last 20 years that land application of both activated and digested sludge has received a renewed interest because of the high energy cost of incineration and the phasing out of ocean dumping. It also helped that the Federal Water Pollution Control Act of 1972 required secondary treatment of sewage which produced cleaner water – and much more sludge. Of the 2 million metric tonnes of municipal sludge generated per day, about one-third is applied to land. Land treatment of sludge can consist of:

1 composting, in which the humification process is accelerated (Table 17.3) (Inbar *et al.* 1989);

2 land farming, which incorporates organic wastes into soil to facilitate microbial degradation, physical adsorption, chemical degradation, and volatilization;

3 sludge application to forests, crops, and pastures to improve soil structure and fertility (Boyle *et al.* 1989).

The soil–plant root system provides an effective barrier from all but a few metals that can accumulate in the edible portions of certain crops (Table 17.4) (Chaney 1983). The rate of application to agronomic land is usually dictated by the level of sludge nitrogen, while the total accumulated amount is regulated by metals content (e.g., cadmium). As the water content in primary, secondary, or digested sludge decreases with land application, the microbial biomass shifts from bacterial towards fungal dominance. In soil, bacteria generally degrade compounds that have a high degree of water solubility whereas fungi metabolize chemicals of lower water solubility and greater adsorptivity. The result is a further mineralization of the labile compounds to carbon dioxide, and humification of the remaining organic fraction.

Table 17.4 Metal behavior in soil–plant systems

Translocated to shoots	Phytotoxic	Animal toxic and bioaccumulated in plants	Chelated in root sap
Zn	Zn	Cd	Cu
Cd	Ni	Se	Ni
Mn	Cu	Mo	Co
B	Mn	Co	Pb
Se	Se		Cr
Mo	Mo		Hg

17.7 Wastewater biotechnology

If the physical and chemical parameters are maintained at optimum levels, then the limiting factor to degradation seems to be biological. With the improvement of recombinant DNA techniques, the possibility of "micromanaging" degradative enzymes is dawning. If appropriate genetic information can be introduced into robust microbial competitors, then an expansion of enzymatic capacity could increase the degradation rates of recalcitrant compounds. Inoculating (seeding) wastewater influent with xenobiotic-degrading bacteria could also prevent the build up of biotoxics that could jeopardize effluent quality. Strain improvement and molecular breeding has the potential to enhance microbial degradation of natural compounds without biomass increase, thereby improving overall efficiency of wastewater treatment (Table 17.5).

Areas of biological improvement of wastewater efficiency, suggested by Lindow *et al.* (1989), include enhancing the production of

Table 17.5 Some current limitations with wastewater treatment

I *Aerobic activated sludge treatment*
1 Mixing and aeration is energy intensive
2 Large amounts of sludge are produced
3 Oxygen input and energy needs require continuous management of plant operation
4 Sludge bulking can be caused by nonsettling filamentous bacteria

II *Anaerobic treatment*
1 The temperature needs to be maintained at 30°C (municipal waste water can average 10–20°C)
2 Anaerobes have slow growth rates and are sensitive to toxics
3 An aerobic polishing step is often required to oxidize ammonium to nitrate to facilitate a subsequent denitrification step

flocculating biopolymers, improving the oxygen scavenging ability of beneficial floc-forming bacteria, and using immobilized cells containing metal-sequestering peptides that can recover toxic metals from wastewater. Augmenting the production of extracellular oxidases and microbial emulsifiers could also increase the rates of degradation of recalcitrant and adsorbed organics. Enhanced bioengineering of waste water will depend on the survival of introduced microbes, or their immobilized enzymes, under fluctuating environmental conditions. The resolution of problems associated with the heterogeneous spatial distribution of pollutants, nutrients, oxygen, cosubstrate, and microorganisms will also be necessary. But the ultimate decision to proceed with bioengineering will be based on its economic competitiveness with established wastewater technology.

Acknowledgment

Partial support was supplied by grant #199-61-12-22 from the National Aeronautics and Space Administration.

References

Alexander M. (1981) Biodegradation of chemicals of environmental concern. *Science* **211**, 132–138.

Atlas R.M. & Bartha R. (1987) *Microbial Ecology, Fundamentals and Applications*. Benjamin Cummings, Menlo Park.

Boyle M., Frankenberger W.T. & Stolzy L.H. (1989) The influence of organic matter on soil aggregation and water infiltration. *J. Prod. Agric.* **2**, 290–299.

Boyle M., Ford T., Mitchell R. & Maki J. (1990) Survival of pathogenic bacteria under nutrient starvation conditions. *SAE International Conference on Environmental Systems, Williamsburg, VI.* July 9–12, 1990.

Bryers J.D. & Characklis W.G. (1990) Biofilms in water and wastewater treatment. In Characklis W.G. & Marshall K.C. (eds) *Biofilms*, pp. 671–696. John Wiley & Sons, New York.

Bull A.T. (1980) Biodegradation: some attitudes and strategies of microorganisms and microbiologist. In Ellwood D.C., Hedger J.N., Latham M.J., Lynch J.M. & Slater J.H. (eds) *Contemporary Microbial Ecology*, pp. 107–136. Academic Press, New York.

Chaney R.L. (1983) Plant uptake of inorganic waste constituents. In Parr J.F., Marsh P.B. & Kla J.M. (eds) *Land Treatment of Hazardous Wastes*, pp. 50–76. Noyes Data Co.

Gadd G.M. & Griffiths A.J. (1978) Microorganisms and heavy metal toxicity. *Microb. Ecol.* **4**, 303–317.

Gould M.S. & Genetelli E.J. (1975) Heavy metal complexation behavior in anaerobically digested sludge. *Wat. Res.* **12**, 505–512.

Grant W.D. & Long P.E. (1981) *Environmental Microbiology*. John Wiley & Sons, New York.

Hartenstein R. (1981) Sludge decomposition and stabilization. *Science* **212**, 743–749.

Inbar Y., Chen Y. & Hadar Y. (1989) Solid-state carbon-13 nuclear magnetic resonance and infrared spectroscopy of composted organic matter. *Soil Sci. Soc. Am. J.* **53**, 1695–1701.

Jewell W.J. (1987) Anaerobic sewage treatment. *Environ. Sci. Technol.* **21**, 14–21.

Kononova M.M. (1966) *Soil Organic Matter: its Nature, its Role in Soil Formation and in Soil Fertility*, 2nd edn. Pergamon, New York.

Lake D.L., Kirk P.W.W. & Lester J.N. (1984) Fractionation, characterization and speciation of heavy metals in sewage sludge and sludge amended soil: a review. *J Environ. Qual.* **13**, 175–183.

Lawler D.F. (1986) Removing particles in water and wastewater. *Environ. Sci. Technol.* 20, 856–861.

Lindow S.E., Panopoulos N.J. & McFarland B.L. (1989) Genetic engineering of bacteria from managed and natural habitats. *Science* **244**, 1300–1307.

Martin J.P. & Haider K. (1986) Influence of mineral colloids on turnover of soil organic carbon. In Huang P.M. & Schnitzer M. (eds) *Interactions of Soil Minerals with Natural Organics and Microbes*, pp. 283–304. SSSA Spec. Publ. 17. Soil Sci. Soc. Am., Madison.

McCarty P.L. & Smith D.P. (1986) Anaerobic wastewater treatment. *Environ. Sci. Technol.* 20, 1200–1206.

Parkin G.F. & Owen W.F. (1986) Fundamentals of anaerobic digestion of wastewater sludges. *J. Environ. Engineer.* **113**, 867–920.

Rittmann B.E. (1987) Aerobic biological treatment. *Environ. Sci. Technol.* **21**, 128–136.

Schink B. (1988) Principles and limits of anaerobic degradation: environmental and technological aspects. In Zehnder A.J.B. (ed.) *Biology of Anaerobic Microorganisms*, pp. 771–847. John Wiley & Sons, New York.

Schroeder E.D. (1977) *Water and Wastewater Treatment.* McGraw-Hill, New York.

Sommers L.E. (1977) Chemical composition of sewage sludge and analysis of their potential use as fertilizers. *J. Environ. Qual.* **6**, 225–232.

Sterritt R.M. & Lester J. (1988) *Microbiology for Environmental and Public Health Engineers.* E. & F.N. Spon, New York.

Stevenson F.J. (1982) *Humus Chemistry: Genesis, Composition, Reactions.* John Wiley & Sons, New York.

Waksman S.A. (1938) *Humus: Origin, Chemical Composition and Importance in Nature*, 2nd edn. Williams & Wilkins, Baltimore.

Timothy E. Ford

The microbial ecology of water distribution and outfall systems

18.1 Introduction

Interest in environmental microbiology is motivated in part by the implications to health, agriculture, and industry. In no area is this more clearly demonstrated than in the municipal and industrial uses and disposal of water.

The interactions between anthropogenic and natural systems result in the development of unique microbiological ecosystems. Iron tubercle formation on water distribution and cooling pipes, and the sulfuretum in concrete sewers are perhaps the best examples. The common denominator is the use of source water. Whether fresh, brackish, or marine, a source water contains a community of bacteria, fungi, protozoa, algae, and higher invertebrates that initially contaminate and may thrive in the anthropogenic system.

This chapter will review aspects of the microbiology of water distribution and wastewater outfall, focusing on the unique ecological interactions between microorganisms and anthropogenic systems. It is not the intention of this chapter to stress drinking water microbiology *per se*; there are excellent texts in this area (McFeters 1990a), but rather to emphasize:

1 the implications of natural microbial population development in piping systems, storage tanks, and cooling systems etc., which may result in biofouling, odor and color problems ("dirty water"), health risks, and biocorrosion;

2 the implications of wastewater outfalls on the microbiology of natural systems, e.g., influence of chemical and biological pollution, and physical changes on microbial community structure and function.

18.2 Biofilms and their implications

Biofouling, dirty water, health problems, and biocorrosion are all primarily the direct consequences of microbial film formation on the surface of pipes. Microbial film formation, or in engineering terms,

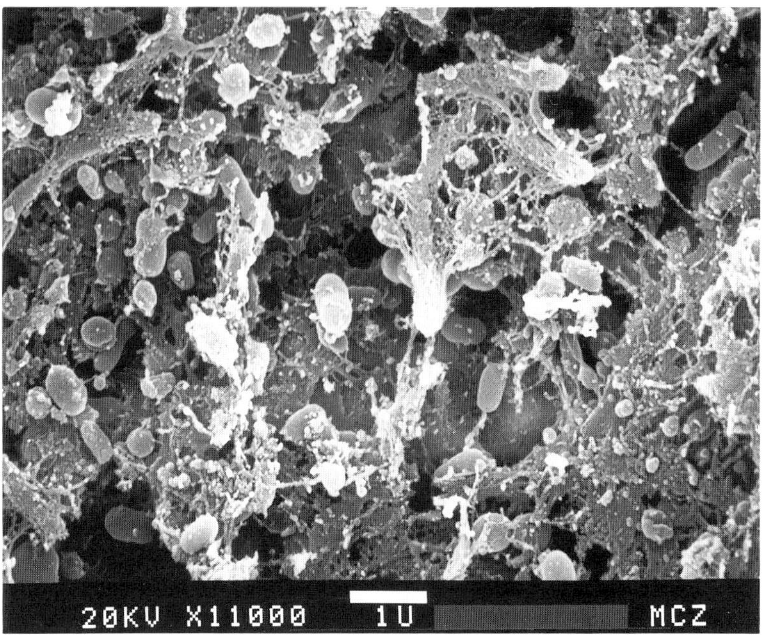

Figure 18.1 Scanning electron micrograph of a complex biofilm on stainless steel.

biofilm formation (see Characklis & Marshall 1990), is caused by adhesion of bacteria to surfaces. The process of microbial attachment and subsequent film formation has been extensively reviewed (Costerton *et al.* 1978; Absolom *et al.* 1983; Baier 1984; Baier *et al.* 1984; Marshall 1984; Savage & Fletcher 1985; Hamilton 1987). Characklis & Marshall (1990) present a comprehensive discussion of the biology, chemistry, and physics of biofilms, with particular emphasis on industrial impacts of biofilm behavior. A biofilm develops due to colonization of a surface by microorganisms (Figure 18.1). The film consists of an accumulation of cells, extracellular products, and inorganic and organic debris. Within a water distribution system, the biofilm acts as an inoculum for the rest of the piping system and provides a protective environment for pathogens (Anderson *et al.* 1990; Boyle *et al.* 1990).

18.2.1 Microbiology

Microbiological analyses of water distribution systems have tended to emphasize indicators of fecal contamination, in particular the coliform group (for reviews see McFeters 1990a). These indicator organisms can be an important component of the water distribution system biofilm. However, they are not central to formation of the film, but rather the biofilm enables them to survive and occasionally proliferate. There appears to be little or no association between the total heterotrophic population of a water distribution system and

Table 18.1 A summary of heterotrophic plate count (HPC) bacteria in a distribution system biofilm. From LeChevallier *et al.* (1987)

Bacteria surface	Water	Zinc floc	Flushed sediment	Iron tubercle	Pipe
Pseudomonas vesicularis	++	++	+		++
Flavobacterium spp.	++	++	+		++
Pseudomonas diminuta	+			+	
Pseudomonas cepacia					+
Pseudomonas pickettii					+
Pseudomonas stutzeri			+		+
Pseudomonas fluorescens			+	+	
Pseudomonas putida				+	
Pseudomonas paucimobilis	+			+	
Pseudomonas maltophilia	+				
Alcaligenes spp.					+
Acinetobacter spp.	+				+
Moraxella spp.	+		+		+
Agrobacterium radiobacter					+
Arthrobacter spp.	+		++	++	+
Corynebacterium spp.				+	+
Bacillus spp.					+
Yeasts					+
CDC group II J	+				
Enterobacter agglomerans					+
Micrococcus spp.	+				

+, Organism present.
++, Predominant organisms.
CDC, Center for Disease Control.

total coliforms (LeChevallier *et al.* 1987; Edberg & Smith 1989). Many microorganisms present in natural aquatic environments will thrive in the water distribution system, becoming part of the biofilm consortium that provides protection from disinfection and cleaning techniques (Nickels *et al.* 1981; Pope *et al.* 1983; Costerton & Lappin-Scott 1989; Anderson *et al.* 1990; Keevil *et al.* 1990). Despite water treatment, regrowth may occur throughout the system, particularly at dead-end sites where biocide residual is lost (Reasoner *et al.* 1989).

LeChevallier *et al.* (1987) characterized biofilms from a distribution system by culture examination. They examined water, zinc floc, flushed sediment, iron tubercle, and pipe surface (summarized in Table 18.1). The approach to examining heterotrophic bacteria using dilute plating media, R_2A (Reasoner & Geldreich 1985), provides data for between-system comparison. However, as with all environmental isolation, plating techniques are by nature specific and will isolate only a small percentage of organisms present. Culture conditions of time, temperature, and concentration of organic nutrients have all been shown to be critical in recovery of bacteria from chlorinated distribution systems (Maki *et al.* 1986). This is particularly true for the iron- and manganese-depositing bacteria which are notoriously difficult to isolate (for reviews see Ghiorse 1984; Jones 1986).

There is considerable microscopic evidence for microbial colonization of distribution systems. Ridgway & Olson (1981) used scanning electron microscopy to show morphological diversity of sessile microbial communities. Based on morphology, they identified *Gallionella*, an iron-oxidizing bacterium, in many samples. In addition, these authors identified "extracellular fibrillar appendages," presumably extracellular polysaccharides, mediating in cellular attachment. Sly *et al.* (1988a,b; 1990) identified *Pedomicrobium manganicum* as an important agent in oxidizing and depositing manganese in high water velocity areas of a distribution system. The consequences of deposition of iron and manganese oxides are tubercle formation, resulting in restricted water flow, biocorrosion, and decreased aesthetic quality as the deposits slough off to form "dirty water." Figure 18.2 shows a distribution pipe heavily encrusted with these deposits. Direct analyses of tubercles have shown large populations of metal-depositing bacteria, among which are *Hyphomicrobium* (Tyler & Marshall 1967a,b), *Gallionella* (Kobrin 1976; Ridgway & Olson 1981; Ridgway *et al.* 1981; Borenstein & Lindsay 1988), and, less frequently, *Sphaerotilis*, *Crenothrix*, *Leptothrix*, and *Siderocapsa*

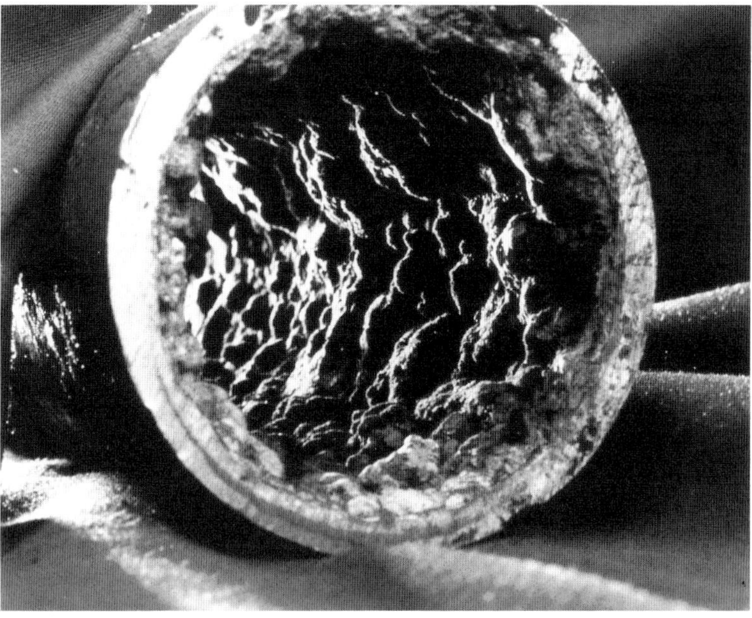

Figure 18.2 Distribution system pipeline heavily encrusted with iron and manganese oxide deposits. From Herro (1991).

(Iverson 1974; Kobrin 1976; Tatnall 1981). Sulfate- and nitrate-reducing bacteria, nitrite oxidizers, and numerous unidentified heterotrophs have also been found (Tuovinen *et al.* 1980; Tuovinen & Hsu 1982).

Keevil *et al.* (1990) reported that nitrite in London's distribution systems, surpassing acceptable concentrations set in EEC Directive 80/778/EEC (1980), was linked to breakdown of the disinfectant monochloramine by ammonia-oxidizing bacteria. Wolfe *et al.* (1990) investigated nitrification in chloraminated distribution systems. Nitrification is beneficial in water treatment through the removal of ammonia; however, if incomplete, nitrite accumulates and rapidly reduces free chlorine. Wolfe *et al.* (1990) found nitrifiers, identified as members of the genus *Nitrosomas*, throughout the system studied. Their data showed seasonal dependence with high numbers in the warm summer months, a relative resistance to disinfection by chloramines and higher numbers in finished water reservoir and pipe sediments than in pipe biofilm.

Pigmented bacteria appear to be a major component of the microbial community in water distribution systems (Maki *et al.* 1986; Reasoner *et al.* 1989). There has been some interest in the use of these organisms as a marker for monitoring water quality, due to their association with oligotrophic, unpolluted waters (Guthrie *et al.* 1974; Reasoner *et al.* 1989). Many of the pigmented forms are from the genus *Flavobacterium* (although pigmented species of *Corynebacterium*, *Mycobacterium*, *Serratia*, *Micrococcus*, *Chromobacterium*, and *Flavobacterium-Cytophaga* have also been isolated). Their predominance in water distribution systems suggests survival of treatment and disinfectant residuals. Reasoner *et al.* (1989) analyzed percentage pigmented bacteria present in the heterotrophic plate count from dead-end fire hydrant sites as 6.6–94%.

Although this review will focus on bacteria, it is important to note that many invertebrates are important components of anthropogenic systems. There are a vast number of free-living protozoa, many of which resist disinfection through cyst formation (see Pennak 1989). Their role in a distribution system biofilm is unknown, but as in other aquatic systems, they may be important in bacterial grazing (e.g., Bloem & Bar-Gilissen 1989). Numerous other invertebrates occur, persist, and grow within distribution lines, including crustaceans, nematodes, flatworms, water mites, and insect larvae (Levy 1990). Many that are less resistant to water treatment, such as the bryozoa, are economically important due to clogging of intake systems (Pennak 1989). In addition, dead, decaying organisms on intake screens or wedged in intake or condenser pipes provide both an important nutrient source and increased disinfectant demand. As in natural aquatic systems, the grazing and settlement behavior of invertebrates is likely to be important in the ecology of the distribution system biofilm.

18.2.2 Health implications of environmental organisms in drinking waters

In potable water systems, biofilm formation presents serious implications for hygiene, odor, and taste. Among naturally occurring biofilm bacteria, we can include pathogenic and opportunistically pathogenic bacteria such as species of *Pseudomonas*, *Campylobacter*, *Clostridium*, *Flavobacterium*, and *Legionella*. Many of the free-living amoebae isolated from chlorinated drinking and swimming waters are pathogenic, e.g., species of *Acanthamoeba*, which cause chronic eye infections, and *Naegleria*, which cause meningoencephalitis. The *Acanthamoeba* may be of particular concern as their cysts appear to be particularly resistant to disinfection (e.g., De Jonckheere & Van de Voorde 1976; Cursons *et al.* 1980; Brandt *et al.* 1989). Although not directly pathogenic, the hartmannellid amoebae may also be important due to an association with *Legionella pneumophila* (see below).

L. pneumophila presents an ideal example of a biofilm pathogen. *Legionella* is present in most natural waters (Fliermans *et al.* 1981) and is an example of one of numerous naturally occurring bacteria that can colonize an anthropogenic piping system. However, within the distribution system it appears to have carved out a unique niche. This is in part due to the fastidious nutrient requirements (high levels of iron and cysteine) that initially made it very difficult to isolate and characterize (States *et al.* 1990). An additional detection problem may be that *L. pneumophila* enters a viable but nonculturable stage when placed in drinking water (see Section 18.2.3) (Hussong *et al.* 1987). As a result, *L. pneumophila* has not always been isolated from water samples during outbreaks of legionnaires' disease. However, it may be one of the most underrated waterborne pathogens (J. Rose personal communication, 1991). According to Bennett *et al.* (1987), 1985 Center for Disease Control estimates of morbidity and mortality associated with *Legionella* were 75 000 and 11 250, respectively. In comparison, they reported numbers for identified water-borne disease of 940 000 and 900, respectively.

From an ecological perspective, outbreaks of legionnaires' disease represent an interesting strategy for the survival of an organism within a competitive environment. Stout *et al.* (1985) describe a scenario in which the water from a hot water storage tank provided conditions to support $>10^4$ *L. pneumophila*. These authors concluded that a number of environmental criteria combine to provide conditions ideal for growth of this bacterium. The combination of sediment, environmental microflora, and hot water (38–50°C) appears to favor the growth of *L. pneumophila*. Environmental microflora are more prevalent when sediment is present due to its nutrient adsorption capacity. Stout *et al.* (1985) consider that stimulation of environmental organisms, both by increased temperature and presence of sediment, results in production of nutrients, such as cysteine, in quantities that can support large *Legionella* populations. Close association between *Legionella* and cyanobacteria (Tison *et al.* 1980; Bohach & Snyder 1982), *Flavobacterium breve* (Wadowsky & Yee 1983), and a number of environmental isolates not identified to species level (Stout *et al.* 1985), have been reported. In addition, *L. pneumophila* has been shown to grow intracellularly in certain strains of free-living amoebae, particularly the *Hartmannella* spp. (e.g., Fields *et al.* 1989; Wadowsky *et al.* 1991).

In agreement with the above scenario is the fact that *Legionella* is seldom recovered without some disturbance of sediments (Stout *et al.* 1982), and outbreaks of legionnaires' disease have been traced to disturbance of stagnant hot water systems (Fisher-Hoch *et al.* 1982).

An important health issue was discussed by Calomiris *et al.* (1984). These authors found that a high percentage of standard plate-count bacteria isolated from a drinking-water system expressed tolerance to metals associated with water treatment and distribution. This tolerance appeared to be associated with multiple antibiotic resistance. Bacteria otherwise considered harmless may be a health threat if they possess resistance factors to antibiotics due to the ready exchange of genetic material (see Chapter 19).

18.2.3 Survival of enteric pathogens

As previously mentioned, biofilms allow survival and proliferation of many of the pathogenic bacteria, viruses, and protozoa by providing a nutrient-rich film, protected from chemical disinfection. Many of the bacteria are referred to as "injured" (for reviews see McFeters 1990b; Singh & McFeters 1990), defined as the sublethal physiological damage caused by stress factors in the aquatic environment (e.g., chlorine). As a result of injury, they do not grow on selective media generally used for their enumeration. However, under suitable conditions injured pathogens may return to full virulence, as shown for *Escherichia coli* (Singh & McFeters 1986) and *Yersinia enterocolitica* (Singh & McFeters 1987). Standard techniques to enumerate pathogens in drinking water may not detect potentially virulent pathogens and other indicators of fecal pollution.

In a related issue, Byrd *et al.* (1991) discussed the idea of "viable but nonculturable" bacteria in drinking water. This hypothesizes that bacteria may be nonculturable, not specifically as the result of injury, but as an adaptation to environmental or nutritional stresses (see Grimes *et al.* 1986). Viable organisms can no longer be cultured on their selective or nonselective media, but are identified by the direct viable-count procedure of Kogure *et al.* (1979), which depends on direct observation of cell elongation. Byrd *et al.* (1991) found evidence of viable but nonculturable states of *Klebsiella pneumoniae*, *Enterobacter aerogenes*, *Agrobacterium tumefaciens*, *Streptococcus faecalis*, and *Micrococcus flavus*. Pathogenicity of these viable cells was not determined.

Other enteric pathogens of interest include protozoa and viruses. The enteric protozoa, *Giardia* and *Cryptosporidium*, survive water treatment by formation of cysts, just as their free-living counterparts. Cyst formation conveys greater disinfection resistance than is found in many bacterial indicators (see Rose 1990). Rose *et al.* (1991a) were able to detect *Cryptosporidium* oocysts in 17% of 36 drinking-water samples that had undergone varying degrees of treatment, including two samples of water that had undergone coagulation, sedimentation, sand filtration, and disinfection.

Viruses in water distribution networks are a ubiquitous problem, particularly in the developing countries (Deetz *et al.* 1984; Keswick *et al.* 1984; Bitton *et al.* 1986). There is epidemiological evidence that hepatitus A, Norwalk agent, and rotaviruses are a frequent cause of water-borne disease (Bitton *et al.* 1986). It appears that water-treatment processes will not completely eliminate viral contamination and that this is probably an unrealistic goal (Payment *et al.* 1985). Viruses are an important part of the microbiology of water-distribution systems. Their ubiquity and long survival associated with particulate material results in less efficient removal in water-treatment processes. Indeed, it has been reported that many drinking-water samples containing infectious viruses meet US coliform standards (Deetz *et al.* 1984). Stetler *et al.* (1984) reported that "it has been repeatedly demonstrated that pollution indicator bacteria are more effectively destroyed by chlorine than are enteric viruses." As a result, waters with a chlorine residual, that satisfy coliform standards for drinking have been found to contain enteric viruses such as hepatitis A (Hejkal *et al.* 1982). It is beyond the scope of this chapter to present a detailed discussion of viruses in treated drinking water. The interested reader is referred to reviews by Bitton *et al.* (1986), Rose & Gerba (1986), and Gerba & Rose (1990).

18.3 Economic implications

The focus of water distribution microbiology on public health concerns has produced a considerable body of literature on pathogen survival and proliferation (see McFeters 1990a; Anderson *et al.* 1990). However, this provides limited information on the condition of the biofilm, the loss of efficiency due to fouling, the risk of system failure due to corrosion, and potential odor and color (dirty water) problems. Economic implications of microbial film formation in anthropogenic systems are enormous. Specific costs due to biocides, antifoulants, anticorrosives, pigging operations, health implications, and customer dissatisfaction cannot be accurately estimated.

Biocorrosion is a major contributor to both water quality and environmental contamination (through spills, etc.). In his review paper, Iverson (1974) summarized some of the early estimates of economic loss due to biocorrosion. For example, costs of deterioration of buried pipeline in the USA were put at between 0.5 and 2 billion dollars per year in 1954 (Greathouse & Wessel 1954). In 1964, Booth suggested that 50% or more of the failure of buried metal in the UK was due to bacterial action. Rates of deterioration are certainly significant, Booth (1964) measured deterioration rates of cast iron pipe in the presence of sulfate reducing bacteria as high as 1389 mg dm^{-2} d^{-1}.

Since the early estimates of economic loss, there have been few calculations of biocorrosion cost, as researchers have been unable to accurately determine a microbial component. Raloff (1985) quoted a National Bureau of Standards statement that a "good portion of metal corrosion in the US, estimated at about 167 billion dollars, is biologically induced." Simply put, accurate estimation of biologically influenced damage cannot be given. It is likely to be vast, but until research enables us to readily distinguish between chemical and biological mechanisms, these estimates must be speculative.

18.4 Microbe–metal interactions

Many of the economic implications of microbial contamination of anthropogenic systems stem from the interactions between microorganisms and metals. Microbe–metal interactions have been reviewed elsewhere (Beveridge & Doyle 1989; Ford & Mitchell 1992; Ford *et al.* in press) and will only be briefly described.

Microbe–metal interactions may be favorable or unfavorable for anthropogenic consideration. Favorable interactions include the use of bacteria and their products in wastewater treatment, bioextraction, and mining practices (Brierly *et al.* 1989). Unfavorable interactions include transformation and mobilization of toxic metals (Ford & Mitchell 1992) and corrosion (Ford & Mitchell 1990). Microbe–metal interactions may be divided into either intracellular or extracellular. Many of these interactions may represent important detoxification steps for bacteria exposed to a metal-rich environment. Within the water distribution system, a number of toxic metals are likely to be present, including lead from old piping and solder, copper from piping, zinc from corrosion inhibitors, and other components from steel piping, fitments, and weldments. A typical detoxification mechanism would be the intracellular accumulation of lead as the phosphate precipitate by a *Citrobacter* sp. suggested by Aickin *et al.* (1979). The mining industry has exploited the ability of certain bacteria to accumulate metals intracellularly, particularly in management of effluent treatment lagoons (Brierly & Brierly 1980).

Extracellular interactions between bacteria and metals are probably of greater importance to the integrity and ecology of the water distribution system. Three processes can be considered of paramount importance:

1 interaction between bacterial cell walls and metals;
2 interaction between bacterial extracellular materials and metals;
3 production of acids and other corrosive metabolites.

Bacterial cell walls possess lipopolysaccharides (LPS) in their outer membranes. These are complex structures, consisting of a hydrophobic, phosphorylated section known as lipid A; a core oligosaccharide section; and a variable o-specific side chain consisting of a number of unusual sugars (Figure 18.3a). The o-specific side chain projects out from the cell membrane and contains different functional groups capable of binding metals. The most abundant are the phosphoryl groups which strongly bind polyvalent metals due to the presence of closely opposed reactive sites (Coughlin *et al.* 1983; Beveridge 1984; Ferris 1989).

Extracellular materials, the structural component of the biofilm, will also bind metals. Interaction between bacterial exopolysaccharides and metals has been the subject of increasing research interest. Most of the original research stems from the importance of biological flocculation of metals in wastewater treatment processes (e.g., Brown & Lester 1979; 1982a,b; Lester *et al.* 1984). As a result of extensive chemical studies of extracellular polysaccharides (EPS) by Sutherland (1972; 1982; 1985) and metal-binding studies by G.G. Geesey and coworkers (Geesey & Mittelman 1985; Geesey *et al.* 1986; 1987; 1988; Geesey & Jang, 1989) and T.E. Ford & coworkers (Ford *et al.* 1987; 1988a; 1990), a clearer picture has emerged of the interactions between bacterial exopolymers and the metal surface. A typical EPS repeating unit is shown in Figure 18.3b. Interaction with metal ions is generally considered a direct consequence of the presence of negatively charged functional groups on the exopolymer. These groups include hydroxyl, phosphate, pyruvate, succinate, and uronate. Binding of positively charged cations can rapidly occur, resulting in equilibrium changes in ion concentrations within the biofilm. These in turn are thought to accelerate metal dissolution from a surface (see Section 18.5).

The ability of organic or inorganic acids produced by microorganisms, including species of *Thiobacillus*, *Serratia*, *Pseudomonas*, *Bacillus*, *Penicillium*, and *Aspergillus*, to extract metals from solid

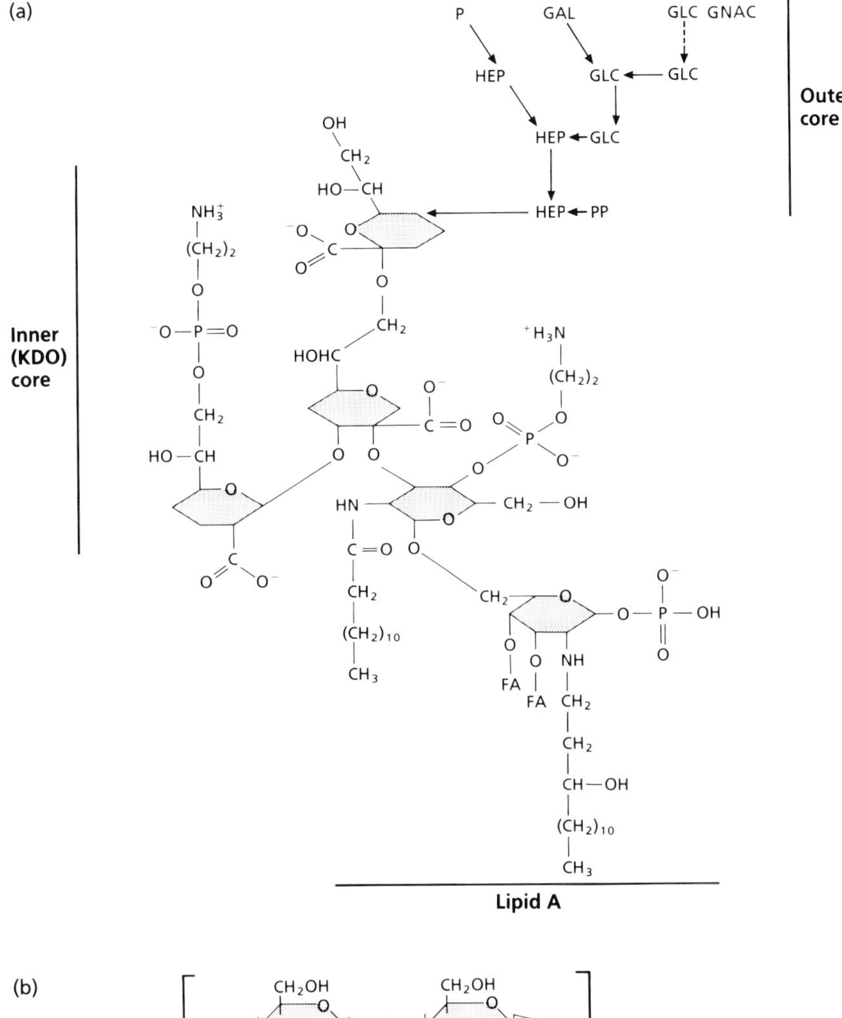

Figure 18.3 Typical structures of bacterial lipopolysaccharides (LPS) and extracellular polysaccharides (EPS). (a) Possible structure of LPS from *Escherichia coli* k-12. (b) Structure of EPS from *Xanthomonas campestris*. (a) From Ferris & Beveridge (1986); (b) from Sutherland (1982); adapted from Jansson *et al.* (1975).

substrates has been extensively used in mining practices (Schinner & Burgstaller 1989). For example, they used citric acid production by a *Penicillium* sp. to selectively extract zinc from industrial waste. Production of these metabolites has considerable implications for corrosion, because acidic conditions may result in solubilization of component metals from a piping system.

18.5 Biocorrosion

There are a number of ways in which surface films can be responsible for accelerated corrosion (for review of these processes see Ford & Mitchell 1990).

1 Heterogeneity (formation of oxygen concentration cells). Patchiness of a biofilm results in differential oxygen concentration on a surface resulting in formation of anodic and cathodic sites.

2 Exopolymer–metal interactions (formation of ion concentration cells). Differential concentrations of metals in the biofilm stimulate corrosion cells by equilibrium changes and by galvanic coupling.

3 Metabolic interaction. Any metabolic process may influence corrosion rates, either by providing an electron sink or an electron source. In particular, extracellular enzymes may be important in electron flow.

4 Organic and inorganic acid production by microorganisms. Solubility of metals increases as pH decreases; in addition, protons provide an electron sink which accelerates the cathodic reaction.

5 Production of other corrosive metabolites, e.g., hydrogen and sulfide. Hydrogen may cause embrittlement or acceleration of stress corrosion cracking; sulfide is implicated in cathodic depolarization – removal of hydrogen from a surface which stimulates the cathodic reaction.

18.5.1 Tubercle formation

We frequently see water piping systems filled with microorganisms, their exopolymers, and inorganic deposits. These deposits may even develop to the extent of actually blocking flow within the pipe (Schmitt 1986), and, at the very least, may substantially reduce efficiency and discolor the water. Underneath the deposits, oxidation of the metal will eventually result in failure. The most commonly occurring form of biocorrosion in metal pipes carrying at least partially oxic waters presents a fascinating and complex ecosystem that results in many, if not all, the biocorrosion processes mentioned above.

The typical development of this form of biocorrosion is manifest by the formation of tubercles (Figure 18.4). These result from chemical or biological deposition of iron and manganese oxides to form deposits or concretions on a metal surface. Biological deposition may result from intracellular or extracellular interaction with the iron and manganese, which may be important in colony morphology or detoxification reactions (Caldwell & Caldwell 1980; Cowen & Silver 1984). In addition, enzymatic oxidation of the metal may occur, with concurrent energy gains to the bacterium, particularly in the case of manganese (Ghiorse 1984). In the case of iron it is considered more usual for ferric iron to be deposited bound to exopolymers or cell walls, which in turn may accelerate autoxidation of ferrous iron.

A mass of bacterial cells and minerals is formed that is bound together by bacterial exopolymers. These in turn limit diffusion of oxygen and nutrients into and out of the tubercle, resulting in steep chemical gradients. Outer layers remain aerobic with a near neutral pH and inner regions become highly reduced and acidic from organic acid production. These conditions favor the growth of sulfate- and nitrate-reducing bacteria, and hence a complex consortia of aerobes and anaerobes is able to coexist within the same tubercle. Recent work by Herro (1991) examined tubercle formation and growth on ferrous alloys. Figure 18.4 is adapted from this work and shows a typical tubercle with chemical and microbiological constituents. Herro (1991) isolated $>10^6$ colony forming units (CFU) aerobes, $>2 \times 10^4$ anaerobes (including $>2 \times 10^3$ sulfate-reducing bacteria), but were unable to isolate iron-depositing bacteria. This may be because of the selectivity of the plating media (as discussed earlier).

18.5.2 *Deterioration of wastewater systems and other concrete structures*

While there is a classic form of biocorrosion that affects steel piping systems, there is also a classic form of concrete corrosion that has had enormous economic implications. Major city sewerage systems have been very badly damaged by acid deterioration, often known as sewer crown corrosion. The microbiology is straightforward and is

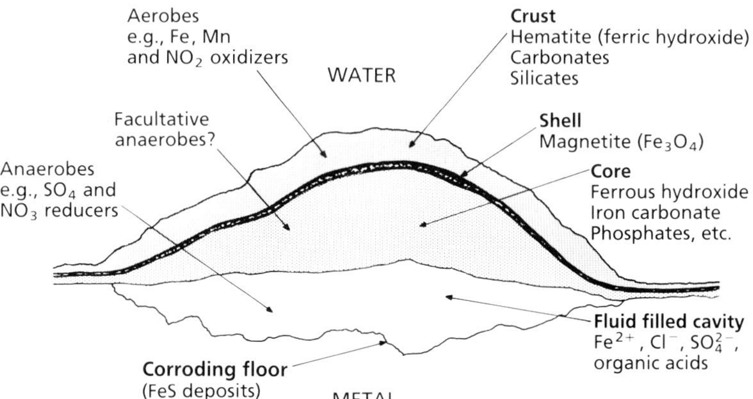

Figure 18.4 Chemical and biological characteristics of a tubercle. Adapted from Herro (1991).

dependent on the sulfur cycle. Early work by Parker (1945a,b) identified the sulfur oxidizing thiobacilli as the probable culprits in concrete degradation. Known as a sulfuretum, the typical process of events has now been well documented (Rigdon & Beardsley 1958; Forrester 1959; Hueck van der Plas 1968; Sand & Bock 1984) and is listed below.

1 Carbon dioxide and hydrogen sulfide (products of sewage decomposition) react with the damp concrete surfaces to form carbonates and calcium sulfate. As a result the surface pH of the concrete drops from approximately 12.5 to about 8.5.

2 Hydrogen sulfide is partially oxidized in air to form thiosulfuric and polythionic acids. These in turn lower the pH further to approximately 7.5.

3 Below pH 10 a number of different bacteria oxidize thiosulfates and polythionates to sulfate. Below pH 9 *Thiobacillus thioparus* becomes active and oxidizes thiosulfate to elemental sulfur.

4 The pH continues to decline, and at about pH 5 *Thiobacillus thiooxidans* oxidizes elemental sulfur to sulfuric acid. The final pH can reach 0.6.

Mansfeld *et al.* (1990) described the interesting case history of failure of the Los Angeles sewer system. Inspection in the 1970s led the county sanitation department to believe that repair or replacement would be unnecessary for several decades to hundreds of years. Less than one decade later, many of the sewers were so badly damaged that reinforcing bars were exposed and corroded. This has been attributed to the limitations placed on strength and toxicity of industrial waste water beginning in 1975 and the 1983 Environmental Protection Agency's Categorical Pretreatment Program for industrial waste discharge which significantly reduced cyanide and toxic metals. It would appear that the radical reduction in toxicity of industrial waste waters discharged into the sewer system has enabled the microbial consortium of sulfate-reducing and sulfide-oxidizing bacteria to reestablish itself.

18.6 Anthropogenic influence on the environment

This is a subject that cannot be comprehensively covered in one chapter. Examples will therefore be chosen to illustrate some of the potential microbiological consequences of wastewater disposal to the environment. In this context, waste water refers not only to the final product of waste treatment facilities, but to irrigation runoff, and once-through cooling systems etc. These can all be defined as anthropogenic influences on the aquatic environment and their potential effects are briefly summarized below:

1 nutrient enrichment (leading to eutrophication, see Chapter 4);

2 addition of toxic chemicals (disinfectants, antifoulants, process by-

products etc. may lead to population kills, selection of resistant species etc.);
3 physical changes, e.g., thermal pollution, salinity, and pH, physical affects of jetted discharge and intake dredging;
4 addition of nonindigenous or genetically altered microorganisms (see Chapter 19).

18.6.1 Waste discharge from sewage treatment and storm drains

One of the most direct ways in which human activities have an impact on the aquatic environment is through the disposal of sewage and other household wastes. From a microbiological perspective this has perhaps the most extensive impact on community structure, eutrophication, and oxygen demand. If we take, for example, enrichment with nitrate and phosphate, concentrations of these nutrients are low in most natural waters unless waste effluent is present. As these nutrients are generally limiting, their addition may stimulate the growth of microorganisms to excessive quantities. This has been shown in a number of artificial enrichment studies (Elwood *et al.* 1981; Peterson *et al.* 1985; Schindler 1985; Perrin *et al.* 1987; Lock *et al.* 1990). According to Guthrie *et al.* (1975), effluents from wastewater treatment plants may contain $>0.5\,mg\,l^{-1}$ nitrates. There are also numerous sources of phosphates: wastewater treatment, laundry and cleaning effluents, runoff from fertilized land, and residues from biological material (Guthrie *et al.* 1975).

The most obvious result of nitrate and phosphate enrichment on a natural aquatic system is a rapid increase in autotrophic production, the algal bloom, and this has been the most consistently measured parameter. However, an increase in autotrophic production is generally closely followed by a large increase in heterotrophic activity. The heterotrophs are stimulated, not only by the inorganic nutrients but by the vast production of algal exudate material (Haack & McFeters 1982; Kaplan & Bott 1982). Considerable increases in heterotrophic activity associated with phosphate enrichment of natural streams have also been demonstrated (Peterson *et al.* 1985; Ford *et al.* 1988b; Lock *et al.* 1990).

Enrichment from wastewater discharge is further compounded by the presence of microorganisms that are not normally associated with natural systems. There have been a number of reports on the survival of pathogens in natural environments in the viable but nonculturable state discussed earlier (see Section 18.2.3) (Grimes *et al.* 1986). However, the enriched conditions of the receiving waters may also allow these organisms to proliferate and, indeed, outcompete the natural populations. The problem of waste enrichment is readily seen. Outbreaks of cholera that appear with devastating consequences, particularly after natural disasters, are in part due to the ability of the bacterium *Vibrio cholerae* to survive and proliferate in sewage-enriched natural waters. An additional area of concern is the ease with which genetic material is exchanged in the environment. This raises questions of transference of antibiotic resistance and pathogenicity factors to natural populations (see Chapter 19) (see reviews by DeFlaun & Levy 1989; Miller & Levy 1989).

Domestic discharges consist of gray and black water. Black water (toilet wastes) in developed countries undergoes some form of municipal treatment (see Chapter 17). Gray water (sources of water other than toilet, e.g., baths, laundry etc.), although presently undergoing some form of treatment with black water, is under consideration in the US for direct re-use in lawn, landscape, and garden irrigation (Rose *et al.* 1991b). In a study to determine the microbial quality of gray water for possible household re-use, Rose *et al.* (1991b) measured 6.1×10^6 CFU ml^{-1} (standard plate count) in combined gray-water samples. Total coliforms averaged 2.8×10^5 CFU ml^{-1} and fecal coliforms ranged from 1.82×10^2 to 7.94×10^4 CFU ml^{-1}. These authors also measured survival, though not growth, of *Salmonella typhimurium*, *S. dysenteriae*, and poliovirus in gray water over a 12-day period. Total coliforms in gray water have been reported to be higher than in toilet water due to higher concentrations of labile nutrients (Brandes 1978). Estimates of gray-water production in the US vary from 29.4 to 59 gallons per person per day, depending on region; in Tucson, Arizona, where water conservation is critical, a family of three produces approximately 3.4×10^4 gallons of gray water per year (Rose *et al.* 1991b). Rose *et al.* (1991b) conclude that treatment is necessary for re-use of gray water in irrigation. Gray-water volumes of this nature will directly affect the quality of the resource. In many areas of the US, the consequences of hundreds of years of uncontrolled discharge are apparent in the decline in aesthetic qualities of many of the aquatic systems. This discharge continues today through use of combined sewer overflows, resulting in untreated gray and black water discharging to the environment during storm events (see Section 18.7).

Treatment of waste water does not necessarily solve problems of microbiological contamination. Table 18.2 compares concentrations of indicator bacteria in pristine, agricultural, and treated waste waters. These authors conclude that quality of treated waste water at its best might be acceptable for swimming water. However, during the summer when quality is poor, 100–400 times dilution would be necessary to meet acceptable swimming water standards (Finnish criteria of <100 thermotolerant coliforms 100 ml^{-1}). Increased disinfection of treated waste water may not be the answer. Chlorination in particular may be harmful to the natural ecosystem (see

Table 18.2 Statistical parameters of the concentrations of indicator bacteria in pristine and agricultural areas and in treated waste water (CFU 100 ml^{-1}). From Niemi & Niemi (1991)

		Sampling site		
Bacterial group	**Parameter**	**Pristine area**	**Agricultural area**	**Treated waste water**
Thermotolerant coliform bacteria	N	49	24	24
	Mean	12	3802	46 863
	Median	0	307	17 800
	Min	0	8	500
	Max	268	38 200	414 300
Presumptive *E. coli*	N	47	24	23
	Mean	12	3610	20 639
	Median	0	283	9100
	Min	0	7	300
	Max	259	38 200	100 000
Presumptive fecal streptococci	N	50	24	23
	Mean	48	1629	35 092
	Median	3.5	233	16 900
	Min	0	41	900
	Max	740	16 900	187 400
Confirmed fecal streptococci	N	39	21	20
	Mean	26	495	35 092
	Median	0	106	16 900
	Min	0	5	900
	Max	609	4 450	62 600

Section 18.6.2). In fact, dechlorination has been recommended for waste water discharged to sensitive marine environments (J. Rose personal communication, 1991).

18.6.2 Cooling-water discharge

In recent years, electric power generation has been a major contributor to environmental change. The influences of cooling-water discharges and fly-ash generation have been the subjects of much research. To understand the environmental impact, Howells (1983) reported that freshwater cooling use in the UK was equivalent to 28% of the rainfall in England and Wales and more than 50% of the total river flow. Water used for cooling thermal power plants is subject to temperature increase, chemical changes from pressure and additives, and biological changes before release to the receiving bodies. In addition, thermal power plant operation results in huge amounts of heavy- and fly-ash waste from coal combustion (Guthrie *et al.* 1982). Population changes in heterotrophic bacteria caused by temperature increase, fly-ash effluent, and pH have been observed (Guthrie *et al.* 1974; 1978; Cherry *et al.* 1974a,b; Larrick *et al.*

1981). Typically, increase of water temperature resulted in increased bacterial numbers but a decrease in diversity, and chemical stresses resulted in changes in community structure (Cherry *et al.* 1974b). The extent of these perturbations is hard to assess. Clearly, these wastes have a greater influence on the microbiology of ponded waters than running waters (Cherry *et al.* 1974b; Larrick *et al.* 1981). However, the continuing increase in energy demands will result in the alteration of more aquatic systems, particularly in developing countries where uncontrolled dumping of wastes in marine environments may occur (Subramanian *et al.* 1990).

Most anthropogenic systems depend to some extent on disinfection. We are most familiar with drinking-water treatment and distribution systems which rely on a chlorine (or other disinfectant) residual to prevent survival of pathogens in the water. Chlorination has also been the disinfectant of choice in water distribution for cooling thermal and nuclear power plants. The use of disinfectants for fouling prevention and microbiological control in cooling systems is a greater risk to the environment than in water distribution for domestic uses. By the time water has been processed through the household, disinfectant residuals are likely to be negligible or non-existent. In fact, Reasoner *et al.* (1989) measured no free chlorine residuals at most dead-end fire hydrants in a water-distribution system. Although no free chlorine may be released, the release of chlorinated organic compounds, particularly in sewage discharge, and their potential as carcinogens or mutagens, is of considerable concern (see Mitchell 1991).

Licina (1990) reviewed the mitigation methods against microbiologically influenced corrosion in nuclear power plants. In addition to chlorine, other oxidizing agents, such as bromine compounds, ozone, and hydrogen peroxide, may be employed as biocides. Non-oxidizing biocides, such as quaternary amines, glutaraldehyde, and isothiazolin, may also be used. When it is considered that a biodispersant, corrosion inhibitor and deposit control agent may also be present, it becomes apparent that the potential range of toxic chemicals discharged is vast. In view of the fact that many of them are specifically designed to be toxic to microorganisms and/or invertebrate settlement, their effects on the localized aquatic environment receiving such discharges demands further study. At present, federal regulations in the US permit free available chlorine in power plant cooling water effluent to average 0.2 ppm with a maximum value of 0.5 ppm (from Licina 1990). However, long-term exposure to 0.01 ppm chlorine (or even less) may be toxic to some aquatic organisms (Glasstone & Jordan 1980).

18.7 System and treatment design

Recent evidence suggests a correlation between consumption of drinking water meeting current microbiological standards and gastrointestinal disease (Payment *et al.* 1991). This study underlines our concerns with drinking-water quality. These authors monitored incidence of gastrointestinal disease among families using tap water and those using tap water passed through a domestic water filter designed to eliminate microbial and chemical contaminants. Water leaving the treatment plant met current microbiological standards (no detectable indicator bacteria, bacterial pathogens, and cultivable human enteric viruses). However, these authors estimated that 35% of gastrointestinal illnesses were attributable to consumption of this water. This data raises serious questions on the effectiveness of water treatment, the adequacy of drinking-water standards, and bacteriological quality of water in the distribution system and domestic connections (Payment *et al.* 1991).

System and treatment design more than any other factors, contribute both to the problems of distribution system biofilms and to the problem of anthropogenic discharge to the environment. Analysis of water distribution systems illustrates the fundamental problem of dead-end sites. On a large scale this is obvious, as shown for the Boston distribution system in Figure 18.5. Consider in addition, the hundreds of thousand individual household hook-ups to this system. As previously discussed, regrowth of coliforms and heterotrophs has been frequently reported in water distribution systems (LeChevallier *et al.* 1987; 1991). Regrowth is associated primarily with loss of chlorine residual, particularly in dead-end sections, although other factors such as assimilable organic carbon (AOC) are considered important (LeChevallier *et al.* 1991). These authors showed that AOC levels in excess of $50\,\mu g\,l^{-1}$ were associated with an increased incidence of coliforms in the distribution water. Increased disinfection results in further problems of disinfectant and disinfectant by-product discharge to the consumer and environment. Alternative disinfectant regimes, e.g., use of ozone, are being considered (Glaze 1987; Wolfe *et al.* 1989). However, although a powerful oxidant and effective disinfectant with few toxic by-products, ozone appears particularly susceptible to high turbidities and is unable to sustain a disinfectant residual throughout a distribution system. Environmentally, however, it would appear to have many advantages over chlorine compounds, particularly for wastewater treatment (e.g., Rice *et al.* 1981).

Production of gray and black water coupled with industrial wastewater discharge produces vast volumes of chemically and microbiologically contaminated water that often exceeds the capacity of

Figure 18.5 The complexities of a typical water distribution system (Boston inner city; bold lines represent the distribution mains).

waste treatment facilities. When treatment capacity is exceeded, the facility has no choice but to discharge directly to the environment. This inadequacy of the system results in increased use of combined storm/sewer networks. Rapid rates of urbanization have resulted in connection of new sewer systems to older systems with limited capacity, resulting in considerable increase in the volume discharged by storm outfalls (Cherrered & Chocat 1990). The result in many large cities is the intermittent, yet significant discharge of untreated domestic and industrial wastes to the environment. The ecological effects of wastes upon receiving waters was the subject of a special edition of *Water Science and Technology* (Aalderink *et al.* 1990). In this edition, effects of nutrient loading and chemical oxygen demand (Lijklema *et al.* 1990), benthic macroinvertebrates (Payne & Hedges 1990), and phyto- and zooplankton (Gast *et al.* 1990) were examined. In all cases ecological disturbances were measured, but long-term effects remain uncertain, particularly in open- or flowing-water bodies.

18.8 Conclusion

This chapter discusses some of the important microbial processes associated with water distribution and outfall. We are limited by our lack of knowledge of mechanisms of interaction between microorganisms and their substrata, their survival strategies against water treatment and disinfection, and their ability to survive and proliferate in a nutrient-deficient environment. In terms of outfall, the database is insufficient to assess accurately the impact on the environment. Extrapolation from small impoundments to open ocean or flowing waters is unrealistic. In order to address the effect of outfall on the microbiological characteristics of an ecosystem, far more sensitive approaches and techniques are needed to identify microorganisms and to measure the significance of changing processes within the ecosystem.

References

Aalderink R.H., Lijklema L. & Ellis J.B. (eds) (1990) Urban storm water quality and ecological effects upon receiving waters. *Wat. Sci. Tech.* **22**, Nos 10/11.

Absolom D.R., Lamberti F.W., Policova Z., Zingg W., Van Oss D.J. & Neumann A.W. (1983) Surface thermodynamics of bacterial adhesion. *Appl. Environ. Microbiol.* **46**, 90–97.

Aickin R.M., Dean A.C.R., Cheetham A.K., Jakeman R.J.B. & Skarnulis A.J. (1979) Electron microscope studies on the uptake of lead by a *Citrobacter* sp. *Microbios Lett.* **9**, 7–16.

Anderson R.L., Holland B.W., Carr J.K., Bond W.W. & Favero M.S. (1990) Effect of disinfectants on *Pseudomonas* colonized on the interior surface of PVC pipes. *Am. J. Public Health* **80**, 17–21.

Baier R.E. (1984) Adhesion in the biologic environment. *Biomat. Med.* **12**, 133–160.

Baier R.E., Meyer A.E., Natiella J.R., Natiella R.R. & Carter J.M. (1984) Surface properties determine bioadhesive outcomes: methods and results. *J. Biomed. Mater. Res.* **42**, 337–355.

Bennett J.V., Holmberg S.D., Rogers M.F. & Solomon S.L. (1987) Infectious and parasitic diseases. *Am. J. Prev. Med.* **3**, 102–114.

Beveridge T.J. (1984) Mechanism of the binding of metallic ions to bacterial walls and the possible impact on microbial ecology. In Reddy C.A. & Klug M.J. (eds) *Current Perspectives in Microbial Ecology*, pp. 601–607. American Society for Microbiology, Washington DC.

Beveridge T.J. & Doyle R.J. (eds) (1989) *Metal Ions and Bacteria.* John Wiley & Sons, New York.

Bitton G., Farrah S.R., Montague C.L. & Akin E.W. (1986) Viruses in drinking water.

Environ. Sci. Technol. **20**, 216–222.

Bloem J. & Bar-Gilissen M.-J.B. (1989) Bacterial activity and protozoan grazing potential in a stratified lake. *Limnol. Oceanogr.* **34**, 297–309.

Bohach G.A. & Snyder I.S. (1982) Cyanobacterial stimulation of growth and oxygen uptake by *Legionella pneumophila. Appl. Environ. Microbiol.* **46**, 528–531.

Booth G.H. (1964) Sulfur bacteria in relation to corrosion. *J. Appl. Bacteriol.* **27**, 174–181.

Borenstein S.W. & Lindsay P.B. (1988) Microbiologically influenced corrosion failure analyses. *Mat. Perf.* **27**, 51–54.

Boyle M., Ford T., Mitchell R. & Maki J. (1990) Survival of pathogenic bacteria under nutrient starvation conditions. *Proc. 20th Intersociety Conf. Environ. Syst.*, Williamsburg, Virginia. SAE Technical Paper No. 901381.

Brandes M. (1978) Characteristics of effluent from gray and black water septic tanks. *J. Wat. Poll. Fed.* **50**, 2547–2559.

Brandt F.H., Ware D.A. & Visvesvara G.S. (1989) Viability of *Acanthamoeba* cysts in ophthalmic solutions. *Appl. Environ. Microbiol.* **55**, 1144–1146.

Brierly C.L., Brierly J.A. & Davidson M.S. (1989) Applied microbial processes for metals recovery and removal from wastewater. In Beveridge T.J. & Doyle R.J. (eds) *Metal Ions and Bacteria*, pp. 359–382. John Wiley & Sons, New York.

Brierly J.A. & Brierly C.L. (1980) Biological methods to remove selected inorganic pollutants from uranium mine wastewater. In Trudinger P.A., Walter M.R. & Ralph B.J. (eds) *Biogeochemistry of Ancient and Modern Environments*, p. 661. Australian Academy of Science, Canberra.

Brown M.J. & Lester J.N. (1979) Metal removal in activated sludge: the role of bacterial extracellular polymers. *Wat. Res.* **13**, 817–837.

Brown M.J. & Lester J.N. (1982a) Role of bacterial extracellular polymers in metal uptake in pure bacterial culture and activated sludge – I. Effects of metal concentration. *Wat. Res.* **16**, 1539–1548.

Brown M.J. & Lester J.N. (1982b) Role of bacterial extracellular polymers in metal uptake in pure bacterial culture and activated sludge – II. Effects of mean cell retention time. *Wat. Res.* **16**, 1549–1560.

Byrd J.J., Huai-Shu X. & Colwell R.R. (1991) Viable but nonculturable bacteria in drinking water. *Appl. Environ. Microbiol.* **57**, 875–878.

Caldwell D.E. & Caldwell S.J. (1980) Fine structure of *in situ* microbial iron deposits. *Geomicrobiol. J.* **2**, 39–53.

Calomiris J.J., Armstrong J.L. & Seidler R.J. (1984) Association of metal tolerance with multiple antibiotic resistance of bacteria isolated from drinking water. *Appl. Environ. Microbiol.* **47**, 1238–1242.

Characklis W.G. & Marshall K.C. (eds) (1990) *Biofilms*. Wiley Interscience, New York.

Cherrered M. & Chocat B. (1990) Development of a methodology to determine the pollution discharged by a combined sewer network. *Wat. Sci. Tech.* **22**, 15–22.

Cherry D.S., Guthrie R.K. & Harvey R.S. (1974a) Temperature influence on bacterial populations in three aquatic systems. *Wat. Res.* **8**, 149–155.

Cherry D.S., Guthrie R.K. & Harvey R.S. (1974b) Bacterial populations of aquatic systems receiving different types of stress. *Wat. Res. Bull.* **10**, 1009–1016.

Costerton J.W. & Lappin-Scott H.M. (1989) Behavior of bacteria in biofilms. *ASM News* **12**, 650–654.

Costerton J.W., Geesey G.G. & Cheng K.-J. (1978) How bacteria stick. *Sci. Am.* **238**, 86–95.

Coughlin R.T., Tonsager S. & McGroarty E.J. (1983) Quantitation of metal cations bound to membranes and extracted lipopolysaccharide from *Escherichia coli. Biochemistry* **22**, 2002.

Cowen J.P. & Silver M.W. (1984) The association of iron and manganese with bacteria on marine macroparticulate material. *Science* **224**, 1340–1342.

Cursons R.T.M., Brown T.J. & Keys E.A. (1980) Effect of disinfectants on pathogenic free-living amoebae: in axenic conditions. *Appl. Environ. Microbiol.* **40**, 62–66.

Deetz T.R., Smith E.M., Goyal S.M., Gerba C.P., Vollet J.J., Tsai L., DuPont H.L. & Keswick B.H. (1984) Occurrence of rota- and enteroviruses in drinking and environmental water in a developing nation. *Wat. Res.* **18**, 567–571.

DeFlaun M.F. & Levy S.B. (1989) Genes and their varied hosts. In Levy S.B. & Miller R.V. (eds) *Gene Transfer in the Environment*, pp. 1–32. McGraw-Hill, New York.

De Jonckheere J. & Van de Voorde H. (1976) Differences in destruction of cysts of pathogenic and nonpathogenic *Naegleria* and *Acanthamoeba* by chlorine. *Appl. Environ. Microbiol.* **31**, 294–297.

Edberg S.C. & Smith D.B. (1989) Absence of association between total heterotrophic and total coliform bacteria from a public water supply. *Appl. Environ. Microbiol.* **55**, 380–384.

EEC Directive 80/778/EEC (1980) Quality of water intended for human consumption. EEC, Brussels.

Elwood J.W., Newbold J.D., Trimble A.F. & Stark R.W. (1981) The limiting role of phosphorus in a woodland stream ecosystem: effects of enrichment on leaf decomposition and primary producers. *Ecology* **82**, 146–158.

Ferris F.G. (1989) Metallic ion interactions with the outer membrane of Gram-negative bacteria. In Beveridge T.J. & Doyle R.J. (eds) *Metal Ions and Bacteria*, pp. 295–323. John Wiley & Sons, New York.

Ferris F.G. & Beveridge T.J. (1986) Site specificity of metallic ion binding in *Escherichia coli* K-12 lipopolysaccharide. *Can. J. Microbiol* **32**, 52–55.

Fields B.S., Sanden G.N., Barbaree J.M., Morrill W.E., Wadowsky R.M., White E.H. & Feeley J.C. (1989) Intracellular multiplication of *Legionella pneumophila* in amoebae isolated from hospital hot water tanks. *Curr. Microbiol.* **18**, 131–137.

Fisher-Hoch S.P., Smith M.G. & Colbourne J.S. (1982) *Legionella pneumophila* in hospital hot water cylinders. *Lancet* **i**, 1073.

Fliermans C.B., Cherry W.B., Orrison L.H., Smith S.J., Tison D.L. & Pope D.H. (1981) Ecological distribution of *Legionella pneumophila. Appl. Environ. Microbiol.* **41**, 9–16.

Ford T.E. & Mitchell R. (1990) The ecology of microbial corrosion. In Marshall K.C. (ed.) *Advances in Microbial Ecology*, vol. 11, pp. 231–262. Plenum, New York.

Ford T.E. & Mitchell R. (1992) Microbial transport of toxic metals. In Mitchell R. (ed.) *New Concepts in Environmental Microbiology*, pp. 83–101. John Wiley & Sons, New York.

Ford T.E., Maki J.S. & Mitchell R. (1987) The role of metal-binding bacterial exopolymers in corrosion processes. *CORROSION/87*, Paper No. 380, Natl Assoc. Corrosion Eng., Houston.

Ford T.E., Maki J.S. & Mitchell R. (1988a) Involvement of bacterial exopolymers in biodeterioration of metals. In Houghton D.R., Smith R.N. & Eggins H.O.W. (eds) *Biodeterioration* 7, pp. 378–384. Elsevier, London.

Ford T.E., Walch M., Mitchell R., Kaufman M.J., Vestal J.R., Ditner S.A. & Lock M.A. (1988b) Microbial film formation on metals in an enriched arctic river. *Biofouling* **1**, 301–311.

Ford T.E., Black J.P. & Mitchell R. (1990) Relationship between bacterial exopolymers and corroding metal surfaces. *CORROSION/90*, Paper No. 110, Natl Assoc. Corrosion Eng., Houston.

Ford T.E., Maki J.S. & Mitchell R. (in press) Metal–microbe interactions. In Gaylarde C. & Videla H. (eds) *Bioextraction and Biodeterioration of Metals*. Cambridge University Press, Cambridge.

Forrester J.A. (1959) Destruction of concrete caused by sulfur bacteria in a purification plant. *Surveyor* **118**, 881–884.

Gast H.F., Suykerbuyk R.E.M. & Roijackers R.M.M. (1990) Urban storm water discharges: effects upon plankton communities. *Wat. Sci. Tech.* **22**, 155–162.

Geesey G.G. & Mittelman M.W. (1985) The role of high-affinity, metal-binding exopolymers of adherent bacteria in microbial-enhanced corrosion. *CORROSION/*

85, Paper No. 297, Natl Assoc. Corrosion Eng., Houston.
Geesey G.G. & Jang L. (1989) Interactions between metal ions and capsular polymers. In Beveridge T.J. & Doyle R.J. (eds) *Metal Ions and Bacteria*, pp. 325–357. John Wiley & Sons, New York.
Geesey G.G., Mittelman M.W., Iwaoka T. & Griffiths P.R. (1986) Role of bacterial exopolymers in the deterioration of metallic copper surfaces. *Mat. Perf.* **25**, 37–40.
Geesey G.G., Iwaoka T. & Griffiths P.R. (1987) Characterization of interfacial phenomena occurring during exposure of a thin copper film to an aqueous suspension of an acidic polysaccharide. *J. Colloid Interface Sci.* **120**, 370–376.
Geesey G.G., Jang L., Jolley J.G., Hankins M.R., Iwaoka T. & Griffiths P.R. (1988) Binding of metal ions by extracellular polymers of biofilm bacteria. *Proc. Internat. Conf. on Wastewater Microbiology*, vol 1. Newport Beach.
Gerba C.P. & Rose J.B. (1990) Viruses in source and drinking water. In McFeters G.A. (ed.) *Drinking Water Microbiology*, pp. 380–396. Springer Verlag, New York.
Ghiorse W.C. (1984) Biology of iron- and manganese-depositing bacteria. *Ann. Rev. Microbiol.* **38**, 515–550.
Glasstone S. & Jordan W.H. (1980) *Nuclear Power and its Environmental Effects.* American Nuclear Society, La Grange Park.
Glaze W.H. (1987) Drinking-water treatment with ozone. *Environ. Sci. Technol.* **21**, 224–230.
Greathouse G.A. & Wessel C.J. (1954) *Deterioration of Materials*. Reinhold, New York.
Grimes D.J., Atwell R.W., Brayton P.R., Palmer L.M., Rollins D.M., Roszak D.B., Singleton F.L., Tamplin M.L. & Colwell R.R. (1986) The fate of enteric pathogenic bacteria in estuarine and marine environments. *Microbiol. Sci.* **3**, 324–329.
Guthrie R.K., Cherry D.S. & Ferebee R.N. (1974) A comparison of thermal loading effects on bacterial populations in polluted and non-polluted aquatic systems. *Wat. Res.* **8**, 143–148.
Guthrie R.K., Cherry D.S. & Singleton F.L. (1975) Effects of nitrate and phosphate concentration on natural aquatic bacterial populations. *Wat. Res. Bull.* **11**, 1131–1136.
Guthrie R.K., Cherry D.S. & Singleton F.L. (1978) Responses of heterotrophic bacterial populations to pH changes in coal ash effluent. *Wat. Res. Bull.* **14**, 803–808.
Guthrie R.K., Davis E.M., Cherry D.S. & Walton J.R. (1982) Impact of coal ash from electric power production on changes in water quality. *Wat. Res. Bull.* **18**, 135–138.
Haack T.K. & McFeters G.A. (1982) Nutritional relationships among microorganisms in an epilithic biofilm community. *Microb. Ecol.* **8**, 115–126.
Hamilton W.A. (1987) Biofilms: microbial interactions and metabolic activities. In Fletcher M., Gray T.R.G. & Jones J.G. (eds) *Ecology of Microbial Communities, Symposia of the Society for General Microbiology 41*, pp. 361–387. Cambridge University Press, Cambridge.
Hejkal, T.W., Keswick B., LaBelle R., Gerba C., Sanchez Y., Dreesman G., Hafkin B. & Melnick J.K. (1982) Viruses in a community water supply associated with an outbreak of gastroenteritis and infectious hepatitis. *J. Am. Water Works Assoc.* **74**, 318–321.
Herro H.M. (1991) Tubercle formation and growth on ferrous alloys. *CORROSION/91*, Paper No. 84. Natl Assoc. Corrosion Eng., Houston.
Howells G.D. (1983) The effects of power station cooling water discharges on aquatic ecology. *Wat. Pollut. Control* **82**, 10–17.
Hueck van der Plas (1968) The microbiological deterioration of porous building materials. *Int. Biodeter. Bull.* **4**, 11–28.
Hussong D., Colwell R.R., O'Brien M., Weiss E., Pearson A.D., Weiner R.M. & Burge W.D. (1987) Viable *Legionella pneumophila* not detectable by culture on agar

media. *Bio/Technology* **5**, 947–950.

Iverson W.P. (1974) Microbial corrosion of iron. In Neilands J.B. (ed.) *Microbial Iron Metabolism*, pp. 475–512. Academic Press, New York.

Jansson P.-E., Kenne L. & Lindberg B. (1975) Structure of the extracellular polysaccharide from *Xanthomonas campestris*. *Carbohydrate Research* **45**, 275–282.

Jones J.G. (1986) Iron transformations by freshwater bacteria. *Adv. Microb. Ecol.* **9**, 149–185.

Kaplan L.A. & Bott T.L. (1982) Diel fluctuations of DOC generated by algae in a piedmont stream. *Limnol. Oceanogr.* **27**, 1091–1100.

Keevil C.W., Mackerness C.W. & Colbourne J.S. (1990) Biocide treatment of biofilms. *Int. Biodeter.* **26**, 169–179.

Keswick B.D., Gerba C.P., DuPont H.L. & Rose J.B. (1984) Detection of enteric viruses in treated drinking water. *Appl. Environ. Microbiol.* **47**, 1290–1294.

Kobrin G. (1976) Corrosion by microbiological organisms in natural waters. *Mat. Perf.* **15**, 38–43.

Kogure K., Simidu, U. & Taga N. (1979) A tentative direct microscopic method for counting living marine bacteria. *Can. J. Microbiol.* **25**, 415–420.

Larrick S.R., Clark J.R., Cherry D.S. & Cairns J. (1981) Structural and functional changes of aquatic heterotrophic bacteria to thermal, heavy, and fly ash effluents. *Wat. Res.* **15**, 875–880.

LeChevallier M.W., Babcock T.M. & Lee R.G. (1987) Examination and characterization of distribution system biofilms. *Appl. Environ. Microbiol.* **53**, 2714–2724.

LeChevallier M.W., Schulz W. & Lee R.G. (1991) Bacterial nutrients in drinking water. *Appl. Environ. Microbiol.* **57**, 857–862.

Lester J.N., Sterrett R.M., Rudd T. & Brown M.J. (1984) Assessment of the role of bacterial extracellular polymers in controlling metal removal in biological waste water treatment. In Grainger J.M. & Lynch J.M. (eds) *Microbiological Methods for Environmental Biotechnology*, pp. 197–217. Academic Press, London.

Levy R.V. (1990) Invertebrates and associated bacteria in drinking water distribution lines. In McFeters G.A. (ed.) *Drinking Water Microbiology*, pp. 224–248. Springer Verlag, New York.

Licina G.J. (1990) Microbiologically influenced corrosion in nuclear power plants – a review of susceptibility and mitigation methods. *CORROSION/90*, Paper No. 527. Natl Assoc. Corrosion Eng., Houston.

Lijklema L., Meuleman A.F.M. & Bosgoed H.M.M. (1990) Fluxes of oxygen equivalents and nutrients across the sediment–water interface after combined sewer overfows. *Wat. Sci. Tech.* **22**, 111–118.

Lock M.A., Ford T.E., Hullar M.A.J., Kaufman M., Vestal J.R., Volk G.S. & Ventullo R.M. (1990) Phosphorus limitation in an arctic river biofilm – a whole ecosystem experiment. *Wat. Res.* **24**, 1545–1549.

Maki J.S., LaCroix S.J., Hopkins B.S. & Staley J.T. (1986) Recovery and diversity of heterotrophic bacteria from chlorinated drinking waters. *Appl. Environ. Microbiol.* **51**, 1047–1055.

Mansfeld F., Shih H., Postyn A., Devinny J., Islander R. & Chen C.L. (1990) Corrosion monitoring and control in concrete sewer pipes. *CORROSION/90*, Paper No. 113. Natl Assoc. Corrosion Eng., Houston.

Marshall K.C. (ed.) (1984) *Microbial Adhesion and Aggregation*. Dahlem Konferenzen, Springer Verlag, New York.

McFeters G.A. (ed.) (1990a) *Drinking Water Microbiology*. Springer Verlag, New York.

McFeters G.A. (1990b) Enumeration, occurrence, and significance of injured indicator bacteria in drinking water. In McFeters G.A. (ed.) *Drinking Water Microbiology*, pp. 478–492. Springer Verlag, New York.

Miller R.V. & Levy S.B. (1989) Horizontal gene transfer in relation to environmental release of genetically engineered microorganisms. In Levy S.B. & Miller R.V. (eds) *Gene Transfer in the Environment*, pp. 405–420. McGraw-Hill, New York.

Mitchell R. (ed.) (1991) *New Concepts in Environmental Microbiology*. John Wiley & Sons, New York.

Nickels J.S., Bobbie R.J., Lott D.F., Martz R.F., Benson P.H. & White D.C. (1981) Effect of manual brush cleaning on biomass and community structure of microfouling film formed on aluminum and titanium surfaces exposed to rapidly flowing seawater. *Appl. Environ. Microbiol.* **41**, 1442–1453.

Niemi R.M. & Niemi J.S. (1991) Bacterial pollution of waters in pristine and agricultural lands. *J. Environ. Qual.* **20**, 620–627.

Parker C.D. (1945a) The corrosion of concrete. 1. The isolation of a species of bacterium associated with the corrosion of concrete exposed to atmospheres containing hydrogen sulphide. *Aust. J. Exp. Biol. Med. Sci.* **23**, 81–90.

Parker C.D. (1945b) The corrosion of concrete. 2. The function of *Thiobacillus concretivorus* (nov. spec.) in the corrosion of concrete exposed to atmospheres containing hydrogen sulphide. *Aust. J. Exp. Biol. Med. Sci.* **23**, 91–98.

Payment P., Trudel M. & Plante R. (1985) Elimination of viruses and indicator bacteria at each step of treatment during preparation of drinking water at seven water treatment plants. *Appl. Environ. Microbiol.* **49**, 1418–1428.

Payment P., Richardson L., Siemiatycki J., Dewar R., Edwardes M. & Franco E. (1991) A randomized trial to evaluate the risk of gastrointestinal disease due to consumption of drinking water meeting current microbiological standards. *Am. J. Public Health* **81**, 703–708.

Payne J.A. & Hedges P.D. (1990) An evaluation of the impacts of discharges from surface water sewer outfalls. *Wat. Sci. Tech.* **22**, 127–135.

Pennak R.W. (1989) *Fresh-Water Invertebrates of the United States*, 3rd edn. John Wiley & Sons, New York.

Perrin C.J., Bothwell M.L. & Slaney P.A. (1987) Experimental enrichment of a coastal stream in British Columbia: effects of organic and inorganic addition on atrophic periphyton production. *Can. J. Fish. Aquat. Sci.* **44**, 1247–1256.

Peterson B.J., Hobbie J.E., Hershey A.E., Lock M.A., Ford T.E., Vestal J.R., McKinley V.L., Hullar M.A.J., Miller M.C., Ventullo R.M. & Volk G.S. (1985) Transformation of a tundra river from heterotrophy to autotrophy by addition of phosphorus. *Science* **229**, 1383–1386.

Pope D.H., Duquette D.J., Johannes A.H. & Wayner P.C. (1983) Microbiologically-influenced corrosion of industrial alloys. *CORROSION/83*, Paper No. 247, Natl Assoc. Corrosion Eng., Houston.

Raloff J. (1985) The bugs of rust. *Science News* **128**, 42–44.

Reasoner D.J. & Geldreich E.E. (1985) A new medium for the enumeration and subculture of bacteria from potable water. *Appl. Environ. Microbiol.* **49**, 1–7.

Reasoner D.J., Blannon J.C., Geldreich E.E. & Barnick J. (1989) Nonphotosynthetic pigmented bacteria in a potable water treatment and distribution system. *Appl. Environ. Microbiol.* **55**, 912–921.

Rice R.G., Evison L.M. & Robson C.M. (1981) Ozone disinfection of municipal wastewater – current state-of-the-art. *Ozone Sci. Eng.* **3**, 239–272.

Ridgway H.F. & Olson B.H. (1981) Scanning electron microscope evidence for bacterial colonization of a drinking water distribution system. *Appl. Environ. Microbiol.* **41**, 274–287.

Ridgway H.F., Means E.G. & Olson B.H. (1981) Iron bacteria in drinking-water distribution systems: elemental analysis of *Gallionella* stalks, using X-ray energy-dispersive microanalysis. *Appl. Environ. Microbiol.* **41**, 288–297.

Rigdon J.H. & Beardsley C.W. (1958) Corrosion of concrete by autotrophs. *Corrosion* **14**, 206–208.

Rose J.B. (1990) Occurrence and control of *Cryptosporidium* in drinking water. In McFeters G.A., (ed.) *Drinking Water Microbiology*, pp. 294–321. Springer Verlag, New York.

Rose J.B. & Gerba C.P. (1986) A review of viruses in treated drinking water. *Curr. Pract. Environ. Sci. Engin.* **2**, 119–141.

Rose J.B., Gerba C.P. & Jakubowski W. (1991a) Survey of potable water supplies for *Cryptosporidium* and *Giardia*. *Environ. Sci. Technol.* **25**, 1393–1400.

Rose J.B., Sun G-S., Gerba C.P. & Sinclair N.A. (1991b) Microbial quality and persistence of enteric pathogens in graywater from various household sources. *Wat. Res.* **25**, 37–42.

Sand W. & Bock E. (1984) Concrete corrosion in the Hamburg sewer system. *Environ. Technol. Lett.* **5**, 517–528.

Savage D.C. & Fletcher M. (eds) (1985) *Bacterial Adhesion.* Plenum, New York.

Schindler D.W. (1985) The coupling of elemental cycles by organisms: evidence from whole-lake chemical perturbations. In Stumm W. (ed.) *Chemical Processes in Lakes*, pp. 225–250. John Wiley & Sons, New York.

Schinner F. & Burgstaller W. (1989) Extraction of zinc from industrial waste by a *Penicillium* sp. *Appl. Environ. Microbiol.* **55**, 1153–1156.

Schmitt C.R. (1986) Anomalous microbiological tuberculation and aluminum pitting corrosion – case histories. In Dexter S.C. (ed.) *Biologically Induced Corrosion*, pp. 69–75. Natl Assoc. Corrosion Eng., Houston.

Singh A. & McFeters G.A. (1986) Recovery, growth, and production of heat-stable enterotoxin by *Escherichia coli* after copper-induced injury. *Appl. Environ. Microbiol.* **51**, 738–742.

Singh A. & McFeters G.A. (1987) Survival and virulence of copper- and chlorine-stressed *Yersinia enterocolitica* in experimentally infected mice. *Appl. Environ. Microbiol.* **53**, 1768–1774.

Singh A. & McFeters G.A. (1990) Injury of enteropathogenic bacteria in drinking water. In McFeters G.A. (ed.) *Drinking Water Microbiology*, pp. 368–379. Springer Verlag, New York.

Sly L.I., Arunpairojana V. & Hodgkinson M.C. (1988a) *Pedomicrobium manganicum* from drinking-water distribution systems with manganese-related "dirty water" problems. *Syst. Appl. Microbiol.* **11**, 75–84.

Sly L.I., Hodgkinson M.C. & Arunpairojana V. (1988b) Effect of water velocity on the early development of manganese depositing biofilm in a drinking-water distribution system. *FEMS Microbiol. Ecol.* **53**, 175–186.

Sly L.I., Hodgkinson M.C. & Arunpairojana V. (1990) Deposition of manganese in a drinking water distribution system. *Appl. Environ. Microbiol.* **56**, 628–639.

States S.J., Wadowsky R.M., Kuchta J.M., Wolford R.S., Conley L.F. & Yee R.B. (1990) *Legionella* in drinking water. In McFeters G.A. (ed.) *Drinking Water Microbiology*, pp. 340–367. Springer Verlag, New York.

Stetler R.E., Ward R.L. & Waltrip S.C. (1984) Enteric viruses and indicator bacteria levels in a water treatment system modified to reduce trihalomethane production. *Appl. Environ. Microbiol.* **47**, 319–324.

Stout J.E., Yu V.L. & Best M.G. (1985) Ecology of *Legionella pneumophila* within water distribution systems. *Appl. Environ. Microbiol.* **49**, 221–228.

Stout J., Yu V.L., Vickers R.M., Zuravleff J., Best M., Brown A., Yee R.B. & Wadowsky R. (1982) Ubiquitousness of *Legionella pneumophila* in the water supply of a hospital with endemic legionnaires' desease. *N. Engl. J. Med.* **306**, 466–468.

Subramanian B., Prabu S.K. & Mahadevan A. (1990) Influence of thermal power station effluents on hydrobiology of seawater. *Wat. Air Soil Poll.* **53**, 131–137.

Sutherland I.W. (1972) Bacterial exopolysaccharides. *Adv. Microbiol. Physiol.* **8**, 143–213.

Sutherland I.W. (1982) Biosynthesis of microbial exopolysaccharides. *Adv. Microbiol. Physiol.* **23**, 79–150.

Sutherland I.W. (1985) Biosynthesis and composition of Gram-negative bacterial extracellular and wall polysaccharides. *Ann. Rev. Microbiol.* **39**, 243–270.

Tatnall R.E. (1981) Case histories: bacteria-induced corrosion. *Mat. Perf.* **20**, 41–48.

Tison D.L., Pope D.H., Cherry W.B. & Fliermans C.B. (1980) Growth of *Legionella pneumophila* in association with blue-green algae (cyanobacteria). *Appl. Environ. Microbiol.* 39, 456–459.
Tuovinen O.H. & Hsu J.C. (1982) Aerobic and anaerobic microorganisms in tubercles of the Columbus, Ohio, water distribution system. *Appl. Environ. Microbiol.* **44**, 761–764.
Tuovinen O.H., Button K.S., Vuorinen A., Carlson L., Mair D.M. & Yut L.A. (1980) Bacterial, chemical, and mineralogical characteristics of tubercles in distribution pipelines. *J. Am. Water Works Assoc.* 72, 626–635.
Tyler P.A. & Marshall K.C. (1967a) Hyphomicrobia – a significant factor in manganese problems. *J. Am. Water Works Assoc.* 59, 1043–1048.
Tyler P.A. & Marshall K.C. (1967b) Microbial oxidation of manganese in hydrooelectric pipelines. *Antonie van Leeuwenhoek* 33, 171–183.
Wadowsky R.M. & Yee R.B. (1983) Satellite growth of *Legionella pneumophila* with an environmental isolate of *Flavobacterium breve. Appl. Environ. Microbiol.* **46**, 1447–1449.
Wadowsky R.M., Wilson T.M., Kapp N.J., West A.J., Kuchta J.M., States S.J., Dowling J.N. & Yee R.B. (1991) Multiplication of *Legionella* spp. in tap water containing *Hartmannella vermiformis. Appl. Environ. Microbiol.* 57, 1950–1955.
Wolfe R.L., Stewart M.H., Liang S. & McGuire M.J. (1989) Disinfection of model indicator organisms in a drinking water pilot plant by using peroxone. *Appl. Environ. Microbiol.* 55, 2230–2241.
Wolfe R.L., Lieu N.I., Izaguirre G. & Means E.G. (1990) Ammonia-oxidizing bacteria in a chloraminated distribution system: seasonal occurrence, distribution, and disinfection resistance. *Appl. Environ. Microbiol.* 56, 451–462.

19

Robert V. Miller

Genetic stability of genetically engineered microorganisms in the aquatic environment

19.1 Introduction

Originally, genetically engineered microorganisms (GEMs) were envisioned to be useful in the medical and industrial disciplines where the GEM would be used to produce a specified chemical product which would then be isolated and used in a purified form (Curtiss 1976; Kirk & Farrell 1987; Timberlake & Marshall 1989). Some success has been achieved in these areas, however, unique safety and risk concerns became evident when these products were proposed for use as drugs or food additives. These concerns usually center around assuring the purity and efficacy of the product.

Recently, the great potential of biotechnology for use in agriculture and for environmental applications has been realized. Genetically modified microbes are being developed to enhance waste-treatment technologies, bioremediate environmental pollutants, to serve as biological pest control agents and fertilizers, for use in regulating corrosion and biofouling in cooling systems and other liquid transport and delivery systems, and for other environmental applications (Ghosal *et al.* 1985; Monticello & Finnerty 1985; Chakrabarty 1986; Schoenen 1986; Forster & Wase 1987; Pooley 1987; Timmis 1987; Timmis *et al.* 1988; Lindow *et al.* 1989; Geesey & White 1990; Sayler *et al.* 1990; Stock *et al.* 1990). In most of these applications, it is the GEM itself that is the desired biotechnological product. Organisms with unique metabolic and physiological potential are applied to a polluted site or crop field to achieve the desired result. Such environmental releases require consideration of additional factors in determining the potential risk associated with use of these GEMs. One of the most important of these is the assessment of the fate of not only the GEM but its engineered genetic sequence as well. The potential for genetic instability is one of the most important factors which must be taken into account in assessing the fate of an engineered sequence released into the environment (Sayre & Miller 1991).

Questions concerning genetic instability can be divided into two general categories (Miller 1988). First are questions concerning the probability of genetic alteration of the engineered sequence while residing in the GEM (intracellular instability). Intracellular alterations of the sequence may lead to an increased or decreased fitness of the host GEM, or may lead to alterations in the level of expression of the product of the engineered sequence. Second are questions which address the potential for horizontal gene transfer among organisms within the environmental population (intercellular instability). Horizontal gene transfer may lead to stabilization of the engineered sequence in the environment even after the original host GEM has been eliminated.

As GEMs released into any environment have the potential of entering aquatic habitats through runoff or transport through groundwater systems, an understanding of the potential for gene transfer and genetic alteration in all environments is fundamental to the assessment of the genetic stability of aquatic populations of bacteria. In this chapter, I will summarize the data which is currently available to us for assessment of both intercellular and intracellular genetic stability in natural populations of microorganisms with an emphasis on aquatic habitats.

19.2 Horizontal gene transfer among bacteria in natural aquatic environments

There are three and possibly four recognized methods of horizontal gene transfer in bacteria. The processes of transformation, conjugation, and transduction have been recognized since the early 1950s and are well described under laboratory conditions (Levy & Miller 1989). Recently, outer-membrane derived vesicles containing DNA have been detected in various Gram-negative bacteria (Dorward & Garon 1990). Such export of DNA within nuclease-resistant vesicles may also allow for gene exchange to occur via membrane fusion. In addition, fusion of hyphae leading to genetic exchange has been recognized in the streptomycetes for several years (Rafii & Crawford 1989).

19.2.1 Transformation

Natural bacterial transformation (Stewart 1989) is a mechanism of horizontal gene transfer in which living bacterial cells acquire new genetic elements directly from their external milieu (Figure 19.1). Given the appropriate environmental stress factors, many bacterial species can develop "natural competence" (Table 19.1), which is a physiological state which allows the absorption of genetic material (DNA) through the cellular envelope. Even though it was the first mechanism of gene transfer discovered in bacteria (Griffith 1928;

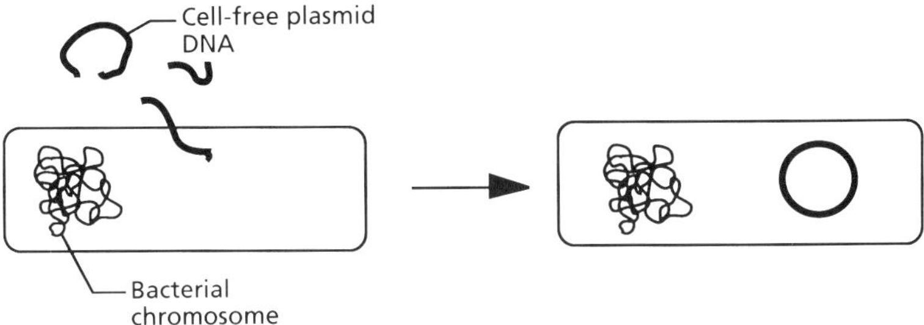

Figure 19.1 Transformation. In natural transformation, cell-free DNA is absorbed through the cell envelope of a "competent" cell. See Stewart (1989) for a complete description of natural transformation.

Table 19.1 A partial list of bacterial genera which demonstrate natural competence. Adapted from Stewart (1989)

Achromobacter	*Halobacterium*	*Pseudomonas*
Acinetobacter	*Methylobacterium*	*Streptococcus*
Anacystis	*Micrococcus*	*Streptomyces*
Azotobacter	*Mycobacterium*	*Synechococcus*
Bacillus	*Moraxella*	*Vibrio*
Haemophilus	*Neisseria*	

Table 19.2 Ecosystems in which transformation has been observed

Environment	**Representative references**
Soil and terrestrial environments	Graham & Istock 1978; Kelley & Reanney 1984; Lorenz *et al.* 1988; Lorenz & Wackernagel 1990
Mammals	Griffith 1928; Avery *et al.* 1944
Marine (estuarine)	Jeffrey *et al.* 1989; Stewart & Sinigalliano 1990

Avery *et al.* 1944), transformation has been observed in only a few environmental settings (Table 19.2). This may be due to the fact that it was generally assumed that DNA would not survive in the environment. However, it has recently been realized that adsorption of DNA to clays and other environmental particulates protect it from degradation (Lorenz *et al.* 1988; Lorenz & Wackernagel 1990). Sand-adsorbed DNA has been shown to be active in transformation (Lorenz & Wackernagel 1990). These data increase the probability that transformation may be a significant method of gene transfer in a variety of natural ecosystems.

The potential of estuarine and freshwater microbial populations to exchange genetic material via transformation has been investigated in some detail (Paul *et al.* 1988a,b; 1989; 1991; Paul & Pichard 1989; Jeffrey *et al.* 1989; Stewart & Sinigalliano 1990; Frischer *et al.* 1990). Seasonal and diel variability in the concentration of dissolved DNA in these environments was found to correlate better with excretion and lysis of cells than with cell growth (Paul *et al.* 1988b). In other words, DNA is released as microbial populations die off. Not surprisingly, the majority of the nucleotides contained in this extracellular DNA are salvaged by estuarine microbial populations for use in nucleic acid synthesis (Paul *et al.* 1988a). Paul *et al.* (1989) measured rapid turnover (~12 hours) of extracellular DNA in subtropical eutrophic and oligotrophic freshwater environments.

Aquatic microbial populations were found to contain specific uptake-binding mechanisms for compounds containing phosphodiester bonds (Paul & Pichard 1989). These mechanisms were shown to be capable of recognizing oligonucleotides as short as dinucleotides. However, in addition, nucleosides, phosphomonoesters, and inorganic phosphate were transported by this mechanism. Paul & Pichard (1989) suggest that nucleic acid-binding mechanisms of this type have evolved in marine and freshwater microorganisms to allow the utilization of extracellular DNA and, perhaps, RNA (which are abundant in these environments), for nucleic acid synthesis. These results suggest that engineered DNA sequences released from the lysis of GEMs might not exist long enough to be transformed intact into other microorganisms.

Even though the majority of nucleotides contained in extracellular DNA may be utilized as building blocks for intracellular metabolism, recent studies have indicated that the absorption of DNA by various marine bacterial species can lead to the stable inheritance of new genetic traits (Jeffrey *et al.* 1989; Stewart & Sinigalliano 1990; Paul *et al.* 1991). Jeffrey *et al.* (1989) isolated a naturally transformable *Vibrio* sp. (designated DI-9) from Tampa Bay, Florida. They found that this bacterium absorbed and stably established the DNA of the broad-host-range plasmid pKT230. This plasmid, which cannot be transferred by conjugation (Bagdasarian *et al.* 1981), was transformed into *Vibrio* sp. DI-9 on both filters and in sterile sediment microcosms. The rate of transformation was enhanced when the exogenous DNA was present as multimeric forms of the plasmid. Transformation frequencies of DI-9 ranged from 3×10^{-9} to 3×10^{-7} transformants per recipient. Growth of plasmid-containing transformants of this strain in nonselective medium resulted in the isolation of cured variants which transformed at significantly higher frequencies ranging from 10^{-8} to 10^{-4} transformants per recipient (Frischer *et al.* 1990).

Recently, Paul *et al.* (1991) utilized one of these high-frequency of transformation (HfT) variants of DI-9 to measure transformation in marine water and sediment microcosms. Using broad-host-range plasmid multimers as the source of transforming DNA, these investigators were able to observe transformation frequencies of 10^{-10}–10^{-6} transformants per recipient in water columns. These results were obtained in both sterile microcosms and those containing the natural microbial community. While these transfer frequencies are low, they were above the spontaneous mutation rates for the selected phenotypes. The authors also confirmed the presence of the plasmid DNA by colony hybridization of the presumptive transformants. Transformation rates in the 10^{-10}–10^{-6} range were also observed in sterile sediment microcosms, but no transformation could be detected when the native community was present in sediment. Stewart & Sinigalliano (1990), using *Pseudomonas stutzeri*, observed transformation of chromosomally encoded rifampicin

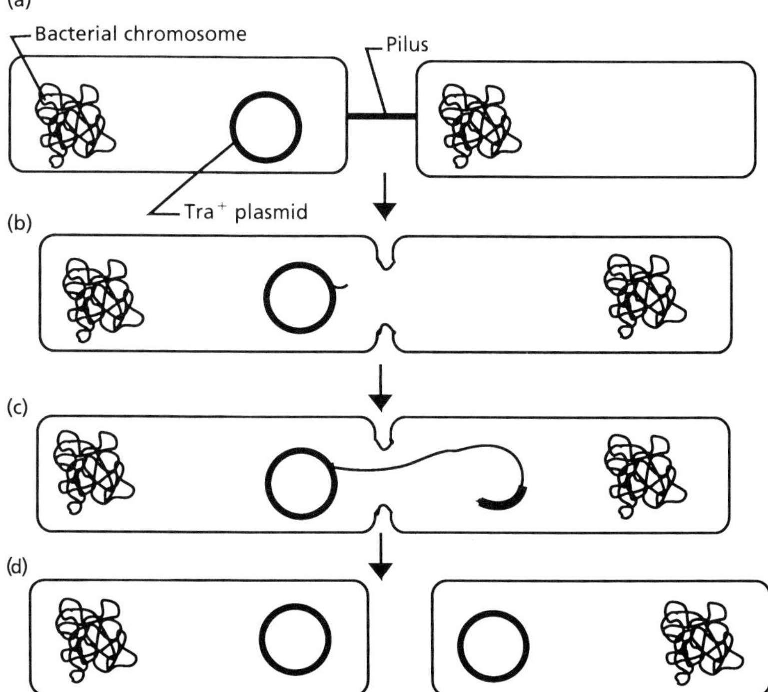

Figure 19.2 Conjugal transfer of plasmid DNA. (a) Initial contact is made between a plasmid donor and recipient cell through interaction between the pilus of the donor cell and the cell surface of the recipient cell. (b) This is followed by fusion of the two cells and the formation of a cytoplasmic bridge. (c) Rolling-circle replication of the plasmid is induced and transfer of a linear copy of the plasmid through the cytoplasmic bridge ensues. (d) After transfer is complete, the cells dissociate, the transferred plasmid circularizes, and genes carried on the plasmid are expressed. The recipient cell is now capable of acting as a donor in subsequent matings. See Ippen-Ihler (1989) for complete discussion. Adapted from Miller (1988).

Table 19.3 A partial list of bacterial genera with known conjugal plasmids. Adapted from McIntire (1991)

Aeromonas	*Clostridium*	*Rhizobium*
Acetobacter	*Corynebacterium*	*Rhodococcus*
Acinetobacter	*Enterobacter*	*Serratia*
Agrobacterium	*Erwinia*	*Staphylococcus*
Alcaligenes	*Haemophilus*	*Streptococcus*
Arthrobacter	*Klebsiella*	*Streptomyces*
Bacillus	*Neisseria*	*Vibrio*
Bacteroides	*Nocardia*	*Yersinia*
Campylobacter	*Proteus*	
Citrobacter	*Pseudomonas*	

resistance in marine sediments containing the autochthonous microbial community.

19.2.2 Conjugation

Conjugation is a parasexual process which allows unidirectional transfer of DNA from a donor cell to a recipient cell of the same or different species (Ippen-Ihler 1989). DNA is transferred through a cytoplasmic bridge from the donor to the recipient cell (Figure 19.2). In most cases, extrachromosomal (i.e., plasmid) DNA is transferred. However, under the appropriate conditions, chromosomal DNA may also be transferred (Ippen-Ihler 1989). Interaction between a donor and recipient cell stimulates a specific kind of DNA replication, termed transfer replication, in which a copy of the plasmid is made in the donor cell for transfer to the recipient cell (Willetts & Skurray 1980). The process results in both the donor and recipient cells containing the extrachromosomal DNA element or the transferred chromosomal DNA. Conjugation has been shown to occur (at least under laboratory conditions) between numerous species of bacteria, many of which are of environmental importance (Table 19.3). In addition, conjugation has been shown to support both trans- and intrakingdom transport of genetic material (Table 19.4) and has been demonstrated in a number of natural ecosystems (Table 19.5).

Conjugation is perhaps the best studied form of horizontal gene transfer in aquatic environments. Significant work has been done in waste treatment facilities as they are potentially ideal for gene transmission (Saye & Miller 1989). They provide a nutrient-rich and protected environment containing high concentrations of particulates (Gealt 1991). The concentrations of bacteria in these facilities are high ($>10^6$ colony forming units (CFU) per milliliter) and can range as high as 10^9 in activated sludge (Blackburn *et al.* 1987; Gealt 1991). Many of these bacteria have been shown to contain conjugative plasmids (Sturtevant & Feary 1969; Grabow *et al.* 1975;

Table 19.4 Conjugal trans- and intrakingdom transfer of genetic material. Adapted from Hirsch (1990)

Donor group	Recipient group
Purple bacteria	Cyanobacteria
Purple bacteria	Flavobacteria
Purple bacteria	Plants
Purple bacteria	Fungi
Purple bacteria	Gram-positive bacteria
Gram-positive bacteria	Purple bacteria

Table 19.5 Ecosystems in which conjugation has been demonstrated

Environment	Representative references
Soil	Stotzky 1989; Klingmüller *et al.* 1990; Mergeay *et al.* 1990; Wellington *et al.* 1990
Rhizosphere	Lynch 1990; van Elsas & Trevors 1990
Plants	Farrand 1989
Fresh water	Gowland & Slater 1984; O'Morchoe *et al.* 1988; Saye & Miller 1989; Bauda *et al.* 1990
Epilithon	Bale *et al.* 1987; 1988b; Fry & Day 1990
Marine	Saye & Miller 1989; Gauthier & Breittmayer 1990
Wastewater treatment facilities	Mach & Grimes 1982; Gealt *et al.* 1985; Mancini *et al.* 1987; Bauda *et al.* 1990
Activated sludge	McClure *et al.* 1990
Insects	Jarrett & Stephenson 1990
Mammals	Roberts 1989

Mach & Grimes 1982; Bell *et al.* 1983; Alcaide & Garay 1984; Gealt *et al.* 1985; McPherson & Gealt 1986; Mancini *et al.* 1987; Trevors *et al.* 1987). In general, these studies have addressed the occurrence and transmission of antibiotic-resistance plasmids in the waste treatment environment. In most cases, it has been assumed that conjugation was the mechanism of genetic exchange leading to the epidemiological characteristics of plasmid dissemination (Saye & Miller 1989). Unfortunately, in the majority of studies the conjugal nature of the transfer was not confirmed. Where an attempt has been made to confirm conjugation, it has sometimes been unsuccessful. An interesting example was documented by Hedges & Jacoby (1980). They found that a plasmid, Rms149, which had originally been classified as transfer proficient, could not be shown to be transferred by conjugation. Additional investigation revealed that the plasmid was transferred from the original environmental isolate by transduction mediated by a coresident prophage (see below).

Recently, several studies have addressed the potential for conjugal transfer in the waste-treatment environment directly. As early as 1982, Altherr & Kasweck (1982) used membrane diffusion chambers immersed in the degritter tank of a sewage-treatment plant to measure gene transfer. They introduced an antibiotic-resistant strain of *Escherichia coli* isolated from the facility to measure transfer of an antibiotic-resistance plasmid (R-plasmid) to an antibiotic-sensitive environmental *E. coli* isolate. In this *in situ* experiment, transfer was observed at significant rates (10^{-6}–10^{-5}) which were only approximately five-fold lower than the rates obtained in laboratory matings between the two bacteria.

In the same year, Mach & Grimes (1982) measured R-plasmid transfer in diffusion chambers incubated in the primary and secondary clarifiers of a wastewater treatment facility. They observed both intra- and interspecies transfer. The transfer rates observed *in situ* were 100-fold lower than those obtained in the laboratory.

Gealt *et al.* (1985) examined the potential for mobilization of transfer-deficient (Tra^-) plasmids by coresident transfer-proficient (Tra^+) plasmids in sterile waste water. They used the Tra^+ plasmid R100-1 to transfer the Tra^- plasmids pBR322 and pBR325 from an *E. coli* strain to *E. coli* and *Enterobacter cloacae* strains isolated from raw waste water. Transconjugants were observed in both matings although the rates of transfer were several orders of magnitude lower than were observed in mating in rich medium (Luria Broth) in the laboratory.

McPherson & Gealt (1986) expanded the study by searching for naturally occurring mobilizing plasmids in wastewater bacteria. They obtained isolates of *E. coli* and *Klebsiella pneumoniae* that contained high molecular weight R-plasmids with the ability to mobilize pBR325. These plasmids were capable of participating in triparental matings (Saye & Miller 1989) which allowed transfer of pBR325 from a laboratory strain of *E. coli* to recipients of either laboratory or wastewater origin.

Triparental mobilization of the Tra^- plasmid pHSV106 containing the herpes simplex thymidine kinase (*tk*) gene has been evaluated in a laboratory-scale, waste treatment facility (Mancini *et al.* 1987; Gealt *et al.* 1988; Gealt 1991). The recombinationally deficient (*recA*) strain HB101 (Miller & Kokjohn 1990) was used as the primary donor, and both laboratory and wastewater isolates were used as mobilizers and recipients in the triparental matings. Significant numbers of transconjugants were found throughout the waste treatment facility with the majority recovered from near the bottom of the primary clarifier (Gealt 1991). These findings suggest that a quiescent setting containing significant levels of particulate surfaces is conducive to horizontal gene transfer via conjugation.

Various parameters have been tested to determine if they affect the frequency of conjugal transfer in these environments (Khalil & Gealt 1987; Mancini *et al.* 1987; Gealt 1991). Nutrient composition and concentration, influent flow rate, and origin of the parental strains did not alter conjugal frequency. However, other factors, such as cationic concentration and composition, extremes of temperature and pH, and presence of detergent, did influence the frequency of transfer.

Conjugation has also been extensively investigated in freshwater environments. Many of these studies have been epidemiological in that the frequency and distribution of R-factors present in these environments have been enumerated without regard for the actual mechanism of transfer (for review see Saye & Miller 1989). Several recent studies have investigated the potential for conjugal transfer in freshwater environments (Gowland & Slater 1984; Bale *et al.* 1987; 1988a,b; O'Morchoe *et al.* 1988). Gowland & Slater (1984) demonstrated conjugal transfer between *E. coli* strains incubated in environmental chambers filled with sterilized pond water and submerged for 2–3 weeks in a freshwater pond. Transconjugants were detected at frequencies of 10^{-7}–10^{-6} transconjugants per donor after 192–360 hr of incubation. These levels were several orders of magnitude lower that the rates observed in continuous-flow chambers incubated under laboratory conditions ($\sim 10^{-2}$).

O'Morchoe *et al.* (1988) studied the transfer of two laboratory plasmids between *Pseudomonas aeruginosa* donor and recipient strains in lake-water-filled chambers suspended in a freshwater lake. Transfer was observed in both the absence and the presence of the indigenous microbial population, although transfer frequencies were significantly reduced when the autochthonous community was present. In this study, the efficiency of transfer of a plasmid known to transfer with greater efficiency on surfaces (R68.45) was compared with the efficiency of transfer of a plasmid which is transferred with equal frequencies in solid and liquid media (FP5). Conjugal transfer rates were consistently 10-fold higher for R68.45 than for FP5 (Miller *et al.* 1990a). These results are consistent with the findings in wastewater treatment facilities and suggest that surfaces may be conducive to gene transfer in aquatic environments. However, when surfaces of the environmental containment chambers used in this study were sampled, no increase in the detectable levels of transfer were observed when compared to the liquid phase of the chamber milieu (O'Morchoe *et al.* 1988). During this study, several factors with the potential for affecting conjugal transfer were evaluated in laboratory simulations and in field studies. The most significant effect was observed when the donor/recipient ratio was altered. As much as 100-fold stimulation in conjugation frequency was obtained when the recipient cells were in excess, compared to experiments in which the donor population was the most abundant (O'Morchoe *et al.* 1988; Miller *et al.* 1990a).

Conjugal transfer of plasmids among *P. aeruginosa* introduced into the natural epilithon has been evaluated in a freshwater microcosm and *in situ* in a Welsh river (Bale *et al.* 1987; 1988a,b; Day *et al.* 1988; Fry & Day 1990). These experiments are important because they are the first demonstration of gene transfer in an unenclosed *in situ* experiment. The epilithon contains a community of organisms enclosed in a slimy layer which develops on the surfaces of stones in aquatic habitats (Fry & Day 1990; see also Chapter 6, this book). It is a particularly nutrient-rich environment in which bacterial concentrations often reach 10^6–10^8 cm^{-2}. For these studies, naturally occurring Tra^+ plasmids of *Pseudomonas* spp. were "captured" from the natural microbiota by mating them with laboratory recipients. Many of these natural Tra^+ plasmids carry determinants for heavy metal and ultraviolet (UV) radiation resistance (Bale *et al.* 1987; 1988b; Rochelle *et al.* 1989; Fry & Day 1990). Plasmids which could be transferred to *P. aeruginosa*, *P. putida*, and *P. fluorescens* were isolated in this fashion (Fry & Day 1990).

The conditions necessary for optimum transfer of these plasmids to *P. aeruginosa* were investigated (Rochelle *et al.* 1989). Factors including temperature, cell density, mating time, and nutrient availability were shown to affect transfer frequency (Rochelle *et al.* 1989; Fry & Day 1990). Consistent with the observations of O'Morchoe *et al.* (1988), as discussed above, one of the most significant effectors of transfer frequency was found to be donor/recipient ratio. Transfer was significant only when recipients were in excess (Rochelle *et al.* 1989; Fry & Day 1990).

The initial plasmid isolations from epilithic bacteria were conducted at 20°C, and the temperature optimum for their transfer was 20°C (Rochelle *et al.* 1989). Further studies were conducted to determine if the isolation temperature selected for plasmids with a corresponding temperature optimum for transfer. Isolation experiments were conducted during the winter when river-water temperature ranged from 7 to 10°C. Isolations were done at 10°C. In these experiments, plasmids with maximum transfer temperatures of 10°C were isolated, and gene transfer occurred well at temperatures as low as 5°C (Fry & Day 1990). These results suggest that plasmids exist in nature which have been adapted for transfer at various environmental temperatures, and that temperature should not be considered a limiting factor when predicting whether conjugal transfer will take place in a specific environment.

Bale *et al.* (1987; 1988b) utilized captured plasmids to measure gene transfer in the epilithon. Two methods were used *in situ* to demonstrate plasmid transfer. In the first, cultures of donor and recipient bacteria were mixed and deposited onto a membrane filter which was subsequently placed face down onto a flat, scrubbed, sterile stone and incubated in the river. In the second method (Bale *et al.* 1988a), donor and recipient bacteria were first deposited onto separate membrane filters and secured onto separate flat stones on which a natural epilithic community was growing. The stones were incubated in the river for 24 hours to allow the donor and recipient bacteria to be incorporated into the epilithon. The filters were then removed, and the stones placed together so the areas covered by the membranes were in contact. After further incubation *in situ*, plasmid transfer between donor and recipient epilithons was evaluated. Comparison of results obtained *in situ* and in microcosms indicated that rates of transfer were as high in the river as they were in the laboratory microcosm. Transfer was observed in river water at temperatures between 5 and 21°C. The rate of transfer increased linearly with temperature (Fry & Day 1990).

Comparatively little is known about conjugal gene transfer in marine environments (Saye & Miller 1989; Gauthier & Breittmayer 1990). Those studies which have been carried out have suggested that gene transfer may take place, but direct proof is still lacking. Stewart & Koditschek (1980) demonstrated transfer of R-plasmids in sterilized sea water and sediments collected from the New York Bight, in the vicinity of the New York City sewer discharge pipes. Transfer of tetracycline and streptomycin resistance to a sensitive strain of *E. coli* was measured in incubations carried out at 10°C. These experiments demonstrated that gene transfer could take place in saline water and at environmentally realistic temperatures. Unfortunately, the investigators did not confirm that transfer was via conjugation, and not one of the other mechanisms of horizontal gene transmission known to exist in bacteria.

Recently, Gauthier & Breittmayer (1990) were able to demonstrate the transfer of the broad-host-range plasmid RP4 between strains of

E. coli in sea-water and marine sediments. They found that transfer was not significantly affected by increasing salinity (≤37%) of the mating milieu. Thus it seems clear that the potential for gene transfer is not eliminated in terrestrial bacteria simply because they find themselves in a saline environment.

The potential for conjugal transfer among marine bacteria is not as clear. Sizemore & Colwell (1977) were able to isolate antibiotic-resistant strains of marine bacteria from harbor, inshore, and off-shore waters which harbored Tra^+ plasmids. Wortman & Colwell (1988) isolated bacteria from the digestive tract of deep-sea amphipods. They divided these isolates into two taxonomic groups identified as *Pseudomonas*-like and *Vibrio*-like. An identical 63 kD plasmid was isolated from both groups suggesting that interspecies transfer had taken place.

The potential for gene transfer between marine species and terrestrial bacteria in a marine environment has been demonstrated. Patt *et al.* (1972) reported that plasmid-encoded salicylate utilization could be transferred from a nonmarine *Pseudomonas* sp. to *Vibrio fisherii*. Transfer of antibiotic-resistance determinants from *E. coli* to several different marine species was also demonstrated in this study. Thus marine species can act as recipients of plasmid DNA transferred from nonmarine organisms.

Some evidence also exists that marine bacteria can act as donors in conjugal mating with terrestrial bacteria. Transfer of mercury-resistance plasmids from marine members of the families Pseudomonadaceae and Vibrionaceae to *E. coli* was demonstrated by Gauthier *et al.* (1985). Sizemore & Colwell (1977) reported the transfer of antibiotic-resistance determinants from several marine species including *Vibrio harveyi*, *V. parahaemolyticus*, *Pseudomonas bathycetes*, and *Alteromonas haloplanktis*. Unfortunately, these plasmids were not stably maintained in *E. coli* and transfer could not be confirmed.

It is clear from the studies carried out to date that the potential for gene transfer via conjugation exists in aquatic environments. This form of gene transfer must be considered in evaluating GEMs for use in these environments. Evidence is also accumulating that conjugal transfer is significant in soils (for review see Stotzky 1989). While the transport of microbes from terrestrial environments into aquatic habitats is not yet well documented, it is conceivable that genetically engineered sequences, introduced into terrestrial habitats could be incorporated into the gene pools of autochthonous aquatic organisms.

19.2.3 Transduction

In the third mechanism of transfer, transduction, DNA is packaged into a viral capsid which mediates its transfer to a recipient bacterial cell (Kokjohn 1989). When a temperate bacteriophage (i.e., bacterial virus) infects a cell, it has two possible fates. In one, the lytic response, the viral genetic information is expressed, first producing copies of the viral genome and then producing viral proteins. These proteins are assembled into protein shells called capsids. Viral-specific nucleic acid is then packaged into the capsid to form a virion. When a number of virions are assembled, the host cell is lysed, phage particles are released, and the cycle of infection and release is repeated. In the alternative life cycle, a temperate phage can establish lysogeny (lysogenic response). In this case, the expression of the viral genome is repressed and a quasi symbiotic relationship is established with the host. The viral genome (or prophage) is maintained in the host and replicated along with the host genome. Daughter copies of the prophage are segregated into the two daughter cells at cell division. Various environmental factors, including UV light, have the ability to activate lysogenic cells to the lytic response. Virulent phages can only carry out the lytic response.

During lytic growth, the viral packaging system will, on occasion, make a mistake and, instead of packaging a viral genome, package bacterial DNA into the viral capsid. Both plasmid and chromosomal DNA can be packaged. Since the infectious process is a characteristic of the protein coat (capsid) and not the DNA itself, this "transducing particle" can infect a new cell, introducing bacterial DNA (instead of viral DNA) into the recipient cell. This DNA can then be established in the cell, producing a transductant (Figure 19.3).

Transduction may be the most widespread mechanism of gene transfer among bacteria. The phages of numerous species of bacteria have been shown to be capable of carrying out transduction (Table 19.6), and it is likely that most viruses (both temperate and virulent) mediate the process under the appropriate conditions (Miller *et al.* 1990b; Kokjohn *et al.* 1991). A number of phages capable of interspecies transfer have been identified (Miller 1991a).

Because of the specificity of the host–parasite interaction explicit in the process, many observers have felt that transduction could not be a viable form of gene transfer in the environment. Nevertheless, it has been observed in a number of habitats (Table 19.7). The recent observation of the high concentrations of bacteriophage particles in many aquatic environments (often as high as 10^{11} particles per milliliter) has spawned new interest in the role of bacterial viruses in the ecology of natural microbial communities (Ewert & Paynter 1980; Bergh *et al.* 1989; Børsheim *et al.* 1990; Bratbak *et al.* 1990; Ogunseitan *et al.* 1990; Proctor & Fuhrman 1990). Not only does

Table 19.6 A partial list of bacterial genera with known transducing phages. Adapted from Kokjohn & Miller (1991)

Achromobacter	*Proteus*
Acidiphilium	*Providencia*
Acinetobacter	*Pseudomonas*
Bacillus	*Rhizobium*
Caulobacter	*Rodococcus*
Citrobacter	*Salmonella*
Corynebacterium	*Serratia*
Desulfovibrio	*Shigella*
Enterobacter	*Staphylococcus*
Escherichia	*Streptococcus*
Klebsiella	*Streptomyces*
Myxococcus	*Vibrio*

gene transfer via transduction seem more probable in light of these observations, but it also seems probable that bacteriophages could be important in regulating microbial cell densities and nutrient cycling in these environments (Miller & Sayler 1991).

In situ demonstrations of transduction in aquatic environments

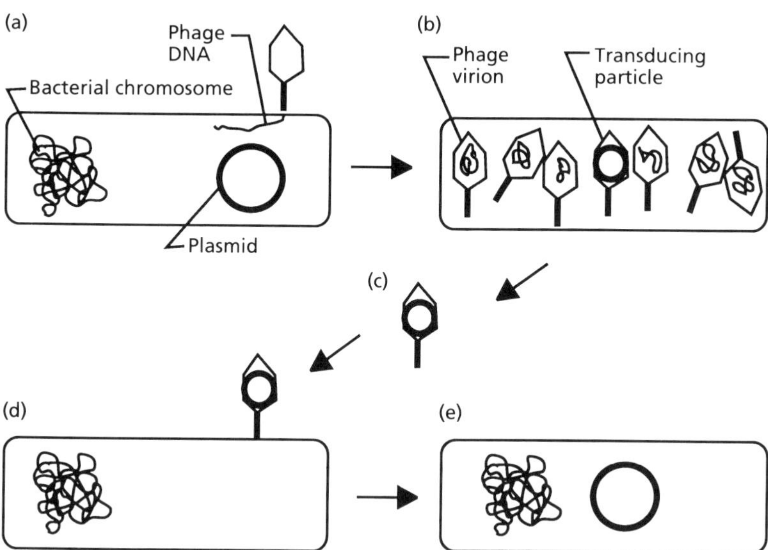

Figure 19.3 Transduction of a plasmid. (a) A generalized transducing phage infects a plasmid-containing cell. (b) The phage genome is expressed leading to synthesis of viral-specific proteins and DNA. The proteins are assembled into capsids and the phage DNA packaged into these capsids producing virions. Occasionally, bacterial DNA may be packaged in place of the viral genome producing a transducing particle. (c) Cell lysis occurs liberating both phage and transducing particles. (d) If a new cell is infected with a transducing particle containing plasmid DNA, the plasmid DNA is injected. (e) The plasmid DNA is established in the host and its genetic potential is expressed. See Miller (1991a) for full discussion. Adapted from Miller (1988).

Table 19.7 Ecosystems in which transduction has been demonstrated

Environment	Representative references
Soil	Germida & Khachatourians 1988; Zeph *et al.* 1988; Zeph & Stotzky 1989
Lake water	Morrison *et al.* 1978; Saye *et al.* 1987; 1990
River water	Amin & Day 1988
Waste treatment facility	Osmond & Gealt 1988
Shellfish	Baross *et al.* 1978
Mammals	Novick & Morse 1967

have been made almost exclusively in freshwater habitats (Morrison *et al.* 1978; Saye *et al.* 1987; 1990; Amin & Day 1988; Saye & Miller 1989; Miller *et al.* 1990a,b). Saye *et al.* (1987) investigated the transduction of a Tra$^-$, mobilization-deficient (Mob$^-$) plasmid, Rms149, in environmental chambers incubated in a freshwater lake. Transduction was observed in chambers which contained sterilized lake water and in chambers in which the autochthonous microbial community was present. These investigators explored several paradigms for the source of transducing phages and found that the greatest number of transductants were recovered when a lysogenic strain served as both the source of the transducing phage and as the recipient of the transduced DNA (Saye & Miller 1989). The donor of the plasmid DNA was a nonlysogen. This paradigm was probably most effective not because the rate of transduction was higher, but because the immunity to phage killing imparted by the resident prophage (Kokjohn 1989; Miller 1991a) increased the probability of recovery (identification) of the transductants.

Transduction of chromosomal genes has also been demonstrated in freshwater environments. Morrison *et al.* (1978) observed transduction of chromosomally encoded streptomycin resistance in flow-through chambers suspended in a freshwater lake. Transduction frequencies (transductants per phage particle) were equivalent to those observed in the laboratory. Similar results were obtained by Amin & Day (1988) who measured transduction in the epilithon of river stones submerged in the River Taff in southern Wales. Saye *et al.* (1990) measured the frequency of cotransduction of linked chromosomal genes in environmental test chambers suspended in Fort Loudon Lake in eastern Tennessee. Transductants of lysogenized strains were recovered at frequencies 10- to 100-fold higher than transductants of nonlysogenic parents. Reciprocal exchange of chromosomal alleles was observed between the two parental strains.

Recently, we have been evaluating the ability of natural freshwater microbial communities to act as recipients in transduction (Miller *et al.* 1990b). For these studies, we isolated a generalized transducing phage, UT1 (Ogunseitan *et al.* 1990) from Fort Loudon Lake. This phage was used as a vector for transduction of Rms149. Transducing particles were prepared from a laboratory strain of *P. aeruginosa* and used to transduce a heterogeneous population of bacteria isolated from the same lake. In these experiments, frequencies of plasmid transduction between 10^{-6} and 10^{-4} transductants per recipient were observed.

In these studies, several factors were shown to affect the levels of transductants recovered from *in situ* incubations (Saye *et al.* 1987; 1990; Saye & Miller 1989; Miller *et al.* 1990a; Miller & Sayler 1991). First, the presence of the natural community was shown to lower the numbers of transductants recovered. It is not yet clear whether this is due to unrelated factors, such as competition between the introduced and natural bacteria, or related factors, such as competitive inhibition by natural bacteria of the binding of bacteriophages to the introduced recipient bacteria. Second, the ratio of donor-to-recipient cells present in the test chamber affected the number of transductants detected. Unlike conjugation (O'Morchoe *et al.* 1988), transduction peaked at ratios close to 1:1. Third, the phage/bacterium ratio produced in the test system was found to correlate with the transduction frequency. No transductants were observed in systems where this ratio fell below 10^{-3} phage per bacterium.

Transduction has been observed in other aquatic environments as well. Baross *et al.* (1974) detected bacteriophage P4-mediated transduction of the genetic determinant encoding agarase activity between strains of *V. parahaemolyticus* infected into the gills of oysters during incubation in a marine aquarium containing sterilized sea water. Coliphage-mediated transduction has also been documented in synthetic waste water (Osmond & Gealt 1988). In these studies both laboratory strains of *E. coli* and isolates from sewage-treatment facilities were used as recipients. Transduction of plasmid Rms149 has been detected in microcosms designed to simulate freshwater swamps (R.B. Coffin & R.V. Miller, unpublished). It would appear that transduction has the potential to mediate gene transfer in natural populations of bacteria.

Many investigators originally felt that gene transfer in the environment would proceed via conjugation if it occurred at all. Because this process requires an extrachromosomal DNA element (the plasmid), one of the scenarios which has often been promoted as a mechanism for reducing risk of gene transfer in natural habitats, is to stabilize engineered DNA in GEMs by introducing it into the host's genome. The probability of transfer could then be reduced even further by making that host recombinationally deficient (Miller & Kokjohn 1990). While these operations will certainly reduce transfer via conjugation, the process of packaging host DNA into viral capsids does not require an homologous recombinational event within the donor cell (although it may require recombination to establish the sequence in the recipient cell). Hence, the possibility of gene transfer via transduction (or transformation for that matter) is not altered by introducing the engineered sequences into the chromosome of a recombinationally deficient host.

19.2.4 Models for gene exchange

Several models for gene transfer have been developed. Most are designed to describe conjugal transfer of genetic material. Levin *et al.* (1979) described conjugational exchange in *E. coli* suspended in a liquid environment, by a simple mass-action model. This model assumes that the rate of transconjugant production is regulated by the probability of contact between a donor and recipient cell by simple Brownian motion. It assumes that the bacteria are passive in the process and that the frequency of conjugation is proportional to the product of the donor and recipient concentrations. While evidence is mounting that donor–recipient recognition may actually be an active process (Clewell *et al.* 1987), Levin's model allows the calculation of a transfer rate constant which is a good approximation of the process in liquid cultures where growth is exponential and cells are equally distributed in space. This model does not hold in environments where growth density is heterogeneous and patchy, such as occurs on solid surfaces. Simonsen (1990) addressed this problem to some degree by developing a model for the dynamics of plasmid transfer through bacterial populations immobilized on solid surfaces.

Freter *et al.* (1983) refined Levin's mass-action model and studied the transmission of plasmids through populations of *E. coli* in the gut of gnotobiotic mice and in continuous-flow cultures. Knudsen *et al.* (1987) adapted the model to allow use of laboratory data to predict conjugal transfer in *P. cepacia* populations present in the phyllosphere and rhizosphere. Levin (1986) has developed a more realistic form of this model which takes into account various factors such as:

1 transitory derepression of conjugative pili synthesis;
2 restriction and modification;
3 periodic selection;

4 group selection which may influence conjugal plasmid maintenance in a microbial community.

An interesting form of this model has been developed that attempts to describe the consequences of conjugation in controlling the actions of bacterial populations in biological processes of the type which might be carried out in bioreactors (Rittmann *et al.* 1990; Smets *et al.* 1990). This model is interesting because gene transfer is treated in terms of biomass (grams per milliliter) as opposed to cell concentration (cells per milliliter). Since conjugation depends on cell-to-cell contact, this model seems most appropriately applied to environments where the mass of a cell is known or at least constant. As cell volume and, hence, the number of cells per gram biomass changes dramatically with growth conditions (i.e., starvation as opposed to exponential growth), it is questionable whether this model will be useful for evaluating gene transfer in natural habitats.

Simonsen *et al.* (1990) developed a further extension of the Levin model which tries to take into account environmental factors potentially affecting conjugation rates in aquatic environments. This model relies on data which is easier to obtain than those required for application of the original Levin model.

Our laboratory has been working on a chemostatic model designed to evaluate the effects of transduction on gene-pool diversity in aquatic bacterial populations (Replicon & Miller 1990). Our studies demonstrate that (1) transduction can act to stabilize a genotype in a bacterial population from which it would otherwise be lost, and (2) factors other than simple mass-action considerations control the rate of transduction in aquatic environments.

19.3 Mechanisms of intracellular genetic alteration

19.3.1 *Mutation*

Mutation is the alteration of information contained in the genetic material by sporadic changes in the DNA base sequence (Miller & Levy 1989). It has the potential to be a major source of genetic variation in the environment (Reanney 1976; Saye & Miller 1989). Mutational events can alter both the fitness and efficacy of a GEM. Not only can they lead to the elimination of specific phenotypic activities, but they can activate cryptic genes, producing unsuspected phenotypes in the organism (Kaiser & Murray 1980; Downs & Roth 1987; Parker *et al.* 1988). The frequency of mutation can be increased by various substances (mutagens) which damage DNA. The presence of these damaging agents in the cellular milieu often activates various repair processes which are mutagenic (Walker 1984; Sedgwick 1986; Miller & Levy 1989). There are many plant-derived compounds and xenobiotics which have the potential for increasing mutation rates (Miller 1991b). Thus, utilizing GEMs for their intended purposes in agriculture and bioremediation will expose them to compounds which can induce systems of genetic instability.

Various *in situ* studies (Miller & Levy 1989; Saye & Miller 1989; Miller *et al.* 1990a) have shown that environmental stress can act to increase mutation frequency. In our own studies, we determined that the mutation frequency for several genes of *P. aeruginosa* was around 10^{-7} mutations per cell generation in microcosms designed to mimic freshwater environments. However, when we measured the frequencies of these same genes *in situ*, we found that they increased rapidly to 10^{-4} mutations per gene per generation (about 1000-fold higher than was observed in the microcosms). From these kinds of data one must conclude, first, that mutation should be considered a potentially significant factor in assessing gene stability in the environment, and second, that environmental gene stability cannot always be modeled from laboratory, or even microcosm, data.

19.3.2 *Transposition*

Transposition is the movement of a genetic element from one replicon to another without the requirement for a homologous recombinational event (Berg 1989; Berg & Howe 1989). In other words, the process increases the frequency of the element within the cell by introducing it into new replicons. Transposons (Tn elements) are usually made up of an insertion sequence (IS element) bounded by inverted repeats and which contain genes for a transposase (a DNA replicase) and other activities which allow both the replication of the Tn element and its transfer. In addition to the IS element, Tn elements contain one or more vegetative genes, which may encode antibiotic- or heavy metal-resistance functions, or catabolic genes (Figure 19.4). These elements are now being used to produce recombinant organisms by inserting engineered genes into the transposon, either in place of or in addition to the antibiotic-resistance genes. These constructs are then inserted into the host genome. Following incorporation, the transposon is inactivated to ensure that further transposition does not take place.

One concern associated with this technology is the creation of new mutations which may alter the phenotype of the host in unsuspected

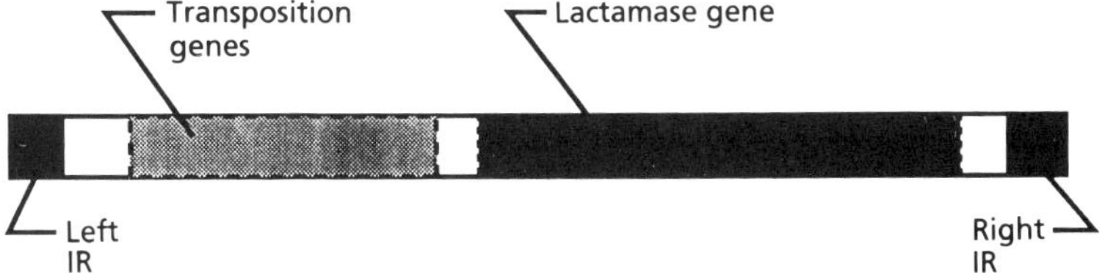

Figure 19.4 A map of Tn3. The Tn element is made up of an IS element containing the transposition genes and a gene for β-lactam resistance (lactamase). The element is bounded by inverted repeats (IR) of DNA sequences. See Berg (1989) for complete discussion.

ways (Sayre & Miller 1991). The transposition event may inactivate genes by inserting into the middle of a gene in the target DNA (Berg & Howe 1989). For instance, transpositional insertion may inactivate a repressor gene allowing expression of a particular biosynthetic or catabolic process which is not normally active and which may not have been suspected in the host. Alternatively, outward-acting promoters within the transposon may activate silent genes which were likewise unsuspected in the host (Parker & Hall 1990a,b).

Naturally occurring transposons also have the potential to produce mutations under environmental conditions (Parker & Hall 1990a,b; Wood *et al.* 1990). These include not only traditional transposons but bacteriophages as well. Numerous bacterial viruses which replicate by transposition have been isolated from natural habitats (Akhverdian *et al.* 1985; Kilbane & Miller 1988). Certainly the best studied of these bacterial viruses is phage Mu of *E. coli* (Harshey 1988). This phage received its name because the transposition of its prophage into the host genome during lysogen establishment frequently produces recognizable mutations.

Recently, studies on *E. coli* have identified retronphages which have characteristics similar to retroviruses of eukaryotic origin (Inouye *et al.* 1991). The prophages of these viruses (called retrons) are often cryptic and require a helper phage in order to produce a virion. Retrons have been found integrated into the chromosome of many isolates of *E. coli* and in myxobacteria as well. Their potential as an environmental source of transposon-derived mutation is yet to be evaluated.

The environment may act on transposons to increase or decrease their rate of movement. Transposition of Tn501, which carries resistance to mercury, is stimulated when mercury is present in the cellular milieu (Sherratt 1989). Likewise, erythromycin stimulates transposition of Tn551 when the host cell is exposed to this antibiotic (Murphy 1989).

O'Morchoe *et al.* (1988) measured transfer of a plasmid *in situ* in a freshwater environment by confirming its presence in the recipient both genetically and physically. They isolated the DNA from transconjugants and asked whether the restriction endonuclease cleavage pattern characteristic of the plasmid was present. In these studies, the plasmid R68.45, which contains several transposable elements, was used. In microcosm studies, they found less than 1% of the plasmids which had been transferred to a new cell showed any alterations (deletions, rearrangements, or insertions) which might suggest that transposition had taken place. However, when transconjugants from the *in situ* incubations were examined, approximately 50% of the

Table 19.8 Factors affecting gene transfer which should be considered in the design of GEMs for environmental uses. Adapted from Miller & Levy (1989)

Factor	Elements
1 The engineered sequence	Genetic elements involved Their product Their molecular regulation Construction through use of a Tn element
2 Vector	Plasmid or phage Modes of transfer
3 Host	Adaptability Mechanism of DNA repair Mutability Presence of Tn elements in genome
4 Environmental factors[1]	Microbial competition Water content of the environment pH Temperature Ionic concentration Surfaces (i.e., particulate matter)
5 Biological factors[1]	Plasmid incompatibility Transfer inhibition Restriction-modification Phenotypic expression Make-up of natural microbial community

[1] For a full discussion of these factors see Miller & Levy (1989) and Miller (1991c).

transferred plasmids contained one or more deletions, rearrangements, or insertions. Numerous factors present in the environment may act to increase the frequency of transposition *in situ*. Therefore, transposition rates should be investigated in the appropriate environment. Laboratory results may be insufficient to certify a Tn element as inactivated.

19.4 Conclusion

Five critical factors should be considered in developing a GEM for environmental use (Table 19.8). Many of these will affect the final probability for gene transfer and stability when the engineered organism is released into the environment. By carefully analyzing each of these factors, an organism can be designed with low risk and high efficacy. In this chapter, I have reviewed the data which suggest that it is important to consider genetic stability in development of organisms for use in environmental processes. Horizontal gene transfer and gene stability must be evaluated in assessing risk associated with the use of GEMs in the environment. Because of the numerous kinds of environmental factors which may affect these processes, we will most likely have to consider each GEM as a special case. We will have to look at the engineered sequence, the physiology of the host organism, its genetics, and the ecosystem into which it will be released. For instance, as microorganisms are applied to problems in environmental toxicology, we are discovering that while the methods for determining the toxicity of chemical substances are reasonably well understood, we currently have very little data to suggest that these risk-assessment models can be applied to biotechnological organisms.

Acknowledgments

I wish to thank my colleagues who shared unpublished observations with me.

This endeavor was supported in part by cooperative agreement No. CR820060 from the Gulf Breeze Environmental Research Laboratory of the US Environmental Protection Agency.

References

Akhverdian V.Z., Khrenova E.A., Revlets M.A., Gerasimova T.V. & Krylov V.N. (1985) The properties of transposable phages of *Pseudomonas aeruginosa* belonging to two groups distinguished by DNA–DNA homology. *Genetika* **21**, 735–747.

Alcaide E. & Garay E. (1984) R-plasmid transfer in *Salmonella* spp. isolated from wastewater and sewage-contaminated surface waters. *Appl. Environ. Microbiol.* **48**, 435–438.

Altherr M.R. & Kasweck K.L. (1982) *In situ* studies with membrane diffusion chambers of antibiotic resistance transfer in *Escherichia coli. Appl. Environ. Microbiol.* **44**, 838–843.

Amin M.K. & Day M.J. (1988) Donor and recipient effects on transduction frequency *in situ. REGEM 1 Program*, p. 11. REGEM, Cardiff.

Avery O.R., MacLeod C.M. & McCarty M. (1944) Studies on the chemical nature of the substance inducing transformation in pneumococcal types. *J. Exp. Med.* **79**, 137–159.

Bagdasarian M., Lurz R., Ruckert B., Franklin F.C.H., Bagdasarian M., Frey J. & Timmis K.N. (1981) Specific purpose plasmid cloning vectors II. Broad host range, high copy number, RSF1010 derived vectors, and a host–vector system for gene cloning in *Pseudomonas. Gene* **16**, 237–247.

Bale M.J., Fry J.C. & Day M.J. (1987) Plasmid transfer between strains of *Pseudomonas aeruginosa* on membrane filters attached to river stones. *J. Gen Microbiol.* **133**, 3099–3107.

Bale M.J., Day M.J. & Fry J.C. (1988a) Novel method for studying plasmid transfer in undisturbed river epilithon. *Appl. Environ. Microbiol.* **54**, 2756–2758.

Bale M.J., Fry J.C. & Day M.J. (1988b) Transfer and occurrence of large mercury resistance plasmids in river epilithon. *Appl. Environ. Microbiol.* **54**, 972–978.

Baross J.A., Liston J. & Morita R.Y. (1974) Some implications of genetic exchange among marine vibrios, including *Vibrio parahaemolyticus*, naturally occurring in the Pacific oyster. In Fujino T., Sakaguchi G., Sakazaki R. & Takeda Y. (eds) *International Symposium on Vibrio parahaemolyticus*, pp. 129–137. Saikon Pub. Col., Tokyo.

Baross J.A., Liston J. & Morita R.Y. (1978) Incidence of *Vibrio parahaemolyticus* bacteriophages and other *Vibrio* bacteriophages in marine samples. *Appl. Environ. Microbiol.* **36**, 492–499.

Bauda P., Menon P., Block J.C., Lett M.C., Roux B. & Hubert J.C. (1990) Laboratory standardized biofilms as a tool to investigate genetic transfer in water systems. In Fry J.C. & Day M.J. (eds) *Bacterial Genetics in Natural Environments*, pp. 81–88. Chapman and Hall, London.

Bell J.B., Elliott G.E. & Smith D.W. (1983) Influence of sewage treatment and urbanization on selection of multiple resistance in fecal coliform populations. *Appl. Environ. Microbiol.* **46**, 227–232.

Berg D.E. (1989) Transposable elements in prokaryotes. In Levy S.B. & Miller R.V. (eds) *Gene Transfer in the Environment*, pp. 99–137. McGraw-Hill, New York.

Berg D.E. & Howe M.M. (1989) *Mobile DNA*. American Society for Microbiology, Washington DC.

Bergh Ø., Børsheim K.Y., Bratbak G. & Heldal M. (1989) High abundance of viruses found in aquatic environments. *Nature* **340**, 467–468.

Blackburn J.W., Jain R.K. & Sayler G.S. (1987) Molecular microbial ecology of a naphthalene-degrading genotype in activated sludge. *Environ. Sci. Technol.* **21**, 884–890.

Børsheim K.Y., Bratbak G. & Heldal M. (1990) Enumeration and biomass estimation of planktonic bacteria and viruses by transmission electron microscopy. *Appl. Environ. Microbiol.* **56**, 352–356.

Bratbak B., Heldal M., Norland S. & Thingstad T.F. (1990) Viruses as partners in spring bloom microbial trophodynamics. *Appl. Environ. Microbiol.* **56**, 1400–1405.

Chakrabarty A.M. (1986) Genetic engineering and problems of environmental pollution. In Rehm H.-J. & Reed G. (eds) *Biotechnology: A Comprehensive Treatise in 8 Volumes*, vol. 8, pp. 515–530. VCH Verlagsgesellschaft, Weinheim.

Clewell D.B., Ehrenfeld E.E., Kessler R.E., Ike Y., Frank A.E., Madion M., Shaw J.H., Wirth R., An R., Mori M., Kitada C., Fujino M. & Suzuki A. (1987) Sex-pheromone systems in *Streptococcus faecalis*. In Levy S.B. & Novick R.P. (eds) *Antibiotic Resistance Genes: Ecology, Transfer, and Expression*, Banbury Report No. 24, pp. 131–142. Cold Spring Harbor Laboratory, Cold Spring Harbor.

Curtiss R. III. (1976) Genetic manipulation of microorganisms: potential benefits and biohazards. *Annu. Rev. Microbiol.* **30**, 507–533.

Day M.J., Bale M.J. & Fry J.C. (1988) Plasmid transfer in a freshwater environment. In Folse J. & Covello V.T. (eds) *Safety Assurance for Environmental Introductions of Genetically Engineered Organisms, Nato ASI Series*, vol. G18, pp. 181–197. Springer Verlag, Berlin.

Dorward E.W. & Garon C.F. (1990) DNA is packaged within membrane-derived vesicles of Gram-negative but not Gram-positive bacteria. *Appl. Environ. Microbiol.* **56**, 1960–1962.

Downs D.M. & Roth J.R. (1987) A novel P22 prophage of *Salmonella typhimurium*. *Genetics* **117**, 367–380.

Ewert D.L. & Paynter M.J.B. (1980) Enumeration of bacteriophages and host bacteria in sewage and the activated sludge treatment process. *Appl. Environ. Microbiol.* **39**, 576–583.

Farrand S.K. (1989) Conjugal transfer of bacterial genes on plants. In Levy S.B. & Miller R.V. (eds) *Gene Transfer in the Environment*, pp. 261–285. McGraw-Hill, New York.

Forster C.F. & Wase D.A.J. (1987) Biopossibilities: the next few years. In Forster C.F. & Wase D.A.J. (eds) *Environmental Biotechnology*, pp. 438–448. Ellis Horwood, Chichester.

Freter R., Freter R.R. & Brickner H. (1983) Experimental and mathematical models of *Escherichia coli* plasmid transfer *in vitro* and *in vivo*. *Infect. Immun.* **39**, 60–84.

Frischer M.F., Thurmond J.M. & Paul J.H. (1990) Natural plasmid transformation in a high frequency of transformation marine *Vibrio* strain. *Appl. Environ. Microbiol.* **56**, 3439–3444.

Fry J.C. & Day M.J. (1990) Plasmid transfer in the epilithon. In Fry J.C. & Day M.J.

(eds) *Bacterial Genetics in Natural Environments*, pp. 55–80. Chapman and Hall, London.

Gauthier M.J. & Breittmayer V.A. (1990) Gene transfer in marine environments. In Fry J.C. & Day M.J. (eds) *Bacterial Genetics in Natural Environments*, pp. 100–110. Chapman and Hall, London.

Gauthier M.J., Cauvin R. & Breittmayer J.-P. (1985) Influence of salts and temperature on the transfer of mercury resistance from a marine pseudomonad to *Escherichia coli*. *Appl. Environ. Microbiol.* **50**, 38–40.

Gealt M.A. (1991) Gene transfer in waste treatment. In Levin M., Seidler R. & Rogul M. (eds) *Microbial Ecology: Principles, Methods, and Application in Environmental Biotechnology*, pp. 327–343. McGraw-Hill, New York.

Gealt M.A., Chai M.C., Alpert K.B. & Boyer J.C. (1985) Transfer of plasmids pBR322 and pBR325 in wastewater from laboratory strains of *Escherichia coli* to bacteria indigenous to the waste disposal system. *Appl. Environ. Microbiol.* **49**, 839–841.

Gealt M.A., Cleaveland P. & Vettese M. (1988) Transfer of genetically engineered DNA sequences (GEDS) into indigenous wastewater bacteria in a laboratory-scale treatment facility, Abst. No. N-61, p. 254. *Abst. Annual Meet. Am. Soc. Microbiol.*, Miami Beach.

Geesey G.G. & White D.C. (1990) Determination of bacterial growth and activity at solid–liquid interfaces. *Annu. Rev. Microbiol.* **44**, 579–602.

Germida J.J. & Khachatourians G.G. (1988) Transduction of *Escherichia coli* in soil. *Can. J. Microbiol.* **34**, 190–193.

Ghosal D., You I.E., Chatterjee D.E. & Chakrabarty A.M. (1985) Microbial degradation of halogenated compounds. *Science* **228**, 135–142.

Gowland P.C. & Slater J.H. (1984) Transfer and stability of drug resistance plasmids in *Escherichia coli* K12. *Microb. Ecol.* **10**, 1–13.

Grabow W.O.K., Prozesky O.W. & Burger J.S. (1975) Behavior in a river and dam of coliform bacteria with transferable or non-transferable drug resistance. *Wat. Res.* **9**, 777–782.

Graham J.B. & Istock C.A. (1978) Genetic exchange in *Bacillus subtilis* in soil. *Mol. Gen. Genet.* **166**, 287–290.

Griffith F. (1928) Significance of pneumococcal types. *J. Hyg.* **27**, 113–159.

Harshey R.M. (1988) Phage Mu. In Calender R. (ed.) *The Bacteriophages*, vol. 1, pp. 193–234. Plenum, New York.

Hedges R.W. & Jacoby G.A. (1980) Compatibility and molecular properties of plasmid Rms 149 in *Pseudomonas aeruginosa* and *Escherichia coli*. *Plasmid* **3**, 1–6.

Hirsch P.R. (1990) Factors limiting gene transfer in bacteria. In Fry J.C. & Day M.J. (eds) *Bacterial Genetics in Natural Environments*, pp. 31–40. Chapman and Hall, London.

Inouye S., Sunshine M.G., Six E.W. & Inouye M. (1991) Retronphage øR73: An *E. coli* phage that contains a retroelement and integrates into a tRNA gene. *Science* **252**, 969–971.

Ippen-Ihler K. (1989) Bacterial conjugation. In Levy S.B. & Miller R.V. (eds) *Gene Transfer in the Environment*, pp. 33–72. McGraw-Hill, New York.

Jarrett P. & Stephenson M. (1990) Plasmid transfer between strains of *Bacillus thuringiensis* infecting *Galleria mellonella* and *Spodoptera littoralis*. *Appl. Environ. Microbiol.* **56**, 1608–1614.

Jeffrey W.H., Paul J.H. & Stewart G.J. (1989) Natural transformation of a marine *Vibrio* sp. by plasmid DNA. *Microb. Ecol.* **19**, 259–268.

Kaiser K. & Murray N.E. (1980). On the nature of *sbcA* mutations in *E. coli* K12. *Mol. Gen. Genet.* **179**, 555–563.

Kelley W.J. & Reanney D.C. (1984) Mercury resistance among soil bacteria: ecological transformability of genes encoding resistance. *Soil Biol. Biochem.* **16**, 1–8.

Khalil T.A. & Gealt M.A. (1987) Temperature, pH, and cations affect the ability of

Escherichia coli to mobilize plasmids in L broth and synthetic wastewater. *Can. J. Microbiol.* **33**, 733–737.

Kilbane J.J. & Miller R.V. (1988) Molecular characterization of *Pseudomonas aeruginosa* bacteriophages: identification and characterization of the novel virus B86. *Virology* **164**, 193–200.

Kirk T.K. & Farrell R.L. (1987) Enzymatic "combustion": the microbial degradation of lignin. *Annu Rev. Microbiol.* **41**, 465–505.

Klingmüller W., Dally A., Fentner C. & Steinlein M. (1990) Plasmid transfer between soil bacteria. In Fry J.C. & Day M.J. (eds) *Bacterial Genetics in Natural Environments*, pp. 133–151. Chapman and Hall, London.

Knudsen G., Walter M., Porteous L., Prince V., Armstrong J. & Seidler R. (1987) Predictive model of conjugative plasmid transfer; in the rhizosphere and phyllosphere. *Appl. Environ. Microbiol.* **54**, 343–347.

Kokjohn T.A. (1989) Transduction: mechanism and potential for gene transfer in the environment. In Levy S.B. & Miller R.V. (eds) *Gene Transfer in the Environment*, pp. 73–98. McGraw-Hill, New York.

Kokjohn T.A. & Miller R.V. (1992) Gene transfer in the environment: transduction. In Fry J.C. & Day M.J. (eds) *Release of Genetically Engineered and Other Microorganisms*, pp. 54–81. Cambridge University Press, Cambridge.

Kokjohn T.A., Sayler G.S. & Miller R.V. (1991) Attachment and replication of *Pseudomonas aeruginosa* bacteriophages under conditions simulating aquatic environments. *J. Gen Microbiol.* **137**, 661–666.

Levin B. (1986) The maintenance of plasmids and transposons in natural populations of bacteria. In Levy S.B. & Novick R.P. (eds) *Antibiotic Resistance Genes: Ecology, Transfer, and Expression*, Banbury Report No. 24, pp. 57–70. Cold Spring Harbor Laboratory, Cold Spring Harbor.

Levin B., Stewart F. & Rice V. (1979) The kinetics of conjugative plasmid transmission: fit of a simple mass action model. *Plasmid* **2**, 247–260.

Levy S.B. & Miller R.V. (1989) *Gene Transfer in the Environment*. McGraw-Hill, New York.

Lindow S.E., Panopoulos N.J. & McFarland B.L. (1989) Genetic engineering of bacteria from managed and natural habitats. *Science* **244**, 1300–1307.

Lorenz M.G. & Wackernagel W. (1990) Natural genetic transformation of *P. stutzeri* by sand-adsorbed DNA. *Arch. Microbiol.* **154**, 380–385.

Lorenz M.G., Aardema B.W. & Wackernagel W. (1988) Highly efficient genetic transformation of *Bacillus subtilis* attached to sand grains. *J. Gen. Microbiol.* **134**, 107–112.

Lynch J.M. (1990) The potential for gene exchange between rhizosphere bacteria. In Fry J.C. & Day M.J. (eds) *Bacterial Genetics in Natural Environments*, pp. 172–181. Chapman and Hall, London.

Mach P.A. & Grimes D.J. (1982) R plasmid transfer in a wastewater treatment plant. *Appl. Environ. Microbiol.* **44**, 1395–1403.

Mancini P., Ferteis S., Nave D. & Gealt M.A. (1987) Mobilization of plasmid pHSV106 from *Escherichia coli* HB101 in a laboratory-scale waste treatment facility. *Appl. Environ. Microbiol.* **53**, 665–671.

McClure N.C., Fry J.C. & Weightman A.J. (1990) Gene transfer in activated sludge. In Fry J.C. & Day M.J. (eds) *Bacterial Genetics in Natural Environments*, pp. 111–129. Chapman and Hall, London.

McIntire S.A. (1991) Analysis of conjugation in bacteria. In Levin M., Seidler R. & Rogul M. (eds) *Microbial Ecology: Principles, Methods, and Application in Environmental Biotechnology*, pp. 199–228. McGraw-Hill, New York.

McPherson P. & Gealt M.A. (1986) Isolation of indigenous wastewater bacterial strains capable of mobilizing plasmid pBR325. *Appl. Environ. Microbiol.* **51**, 904–909.

Mergeay M., Springael D. & Top E. (1990) Gene transfer in polluted soils. In Fry J.C. & Day M.J. (eds) *Bacterial Genetics in Natural Environments*, pp. 152–171.

Chapman & Hall, London.

Miller R.V. (1988) Potential for transfer and establishment of engineered genetic sequences in the environment. In Hodgson J. & Sugden A.M. (eds) *Planned Release of Genetically Engineered Organisms (Trends in Biotechnology and Trends in Ecology and Evolution Special Publication)*, pp. S23–S27. Elsevier, Cambridge.

Miller R.V. (1991a) Methods for evaluating transduction: an overview with environmental considerations. In Levin M., Seidler R. & Rogul M. (eds) *Microbial Ecology: Principles, Methods, and Application in Environmental Biotechnology*, pp. 229–251. McGraw-Hill, New York.

Miller R.V. (1991b) Increasing levels of environmental mutagens: potential for affecting viral evolution and pathogenicity – a speculative review. *Environ. Carcinogenesis Rev.* **C8**, 89–137.

Miller R.V. (1991c) Methods for the evaluation of gene transfer in the environment; general considerations. In Levin M., Seidler R. & Rogul M. (eds) *Microbial Ecology: Principles, Methods, and Application in Environmental Biotechnology*, pp. 141–159. McGraw-Hill, New York.

Miller R.V. & Levy S.B. (1989) Horizontal gene transfer in relation to environmental release of genetically engineered microorganisms. In Levy S.B. & Miller R.V. (eds) *Gene Transfer in the Environment*, pp. 405–420. McGraw-Hill, New York.

Miller R.V. & Kokjohn T.A. (1990) General microbiology of recA: environmental and evolutionary significance. *Annu. Rev. Microbiol.* **44**, 365–394.

Miller R.V. & Sayler G.S. (1991) Bacteriophage–host interactions in aquatic systems. In Wellington E.M. & Van Elsas J.D. (eds) *Genetic Interactions Between Microorganisms in the Microenvironment*, pp. 176–193. University of Manchester Press, Manchester.

Miller R.V., Kokjohn T.A. & Sayler G.S. (1990a) Environmental and molecular characterization of systems which affect genome alteration in *Pseudomonas*. In Silver S., Chakrabarty A.M., Iglewski B. & Kapland S. (eds) *Pseudomonas: Biotransformations, Pathogenesis, and Evolving Biotechnology*, pp. 252–268. American Society for Microbiology, Washington DC.

Miller R.V., Ripp S. & Ogunseitan O.A. (1990b) A naturally-occurring generalized-transducing bacteriophage of *Pseudomonas aeruginosa* isolated from a freshwater lake, Abst. No. Q-127, p. 309. *Abst. Annu. Meet. Am. Soc. Microbiol.*, Anaheim.

Monticello D.J. & Finnerty W.R. (1985) Microbial desulfurization of fossil fuels. *Annu. Rev. Microbiol.* **39**, 371–389.

Morrison D., Miller R.V. & Sayler G.S. (1978) Frequency of F116 mediated transduction of *Pseudomonas aeruginosa* in a natural freshwater environment. *Appl. Environ. Microbiol.* **36**, 724–730.

Murphy E. (1989) Transposable elements in Gram-positive bacteria. In Berg D.E. & Howe M.M. (eds) *Mobile DNA*, pp. 269–288. American Society for Microbiology, Washington DC.

Novick R.P. & Morse S.I. (1967) *In vivo* transmission of drug resistance factors between strains of *Staphylococcus aureus*. *J. Exp. Med.* **125**, 45–59.

Ogunseitan O.A., Sayler G.S. & Miller R.V. (1990) Dynamic interaction of *Pseudomonas aeruginosa* and bacteriophages in lake water. *Microb. Ecol.* **19**, 171–185.

O'Morchoe S., Ogunseitan O., Sayler G.S. & Miller R.V. (1988) Conjugal transfer of R68.45 and FP5 between *Pseudomonas aeruginosa* in a natural freshwater environment. *Appl. Environ. Microbiol.* **54**, 1923–1929.

Osmond M.A. & Gealt M.A. (1988) Wastewater bacteriophages transduce genes from the chromosome and a recombinant plasmid, Abst. No. N-62, p. 254. *Absts. Annu. Meet. Am. Soc. Microbiol.*, Miami Beach.

Parker L.L. & Hall B.G. (1990a) Characterization and nucleotide sequence of the cryptic *cel* operon of *Escherichia coli* K12. *Genetics* **124**, 455–471.

Parker L.L. & Hall B.G. (1990b) Mechanisms of activation of the cryptic *cel* operon of *Escherichia coli* K12. *Genetics* **124**, 473–482.

Parker L.L., Betts P.W. & Hall B.G. (1988) Activation of cryptic gene expression by excision of a DNA fragment. *J. Bacteriol.* **170**, 218–222.

Patt T., Nealson K. & Eberhard C. (1972) Promiscuous transfer of genetic information from nonmarine to marine bacteria. *Biol. Bull.* **143**, 473.

Paul J.H. & Pichard S.L. (1989) Specificity of cellular DNA-binding sites of microbial populations in a Florida reservoir. *Appl. Environ. Microbiol.* **55**, 2798–2801.

Paul J.H., Deflaun M.F. & Jeffrey W.H. (1988a) Mechanisms of DNA utilization by estuarine microbial populations. *Appl. Environ. Microbiol.* **54**, 1682–1688.

Paul J.H., Deflaun M.F., Jeffrey W.H. & David A.W. (1988b) Seasonal and diel variability in dissolved DNA and in microbial biomass and activity in a subtropical estuary. *Appl. Environ. Microbiol.* **54**, 718–727.

Paul J.H., Jeffrey W.H., David A.W., Deflaun M.R. & Cazares L.H. (1989) Turnover of extracellular DNA in eutrophic and oligotrophic freshwater environments of southwest Florida. *Appl. Environ. Microbiol.* **55**, 1823–1828.

Paul J.H., Frischer M.E. & Thurmond J.M. (1991) Gene transfer in marine water column and sediment microcosms by natural plasmid transformation. *Appl. Environ. Microbiol.* **57**, 1509–1515.

Pooley F.D. (1987) Mineral leaching with bacteria. In Forster C.F. & Wase D.A.J. (eds) *Environmental Biotechnology*, pp. 114–134. Ellis Horwood, Chichester.

Proctor L.M. & Fuhrman J.A. (1990) Viral mortality of marine bacteria and cyanobacteria. *Nature* **343**, 60–62.

Rafii F. & Crawford D.L. (1989) Gene transfer among *Streptomyces*. In Levy S.B. & Miller R.V. (eds) *Gene Transfer in the Environment*, pp. 309–345. McGraw-Hill, New York.

Reanney D. (1976) Extrachromosomal elements as possible agents of adaptation and development. *Bacteriol. Rev.* **40**, 552–590.

Replicon J. & Miller R.V. (1990) Modeling the potential for transduction to stabilize a foreign genotype within an established microbial community, Abst No. W75–001, p. 117. *Abst. VIIIth Internat. Cong. Virol.*, Berlin.

Rittmann B.E., Smets B.R. & Stahl D.A. (1990) The role of genes in biological processes. Part 1 of a two-part article. *Environ. Sci. Technol.* **24**, 23–29.

Roberts M.C. (1989) Gene transfer in the urogenital and respiratory tract. In Levy S.B. & Miller R.V. (eds) *Gene Transfer in the Environment*, pp. 347–375. McGraw-Hill, New York.

Rochelle P.A., Fry J.C. & Day M.J. (1989) Factors affecting conjugal transfer of plasmids encoding mercury resistance from pure cultures and mixed natural suspensions of epilithic bacteria. *J. Gen. Microbiol.* **135**, 409–424.

Saye D.J. & Miller R.V. (1989) The aquatic environment: consideration of horizontal gene transmission in a diversified habitat. In Levy S.B. & Miller R.V. (eds) *Gene Transfer in the Environment*, pp. 223–259. McGraw-Hill, New York.

Saye D.J., Ogunseitan O., Sayler G.S. & Miller R.V. (1987) Potential for transduction of plasmids in a natural freshwater environment: effect of plasmid donor concentration and a natural microbial community on transduction in *Pseudomonas aeruginosa*. *Appl. Environ. Microbiol.* **53**, 987–995.

Saye D.J., Ogunseitan O.A., Sayler G.S. & Miller R.V. (1990) Transduction of linked chromosomal genes between *Pseudomonas aeruginosa* during incubation *in situ* in a freshwater habitat. *Appl. Environ. Microbiol.* **56**, 140–145.

Sayler G.S., Hooper S.W., Layton A. & King J.M.H. (1990) Catabolic plasmids for environmental applications: a review. *Microb. Ecol.* **19**, 1–20.

Sayre P. & Miller R.V. (1991) Bacterial mobile genetic elements: Importance in assessing the environmental fate of genetically engineered sequences. *Plasmid* **26**, 151–171.

Schoenen D. (1986) Microbial growth due to materials used in drinking water systems. In Rehm H.-J. & Reed G. (eds) *Biotechnology: A Comprehensive Treatise in 8 Volumes*, vol. 8, pp. 627–647. VCH Verlagsgesellschaft, Weinheim, Germany.

Sedgwick S.G. (1986) Stability and change through DNA repair. In Kirkwood T.B.L.,

Rosenberg R.F. & Galas D.J. (eds) *Accuracy in Molecular Processes. Its Control and Relevance to Living Systems*, pp. 233–289. Chapman and Hall, Cambridge.

Sherratt D. (1989) Tn3 and related transposable elements: site-specific recombination and transposition. In Berg D.E. & Howe M.M. (eds) *Mobile DNA*, pp. 163–184. American Society for Microbiology, Washington DC.

Simonsen L. (1990) Dynamics of plasmid transfer on surfaces. *J. Gen. Microbiol.* **136**, 1001–1007.

Simonsen L., Gordon D.M., Stewart F.M. & Levin B.R. (1990) Estimating the rate of plasmid transfer: an end-point method. *J. Gen. Microbiol.* **136**, 1241–1249.

Sizemore R.K. & Colwell R.R. (1977) Plasmids carried by antibiotic-resistant marine bacteria. *Antimicrob. Agents Chemother.* **12**, 373–382.

Smets B.F., Rittmann B.E. & Stahl D.A. (1990) The role of genes in biological processes. Part 2 of a two-part article. *Environ. Sci. Technol.* **24**, 162–169.

Stewart G.J. (1989) The mechanism of natural transformation. In Levy S.B. & Miller R.V. (eds) *Gene Transfer in the Environment*, pp. 139–164. McGraw-Hill, New York.

Stewart G.J. & Sinigalliano C.D. (1990) Detection of horizontal gene transfer by natural transformation in native and introduced species of bacteria in marine and synthetic sediments. *Appl. Environ. Microbiol.* **56**, 1818–1824.

Stewart K.R. & Koditschek L. (1980) Drug-resistance transfer in *Escherichia coli* in the New York Bight sediments. *Mar. Poll. Bull.* **11**, 130–133.

Stock C.A., Mcloughlin T.J., Klein J.A. & Adang M.J. (1990) Expression of a *Bacillus thuringiensis* crystal protein gene in *Pseudomonas cepacia* 526. *Can. J. Microbiol.* **36**, 879–884.

Stotzky G. (1989) Gene transfer among bacteria in soil. In Levy S.B. & Miller R.V. (eds) *Gene Transfer in the Environment*, pp. 165–222. McGraw-Hill, New York.

Sturtevant A.B. Jr & Feary T.W. (1969) Incidence of infectious drug resistance among lactose-fermenting bacteria isolated from raw and treated sewage. *Appl. Environ. Microbiol.* **18**, 918–924.

Timberlake W.E. & Marshall M.A. (1989) Genetic engineering of filamentous fungi. *Science* **244**, 1313–1317.

Timmis K.N. (1987) Prospects for laboratory engineering of bacteria to degrade pollutants. In Omen G.S. (ed.) *Environmental Biotechnology: Reducing Risks from Environmental Chemicals through Biotechnology*, pp. 72–76. Plenum, New York.

Timmis K.N., Warren G.J., Whitely H.R., Gutterson N., Gasson M.J. & Collins C.H. (1988) Round Table 1: Applications. In Sussman M., Collins C.H., Skinner F.A. & Stewart-Tull D.E. (eds) *The Release of Genetically-engineered Micro-organisms*, pp. 193–206. Academic Press, London.

Trevors J.T., Barkay T. & Bourquin A.W. (1987) Gene transfer among bacteria in soil and aquatic environments: a review. *Can. J. Microbiol.* **33**, 191–198.

Van Elsas J.D. & Trevors J.T. (1990) Plasmid transfer to indigenous bacteria in soil and rhizosphere: problems and perspectives. In Fry J.C. & Day M.J. (eds) *Bacterial Genetics in Natural Environments*, pp. 188–199. Chapman and Hall, London.

Walker G.C. (1984) Mutagenesis and inducible responses to deoxyribonucleic acid damage in *Escherichia coli*. *Microbiol. Rev.* **48**, 60–93.

Wellington E.M.H., Cresswell M., Herron P.R., Clewlow L.J., Saunders V.A. & Wipat A. (1990) Gene transfer between streptomycetes in soil. In Fry J.C. & Day M.J. (eds) *Bacterial Genetics in Natural Environments*, pp. 216–230. Chapman and Hall, London.

Willetts N. & Skurray R. (1980) The conjugation system of F-like plasmids. *Annu. Rev. Genet.* **14**, 41–76.

Wood M.S., Lory C. & Lessie T.G. (1990) Activation of the *lac* genes of Tn951 by insertion sequences from *Pseudomonas cepacia*. *J. Bacteriol.* **172**, 1719–1724.

Wortman A.T. & Colwell R.R. (1988) Frequency and characteristics of plasmids in bacteria isolated from deep-sea amphipods. *Appl. Environ. Microbiol.* **54**, 1284–1288.

Zeph L.R. & Stotzky G. (1989) Use of a biotinylated DNA probe to detect bacteria transduced by bacteriophage P1 in soil. *Appl. Environ. Microbiol.* 55, 661–665.

Zeph L.R., Onaga M.A. & Stotzky G. (1988) Transduction of *Escherichia coli* by bacteriophage P1 in soil. *Appl. Environ. Microbiol.* **54**, 679–699.

Index

à rendre le	rendu le
11. 05. 01	prolongé →
28/05/01	